Cluster Beam Deposition of Functional Nanomaterials and Devices

Frontiers of Nanoscience

Series Editors
Richard E. Palmer
Nanomaterials Laboratory,
College of Engineering,
Swansea University, UK

Roy L. Johnston
School of Chemistry,
University of Birmingham, UK

Vol. 1 **Nanostructured Materials** edited by
Gerhard Wilde

Vol. 2 **Atomic and Molecular Manipulation** edited by
Andrew J. Mayne and Gérald Dujardin

Vol. 3 **Metal Nanoparticles and Nanoalloys** edited by
Roy L. Johnston and J.P. Wilcoxon

Vol. 4 **Nanobiotechnology** edited by
Jesus M. de la Fuente and V. Grazu

Vol. 5 **Nanomedicine** edited by
Huw Summers

Vol. 6 **Nanomagnetism: Fundamentals and Applications** edited by
Chris Binns

Vol. 7 **Nanoscience and the Environment** edited by
Jamie R. Lead and Eugenia Valsami-Jones

Vol. 8 **Characterization of Nanomaterials in Complex Environmental and Biological Media** edited by
Mohammed Baalousha and Jamie R. Lead

Vol. 9 **Protected Metal Clusters: From Fundamentals to Applications** edited by
Tatsuya Tsukuda and Hannu Häkkinen

Vol. 10 **Structure and Properties of Nanoalloys** edited by
Riccardo Ferrando

Vol. 11 **Materials and Processes for Next Generation Lithography** edited by
Alex P. G. Robinson and Richard A. Lawson

Vol. 12 **Computational Modelling of Nanoparticles** edited by
Stefan T. Bromley and Scott M. Woodley

Vol. 13 **Self-Assembly of Nano- and Micro-structured Materials using Colloidal Engineering** edited by
Dwaipayan Chakrabarti, Stefano Sacanna

Vol. 14 **Nanostructured Thin Films: Fundamentals and Applications** edited by
Maria Benelmekki and Andreas Erbe

Frontiers of Nanoscience

Cluster Beam Deposition of Functional Nanomaterials and Devices

Edited by

Paolo Milani

Interdisciplinary Centre for Nanostructured Materials and Interfaces (CIMaINa) and Department of Physics, Università degli Studi di Milano, Milan, Italy

Mukhles Sowwan

Nanoparticles by Design Unit, Okinawa Institute of Science and Technology (OIST), Okinawa, Japan

ELSEVIER

Elsevier
Radarweg 29, PO Box 211, 1000 AE Amsterdam, Netherlands
The Boulevard, Langford Lane, Kidlington, Oxford OX5 1GB, United Kingdom
50 Hampshire Street, 5th Floor, Cambridge, MA 02139, United States

Notices
Knowledge and best practice in this field are constantly changing. As new research and experience broaden our understanding, changes in research methods, professional practices, or medical treatment may become necessary.

Practitioners and researchers must always rely on their own experience and knowledge in evaluating and using any information, methods, compounds, or experiments described herein. In using such information or methods they should be mindful of their own safety and the safety of others, including parties for whom they have a professional responsibility.

To the fullest extent of the law, neither AACCI nor the Publisher, nor the authors, contributors, or editors, assume any liability for any injury and/or damage to persons or property as a matter of products liability, negligence or otherwise, or from any use or operation of any methods, products, instructions, or ideas contained in the material herein.

Library of Congress Cataloging-in-Publication Data
A catalog record for this book is available from the Library of Congress

British Library Cataloguing-in-Publication Data
A catalogue record for this book is available from the British Library

ISBN: 978-0-08-102515-4
ISSN: 1876-2778

For information on all Elsevier publications visit our website at
https://www.elsevier.com/books-and-journals

Publisher: Matthew Deans
Acquisitions Editor: Simon Holt
Editorial Project Manager: Joshua Mearns
Production Project Manager: Anitha Sivaraj
Cover Designer: Roger Mark

Typeset by TNQ Technologies

Contents

Contributors

Moheb Abdelaziz Nanochemistry and Nanoengineering, Department of Chemistry and Materials Science, School of Chemical Engineering, Aalto University, Espoo, Finland

Francesco Banfi Université de Lyon, CNRS, Université Claude Bernard Lyon 1, Institut Lumiére Matiére, Villeurbanne, France

Giulio Benetti Interdisciplinary Laboratories for Advanced Materials Physics (i-LAMP) and Dipartimento di Matematica e Fisica, Università Cattolica del Sacro Cuore, Brescia, Italy; Department of Pathology and Diagnostics—Medical Physics Unit, University Hospital of Verona, P.le Stefani 1, Verona, Italy

Renheng Bo Research School of Engineering, The Australian National University, Canberra, ACT, Australia

Emanuele Cavaliere Interdisciplinary Laboratories for Advanced Materials Physics (i-LAMP) and Dipartimento di Matematica e Fisica, Università Cattolica del Sacro Cuore, Brescia, Italy

Eric Danielson Nanoparticles by Design Unit, Okinawa Institute of Science and Technology, Okinawa, Japan

Marcel Di Vece CIMAINA and Department of Physics, Università degli Studi di Milano, Milano, Italy

Mady Elbahri Nanochemistry and Nanoengineering, Department of Chemistry and Materials Science, School of Chemical Engineering, Aalto University, Espoo, Finland

Luca Gavioli Interdisciplinary Laboratories for Advanced Materials Physics (i-LAMP) and Dipartimento di Matematica e Fisica, Università Cattolica del Sacro Cuore, Brescia, Italy

Panagiotis Grammatikopoulos Okinawa Institute of Science and Technology (OIST) Graduate University, Okinawa, Japan

Avik Halder Materials Science Division, Argonne National Laboratory, Argonne, IL, United States

Marta Haro Nanoparticles by Design, OIST, Onna-son, Okinawa, Japan

Mehdi Keshavarz Hedayati Department of Engineering, Durham University, Durham, United Kingdom

Shahin Homaeigohar Nanochemistry and Nanoengineering, Department of Chemistry and Materials Science, School of Chemical Engineering, Aalto University, Espoo, Finland

Sushant Kumar Department of Chemical and Biochemical Engineering, Indian Institute of Technology Patna, Patna, India

Ondřej Kylián Department of Macromolecular Physics, Faculty of Mathematics and Physics, Charles University, Prague, Czech Republic

Paolo Milani Interdisciplinary Centre for Nanostructured Materials and Interfaces (CIMaINa) and Department of Physics, Università degli Studi di Milano, Milan, Italy

Vladimir N. Popok Department of Materials and Production, Aalborg University, Aalborg, Denmark

Alexander James Porkovich Nanoparticles by Design Unit, Okinawa Institute of Science and Technology, Okinawa, Japan

Tommaso Santaniello Interdisciplinary Centre for Nanostructured Materials and Interfaces (CIMaINa) and Department of Physics, Università degli Studi di Milano, Milan, Italy

Carsten Schulte Interdisciplinary Centre for Nanostructured Materials and Interfaces (CIMaINa) and Department of Physics, Università degli Studi di Milano, Milan, Italy

Jeffrey E. Shield Department of Mechanical & Materials Engineering, Nebraska Center for Materials and Nanoscience, University of Nebraska-Lincoln, Lincoln, NE, United States

Antonio Tricoli Research School of Engineering, The Australian National University, Canberra, ACT, Australia

Stefan Vajda Institute for Molecular Engineering, The University of Chicago, Chicago, IL, United States; Department of Nanocatalysis, J. Heyrovský Institute of Physical Chemistry, Czech Academy of Sciences, Prague, Czech Republic

Preface

When you look at the world with knowledge,
you realize that things are unchangeable
and at the same time are constantly being transformed.

Yukio Mishima

Nanoparticles employed as additives in industrial processes can be considered as a commodity. The large-scale production of carbon nanoparticles as rubber additives for tires or titania nanoparticles for pigmentary and food applications represents a paradigm.

Clusters, a particular class of nanoparticles ranging in size from a few atoms to thousands of atoms, have received, since the advent of nanotechnology, particular attention as possible elemental building blocks of new materials with unprecedented structural and functional properties. Clusters have dimensions such that the addition or removal of every single atom can make a substantial difference in the properties of the aggregate, and this prompted the hope of building a new periodic table based on "super-atomic" units. Unfortunately, this dream does not have a practical counterpart: the mass production, manipulation, and assembling of clusters with an exact number of components assembled in a well-determined and reproducible structure remain unachievable after several decades of efforts.

On the other hand, another paradigm is emerging based on cluster beam deposition (CBD) as a scalable tool to produce materials and devices where the presence of clusters can impart superior properties and performances without the need of bulk production and size selection. CBD is based on gas phase routes for the production and assembling of clusters and played a prominent role in fundamental studies; since one decade it has been recognized that CBD is a versatile scalable approach for the production of devices with a high level of control on their structural and functional properties.

CBD allows the top-down/bottom-up integration for the production of systems such as memory devices, optical devices, sensors, energy storage and catalytic devices, passive photonic networks, and lab-on-chip diagnostic devices. This thanks to the capability of CBD of sorting nanoobjects in terms of size or geometry classification, the ability to transport and deliver them onto substrates, and the ability to control their position.

This book presents and describes some of the most advanced achievements obtained by the use of CBD to assemble complex materials and provides a detailed description of different technological solutions in order to make available to a large audience of scholars and professionals the CBD technology. The book is organized in different sections devoted to catalysis and energy applications (Section I), electronic, optical, and magnetic devices (Section II), systems for cell biology and biomedical applications (Section III), and scale-up of CBD in view of 3D printing and rapid prototyping (Section IV).

1

Synchrotron characterization of clusters for catalysis[a]

Avik Halder[1], Stefan Vajda[2,3,*]

[1]MATERIALS SCIENCE DIVISION, ARGONNE NATIONAL LABORATORY, ARGONNE, IL, UNITED STATES; [2]INSTITUTE FOR MOLECULAR ENGINEERING, THE UNIVERSITY OF CHICAGO, CHICAGO, IL, UNITED STATES; [3]DEPARTMENT OF NANOCATALYSIS, J. HEYROVSKÝ INSTITUTE OF PHYSICAL CHEMISTRY, CZECH ACADEMY OF SCIENCES, PRAGUE, CZECH REPUBLIC

[*]CORRESPONDING AUTHOR

Chapter outline

[a]"The submitted manuscript has been created by UChicago Argonne, LLC as Operator of Argonne National Laboratory ("Argonne") under Contract No. DE-AC02-06CH11357 with the U.S. Department of Energy. The U.S. Government retains for itself, and others acting on its behalf, a paid-up, nonexclusive, irrevocable worldwide license in said article to reproduce the article, prepare derivative works, distribute copies to the public, and perform and display publicly, by or on behalf of the Government. The Department of Energy will provide public access to these results of federally sponsored research in accordance with the DOE Public Access Plan."

Cluster Beam Deposition of Functional Nanomaterials and Devices. https://doi.org/10.1016/B978-0-08-102515-4.00001-5

1.1 Introduction

Clusters consist of particles ranging from sizes as small as a dimer up to few nanometers in size. These small particles have a large surface-to-volume ratio along with a significantly increased fraction of uncoordinated atoms as their size shrinks. Clusters can exhibit a range of fascinating reactive, optical, electronic, and magnetic properties, which are not observed in their bulk phase. Over the past decades, clusters have been widely used in studies of heterogeneous catalytic reactions [1–3]. It is known that the catalytic reactions take place on the surface of the catalysts, which implies that the core atoms in bulkier particles are not exposed to the reactants, and thus do not get directly involved in the reaction. In the case of expensive metals this may result in an ineffective usage of a large fraction of metal loading and increased costs. Small clusters made of a handful of atoms are often nicknamed "particles with no inside" and possess the maximum number of surface atoms, thus offering potential advantages for the optimal use of the metal. Recent studies with small mono- and bimetallic cluster catalysts have also shown that more expensive metals can be replaced in full or in part by a cheaper metal, while maintaining good performance just by the virtue of the right combination of cluster size, cluster composition, and substrate.

As we will discuss select reactions along the chapter, we will see that the activity and selectivity for a reaction, as well as the stability of the clusters and the evolution of the oxidation state of the cluster's metal components (characteristics monitored using synchrotron-based techniques), can change drastically with the addition/removal/substitution of one single atom within the cluster. Using size-selected cluster deposition, it is possible to prepare catalysts which are composed of pure or alloy clusters of a single specific size and composition, with atomic precision. By exerting control over the size and composition of a catalyst, one has the tuning knobs to control the activity and selectivity of the cluster-based catalyst.

Most of the standard characterization techniques like electron microscopy, atomic force microscopy, Raman spectroscopy, or X-ray diffraction are typically not suited to characterize samples with subnanometer-sized clusters and nanogram loadings, especially under realistic reaction conditions. Dedicated synchrotron-based techniques, such as those developed at Argonne National Laboratory, enable ex situ and in situ characterization of such samples. Due to the high flux of X-rays available at the beamlines, synchrotron radiation has become a valuable tool for investigating clusters. In studies of heterogeneous gas-phase reactions, a fixed-bed, continuous flow catalysis reactor can be used, in which GISAXS (grazing incidence small angle X-ray scattering), GIXAS (grazing incidence X-ray absorption spectroscopy), and TPR (temperature-programmed reaction) data are simultaneously collected to probe the chemical and structural changes of cluster catalysts under catalytic reaction conditions [4]. Typically the monodispersed cluster samples are prepared as a fraction of a single atomic monolayer equivalent coverage in order to maintain the size specificity of the particles. GISAXS technique is used to detect changes in the shape, size, and morphology of nanoparticles, while in the case of subnanometer clusters it is primarily used to monitor the sintering resistance/ possible aggregation of small clusters during the course of the reaction. GIXAS is used to follow the evolution of the oxidation state of the individual metal components of the cluster with element specificity, and TPR measures the product formation during the applied temperature ramp.

A wide range of heterogeneous catalysis reactions has been investigated on size-selected clusters using this in situ technique, which in close synergy with computational studies provided fundamental insights into the reaction mechanisms of O_2 activation, C–H bond cleavage, C–C bond scission, partial oxidation of hydrocarbons, CO oxidation, conversion of CO and CO_2, etc. This technique has also been successfully used for probing nanocatalyst prepared by conventional wet-chemical methods, or for ex situ characterization of cluster samples used in electrochemical reactions before and after the reaction. Some of the electrochemical reactions for which these synchrotron-based techniques contributed useful insights were the oxygen evolution reaction (OER) or CO_2 reduction, for example. This chapter is not intended to deliver an exhaustive review of all synchrotron-based techniques that have been published on this topic, nor a comprehensive review of all systems and reactions studied on clusters at light sources. Instead, a select set of examples of synchrotron studies of clusters in heterogeneous catalytic and electrocatalytic reactions performed at Advanced Photon Source (beamlines 12-ID-C and 12-BM) and some other beamlines will be presented, on cluster samples prepared at the Argonne National Laboratory. For those readers who are not familiar with physical cluster preparation techniques, a brief introduction is included.

1.2 Size-selected cluster synthesis

Metal clusters are produced in a magnetron sputtering source based on the original design of Haberland [5,6], using argon ions for sputtering. The schematics of the cluster apparatus is shown in Fig. 1.1A with the main components of its beamline in Fig. 1.1B [7]. The sputtered ions are cooled inside a liquid nitrogen cooled chamber by collisions with the buffer gas helium which is cofed with argon into the sputtering source. The size distribution of the clusters formed in the source can be controlled by tuning the length of aggregation region, the Ar to He ratio, the gas flow, and sputtering power applied. The targets used can be made of pure metals, as well as alloys to prepare monometallic or mixed clusters. The cluster ions emerging from the source are focused

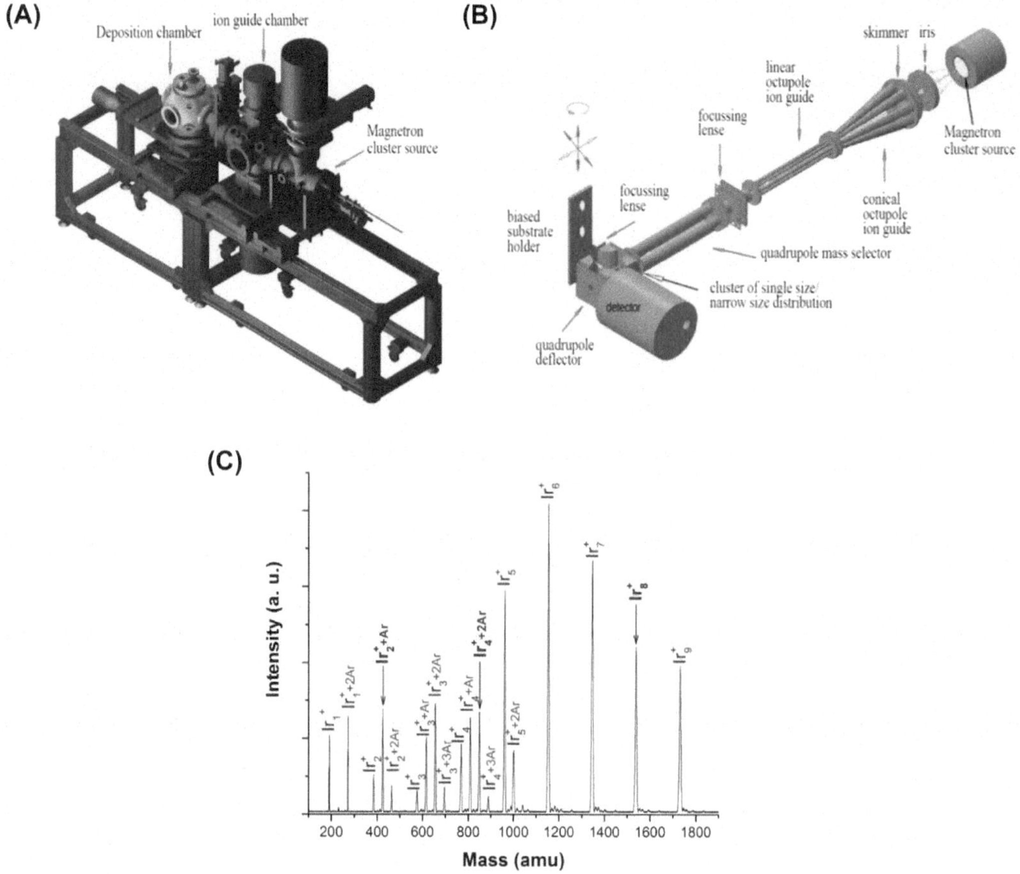

FIGURE 1.1 Size-selected cluster deposition. (A) Magnetron sputtering–based cluster deposition system at Argonne National Laboratory. From right to left are the magnetron cluster source chamber, ion guide chamber (which has a conical octupole followed by a linear octupole guide), and deposition chamber (which has a quadrupole mass filter and a deposition assembly). (B) Schematic of the internal components (ion optics) in the cluster deposition system. (C) Mass spectrum of cationic Ir clusters produced in the 2000 amu range. Each mass peak corresponds to one specific composition as marked within the spectra. *(A–B) Reproduced with permission from Yin et al., J. Chem. Phys. 140 (2014) 174201. Copyright 2014. AIP Publishing LLC. (C) Reproduced with permission from Halder et al., J. Phys. Chem. C 122 (2018) 9965. Copyright 2018, American Chemical Society.*

using a sequential conical and linear octupole ion guide assembly made of Extrel hardware components. The mass selection occurs within a quadrupole filter (Extrel, 19 mm rods, 440 kHz). An example of Ir mass spectra is shown in Fig. 1.1C [8]. The selected cluster ions pass through a home-built quadrupole bender and are subsequently deposited onto a conductive substrate, which is mounted perpendicular to the beam. Using the bender allows to extract the mass-selected cluster ions from the neutral clusters present in the cluster beam and to avoid the codeposition of neutral, not size-selected species.

The deposition currents attained are between 0.5 and 5 nA leading to deposition times of about 15–60 min per sample, depending on cluster size and the desired surface coverage. A typical sample is made with 0.1–0.2 atomic monolayer equivalent surface coverage (ML) with the metal, which corresponds to about 10^{13} clusters, deposited within a spot of diameter of 8 mm. The clusters are softly landed with kinetic energy of less than 1 eV/atom onto the substrate in order to avoid fragmentation of the clusters upon impact. Typical supports used for cluster deposition are of technological relevance, such as oxides (reducible and nonreducible) prepared by atomic layer deposition (ALD) or carbon-based, such as amorphous carbon or ultrananocrystalline diamond (UNCD).

1.3 Synchrotron techniques for in situ and ex situ catalytic studies of clusters

The schematics of the cell reactor used for the in situ and ex situ characterization of supported clusters is shown in Fig. 1.2A [4]. The customized design equips the cell to perform combined in situ TPR/GIXAS/GISAXS characterization. The key components of the cell are: (1) feedthroughs at the top of the cell for gas inlet and outlet—the gas outlet carries with it the products formed during the reaction, which are analyzed on a mass spectrometer (Pfeiffer Prisma 200), (2) feedthroughs at the side of the cell are for cooling the cell body, (3) sample heater, a boron nitride–based ceramic heater (Momentive Performance Materials, HTR1001), (4) the internal volume of the cell is approximately 25 cm^3, (5) the windows of the reactor are made out of a 25 μm-thick Kapton film for the penetration of the X-rays.

The combined TPR/GIXAS/GISAXS experiment is illustrated in Fig. 1.2B [9]. The products' signal (TPR) is collected on a differentially pumped mass spectrometer shown at the right in Fig. 1.2C. A 2048 × 2048 pixel MarCCD detector (see Fig. 1.2C) records the 2-dimensional GISAXS images. The 2D images are then processed by taking horizontal and vertical cuts and the cuts are analyzed to obtain the size distribution using Nika and Irena softwares, for example [10,11]. GIXAS spectra, in fluorescence mode, are typically collected on a 4-element Vortex detector mounted on the horizontal plane perpendicular to the beam (a picture of the detector used is shown in Fig. 1.2D). For the analysis of the GIXAS data analysis multiple packages can be used, for example ATHENA and ARTEMIS programs in the IFEFFIT interactive software package.

1.4 Characterization of heterogeneous catalysts made of monometallic clusters

1.4.1 Oxidation and reduction of Cu clusters

Tuning the oxidation state of Cu clusters offers a potential utility to optimize product yield and selectivity. A combined study using first principles calculations and XANES

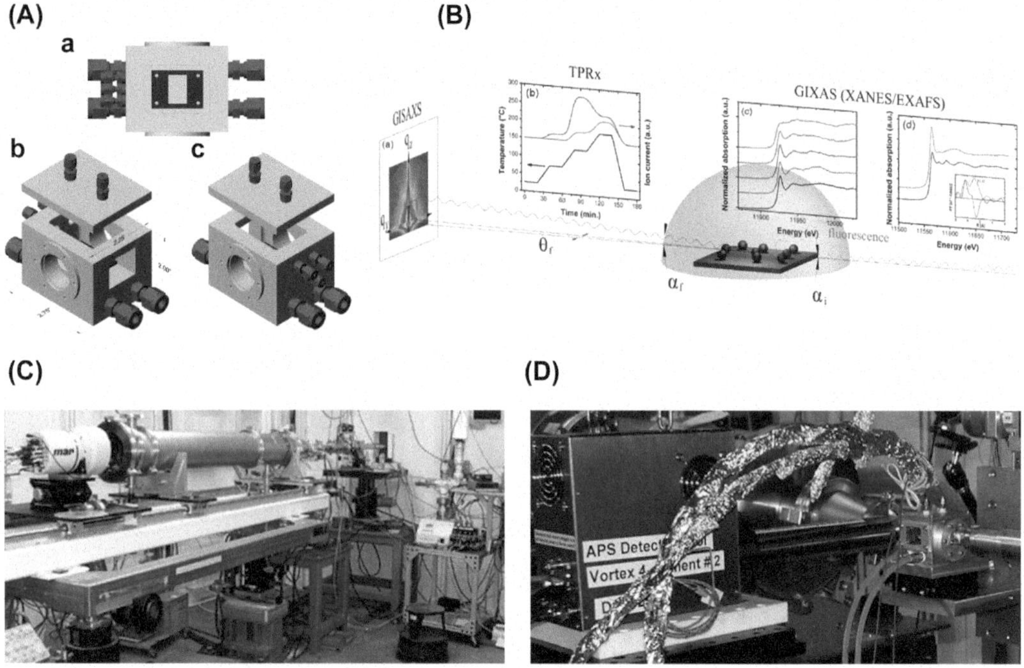

FIGURE 1.2 In situ TPR/GISAXS/GIXAS setup. (A) Three-dimensional schematic of the cell reactor used for in situ temperature-programmed reaction (TPR) studies with grazing-incidence X-ray scattering (GISAXS) and grazing-incidence X-ray absorption (GIXAS) characterization. Three different orientations demonstrate: (a) Top view: sample and heater within the cell. (b) Rectangular and circular openings for entry and exit of X-ray beam. (c) Top lid containing two 1/16 in. gas feedthroughs; three 1/16 in. feedthroughs on the side for electric and thermocouple connections; two 1/4 in. Swagelok connectors for cooling water circulation in the lower part of the cell. (B) Schematic of the combined TPR reaction, and X-ray characterization (GISAXS and GIXAS). The insets illustrate a set of representative data collected: (A) two-dimensional in situ GISAXS pattern which allows to determine evolution in the particle size and shape during a reaction and hence to study particle stability, (B) monitoring of product formation during the reaction as a function of time and temperature, (C) XANES spectra collected during an in situ study for reduction of small gold nanoparticles, and (D) EXAFS scan on Pt_{10} clusters which revealed fully oxidized Pt. (C) The TPR setup at the 12-ID-C beamline at the Advanced Photon Source. A MAR (or Gold, Platinum) CCD detector used to collect the GISAXS images (D) Vortex 4-element detector at the 12-ID-C beamline for GIXAS data collection. *(A) Reproduced with permission from Lee et al., Nucl. Instrum. Methods Phys. Res. A 649 (2011) 200–203. Copyright 2011 Elsevier. (B) Reproduced with permission from Wyrzgol et al., Phys. Chem. Chem. Phys. 12 (2010) 5585–5595. Copyright 2010, Royal Society of Chemistry.*

measurements of size-selected Cu clusters (Cu_4, Cu_{12}, and Cu_{20}) supported on hydroxylated amorphous alumina reveals the size-dependent trend in the oxidizability of the supported clusters and the significant modifications that it undergoes when compared with the gas-phase clusters. The ground state structures of metallic and oxidized clusters were obtained in two steps, where candidate structures were first shortlisted using empirical potentials and simulated annealing, and then further refined using density functional theory implemented in the VASP code [12,13]. Fig. 1.3A–F shows the ground state structures for the elemental clusters (Cu_n) and

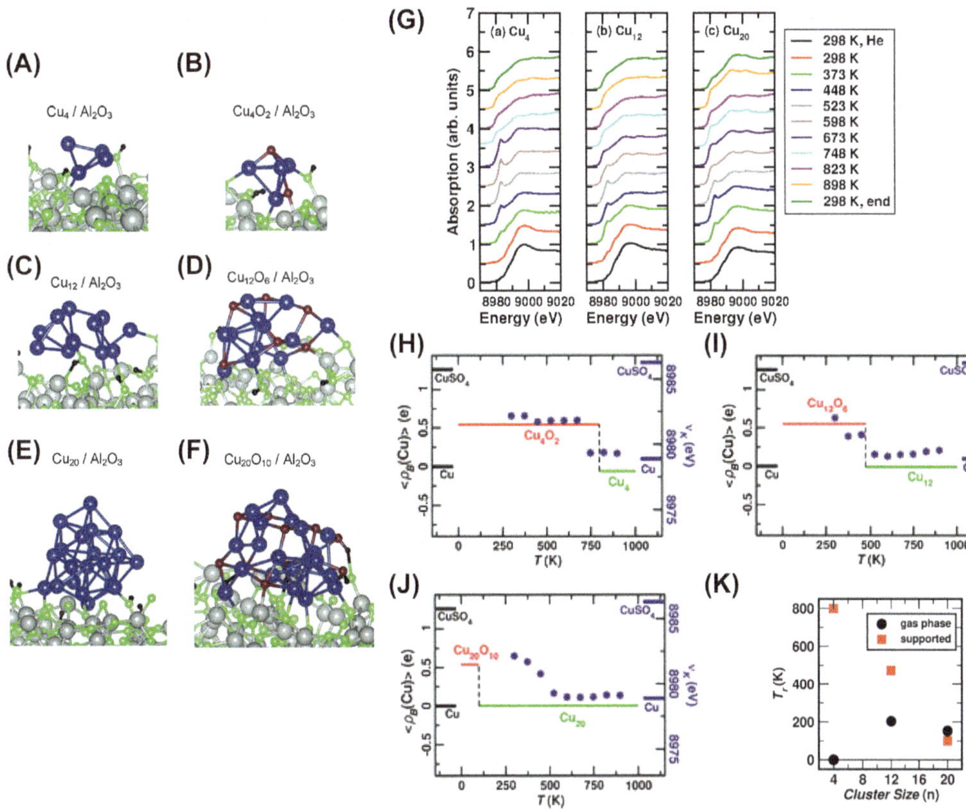

FIGURE 1.3 Oxidation of Cu clusters. Lowest energy structures of (A) Cu_4/Al_2O_3, (B) Cu_4O_2/Al_2O_3, (C) Cu_{12}/Al_2O_3, (D) $Cu_{12}O_6/Al_2O_3$, (E) Cu_{20}/Al_2O_3, (F) $Cu_{20}O_{10}/Al_2O_3$. The gray, black, and light green spheres represent Al, H, and O atoms in the support, respectively; the blue and red spheres represent the Cu and O atoms in the cluster, respectively. (G) XANES spectra for Cu_4, Cu_{12}, and Cu_{20} clusters on Al_2O_3 upon heating in a H_2-rich environment. (H) (J) a comparison of experimental and theoretical results for size-dependent reduction of $Cu_nO_{n/2}$ clusters on HAA support. Violet dots refer to the XANES Cu K-edge energies and $\rho_B(Cu)$ is averaged Bader charge which provides a theoretical estimate for the Bader charge (horizontal red or green lines) over all the Cu atoms in the cluster measured for $p(H_2)/p(H_2O) = 100$. *Figures reproduced with permission from Mammen et al., Eur. J. Inorg. Chem. 2018 (2018) 16–22. Copyright 2018, John Wiley and Sons.*

oxidized clusters ($Cu_nO_{n/2}$) compositions. The reaction studied represents oxidation/reduction in these clusters [14].

The Cu clusters are initially oxidized due to having been exposed to air as seen from the XANES spectra collected in a He environment for each cluster and is shown in Fig. 1.1G. Thereafter H_2 is introduced and the catalysts are heated from 298 to 898 K, with temperature steps of 75 K. H_2 (3.5% in He) was introduced at a flow rate of 30 sccm with the pressure of the cell being kept at 1.2 bar (900 Torr). The samples are heated at a rate of 10 K/min while ramping between different temperatures, to assure stabilization of conditions. At every finite temperature 30 min are spent, during which the GISAXS images and GIXANES spectra (measured at the Cu K-edge) are collected.

GISAXS, performed at 8.9 keV, confirmed that there has been no sintering of the clusters during the heat ramp. The position of the Cu K-edge at each temperature derived from the XANES spectra is plotted in blue stars in Fig. 1.3H–J. A clear transition is observed in the edge position during the reduction process, as the clusters transition from $Cu_nO_{n/2}$ to Cu_n. The solid red and green lines refer to the average Bader charge calculated for the average stable phase of the cluster at a given temperature. For each stoichiometry, the Bader charge is calculated on each Cu atom by analyzing the topological properties of the electronic charge density using the Bader prescription. The transition temperatures for the clusters on the amorphous alumina support and in the gas phase are compared in Fig. 1.3K. Calculations show that for $n = 4$, the gas-phase cluster is in the elemental phase over the entire temperature range, and we therefore define transition temperature $T_r = 0$ K. It is seen from the calculated results as plotted in Fig. 1.3K that for the subnanometer-sized clusters in gas phase smaller Cu clusters are more resistant to oxidation. In stark contrast to the gas-phase clusters, the supported Cu_4 cluster is found to be more easily oxidized than Cu_{12} and Cu_{20}. For $n = 20$, the presence of the support lowers the transition temperature, and hence hinders oxidation. However, in case of $n = 4$ and 12, the results are counterintuitive; as one might expect that the Cu atoms would be relatively inaccessible to O atoms due to the presence of support and make oxidation more difficult relative to gas phase. The study shows that the degree of oxidation for the Cu catalysts can be tuned by not only varying the size of the clusters, but also the chemical nature of the support they are deposited on.

1.4.2 Gas adsorption on supported small silver clusters

A combined computational and experimental XANES study was performed on size-selected Ag_4 and Ag_{16} clusters deposited on a titania film to study the potential of subnanometer Ag clusters in CO oxidation [15]. A real-space multiple scattering method (FEFF code) was used to calculate Ag K-edge XANES spectra for cluster-ligand geometries obtained from density-functional calculations [16]. Theoretical XANES spectra for the bare Ag_4 and Ag_{16} clusters, as well as those with adsorbed oxygen or carbon monoxide, are shown in Fig. 1.4A. Ag_4 XANES spectra reveal two peaks and a prepeak shoulder within 40 eV near edge. The difference in energy of two calculated peaks is ca. 30 eV. Interestingly, molecular oxygen adsorption only slightly changes the XANES pattern. In contrary, dissociative oxygen adsorption makes the first peak more prominent. In case of Ag_{16} cluster, the first two peaks are notably sharper than those of the Ag_4 cluster; also the spectra are very similar for Ag_{16}, Ag_{16}_O_2, Ag_{16}_2O, and Ag_{16}_CO, since the adsorption of a single molecule does not significantly alter the electronic state of Ag atoms in the larger Ag_{16}.

The experimental Ag K-edge spectra collected on the ALD titania-supported Ag_4 and Ag_{16} clusters in air and in CO atmospheres are shown in Fig. 1.4B. The first absorption peak for Ag_4@TiO_2_air and Ag_4@TiO_2_CO appeared at 25.526 keV, while the second peak

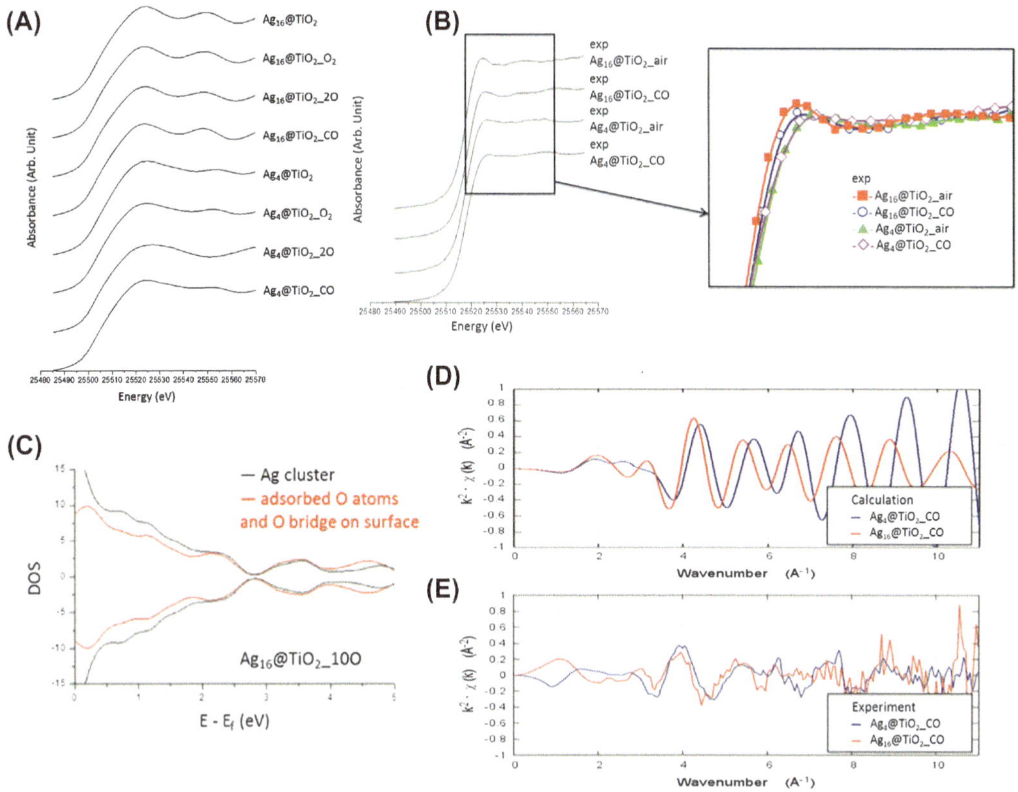

FIGURE 1.4 Reactivity of size-selected Ag clusters. (A) Ag K-edge XANES spectra derived theoretically for Ag_4 and Ag_{16} clusters deposited on TiO_2 (111) surface, with molecularly adsorbed O_2, dissociatively adsorbed 2O, and adsorbed CO molecule. (B) Ag K-edge XANES spectra measured experimentally for Ag_4 and Ag_{16} clusters deposited on TiO_2 surface in air and CO environments. (C) Partial densities of states of titania-supported Ag_{16} cluster with 10 adsorbed oxygen atoms. EXAFS for supported Ag_4 and Ag_{16} clusters in CO gas environment obtained (D) theoretically and (E) experimentally. *Figures reproduced with permission from Chen et al., J. Phys. Chem. C 121 (2017) 6614–6625. Copyright 2017, American Chemical Society.*

is at 25.551 keV; the energy difference of these two peaks is 25 eV. For Ag_{16}@TiO_2_air and Ag_{16}@TiO_2_CO the first absorption peak appeared at 25.524 keV, while the second peak at 25.546 eV; the difference of these two peaks is 22 eV. A comparison with the theoretical spectra from Fig. 1.4A shows a prominent first peak for Ag_{16}@TiO_2, thus indicating dissociative adsorption of oxygen in air. The absence of such prominent peak in case of Ag_4@TiO_2_air is explained with Ag_4@TiO_2_O_2 being energetically more stable than Ag_4@TiO_2_2O. This leads to a conclusion that the oxygen molecule would not dissociate on Ag_4@TiO_2 clusters at room temperature. The calculated spectrum of Ag_4@TiO_2_CO has a shape almost identical with the experimental spectrum of Ag_4@TiO_2_air. This is because Ag_4 on TiO_2 presumably does not considerably change its charge state in air and carbon monoxide environments.

The origin of XANES features lies in the excitation of core electrons to unoccupied electronic states. The projected density of states (pDOS) of unoccupied states of the Ag K-edge XANES for Ag_{16}@TiO_2 is shown in Fig. 1.4C. Beyond Fermi level of Ag the pDOS is mainly attributed to the mixing of 3d orbitals with 4s orbitals. One should note that the orbitals of the Ag atoms hybridize with O atoms (including with the bridging oxygens of TiO_2 surface), the pDOS of Ag and O correlate with each other. In Fig. 1.4C, Ag_{16}@TiO_2_100 display large population near Fermi level, which resulted in the prominent first peak in Ag K-edge spectra. Exposure of the Ag_{16}@TiO_2 to carbon monoxide reduced silver in the cluster, reflected in the greatly reducing intensity of the first absorption peak.

Experimental and calculated extended X-ray adsorption fine structure (EXAFS) spectra of Ag_4@TiO_2 and Ag_{16}@TiO_2 in CO gas environment are shown in Fig. 1.4D and E, respectively. The spectra are comparable in the wavenumber region of 0–6/Å; the distinguishable peaks at 2 and 3/Å in the calculated spectra correspond to the experimental peaks at 1 and 3/Å. The peaks in case of Ag_{16} are shifted slightly to lower wavenumbers in comparison with Ag_4. The overall features are similar to the calculated and measured spectra and they diverge beyond 6/Å.

In situ XAS spectra taken for Ag_4@TiO_2 and Ag_{16}@TiO_2 under air and CO supports reveal marked differences in the properties of these two clusters. From the XAS spectra collected for the clusters exposed to air and CO, it could be inferred that Ag_{16} is more congenial toward O_2 dissociation as well as CO reduction compared to Ag_4.

1.4.3 CO_2 conversion on copper clusters

Copper catalysts have been extensively studied to convert CO_2 to methanol (CH_3OH) through hydrogenation ($CO_2 + 3H_2 \rightarrow CH_3OH + H_2O$). The understanding of the reaction mechanism and exploring the nature of the active sites in Cu catalysts, including cluster size, oxidation state, and support effects still remain an area of active research. Catalysts were fabricated with three monodispersed Cu cluster sizes: Cu_3, Cu_4, and Cu_{20}, deposited on amorphous alumina film. The strongly size-dependent activity for these samples is shown in Fig. 1.5A [17]. The production rate of methanol was monitored at $m/z = 31$ using a mass spectrometer. The turnover rate (TOR) was calculated as the number of methanol molecules formed per second over the total number of copper atoms in the sample. The activity of the Cu_4/Al_2O_3 catalyst exceeded the activity of the Cu_3/Al_2O_3 and Cu_{20}/Al_2O_3 samples, as shown in Fig. 1.5A. Under a constant flow of 20 sccm of a gas mixture of 1% CO_2 and 3% H_2 seeded in helium, at a total pressure of 1.25 atm, the methanol production by Cu_4/Al_2O_3 reached its maximum of around 225°C, with a turnover rate of about $4 \cdot 10^{-4}$ methanol molecules produced per copper atom per second. Above 325°C, methane formation was also observed.

GIXANES results obtained on the three cluster samples at the Cu K-edge are summarized in Fig. 1.5B and C showing a prominent change in the copper XANES feature that is observed as a function of temperature. Also shown in Fig. 1.5B are the bulk XANES

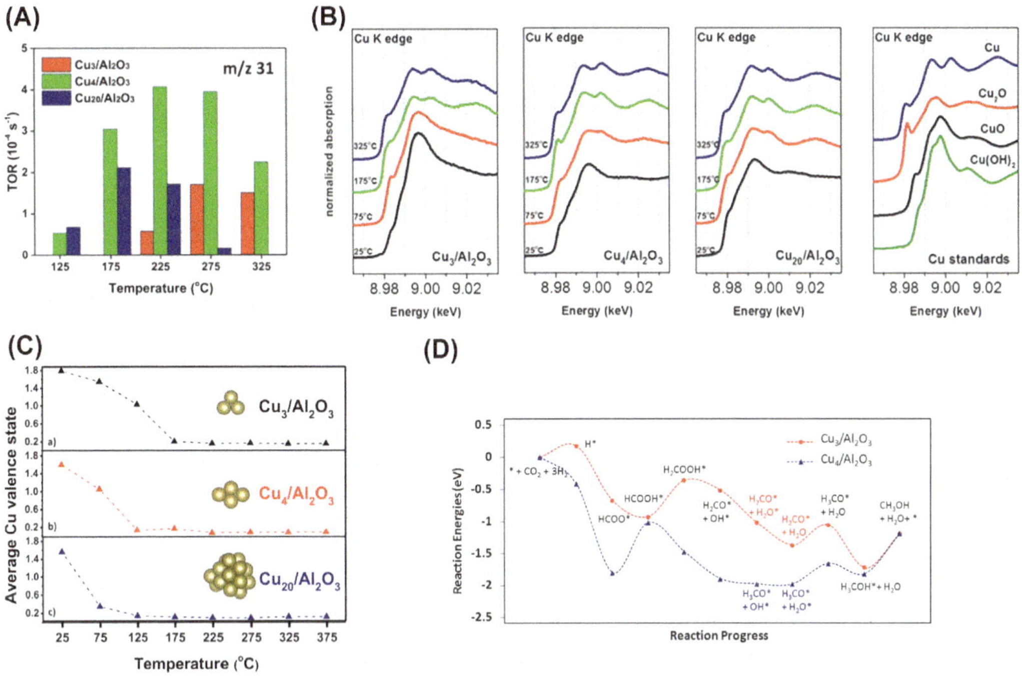

FIGURE 1.5 CO_2 conversion to methanol by Cu_n/Al_2O_3 clusters. (A) Activity toward methanol formation on Cu_3/Al_2O_3 (*red*), Cu_4/Al_2O_3 (*green*), and Cu_{20}/Al_2O_3 (*blue*) catalysts. The Cu_n catalysts were heated from 25 C up to 325 C in steps of 50 C under reaction conditions. (B) experimental XANES spectra obtained for Cu_3/Al_2O_3, Cu_4/Al_2O_3, and Cu_{20}/Al_2O_3 during the heat ramp, and also the bulk Cu standards (C) Average Cu charge state of single-size copper catalysts with increasing reaction temperatures for Cu_3/Al_2O_3, Cu_4/Al_2O_3, and Cu_{20}/Al_2O_3. (D) Comparison of calculated reaction pathways of Cu_3 and Cu_4 clusters on hydroxylated Al_2O_3. *Figures reproduced with permission from Yang et al., J. Phys. Chem. C 121 (2017) 10406. Copyright 2017, American Chemical Society.*

standards of Cu, which were used to determine oxidation state of copper in the clusters by linear combination analysis (LCF) of the in situ spectra. The evolution of the oxidation state of copper derived therein is shown in Fig. 1.5C, which implies that the reduced phase plays a crucial role in the methanol synthesis from CO_2.

DFT calculations for the reaction pathway for Cu_3/Al_2O_3 and Cu_4/Al_2O_3 are shown in Fig. 1.5D, which elucidates the higher activity of Cu_4 relative to Cu_3 clusters, as well as bulk Cu. The activation barrier for conversion of CO_2 to CH_3OH was found to be considerably low due to the strong binding of the adsorbates on the undercoordinated atoms of the copper clusters. This results in an energetically lower-lying reaction pathway and a lower barrier. Cu in Cu_3/Al_2O_3 lies at a higher charge state compared with Cu_4/Al_2O_3, which leads to a weaker binding with intermediates, and ultimately responsible for its lower per atom activity.

The in situ GIXANES proved as critical for the determination of the oxidation state of copper in the working catalysts, while simultaneously collected GISAXS (at 8.9 keV) data did not show any indication of cluster agglomeration into large particles. Clear size dependence for the reduction of Cu_n clusters under reaction atmosphere could be identified, with the reduction temperature decreasing with increasing cluster size: $Cu_3/Al_2O_3 > Cu_4/Al_2O_3 > Cu_{20}/Al_2O_3$. This enables a nice correlation between the active phase of the clusters with rise in activity for the methanol and methane production.

1.4.4 Fischer-Tropsch synthesis on small oxidized cobalt clusters

Fischer-Tropsch (FT) process is one of the most applied techniques for gas-to-liquid (GTL) conversion which is based on the catalytic conversion of CO and H_2 to long-chain hydrocarbons. Alumina-supported cobalt nanoparticles are well-known for their high catalytic activity and selectivity in the FT reaction. There has been considerable interest in recent years to understand the effect of particle size and support in order to develop more active FT catalysts. Here we summarize results of in situ investigation of $Co_{4\pm1}$ and $Co_{27\pm5}$ clusters supported on metal oxide (Al_2O_3, MgO) and carbon (ultrananocrystalline diamond, UNCD) supports under low-pressure CO conversion conditions, with focus on the XANES results measured at the Co K-edge [18].

The XANES of as prepared Co cluster samples on different supports (measured under a continuous flow of 30 sccm gas mixture of H_2:CO:He = 1:0.5:98.5, $p_{total} = 1.07$ bar, $p_{H2} = 0.01$ bar, $p_{CO} = 0.005$ bar) along with the spectra of bulk Co standards are shown in Fig. 1.6A. The comparison of the spectra indicates the presence of a CoO phase in the as made samples. Fig. 1.6B shows in situ XANES results obtained for $Co_{4\pm1}$ clusters supported on alumina. XANES from a set of three samples of $Co_{27\pm5}$ clusters supported on alumina, magnesia, and UNCD are shown in Fig. 1.6C,D and F, respectively, as a function of temperature during reaction. The different extent of changes in the spectra of the individual samples with increasing reaction temperature is illustrated in Fig. 1.6F. It is worth noting, from Fig. 1.6B–E, that the evolution of the spectra of the individual samples are quite distinct, reflecting cluster size as well as support dependency.

Fig. 1.6B shows that XANES of $Co_{4\pm1}/Al_2O_3$ recorded at 175°C and 225°C show a significant decrease in white line intensity as well as a slight white line shift to lower energies indicating partial reduction. The changes are quite identical to $Co_{27\pm5}/Al_2O_3$. On the other hand the spectra of the MgO-supported $Co_{27\pm5}$ clusters remain practically unchanged under the applied reaction conditions indicating a stable oxidation state. $Co_{27\pm5}$ clusters, supported on UNCD exhibited the most pronounced shift toward lower energy and the biggest drop in white line intensity during the reaction, thus indicating the highest degree of reduction. The degree of oxidation among the clusters investigated at 225°C can be ranked as $Co_{27\pm5}$/UNCD $< Co_{4\pm1}/Al_2O_3 \sim Co_{27\pm5}/Al_2O_3 < Co_{27\pm5}$/MgO.

The mass spectrometer signal of methane from Co_4 was used as a reference (Fig. 1.6F) in comparing catalytic activity and selectivity of the studied catalysts, revealing a significant increase in the catalytic activity for 27 atom Co clusters. $Co_{27\pm5}$/UNCD is the most active and also the most selective toward the formation of longer-chain

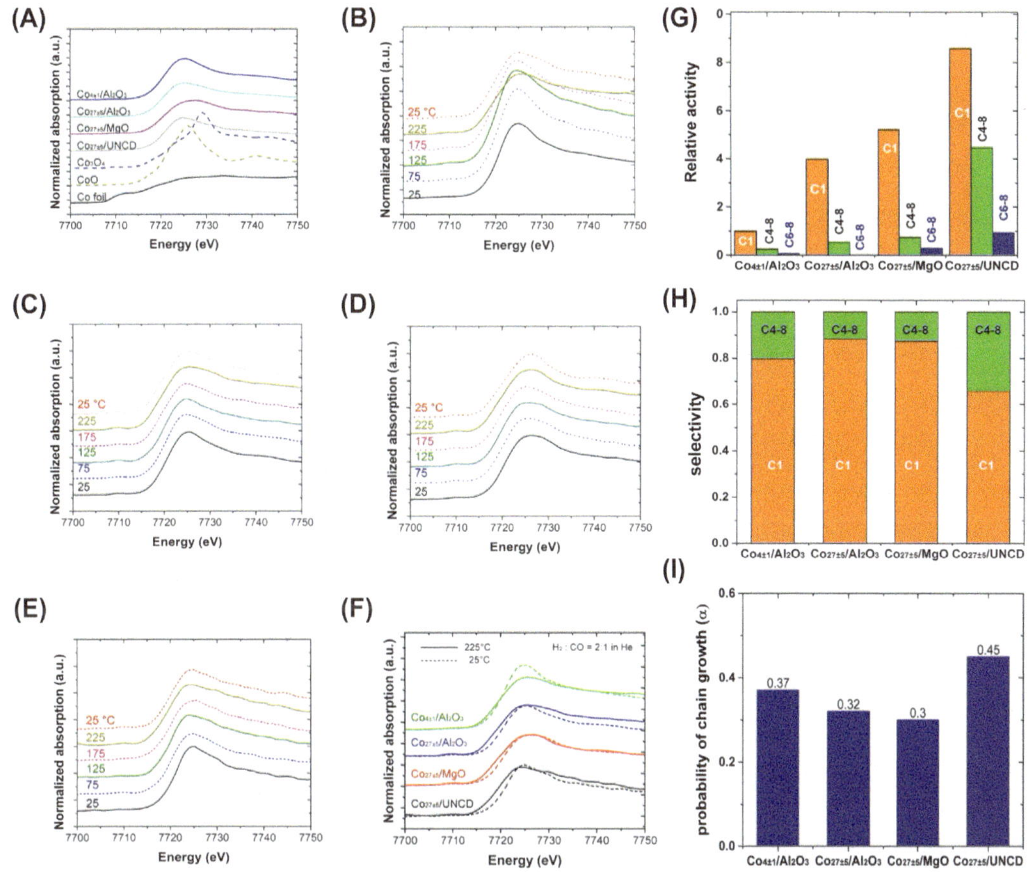

FIGURE 1.6 Fischer-Tropsch Synthesis using Co cluster catalysts. (A) XANES of the as-made $Co_{4\pm1}$ and $Co_{27\pm5}$ cluster samples collected under vacuum at 25°C. Reference spectra of bulk Co standards (Co, CoO, and Co_3O_4) are also shown for comparison. In situ XANES spectra of (B) $Co_{4\pm1}/Al_2O_3$ and (C) $Co_{27\pm5}/Al_2O_3$, (D) $Co_{27\pm5}$/MgO, and (E) $Co_{27\pm5}$/UNCD with temperature ramp (from bottom to top: 25°C, start, 75°C, 125°C, 175°C, 225°C, and 25°C, after reaction) (F) Summary plot of spectra obtained at 25°C and 225°C. (G) Cospecific activity of $Co_{4\pm1}/Al_2O_3$, $Co_{27\pm5}/Al_2O_3$, $Co_{27\pm5}$/MgO, and $Co_{27\pm5}$/UNCD at 225°C. (H) Selectivity ratio of C1 and C4–8. The reactivity was compared on the basis of C1 (methane) formation rate of $Co_{4\pm1}$. The reactivity of various cluster catalysts were normalized by the number of total deposited Co atoms. (I) Probability of chain growth derived based on ASF distribution model at 225°C as the function of cluster size and supports at 225°C. *Figures reproduced with permission from Lee et al., J. Phys. Chem. C 119 (2015) 11210. Copyright 2015, American Chemical Society.*

hydrocarbons as seen from Fig. 1.6G and H. The comparison of the activity and selectivity at 225°C shows that the cumulative activity of the observed products increases in the order $Co_{4\pm1}/Al_2O_3 < Co_{27\pm5}/Al_2O_3 < Co_{27\pm5}/MgO < Co_{27\pm5}/UNCD$. The selectivity toward longer-chain hydrocarbons is best for $Co_{27\pm5}$/UNCD catalysts, which produce the highest fraction of C4+ hydrocarbons (35%) followed by $Co_{4\pm1}/Al_2O_3$ (20%), $Co_{27\pm5}/Al_2O_3$, and $Co_{27\pm5}$/MgO catalysts (12%). It is important to note that in this study low pressure and short contact time compared to real FT conditions were applied, which results in higher selectivity toward low molecular weight hydrocarbons. Mass fragment pattern analysis based on the Anderson–Schulz–Flory (ASF) [19] distribution model showed that under these conditions the probability of chain growth (α) on subnanometer cobalt clusters was $\alpha = 0.3–0.45$ (see Fig. 1.6I), which is lower than those reported for conventional high-pressure FT catalyst ($\alpha = 0.85–0.95$) [20].

In situ XANES cobalt clusters deposited on oxide and carbon supports revealed significant differences in the oxidation state of the clusters affected by the size of the clusters and the nature of the support, which could have implications for the use of such cluster-support combinations in a range of reactions as discussed in this case for FT processes.

1.4.5 Propylene epoxidation on small silver clusters and aggregates assembled of clusters

Ag-based catalysts of size ranging from subnanometer to nanosizes have been found to be active toward direct propylene epoxidation to propylene oxide at low temperature. The catalyst was fabricated by depositing Ag_3 clusters on a thin amorphous alumina support film and its performance probed under a gas mixture of 0.67% propylene and 0.33% oxygen seeded in He [21]. A constant gas flow of 30 sccm was maintained within the reactor kept at a pressure of 133 kPa. The stability of the clusters was monitored in situ by GISAXS measured at 12 keV, with the result shown in Fig. 1.7A and B. The subnanometer clusters start to sinter above 120°C and form stable aggregates at higher temperatures. The dashed line marks the onset of sintering. Accompanying ultrahigh-vacuum and in situ X-ray photoelectron spectroscopy (XPS) characterization were carried out on these samples as well (not shown here), revealing partially oxidized silver. As seen from Fig. 1.7C and D the propylene oxide formation started at a temperature as low as 50°C, accompanied by the production of acrolein. At higher temperatures, combustion prevailed with CO_2 as the dominating product. From DFT calculations (see Fig. 1.7E) it was seen that Ag_3–alumina support interface plays a crucial role in the efficient activation of molecular oxygen on the Ag_3 cluster. The role of atop and interfacial oxygen atoms in the epoxidation versus acrolein/CO_2 channel was also studied.

The silver aggregates forming during the reaction mentioned above showed strikingly different performance and exhibited superior selectivity toward propylene oxide at temperatures up to 120°C as shown in Fig. 1.7F and G. The rate of PO formation was ~1 molecule/Ag atom/s, which was much higher than on other reported Ag catalyst. The O_2 dissociation and propylene epoxidation apparent barriers at the interface between the Ag aggregate and the alumina support are calculated to be 0.2 and 0.2–0.4 eV, respectively, as shown in Fig. 1.7H. The barriers are somewhat lower than those on sites away from the interface. The dissociation mechanism at the interface is similar to what was previously found for the silver trimer on alumina and can account for the high activity observed for the direct oxidation of propylene on the Ag aggregates.

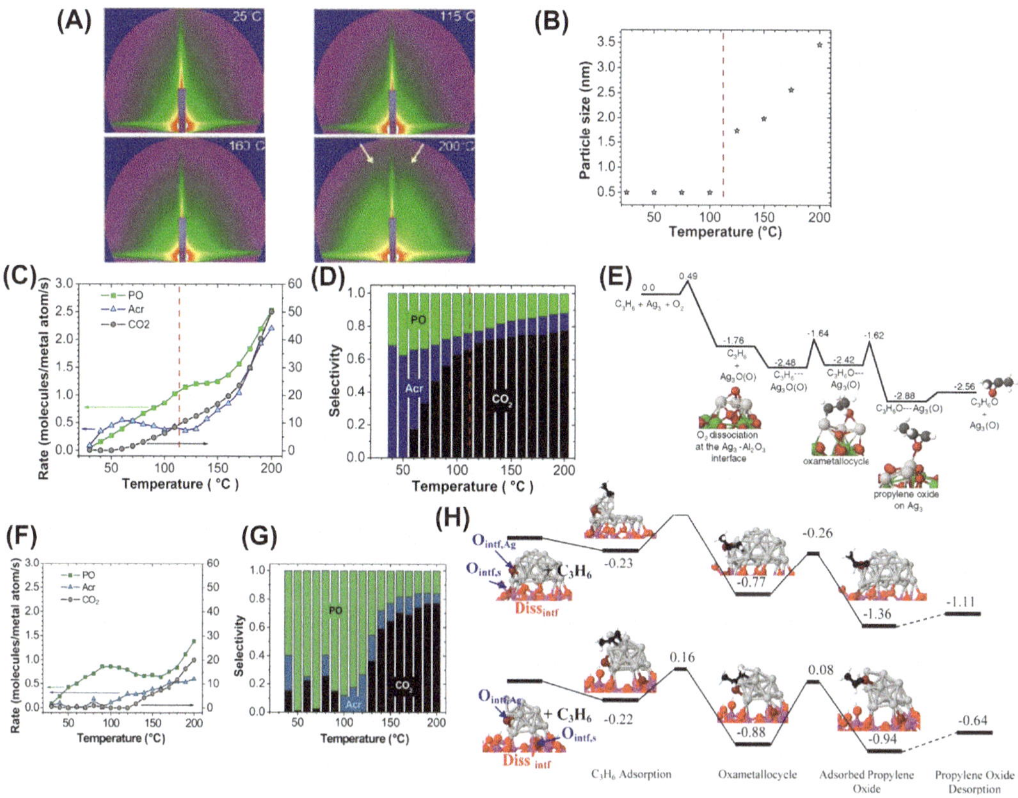

FIGURE 1.7 Propylene oxidation using Ag_3 clusters and Ag aggregates supported on thin ALD-alumina films. (A) In situ GISAXS patterns of the Ag_3 sample at various temperatures. With the used color scheme, the GISAXS pattern is at room temperature dominated by scattering from the alumina support, while the analysis of the 2-dimensional GISAXS image acquired at 200°C reveals 2 nm-tall silver aggregates with 4 nm diameter. The arrows highlight the typical GISAXS pattern arising from small nanoparticles. (B) Average cluster size determined by in situ GISAXS which reveals the onset of sintering of Ag_3 clusters above 120°C. (C) TOR of propylene oxidation on Ag_3 sample toward propylene oxide (PO), acrolein (Acr), and CO_2 calculated per surface silver atom. (D) Selectivity on Ag_3 sample of PO, Acr, and CO_2 as a function of reaction temperature, showing the highest selectivity toward PO and acrolein at lower temperatures. The red vertical dashed lines in plots (B–D) indicate the onset temperature of the sintering of the Ag_3 clusters. (E) Energies of intermediates and transition states in the mechanism for epoxidation of propylene on Ag_3/Al_2O_3 from DFT calculations. (F) TOR of propylene oxidation on sintering-resistant Ag aggregates that formed from Ag_3 clusters during the first reaction cycle. TOR of 1.4/s for PO at 200°C is in agreement with the activity data obtained after annealing of the sample at 200°C. (G) Comparable selectivities observed at 200°C for the Ag aggregates indicate the high stability at ~ 3.5 nm. (H) Reaction energy profiles of propylene epoxidation on $O_{intf,Ag}$ sites of $Diss_{intf}$ and $Diss'_{intf}$ structures. Silver, oxygen, aluminum, and hydrogen atoms are shown in silver, red, magenta, and white, respectively. The two oxygen atoms from the O_2 molecule are shown in dark red for clarity. *(A–G) Reproduced with permission from Lei et al., Science 328 (2010) 224–228. Copyright 2010, American Association for the Advancement of Science. (H) Reproduced with permission from Cheng et al., ACS Catal. 4 (2014) 32. Copyright 2014, American Chemical Society.*

In situ GISAXS revealed the change in the morphology of the deposited Ag trimers during the reaction. The trimers stayed as individual particles at low temperatures and formed aggregates at higher temperatures, the latter exhibiting excellent selectivity toward propylene oxide.

1.4.6 Oxidative dehydrogenation of cyclohexane on nm-size cobalt oxide particles

Co nanoparticles with mean size of 6 and 12 nm made by wet chemistry and dispersed on alumina support were used as catalyst to study oxidative dehydrogenation of cyclohexane [22]. A flow of 30 and 1.2 sccm for cyclohexane (4000 ppm in He) and O_2 (100%), respectively, was maintained in the cell reactor, which produces a cyclohexane:oxygen ratio of 1:10. The pressure within the cell was controlled at 800 Torr. TEM images of the catalyst surface with the 6 and 12 nm particles are shown in Fig. 1.8A and B, respectively. In situ XANES spectra measured at the Co K-edge for the 12 nm sample at select temperatures along with the spectra of bulk Co standards are shown in Fig. 1.8C and D, respectively. The corresponding derivatives of the XANES spectra for the Co nanoparticles and of bulk Co standards are shown in Fig. 1.8E and F. The spectral features are indicative of the presence of Co_3O_4 phase in the catalyst throughout the course of the reaction. Though the XANES spectra initially show slightly different characteristics—the rise of the spectrum to the peak at 7728 eV at initial 25°C is smoother than the spectrum at 250°C, and also the feature in the starting 25°C spectrum at 7746 eV is more prominent than in the spectra obtained at higher temperatures. This difference at low temperatures was due to the presence of a hydroxide phase in the as prepared catalyst as confirmed by ex situ XRD (not shown). The transformation from a mixture of Co_3O_4 and $Co(OH)_2$ to pure Co_3O_4 is more evident on comparison with the XANES derivatives at Fig. 1.8E and F. No change in the height of the particles was observed with in situ GISAXS (measured at 8 keV); however, assembly formation at high temperature and possible disintegration on cooling were observed during the reaction. The changes in the particle diameter for the 6 and 12 nm samples are shown in Fig. 1.8G and H. At room temperature, the 6 nm sample shows a bimodal distribution of particles with peaks centered at 6.1 and 9.8 nm. The changes in the distribution with temperature were minimal with slight broadening seen at higher temperatures. In case of 12 nm the evolution was much different with the initial size distribution showing 11.7 nm particles; a significant rise to larger assemblies was seen with increase in temperature. The extent of assembly is much larger for the 12 nm particles relative to the 6 nm particles, most likely caused by increased mobility of the as prepared –OH terminated particles in case of the former.

The catalyst produce a wide spectrum of products including cyclohexene (C_6H_{10}), cyclohexadiene (C_6H_8), and benzene (C_6H_6), however the process is highly temperature and size dependent as seen from Fig. 1.8I and J. The larger 12 nm particle sample is more selective toward the dehydrogenated products, especially at lower temperatures, whereas the smaller 6 nm particles are more prone toward combustion. At higher temperature both catalysts produce higher fraction of benzene, along with cyclohexene and CO_2.

The in situ XANES spectra collected for the two different size nanoparticles in the course of the reaction showed that, after the initial low temperature effect, none of the

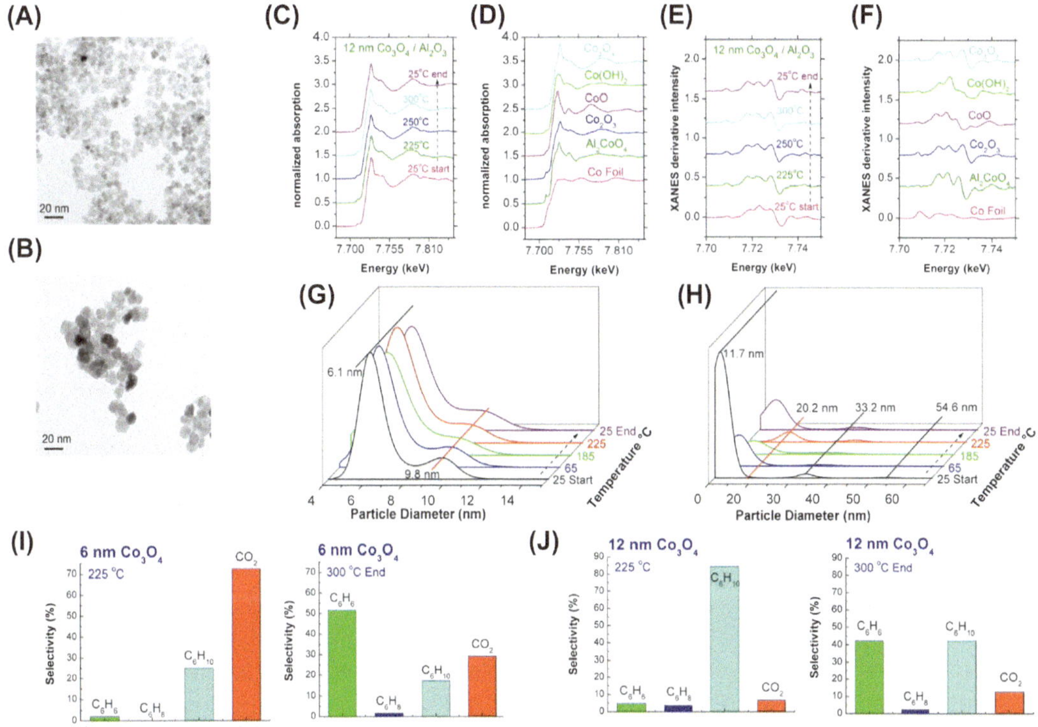

FIGURE 1.8 Oxidative dehydrogenation of cyclohexane on cobalt oxide nanoparticles. TEM image of as synthesized cobalt oxide nanoparticles of two main sizes, 6 nm (A) and 12 nm (B), (C) Co K-edge XANES spectra of the 12 nm Co_3O_4 nanoparticles under ODH conditions during the temperature ramp (from bottom to top: 25°C, start, 225°C, 250°C, 300°C, and 25C, end). (D) XANES spectra of cobalt standards Co metal, Al_2CoO_4, Co_2O_3, CoO, $Co(OH)_2$, and Co_3O_4. The corresponding derivatives of the spectra are shown in (E) and (F). Temperature-dependent evolution of the horizontal size distribution of the 6 nm (G) and 12 nm (H) particles and their assemblies during the reaction. The lines in the plots indicate the main sizes identified by the analysis of the GISAXS data. Carbon-based selectivities for the carbonaceous products for the ODH of cyclohexane by Co_3O_4 6 nm nanoparticles at (I) 225°C, and after 1 h of reaction at 300°C. The same for 12 nm Co_3O_4 nanoparticles is shown in (J). *Figures reproduced with permission from Tyo et al., ACS Catal. 2 (2012) 2409–2423. Copyright 2012, American Chemical Society.*

clusters changes its composition in the course of the reaction, and hence the stability of the clusters as far as their oxidation state is evident. On the other hand the GISAXS data collected on the catalyst surface show that the 6 nm particles are more stable in their size under ODH conditions than the 12 nm size particles.

1.4.7 Evolution of size-selected cobalt nanocluster–based catalysts under oxidative dehydrogenation conditions of cyclohexane

This study involved the elucidation of evolution of the structure, stoichiometry/composition, and morphology of gas-phase-produced size-selected cobalt nanoparticles in the ~1–20 nm size range under cyclohexane ODH conditions, deposited on alumina and UNCD support, on nanoclusters which were exposed to the ambient after deposition [23]. A flow of 30 and 1.2 sccm for cyclohexane (4000 ppm in He) and O_2 (100%), respectively, was maintained into the cell reactor, which produces a cyclohexane:oxygen ratio of 1:10. The pressure within the cell was controlled at 800 Torr. Fig. 1.9A shows the evolution of in situ XANES spectra measured at the Co K-edge of the larger alumina-supported clusters on Al_2O_3 (L_1) under reaction conditions for oxidative dehydrogenation of cyclohexane. As shown in Fig. 1.9B under oxygen-rich conditions the fraction of CoO and Co_3O_4 increases at high temperature and the metallic phase vanishes; Co_3O_4 is the most dominating phase at 300°C. The average valence state of the particles increases from 1 at the beginning of the experiment to 2.5 at the end of the reaction cycle. TEM images (bright field) before reaction are shown in Fig. 1.9C with the clusters having a darker core and a brighter shell. The brighter shell is caused by partial oxidation at the surface due to exposure to air after deposition. After treatment, the core appears bright and is surrounded by a darker area as shown in Fig. 1.9D. The outermost region again is brighter. The image demonstrates an example of formation of a hollow nanoparticle with its origin lying in the nanoscale Kirkendall effect. The horizontal cuts of the GISAXS images collected at 8 keV from a sample L_2, which are composed of 22 nm particles, showed a subtle change occurring around 250°C as shown in Fig. 1.9E. The analysis of GISAXS data revealed that the single size distribution seen for the particles at low temperature evolves into a bimodal distribution as shown in Fig. 1.9F. The lateral size for the two peaks is found to be both larger and smaller than d_i as seen from Fig. 1.9F, which is consistent with the observation of a core-shell formation from the TEM images and the two diameters obtained from GISAXS representing the inner and outer diameters of the transformed particles. Similar observation was also made for L_1 sample (not shown). In case of smaller clusters, only a small fraction of the initial CoO phase was converted to Co_3O_4, so in the final stage the particles consist of about 75% CoO and 25% Co_3O_4 as shown in Fig. 1.9G and H. The average valence state also does not show any change and stays around 2.2. Also from the TEM images shown in Fig. 1.9I and J for the nanoparticles imaged before and after reaction one cannot see any observable differences. These nanosized particles are robust under the reaction conditions.

Fig. 1.9K summarizes the change in particle sizes during the reaction and it can be clearly seen that the overall size of the particle on Al_2O_3 (M_1, L_1, and L_2, black dots) only undergo small size changes compared to UNCD/carbon supported species (L_3, gray dots). The Al_2O_3-supported particles show an isotropic Kirkendall effect, whereas on UNCD, a dramatic lateral size change is found for the particles, which we attribute to an anisotropic form of the Kirkendall effect where a flat disklike structure around the particles has been observed.

A combination of in situ GISAXS and GIXANES study reveals interesting differences in the evolution of morphology and composition of the small, medium, and large cobalt nanoclusters and in identifying the temperature changes that take place depending on

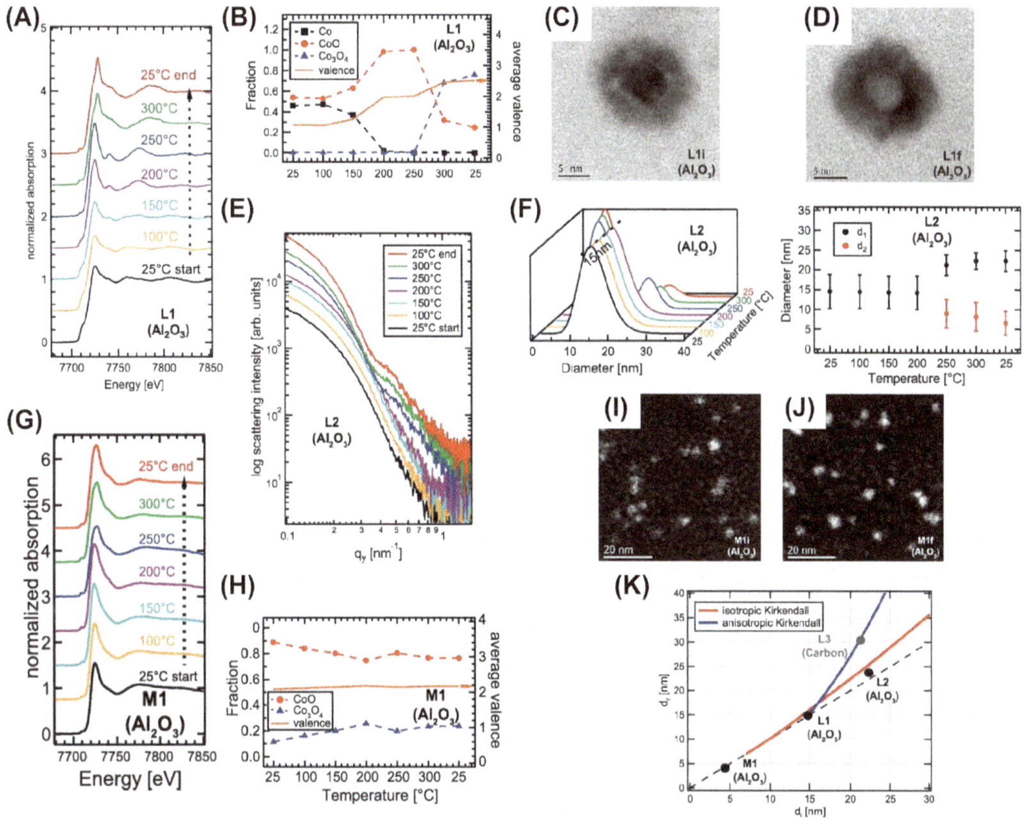

FIGURE 1.9 Structure and morphology evolution of size-selected cobalt nanoclusters under reaction conditions. (A) Normalized GIXANES spectra on Al_2O_3 (L_1, $d_i = 15$ nm) at different temperatures. The spectra are offset vertically for clarity. (B) Composition of clusters on sample L_1 obtained by LCF fitting with bulk Co standards (C) TEM images of L_1 before TPR. All particles show brighter shells compared to the core indicating a partial oxidation at the surface. (D) TEM images of L_1 after TPR. The core appears bright, surrounded by a darker area. The outermost region again shows up brighter. The images indicate hollow particles that are a typical result of the Kirkendall effect. (E) Horizontal cuts taken from the GISAXS images recorded between 25°C and 300°C with Co nanoparticles on alumina (L_2, $d_i = 22$ nm). The spectra are offset vertically for clarity. (F) Horizontal size distributions for the nanoparticles representing a splitting into two size distributions at 250°C. (C) Diameters obtained for L_2. (G) Normalized GIXANES spectra on Al_2O_3 (M_1, $d_i = 4.4$ nm) at different temperatures, and (H) Composition of M_1 sample (I) and (J) are the TEM images of M_1 before and after reaction, respectively. (K) Change of lateral cluster diameter upon reaction. The expected size changes due to the Kirkendall effect based on simple geometric models are also plotted for isotropic (*red*) and anisotropic (*blue*) Kirkendall effect. *Figures reproduced with permission from Bartling et al., ACS Nano 9 (2015) 5984–5998. Copyright 2012, American Chemical Society.*

particle size. It was inferred from detailed experiment that the clusters take completely different structures under reaction conditions on being heated to high temperatures. These, often subtle, differences in particle morphologies and composition can have a profound effect in the catalytic properties of these clusters.

1.4.8 Oxidative dehydrogenation of cyclohexene by oxidized 27-atom cobalt clusters

A study of oxidative dehydrogenation of cyclohexene was performed using subnanometer cobalt clusters, Co_{27}, deposited on alumina (Al_2O_3) and magnesia (MgO) supports [24]. The purpose was to study the effects of oxide supports in the dehydrogenation of cyclohexene under oxygen rich conditions (cyclohexene: oxygen ratio of 1:10, cyclohexene (0.39%) and oxygen (3.90%) in He at a total pressure of 800 torr.). From Fig. 1.10A we can see that the XANES spectra (Co-K edge) of the as-prepared sample and the spectra in reaction gas environment up to 200°C with a dominant mixed phase of CoO and Co_3O_4. A sudden change occurs at 250°C (which was even more pronounced at 300°C) in the Co/MgO sample, which includes forming a sharper peak at 7727 eV and a distinct double-shoulder feature at lower energy. The change in Co/MgO sample is indicative of the transformation to a higher oxidation state, Co_3O_4, along with a possibility of the formation of a mixed phase $Co_xMg_{1-x}O$ with the MgO support. No such change was observed for alumina-supported clusters. GISAXS images collected at 8 keV showed no change in the particle size for Co clusters on the Al_2O_3 substrate (Fig. 1.10B), whereas in case of MgO an assembly formation which is marked by a growth of particle size distribution above 250°C is shown in Fig. 1.10C and D. The TPR results for Co/Al_2O_3 and Co/MgO are shown in Fig. 1.10F. Co/Al_2O_3 shows highest activity and selectivity toward benzene production in the entire temperature region. In the low temperature range when the Co/MgO system still consists of subnanometer size clusters, Co_{27} clearly outperforms Co/Al_2O_3 catalyst, especially, at 150°C and 200°C. At higher temperature Co/MgO forms highly dynamic nanostructures; however, it maintains its high reactivity and seems to adapt its size and morphology to offer optimal active sites for the reaction.

In situ GISAXS/GIXAS study proved instrumental in identifying the size and nature of the working catalysts for ODH of cyclohexene. Co_{27}/Al_2O_3 and Co_{27}/MgO evolved quite differently when exposed to the same reaction conditions. GISAXS results revealed that the sudden jump in the activity of Co_{27} clusters on MgO supports correlated with a transient nanostructure formation from the clusters at high temperatures, and a simultaneous GIXAS measurement revealed intermixing with MgO before the onset of assembly took place.

1.4.9 Hydrogenation of ethylene on Pt nanoparticles

Catalytic activity and characterization of Pt nanoparticles prepared by a wet chemical method supported on a GaN semiconductor substrate were investigated in the hydrogenation of ethylene with deuterium [9]. The smaller particles were 1.8 nm in size, the larger ones were 7 nm. The particle distribution and coverage on the semiconductor substrate were characterized by AFM, with images shown in Fig. 1.11A. During the TPR/GISAXS/GIXAS measurements the reactants used were gases at 1% C_2H_4 in He and 99.7% D_2 flowing at a rate of 50 sccm through the reactor at a pressure of 100 kPa. GISAXS images were collected at 11.5 keV. Under reaction conditions the evolution of the shape

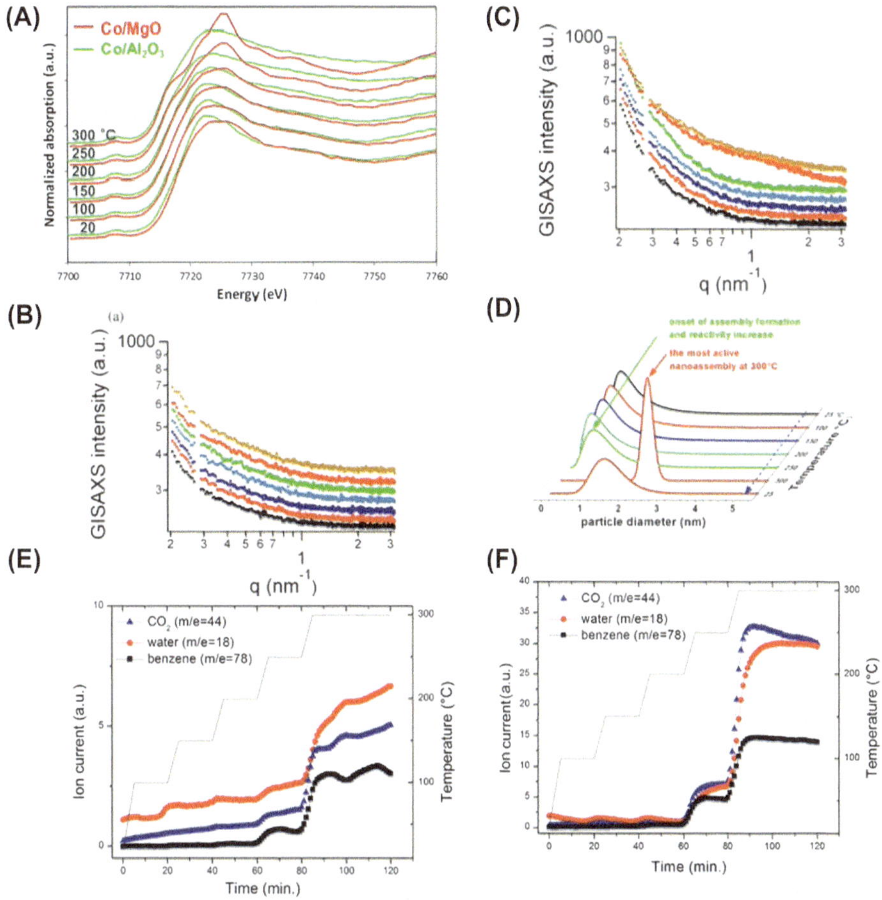

FIGURE 1.10 Oxidative dehydrogenation of cyclohexene on oxidized Co_{27} clusters. (A) Co K-edge XANES spectra of Co/MgO (*red*) and Co/Al_2O_3 (*green*) from bottom to up in the order of reaction temperatures, 20°C, 100°C, 150°C, 200°C, 250°C, and 300°C, respectively (an offset was applied to the curves for better clarity). GISAXS intensity (horizontal line cut) change during the heating cycle, from the bottom 20°C, 100°C, 150°C, 200°C, 250°C, 300°C, and 20°C (after reaction), respectively, for (B) alumina-supported and (C) magnesia-supported cobalt clusters; the spectra are offset for clarity. (D) evolution of particle/nanoassembly size distribution of the magnesia-supported clusters during the reaction. (E) and (F) Background-corrected TPR profile of benzene, water, and carbon dioxide evolution on Co/Al_2O_3 and Co/MgO sample, respectively; the spectra are offset for clarity. *Figures reproduced with permission from Lee et al., Phys. Chem. Chem. Phys. 14 (2012) 9336–9342. Copyright 2012, Royal Society of Chemistry.*

for the small and large Pt particles are shown in Fig. 1.11B and C, respectively. The aspect ratio of the smaller particles became slightly decreased on introducing the reactant mixture, and it recovered to the initial shape after the gas feed was cut off. The larger Pt particles were found more rigid, they did not change their shape during the course of the reaction. GIXANES spectra for the two Pt samples at the Pt L_3-edge are shown in Fig. 1.11D and E. A decrease in the white line intensity is seen for small particles after ethene deuteration, indicative of the reduction Pt. No change was observed for the larger Pt nanoparticles which could be due to the low surface-to-volume ratio of the larger particles, which may limit sensitivity to changes taking place at the surface of the particles. A high TOR of deuterated products was measured for the small particles, reaching 22 molecules of deuterated ethylene per Pt atom, which was higher than reported for other known contemporary hydrogenation catalysts. A drop in the activity by about 30% was observed after the first 10 min of which coincides with the region where a fine change in particle shape, without indication of sintering, took place. In contrast, large particles had a constant turnover rate of 12 molecules (total Pt atoms)$^{-1}$ s^{-1}. The results highlight the importance of the size and shape sensitivity of the smaller and larger Pt particles, during the reaction.

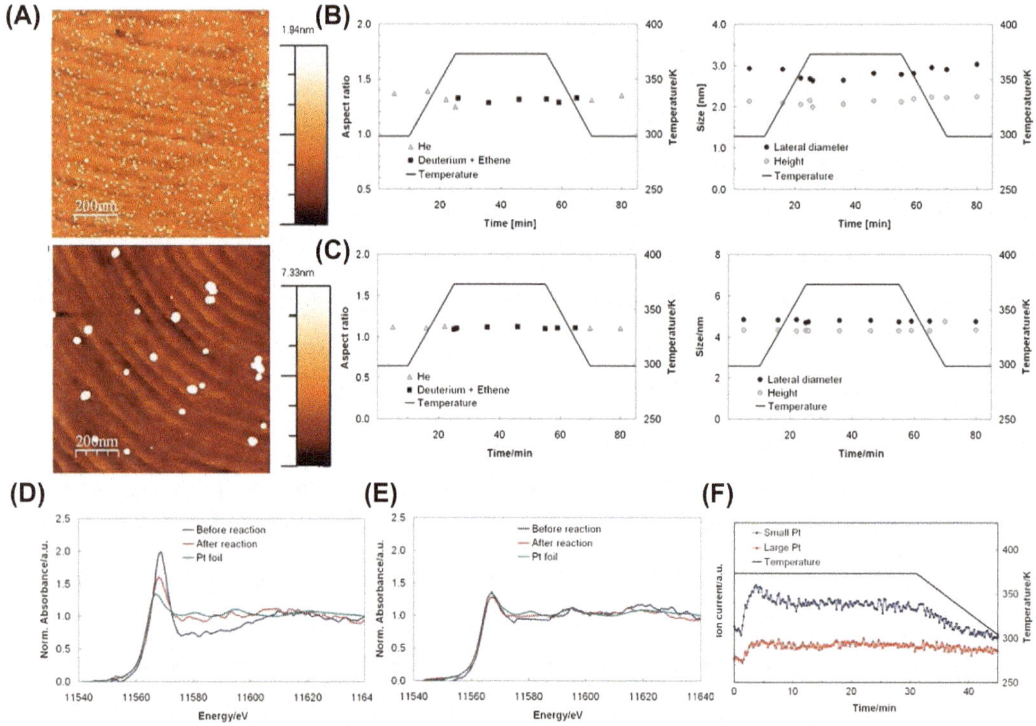

FIGURE 1.11 Platinum nanoparticles in the hydrogenation of ethane. (A) AFM images of spin-coated Pt nanoparticles on n-GaN with particle sizes of 1.8 ± 0.3 nm (*top*) and 6.7 ± 1.1 nm (*bottom*). The root mean square of blank n-GaN was 0.1 nm. (B) Aspect ratios of small Pt nanoparticles on n-GaN during deuterium–ethene reaction and the corresponding lateral diameter and height for small (B) and large (C) Pt nanoparticles. The temperature cycle is plotted as a solid line. Normalized Pt L_3 edge spectra of (D) small and (E) large Pt particles on n-GaN and Pt foil recorded at 298 K in helium before and after deuterium–ethene reaction. (F) Ion current of ethane ($C_2H_4D_2$, m/z = 30) for small and large Pt nanoparticles on n-GaN during deuterium–ethene reaction. The temperature is plotted as solid line. *Figures reproduced with permission from Wyrzgol et al., Phys. Chem. Chem. Phys. 12 (2010) 5585–5595. Copyright 2010, Royal Society of Chemistry.*

GISAXS and GIXANES coupled with TPR proved to be powerful in providing information about size-dependent morphology and oxidation state changes during the hydrogenation of ethylene, showing that the smaller Pt particles changed their shape and oxidation state, while the larger particles were more robust.

1.4.10 Electrochemistry with subnanometer clusters: water oxidation by size-selected palladium clusters

An efficient method for water splitting is a grand scientific challenge. Water oxidation is a key catalytic step for electrical fuel generation. In order to study water oxidation in alkaline conditions, size-selected Pd clusters of three different sizes (Pd_4, Pd_6, and Pd_{17}) supported on UNCD were used to probe the relationship between cluster size and efficacy of the oxygen evolution reaction (OER) [25]. A strong size-dependent activity was observed for the clusters as shown in Fig. 1.12A. It was found that Pd_4 was inactive, whereas Pd_6 and Pd_{17} positioned among the most active electrocatalysts on a per-metal-atom basis, reaching turnover rates of 0.68 and 0.6 oxygen molecules formed per second per Pd atom as shown in Fig. 1.12B. Ex situ GIXANES (measured at Pd $K\alpha_1$ X-ray emission line) and GISAXS (images collected at 24.3 keV) characterization of the samples shown in Fig. 1.12C and D revealed a stable PdO-like composition for all cluster sizes. DFT calculations were employed to calculate the energetics of the underlying reaction steps on gas-phase Pd clusters (Pd_4, Pd_4O_4, Pd_4O_3, and Pd_6O_6 clusters, Fig. 1.12E) to provide a fundamental understanding for the experimental findings. The stark difference in the OER activity observed between Pd_4, Pd_4O_n, and Pd_6O_6 clusters is explained by the activity of the bridging metallic Pd–Pd bond/site present in the Pd_6O_6 driving the reaction, a site which is absent in the smaller Pd_4O_4 cluster.

GISAXS performed ex situ showed a stable morphology of the electrodes and no indication of sintering of clusters, while GIXANES provided information about the oxidation states of clusters that contributed to the understanding of the size effects in their electrocatalytic performance.

1.4.11 Bimetallic/alloy clusters

1.4.11.1 Heterogeneous catalysis with bimetallic clusters: CO oxidation on size- and composition-selected Ag-Pt clusters

Bimetallic Ag-Pt clusters, of two different compositions, Ag_9Pt_2 and Ag_9Pt_3, were deposited on ultrathin amorphous alumina support to study the effect of cluster composition in CO oxidation (CO_{ox}) [26], with the simultaneous monitoring of the stability of the clusters and the evolution of the oxidation state of their components during the course of the reaction. A continuous 30 sccm flow of CO and O_2 seeded in He at a $CO/O_2/He = 1{:}1{:}98$ ratio was maintained within our reactor kept at a pressure of 800 torr and with applied temperature ramp shown in Fig. 1.13A. The TOR was found among the highest reported, amounting to about 250 molecules of CO_2 formed per

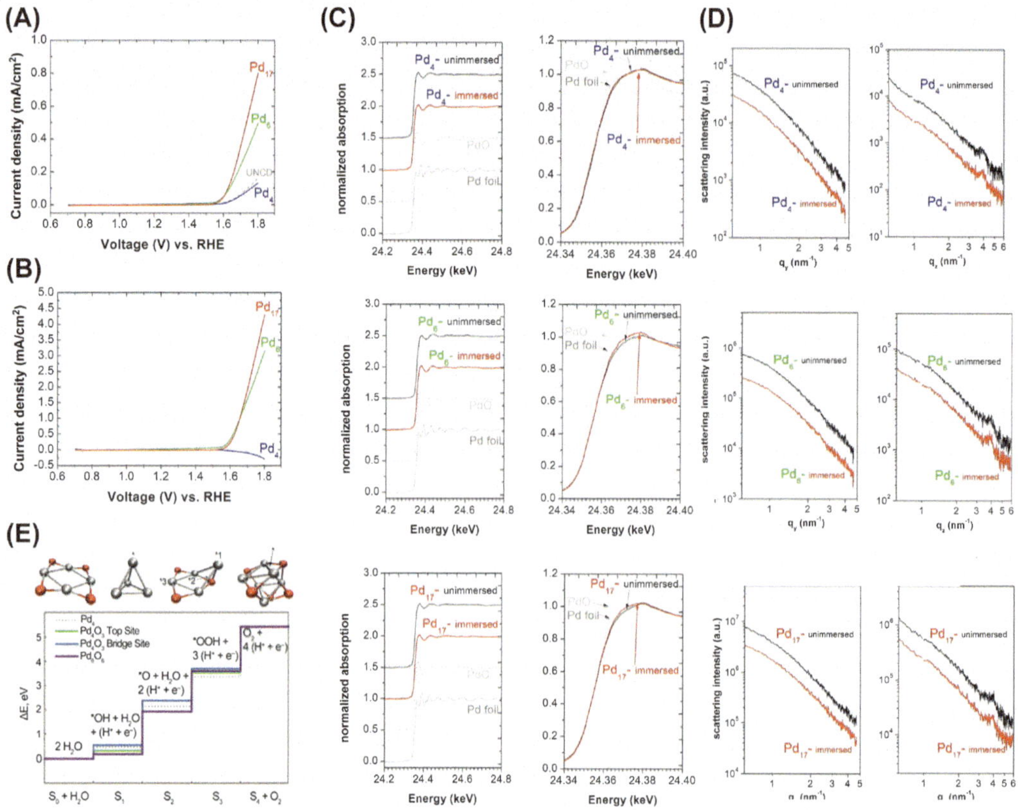

FIGURE 1.12 Pd clusters in oxygen evolution reaction. (A) Performance of size-selected palladium cluster on UNCD under pH 13.6 (1M NaOH) with current densities normalized to the total chip area immersed into the solution blank UNCD support (gray) Pd_4 (blue), Pd_6 (green), and Pd_{17} (red) clusters. (B) Background corrected (i.e., after subtraction of the current of the blank UNCD support) currents normalized for the cluster-covered area (C) Ex situ x-ray absorption spectra for unimmersed (as deposited, black) and immersed and cycled (red) areas of electrodes with Pd clusters. The spectra of bulk standards Pd and PdO are shown for comparison as well. An offset is applied on spectra for clarity. (D) Horizontal (q_y) (left column) and vertical (q_z) (right column) cuts of the GISAXS data, acquired with photons of 24 keV energy, for unimmersed (as deposited) and immersed area for the Pd_4, Pd_6, and Pd_{17} clusters investigated (E) Structures used for calculation of the thermochemistry of the OER reactions ($S_0 \rightarrow S_4$ for changes in free energy) of the Pd_4O_4, Pd_4, Pd_4O_3, and Pd_6O_6 clusters. "*" indicates the reaction sites. Changes in free energy for the intermediates of the electrolysis of water by the subnanometer clusters: Pd_4 (*dashed line*), Pd_4O_3 site 1 (*blue line*), Pd_4O_3 site 2 (*green line*), and Pd_6O_6 (*purple line*) at U = 0 and pH = 0. The products and reactants for each of the four electron transfer reactions are displayed on the $S_0 \rightarrow S_4$ steps. Here the "*" indicates a cluster attached product. *Figures reproduced with permission from Kwon et al., ACS Nano 7 (2013) 5808–5817. Copyright 2013, American Chemical Society.*

second per metal atom. Moreover the AgPt bimetallic clusters were found more stable when compared with monometallic Pt clusters tested under identical reaction conditions [27], showing no signs of deactivation (Fig. 1.13A) The evolution in the XANES features and oxidation state of silver and platinum for Ag_9Pt_2 at the Ag K-edge and Pt L_3-edge is shown in Fig. 1.13B–E, respectively. Under the applied reaction conditions for Ag_9Pt_2/Al_2O_3, the average valence of both Pt and Ag is found to be rather constant at about 0 and 0.6, respectively. GISAXS measurement done at 25.5 keV has shown no sign of particle sintering. Accompanying DFT calculations reveal that the Pt atoms anchor the mixed clusters to the Al_2O_3 support, and also bind and activate two CO molecules, while Ag binds and activates O_2. The proximity of Ag and Pt suppresses poisoning by CO or by carbonate (CO_3) as shown in Fig. 1.13F, in the low temperature regime. Langmuir–Hinshelwood mechanism is proposed at room temperature, which is replaced by Eley–Rideal mechanism at elevated temperatures.

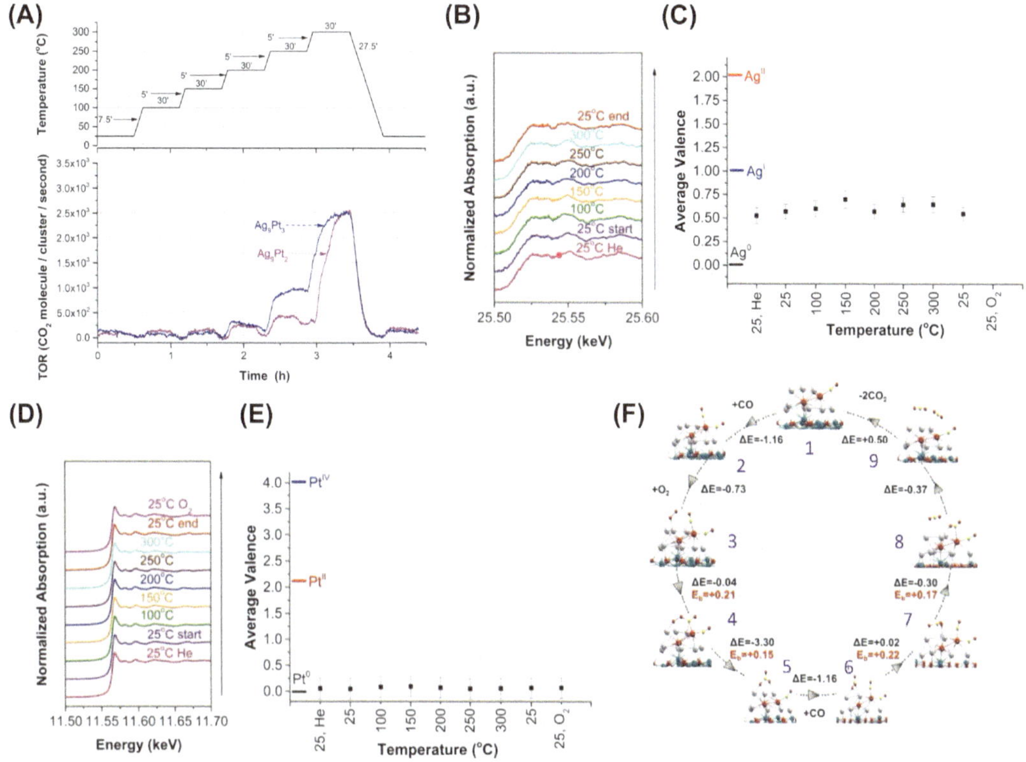

FIGURE 1.13 CO oxidation using bimetallic size - and composition selected Ag-Pt alloy clusters on oxide supports. (A) Top: Temperature ramp applied during the CO oxidation reaction. Bottom: Per cluster turn-over rate of CO_2 formed on Ag_9Pt_2 (pink) and Ag_9Pt_3 (blue) clusters as the function of temperature. (B) and (C) XANES spectra of Ag_9Pt_2 clusters obtained at Ag K-edge with the temperature ramp and the average valence state of silver in alumina supported bimetallic Ag_9Pt_2, respectively. (D) and (E) XANES spectra of Ag_9Pt_2 clusters obtained at Pt L_3-edge with the temperature ramp and the average valence state of silver in alumina supported bimetallic Ag_9Pt_2, respectively (F) Complete COox catalytic cycle on Ag_9Pt_2 supported on the (001) surface of α-alumina. ΔE (*in black*) are reaction energies and E_b (*in red*) are reaction energy barriers. Langmuir-Hinshelwood (LH) steps (2 → 3 → 4 → 5) hold at RT and are replaced by an Eley-Rideal (ER) step at high temperatures. Ag atoms are in gray, Pt atoms are in red, and O and A_l atoms in the substrate are in small dark red and blue spheres, respectively. Energy values are in eV and spin in bohr. *Figures reproduced with permission from Negreiros et al., Angew. Chem., Int. Ed. 57 (2018) 1209. Copyright 2017, Wiley-VCH Verlag GmbH & Co. KGaA.*

In situ GISAXS played a key role in showing that the clusters are stable during prolonged exposure to reactants at high temperatures, while GIXANES turned out to be very helpful in the understanding of the role of oxidation state of the silver and platinum in relation to their role in CO oxidation.

1.4.12 Oxidation states of cobalt and platinum in 3 nm size $Co_{1-x}Pt_x$ nanoclusters under H_2 and $CO + H_2$

Cobalt-based catalysts have been widely used for Fischer-Tropsch Synthesis (FTS). The referred study involved addition of Pt promoter to cobalt, in a well-defined ratio, through the synthesis of mass-selected $Co_{1-x}Pt_x$ clusters and depositing them onto a thin amorphous alumina film [28]. The particles had a diameter of about 2.9 nm and a composition encompassing Pt fractions of $x = 0$, 0.25, 0.5, 0.75, and 1. Results from GIXAS characterization of the samples performed after extended exposure of samples to air are shown in Fig. 1.14A and C, for the Co (K-edge) and Pt (L_3-edge), respectively, indicate oxidized Co and Pt components. Linear combination fitting (LCF) using bulk Co standards reveals the coexistence of Co^{2+} (CoO) and Co^{3+} (Co_3O_4) phases (see Fig. 1.14B). In pure Co clusters the fractions of CoO and Co_3O_4 make up 63% and 37% respectively. In the sample containing 25% Pt ($x = 0.25$), the fraction of the Co_3O_4 component decreases to about 14%. Further increase in the Pt/Co ratio yielded a gradual shift in the Co oxidation state toward the Co_3O_4 phase as seen from Fig. 1.14E. $Co_{0.75}Pt_{0.25}$ exhibits the most prominent Pt^{4+} phase, and no Pt^{2+} component. The $Co_{0.5}Pt_{0.5}$ cluster shows the highest fraction of the Pt^{2+} state, and the $Co_{0.25}Pt_{0.75}$ and Pt clusters contain the highest fractions of Pt^0 component (see Fig. 1.14D). Platinum component within the bimetallic clusters was found in a reduced form or was readily reducible under H_2 in all $Co_{1-x}Pt_x$ clusters, except that with $x = 0.5$, where a complete reduction occurred only under $CO + H_2$ at 25°C. The clusters were first pretreated with 3.5% H_2 balanced with He at 900 torr pressure; followed by treatment with a mixture of 1% CO and 2% H_2 balanced with He at the same pressure. Pt retained its metallic state in all clusters up to 225°C under CO and H_2 mixture as seen from Fig. 1.14F. In contrast, except in the $Co_{0.25}Pt_{0.75}$ cluster, where cobalt seemed to form the core of the particles, the cobalt component was found to undergo a stepwise change during pretreatment with hydrogen. The experiment reveals that by changing the ratio of the metal components and reaction conditions it is possible to control the oxidation state (and in some case the structure) of nanoalloys.

In situ XANES results revealed the interesting feature that the composition dependence for both Co and Pt in the Co-Pt clusters changed nonlinearly with particle composition and temperature. GISAXS images collected at 11.6 and 7.8 keV helped to confirm that the particles don't sinter under the applied conditions.

The shown examples of applications of synchrotron-based characterization were aimed to highlight the impact and importance of such techniques in the

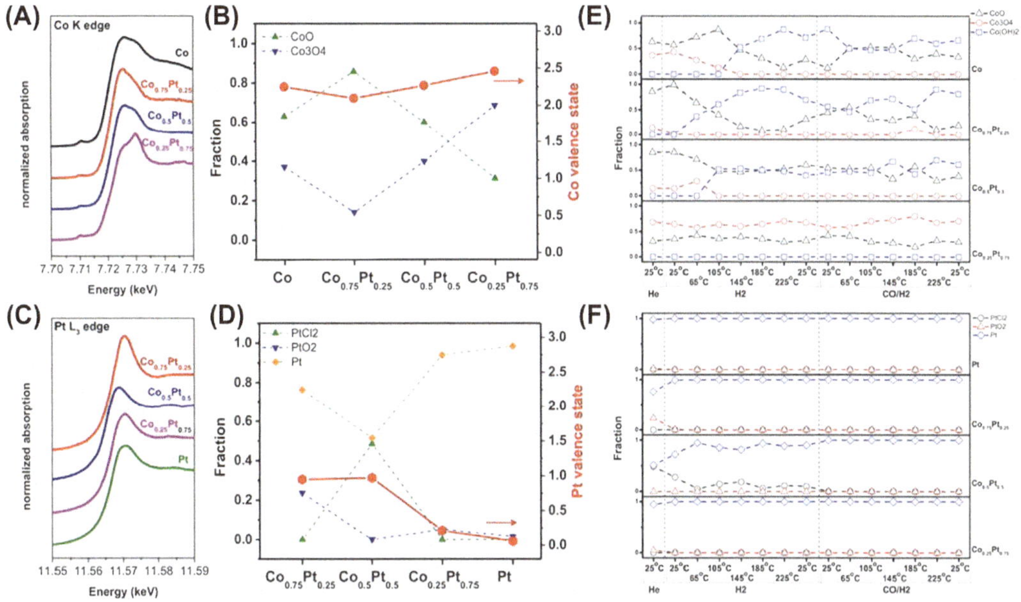

FIGURE 1.14 TPR study with Co-Pt alloy nanoclusters during heating under syngas. XANES spectra and its analysis of as-made $Co_{1-x}Pt_x$ clusters on an Al_2O_3 thin film after aging in the air: (A) Co K-edge XANES spectra of Co, $Co_{0.75}Pt_{0.25}$, $Co_{0.5}Pt_{0.5}$, and $Co_{0.75}Pt_{0.25}$. (B) Fractions of Co components from LCF results using CoO and Co_3O_4 bulk standards and average Co valence state plotted in red. (C) Pt L_3-edge XANES spectra of $Co_{0.75}Pt_{0.25}$, $Co_{0.5}Pt_{0.5}$, $Co_{0.75}Pt_{0.25}$, and Pt. (D) Fractions of Pt components from LCF results using $PtCl_2$ and PtO_2 bulk standards and average Pt valence state plotted in red. (E) Fractions of Co components from LCF using CoO, $Co(OH)_2$, and Co_3O_4. (F) Fractions of Pt components from LCF of the Pt edge using $PtCl_2$ and PtO_2. *Figures reproduced with permission from Yang et al., J. Phys. Chem. C 120 (2016) 21496–21504. Copyright 2016, American Chemical Society.*

identification of the nature of the working catalytic sites in gaining fundamental understanding of catalytic processes and also recipes for the design of new classes of catalytic materials.

Acknowledgment

The authors acknowledge the support by the U.S. Department of Energy, Office of Science, Office of Basic Energy Sciences Materials Science and Engineering under Contract No. DE-AC-02-06CH11357. S.V. also acknowledges support from the European Union's Horizon 2020 research and innovation programme under grant agreement No 810310, which corresponds to the J. Heyrovsky Chair project ("ERA Chair at J. Heyrovský Institute of Physical Chemistry AS CR – The institutional approach towards ERA") during the finalization of the paper. The funders had no role in the preparation of the article.

References

[1] A. Halder, L.A. Curtiss, A. Fortunelli, S. Vajda, Perspective: size selected clusters for catalysis and electrochemistry, J. Chem. Phys. 148 (2018) 110901. https://doi.org/10.1063/1.5020301.

[2] E.C. Tyo, S. Vajda, Catalysis by clusters with precise numbers of atoms, Nat. Nanotechnol. 10 (2015) 577. https://doi.org/10.1038/nnano.2015.140.

[3] S. Vajda, M.G. White, Catalysis applications of size-selected cluster deposition, ACS Catal. 5 (2015) 7152–7176. https://doi.org/10.1021/acscatal.5b01816.

[4] S. Lee, B. Lee, S. Seifert, S. Vajda, R.E. Winans, Simultaneous measurement of X-ray small angle scattering, absorption and reactivity: a continuous flow catalysis reactor, Nucl. Instrum. Meth. A. 649 (2011) 200–203. https://doi.org/10.1016/j.nima.2010.12.172.

[5] H. Haberland, et al., Filling of micron-sized contact holes with copper by energetic cluster impact, J. Vac. Sci. Technol. A 12 (1994) 2925–2930. https://doi.org/10.1116/1.578967.

[6] J. Kraft, O. Rattunde, O. Rusu, A. Häfele, H. Haberland, Thin films from fast clusters: golden TiN layers on a room temperature substrate, Surf. Coat. Technol. 158–159 (2002) 131–135. https://doi.org/10.1016/S0257-8972(02)00234-7.

[7] C. Yin, E. Tyo, K. Kuchta, B.v. Issendorff, S. Vajda, Atomically precise (catalytic) particles synthesized by a novel cluster deposition instrument, J. Chem. Phys. 140 (2014) 174201. https://doi.org/10.1063/1.4871799.

[8] A. Halder, et al., Water oxidation catalysis via size-selected iridium clusters, J. Phys. Chem. C 122 (2018) 9965–9972. https://doi.org/10.1021/acs.jpcc.8b01318.

[9] S.A. Wyrzgol, et al., Combined TPRx, in situ GISAXS and GIXAS studies of model semiconductor-supported platinum catalysts in the hydrogenation of ethene, Phys. Chem. Chem. Phys. 12 (2010) 5585–5595. https://doi.org/10.1039/B926493K.

[10] J. Ilavsky, Nika: software for two-dimensional data reduction, J. Appl. Crystallogr. 45 (2012) 324–328. https://doi.org/10.1107/S0021889812004037.

[11] J. Ilavsky, P.R. Jemian, Irena: tool suite for modeling and analysis of small-angle scattering, J. Appl. Crystallogr. 42 (2009) 347–353. https://doi.org/10.1107/S0021889809002222.

[12] G. Kresse, J. Hafner, Ab initio molecular-dynamics simulation of the liquid-metal–amorphous-semiconductor transition in germanium, Phys. Rev. B 49 (1994) 14251–14269. https://doi.org/10.1103/PhysRevB.49.14251.

[13] G. Kresse, J. Furthmüller, Efficient iterative schemes for ab initio total-energy calculations using a plane-wave basis set, Phys. Rev. B 54 (1996) 11169–11186. https://doi.org/10.1103/PhysRevB.54.11169.

[14] M. Nisha, et al., Reversing size-dependent trends in the oxidation of copper clusters through support effects, Eur. J. Inorg. Chem. 2018 (2018) 16–22. https://doi.org/10.1002/ejic.201701355.

[15] P.-T. Chen, et al., Size-selective reactivity of subnanometer Ag_4 and Ag_{16} clusters on a TiO_2 surface, J. Phys. Chem. C 121 (2017) 6614–6625. https://doi.org/10.1021/acs.jpcc.6b11375.

[16] J.J. Rehr, R.C. Albers, Theoretical approaches to x-ray absorption fine structure, Rev. Mod. Phys. 72 (2000) 621–654. https://doi.org/10.1103/RevModPhys.72.621.

[17] B. Yang, et al., Copper cluster size effect in methanol synthesis from CO_2, J. Phys. Chem. C 121 (2017) 10406–10412. https://doi.org/10.1021/acs.jpcc.7b01835.

[18] S. Lee, B. Lee, S. Seifert, R.E. Winans, S. Vajda, Fischer–tropsch synthesis at a low pressure on subnanometer cobalt oxide clusters: the effect of cluster size and support on activity and selectivity, J. Phys. Chem. C 119 (2015) 11210–11216. https://doi.org/10.1021/jp512157d.

[19] R.B. Anderson, R.A. Friedel, H.H. Storch, Fischer-tropsch reaction mechanism involving stepwise growth of carbon chain, J. Chem. Phys. 19 (1951) 313–319. https://doi.org/10.1063/1.1748201.

[20] S.T. Sie, in: Reviews in Chemical Engineering, vol. 14, 1998, p. 109.

[21] Y. Lei, et al., Increased silver activity for direct propylene epoxidation via subnanometer size effects, Science 328 (2010) 224–228. https://doi.org/10.1126/science.1185200.

[22] E.C. Tyo, et al., Oxidative dehydrogenation of cyclohexane on cobalt oxide (Co3O4) nanoparticles: the effect of particle size on activity and selectivity, ACS Catal. 2 (2012) 2409–2423. https://doi.org/10.1021/cs300479a.

[23] S. Bartling, et al., Pronounced size dependence in structure and morphology of gas-phase produced, partially oxidized cobalt nanoparticles under catalytic reaction conditions, ACS Nano 9 (2015) 5984–5998. https://doi.org/10.1021/acsnano.5b00791.

[24] S. Lee, et al., Oxidative dehydrogenation of cyclohexene on size selected subnanometer cobalt clusters: improved catalytic performance via evolution of cluster-assembled nanostructures, Phys. Chem. Chem. Phys. 14 (2012) 9336–9342. https://doi.org/10.1039/C2CP40162B.

[25] G. Kwon, et al., Size-dependent subnanometer Pd cluster (Pd_4, Pd_6, and Pd_{17}) water oxidation electrocatalysis, ACS Nano 7 (2013) 5808–5817. https://doi.org/10.1021/nn400772s.

[26] F.R. Negreiros, et al., Bimetallic Ag-Pt sub-nanometer supported clusters as highly efficient and robust oxidation catalysts, Angew. Chem. Int. Ed. 57 (2018) 1209–1213. https://doi.org/10.1002/anie.201709784.

[27] C. Yin, et al., Alumina-supported sub-nanometer Pt_{10} clusters: amorphization and role of the support material in a highly active CO oxidation catalyst, J. Mater. Chem. 5 (2017) 4923–4931. https://doi.org/10.1039/C6TA10989F.

[28] B. Yang, et al., Temperature-dependent evolution of the oxidation states of cobalt and platinum in Co1−xPtx clusters under H_2 and CO + H_2 atmospheres, J. Phys. Chem. C 120 (2016) 21496–21504. https://doi.org/10.1021/acs.jpcc.6b06483.

2

From cluster design to energy storage device engineering

Marta Haro*

NANOPARTICLES BY DESIGN, OIST, ONNA-SON, OKINAWA, JAPAN
**CORRESPONDING AUTHOR*

Chapter outline

2.1 Outline

In this chapter, we are going to discuss the work performed in engineering energy storage devices using cluster beam deposition (CBD) technique. In particular, devices based on electrochemical energy storage, which are based on the charge accumulation in two electrodes, and charge movement between these two electrodes through an electrolyte that allows the transfer of ions while provides electrical insulation. To the date, CBD technique has been used for the design of electrodes, but it could also be used for the electrolyte. The works, found by the author, reporting electrochemical energy storage devices using CBD are pointed in Fig. 2.1 following the timeline. These works [1–12] are classified according to the electrochemical energy storage mechanism and the role of the clusters in the system.

First, we distinguish roughly two main groups depending on the particular mechanism to accumulate the charge in the electrode, supercapacitors (SCs) and batteries [13]. Simplified, in the case of the SCs, the energy storage is based on the accumulation of

Cluster Beam Deposition of Functional Nanomaterials and Devices. https://doi.org/10.1016/B978-0-08-102515-4.00002-7

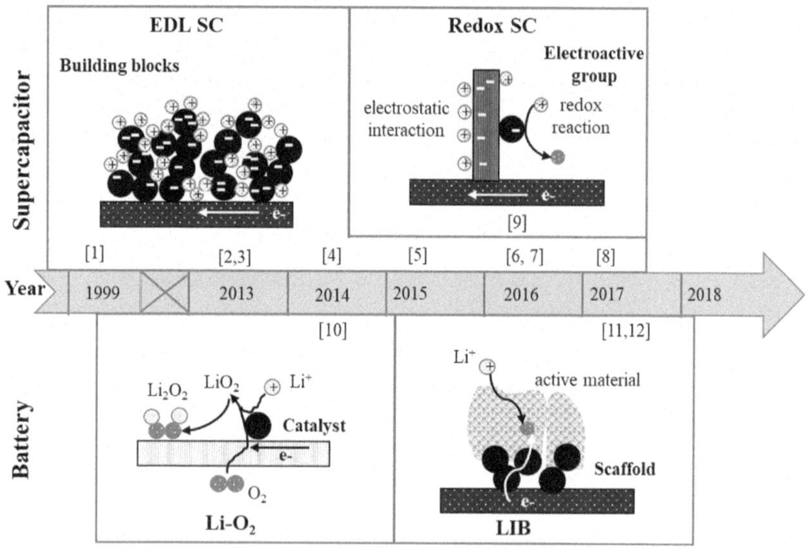

FIGURE 2.1 Timeline of the publications of electrochemical energy storage devices incorporating clusters deposited by cluster beam deposition (CBD). The publications are classified in four different groups: electrical double-layer supercapacitor (EDL SC), redox SC, Li–O_2 battery, and Li-ion battery (LIB). For each group, a diagram is shown of the energy storage mechanism and the role of the clusters, represented by black spheres.

ionic species at the electrode–electrolyte interface, producing a net charge accumulation at the interface, known as electrical double layer (EDL) that depends on the applied voltage. While, in batteries, the energy storage mechanism involves the charge transfer (CT) of ions between the electrode–electrolyte interface, and the formation of new chemical bonds in the material of the electrode depending on the applied voltage. Within this classification, we can collect the reported studies in four groups, detailed below.

Most of the work performed by CBD in SCs is related with the design of nanostructured electrodes (ns-electrodes) made of graphitic carbon as electronically conductor material [1–8]. Fig. 2.1 shows a scheme of one of the electrodes. On the top of a collector (electron conductor), the clusters (black spheres) are deposited by CBD. In the scheme, the clusters maintain their individuality due to the soft landing (SL) and, as a result, their properties due to the so-called "memory effect" [14]. Therefore, the clusters can be seen as "building blocks," in which their composition, size, and interactions between them determine the fundamental properties of the nanostructured material. In this case, energy is stored at the electrode–electrolyte interface by electrostatic interactions between the ions in the electrolyte and the polarized electrode without any redox reaction (i.e., no Faradaic current). This group is classified as electrical double-layer supercapacitor (EDL SC), and the role of clusters is as building blocks that are polarized under an applied voltage. Besides a proof-of-concept reported in 1999, this work extends along the last 5 years, tailoring the properties of the electrode and the fabrication of flexible microsupercapacitors that are integrated in more complex systems.

In 2016, a work of a redox SC using CBD was reported [9]. Redox SCs (or pseudocapacitors) store the energy also faradaically by electron CT between functional groups present on the electrode surface and ions in the electrolyte [15]. This is represented in Fig. 2.1, which shows a carbon nanotube (CNT) on the collector, functionalized with a cluster (black sphere) deposited by CBD. For one side, there is an important EDL capacitance due to the high surface of the electrode and the good electronic conductivity of CNTs. For the other side, the capacitance is increased by faradaic current due to the redox reactions between the clusters deposited on the surface of the CNTs and the ions of the electrolyte. Therefore, the clusters in the redox SC participate actively as electroactive groups by CT with electrolyte ions, and cluster distribution and aggregation state determine the pseudocapacitance of the system. Since the redox groups are on the surface of the electrode, the electrochemical response of redox SCs is faster than batteries but with increased energy compared to EDL SCs.

In the case of batteries, all the works found are related with Li^+ ions. We can distinguish between LIBs and $Li{-}O_2$ batteries since the mechanism of energy storage is different.

In 2017, two independent works on LIBs using CBD technique were published almost at the same time [11,12]. Furthermore, the role of the clusters in both studies is the same, as scaffolds to nanostructure the electroactive material that can incorporate Li^+ ions inside its structure. The scheme shows the clusters (black spheres) deposited on the top of the collector. As in the case of EDL SCs, the clusters maintain their individuality generating a film with plenty of voids, which determine the structure of the following atomic thin-film growth by sputtering. Then, with a cluster film a few nanometers thick, it is possible to tune the properties of the much thicker (tens of nanometers) amorphous film deposited on the top. This thin film is the active material that can incorporate Li^+ ions inside its structure, earning an electron for each Li^+ ion, while the clusters are not actively involved in the storage of Li^+ ions. Then, clusters realize a structural role, which determines the mechanical and electrochemical properties of the electrode.

The last group reported is $Li{-}O_2$ batteries with a study performed in 2014 [10]. In this case, the energy is not stored in the bulk of the electrode as LIBs but in the chemical bonds of Li_2O_2 molecules generated in the battery discharge process between O_2 molecules, Li^+ ions present in the electrolyte, and electrons transported in the electrode. Fig. 2.1 shows the electrode as a mesh through which O_2 gas molecules can pass and arrive to the electrolyte. On the surface of the electrode, there are the clusters that act as catalysts of the reaction between the O_2 molecules, Li^+ ions, and electrons, playing a fundamental role on the storage mechanism of the battery.

In general, we can see that the incorporation of the CBD technique in the design of electrodes for electrochemical energy storage is slowly increasing in the last few years. Also the variety of systems studied has increased, starting with SCs and now integrating other technologies such as redox SC, LIB, and Li–O_2 battery. In each group, the role of the clusters is different, and there are still many possibilities to take advantage of CBD for the design of advanced energy storage systems. Each group summarized herein is explained in detail below in the points 3.2 to 3.5. In point 3.6, some conclusions and perspectives on this field of research are presented.

2.2 Electrical double-layer capacitors

The energy storage in EDL SCs is based on the polarization of the electrode and the resultant accumulation of ions of opposite charge at the electrode–electrolyte interface, producing a net charge accumulation known as double-layer capacitance, Cdl, because the interface charging occurs similarly to an electrical capacitor. Cdl measures the ability of the interface to accumulate ionic species in the vicinity of the surface in function of the applied voltage and depends on the dielectric properties of the interface according to the expression:

$$C_{dl} = \frac{\varepsilon\varepsilon_0}{d} A \tag{2.1}$$

where ε accounts for the effective dielectric constant, ε_0 is the vacuum permittivity, d represents the effective charge separation, and A, the area of the electrode. Ions pile up at the interface producing thin space charge layers with net separation in the range of 1–10 Å that implies Cdl values in the order of 10–100 mF/cm^2, depending on the permittivity. The obvious strategy to increase the value of the total double-layer capacitor is to increase the interface area by means of porous matrices that allow electrolyte penetration and the formation on an extended double layer, being possible to increase the capacitance by several orders of magnitude [16].

The most used materials in SC technology are based on carbon, as activated carbons, carbon fibers, nanotubes or gels (xerogels, cryogels, and aerogels), and graphene. Carbonaceous materials, in the various forms, can have surface areas as large as 2000 m^2/g [17] and good control of the micro- and mesoporosity. Furthermore, the possibility to control the chemical surface allows tailoring the inertness of the electrode to faradaic phenomena or introduce pseduocapacitance. These aspects are well reported in numerous reviews [17–21] and books [22,23]. Herein, we focus on the work performed on nanostructured carbons (ns-C) engineered by supersonic cluster beam deposition (SCBD), which present the advantages of easy fabrication of planar micro-supercapacitors that can be integrated on chip components for miniaturized biomedical devices, sensors, and microelectromechanical systems (MEMS).

Fig. 2.2A shows the principle of operation of the SCBD used for the fabrication of ns-C for SCs [1–8]. SCBD is composed of two ultrahigh-vacuum (UHV) chambers, the first for the cluster nucleation and growth following by a seeded supersonic beam and the second for the cluster deposition on the substrate. The cluster source, hosted in the first chamber, is a pulsed microplasma cluster source (PMCS) [24]. The source is a ceramic body that hosts the target to be vaporized (graphite rod), connected to the negative pole

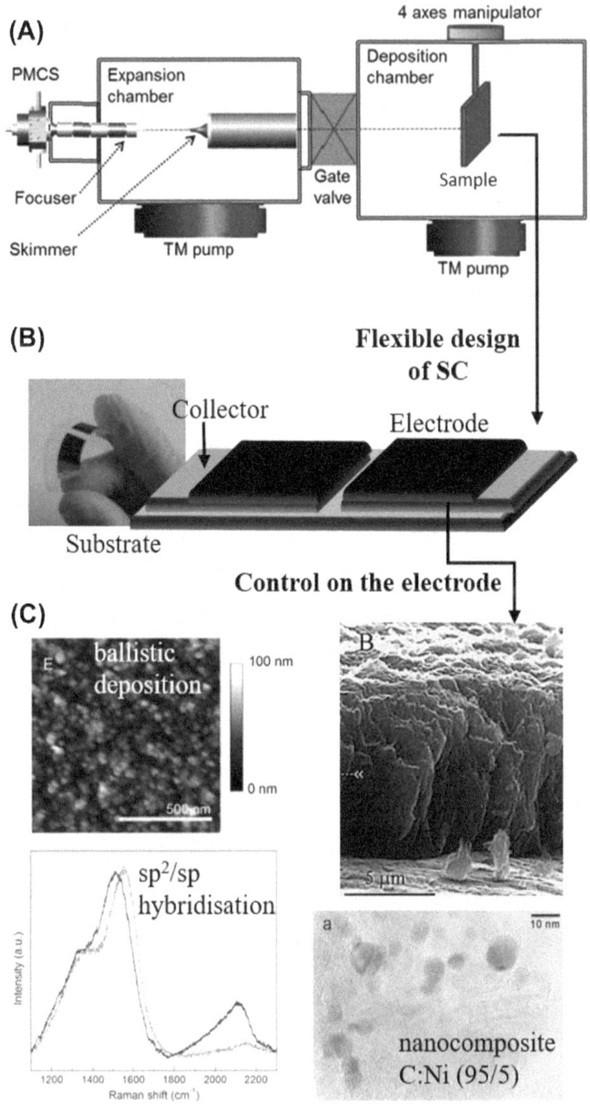

FIGURE 2.2 (A) Schematic representation of an supersonic cluster beam deposition (SCBD) setup coupled to a pulsed microplasma cluster source (PMCS) source for the production of supersonic beams of neutral nanoparticles (NPs). (B) Scheme of a planar supercapacitor (SC), with electrodes deposited on top of the two current collectors, separated by an insulating gap that will be filled by an ionic conductor electrolyte. On the left, a picture of a flexible planar SC device. (C) Different properties of the electrode measured by different techniques: top left: atomic force microscope (AFM) image of the ns-C film deposited on the top of flat sputtered Ni current collector; top-right: scanning electron microscope (SEM) image of the ns-C film in which columnar growth can be seen; bottom-left: Raman spectra of a carbyne-rich ns-C film immersed in ionic liquid (*solid line*) and of a carbyne-depleted ns-C film exposed to air (*dotted line*); bottom-right: high-resolution transmission electron microscope (HRTEM) micrograph of Ni:C composite, the dark features are Ni NPs, which favor the growth of elongated graphitic carbon structures. *(A) Reproduced from F. Soavi, et al., Miniaturized supercapacitors: key materials and .tructures towards autonomous and sustainable devices and systems. J. Power Sources 326 (2016) 717–725.*

of a power source, acting as a cathode. The anode is embedded in the ceramic body. In the back, there is a valve that delivers inert gas (helium) pulses of few hundreds of microseconds at high pressure (typically up to 50 bars) in contrast with the vacuum when the valve is closed (c.10^{-5} bar under working conditions), producing a large pressure difference that causes the formation of a supersonic gas jet directed against the cathode. Before the valve closes, a pulsed high voltage is applied between the electrodes that ionize the gas, generating a plasma confined to an area of less than 1 mm^2, key to ensure high stability and reproducibility. The plasma ablates the material and removes atoms via sputtering. The sputtered carbon atoms thermalize with the buffer gas and condense to form clusters, which are expanded through the source nozzle into the expansion chamber providing a supersonic beam. The supersonic expansion can be exploited in order to control the collimation and dimensions of the beam spot and the cluster mass distribution [25]. The clusters are finally collected on a substrate located on the beam trajectory to form the ns-C film.

This configuration of SCBD allows to fabricate designed SCs by patterning. Fig. 2.2B shows a planar SC configuration with two ns-C electrodes separated by 0.6 mm [3]: the separation between the electrodes and their thickness can be easily controlled by SCBD. An important advantage of this system is that the substrate is at room temperature. Then, any substrate can be used, such as glass [6], flexible polymers [7], or paper [8], with only first sputtering a metallic collector to conduct the electrons. Also the metallic collector can be customized, as, for example, thin film of different metals or nanostructured layers, to favor the electron injection between the collector and the nanostructured film.

Besides high control on the design of SC device, the particular configuration of PMCS and SCBD allows the growth of electrodes with the desired thickness since it can achieve high deposition rates, up to 10 nm/s [27]. The thickness studied for SCs are typically between 100 and 500 nm, but it is possible to obtain films in the order of micrometers, as Fig. 2.2C shows. The obtained structure is typical of a columnar growth with density between 0.5 and 1.4 g/cm^3 in function of the size distribution and kinetic energy of precursor clusters. In the studied ns-C electrodes for SCs, the density is adjusted to 0.5 g/cm^3, providing a BET surface area of 700 m^2/g, with a pore size distribution between 2 and 50 nm, with a maximum at around 3.4 nm. The nanostructure and morphology are a result of random stacking of particles due to deposition-diffusion-

(B) Reproduced from L. Bettini, et al., Flexible, ionic liquid-based micro-supercapacitor produced by supersonic cluster beam deposition. Electrochim. Acta 170 (2015) 57–62. (C) Reproduced from L.G. Bettini, et al., Electrochemical impedance spectroscopy on nanostructured carbon electrodes grown by supersonic cluster beam deposition. J. Nanoparticle Res. 15 (2) (2013) 1429. Reproduced from E. Barborini, et al., Synthesis of carbon films with controlled nanostructure by separation of neutral clusters in supersonic beams. Chem. Phys. Lett. 300 (5–6) (1999) 633–638. Reproduced from L.G. Bettini, et al., Interfacial properties of a carbyne-rich nanostructured carbon thin film in ionic liquid. Nanotechnology 27 (11) (2016) 115403. Reproduced from L.G. Bettini, et al., Nickel nanoparticles effect on the electrochemical energy storage properties of carbon nanocomposite films. Nanotechnology 25 (43) (2014) 435401.

aggregation (DDA) growth at ballistic regime (low diffusion). This deposition growth provides a granular morphology, as the AFM image shows in Fig. 2.2C, with an average grain size of 10 nm.

The ns-C film obtained by SCBD, characterized by Raman spectroscopy, shows the signature of sp^2-hybridized amorphous carbon with the two D and G bands in the 1200–1700/cm^1 spectral region. Nevertheless, at UHV deposition, there exists originally some amount of sp-hybridization (about 30%) that rapidly decays upon exposure to oxygen that reacts with linear carbon structures, polyynes (linear structures alternating single and triple bonds, $-C{\equiv}C-$), and polycumulene (linear structures with identical double bonds, $=C=C=$) species. These species are protected if the ns-C film is directly immersed inside ionic liquid. This effect is observed in the Raman spectra of Fig. 2.2C, in which the C band in the region of 1800–2300/cm^2, associated with the stretching modes of linear sp-hybridized carbon structure (both cumulenes and polyynes), is seen for the ns-C film immersed in ionic liquid while it cannot be observed once the ns-C film is at air [6]. The presence of carbyne units in the ns-C electrode increases the electrical conductivity of the electrode and the capacitance response. Another strategy to increase the electrical conductivity of the ns-C films is embedding metal nanoparticles (NPs) inside the electrode by SCBD. Fig. 2.2C shows the HRTEM micrograph of ns-C film with Ni NPs (5%) obtained by codeposition of both elements by SCBD from two different targets [4]. The results show that the presence of Ni clusters not only increases the electrical conductivity of the electrode, but they also catalyze the C growth in ordered onion-like structures when the film is heated at 300°C in argon atmosphere.

The structural, morphological, and electrical properties determine the behavior of the ns-C films as electrodes, and the mechanical properties are critical for the integration of the electrodes in more complex devices.

The electrode structure and morphology are key to determine the accessible surface for ionic electroadsorption, and the film roughness affects the isoelectric point (IEP) of the film and the EDL capacitance [28]. Also the electron transference between the collector and the electrode and the electrical conductivity within the nanostructured film are key for the electrode performance in the SC. A powerful tool to interpret and discuss the effect of these different parameters on the SC performance is the electrochemical impedance spectroscopy (EIS) using an equivalent circuit model (ECM).

Fig. 2.3A shows a characteristic Nyquist plot obtained with ns-C electrode. The Nyquist plot shows two different features: an arc at high-frequency range (0.1 MHz–10 Hz) and a straight line at lower frequencies than 10 Hz. The high-frequency semicircle is attributable to the collector/electrode and collector/electrolyte interface, while the low-frequency behavior, to the electrode/electrolyte interface. The ECM (Fig. 2.3B) is composed of a resistance, R_S, in series with two parallel combination of resistances, R_1 and R_2, with a constant phase element, CPE_1 and CPE_2. R_S is the real value where the impedance imaginary part is zero and accounts for the ionic conductivity of the electrolyte and the electric conductivity of the electrode. The high-frequency arc is

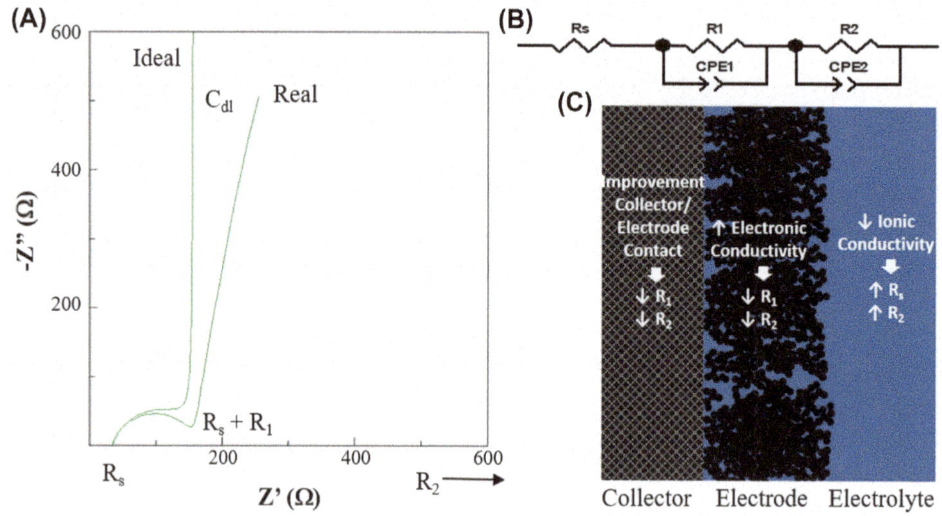

FIGURE 2.3 (A) Nyquist plot obtained from the equivalent circuit model (ECM) represented in (B) for the real behavior, and for the ideal behavior, the arc formed by R_2 and CPE_2 is substituted by C_2. To build the Nyquist plots, the values of R_s, R_1, R_2, CPE_1, and CPE_2 were taken for reference [3] and C_2 calculated from R_2 and CPE_2. (C) Scheme of an ns-C electrode and the effect of several parameters of the electrode design on the resistances of the system.

modeled by the parallel R_1 and CPE_1, which is attributed to a mixed response of the contact interface between the sputtered collector and the ns-C and the EDL formed between the collector and the electrolyte due to the electrolyte infiltration in the porous carbon. The capacitive response of the ns-C electrode/electrolyte interface is modulated by CPE_2, which is identified with the double-layer capacitance, C_{dl}. The observation of a straight line without further features is evidence of the capacitive character and the absence of faradaic processes. Nevertheless, the straight line at low frequencies deviates from the ideal behavior, which should be a straight line with infinite slope, as it is represented in Fig. 2.3A. The deviation from the ideal behavior is caused by the porous nature of the electrode, which is related to the parallel resistance R_2 that is ascribable to the tortuosity of the porosity of the electrode [29] or a leakage current [2] and the superscript α, which is related with the capacitance by the following equation:

$$Z_{CPE} = \frac{1}{Q(i\omega)^{\alpha}} \tag{2.2}$$

In this chapter, we use the ECM represented in Fig. 2.3B, but the reader can find extensive information about ECMs for supercapacitors in Refs. [30,31].

Fig. 2.3C represents an ns-C electrode in contact with the collector in the left side and with the electrolyte on the right. In the figure, the effect of the different parameters studied in Refs. [1–8] to the resistances of the SC system is shown. The nature of the electrolyte, from aqueous solution to liquid ionic, represents a reduction in more than an order of magnitude in ionic conductivity that causes an increase of R_s and R_2, comparison between Refs. [2,3]. On the other hand, the use of liquid ionic increases the electrochemical stability window (ESW) of the electrolyte, which increases the energy storage of the SC according to:

$$E = \frac{1}{2} C V^2 \tag{2.3}$$

$$P = \frac{V^2}{4R_s} \tag{2.4}$$

Where V is the voltage window applied on the SC and C is the capacitance. While R_S is directly related to the power (i.e., the speed that the energy can be delivered) of the SC.

To improve the electronic contact between the metallic collector and the ns-C electrode, two different strategies were tested [2]: (1) the use of a nanostructured collector and (2) mild thermal annealing treatment (300 C during 15 min in air atmosphere). Both strategies reduce R_1 and, particularly, R_2 by an order of magnitude. Finally, it is well known that nanoparticulated films show reduced electrical conductivity. One strategy to increase the conductivity is the fabrication of nanocomposites, as the codeposition of Ni and C NPs [4]. The reduction of R_S, R_1 and R_2, occurs after annealing treatment since Ni catalyzes the C tubular growth. Another strategy is increasing of sp hybridization respect to sp^2. The modification of all these parameters reduces the resistances, while the Cdl maintains practically the same, depending on the available surface area of the electrode, which in the case of the ns-C fabricated by SCBD depends on the film thickness [3], confirming the open porosity. The reduction of the resistance in the system is translated with an increase of power of the SCs as Eq. (2.4) indicates, while the energy is similar according to Eq. (2.3).

Fig. 2.4 shows the electrochemical response of a designed microsupercapacitor with optimized parameters of ns-C electrodes using paper substrate and SCBD technique. The details are described in Ref. [8], and herein we only comment on the design parameters and the electrochemical response. The collector is an ns-Au film (100 nm) deposited by SCBD as the ns-C electrodes of area 0.24 cm^2 and 200 nm thick film. The electrode design and geometry were controlled using shadow masks cut from another sheet of paper without employing any pre- or postdeposition treatment of the samples and producing the electrodes in clean high-vacuum, room temperature conditions. As prepared ns-C electrodes were soaked with 1-ethyl-3-methylimidazolium bis-trifluoromethane-sulfonyl-imide ([Emim][NTf2]) ionic liquid. The encapsulation was performed by blade coating a thin layer of polydimethylsiloxane (PDMS) and curing in oven at 80°C for 60 min. The capacitive behavior of the device is shown in Fig. 2.4A under CV polarization between 0 and 1.8 V showing the characteristic box shape of SCs, acquired at sweep rates of 0.1, 0.5, and 1 V/s. The capacitances are 0.2 F/cm^2 and 6–7 F/cm^3 calculated by:

$$C = \frac{1}{v} \tag{2.5}$$

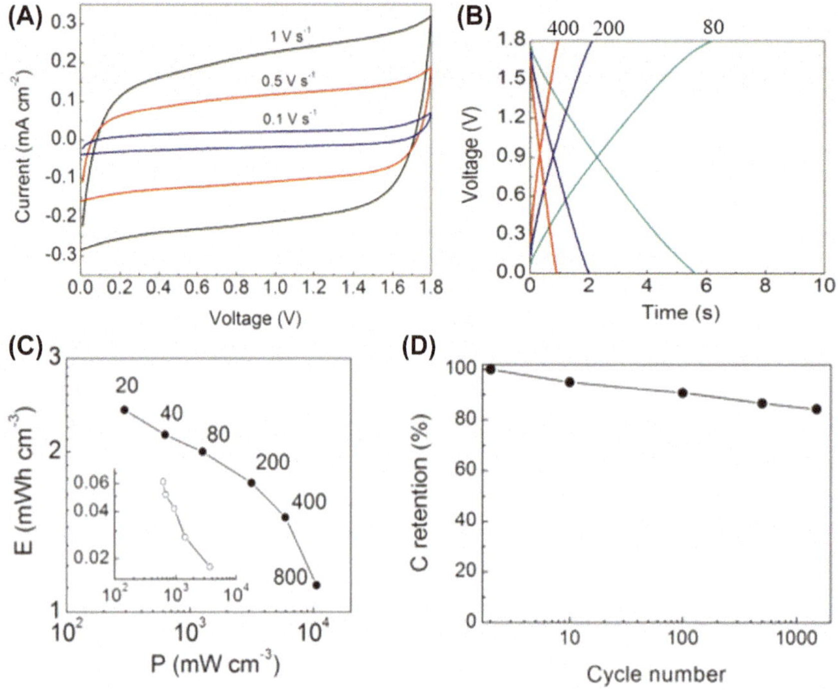

FIGURE 2.4 (A) Electrochemical characterization of a microsupercapacitor printed on paper by supersonic cluster beam deposition (SCBD). (A) CV curves at 0.1, 0.5, and 1 V/s; (B) galvanostatic charge and discharge profiles at 80, 200, and 400 μA/cm^2; (C) Ragone plot of the volumetric energy and power densities (*solid circles*), the labels indicate the current densities in μA/cm^2, together with the volumetric energy and power densities of a MXene-on-paper microsupercapacitor (*open circles*) as estimated from the data reported in Ref. [32]; and (D) capacitance retention as function of cycle number upon galvanostatic cycling at 200 μA/cm^2. *Reproduced from L.G. Bettini, et al., Supersonic cluster beam printing of carbon microsupercapacitors on paper. Flexible and Printed Electronics 2 (2) (2017) 025002.*

where v is the scan speed. The capacitance value agrees with the values obtained by EIS and glavanostatic charge/discharge, plots shown in Fig. 2.4B at different charge/discharge current densities. The ohmic drop (R_S) at the beginning of the discharge points is between 500 and 375 Ω cm^2 for the corresponding discharge currents of 80 and 400 μA/cm^2, leading to a time constant for about 0.1 s. The galvanostatic survey at different current densities was used to build the Ragone plot shown in Fig. 2.4C, where the volumetric energy and power densities are reported. The obtained values for the fabricated microsupercapacitors by SCBD are competitive with other techniques reported in the literature [32], also shown in Fig. 2.4C. Finally, Fig. 2.4D shows the stability of the microsupercapacitor along 1000 cycles, highlighting the suitability of the PDMS technique for the preparation of protecting covers to avoid the electrolyte leakage, the electrical insulation, and the structure protection. This microsupercapacitor can also work at bending conditions. For this purpose, the encapsulated microsupercapacitor was wrapping around a tube of 2 cm of diameter obtaining no significant differences in the electrochemical performances between the flat and bent states.

In summary, SCBD is a versatile technique which allows the design of nanostructured electrodes with tuneable properties toward the electrochemical response. The direct deposition at high vacuum and at room temperature without any binder or posttreatment facilitates the integration of these microsupercapacitors in chips for real applications. This point will be discussed in Section 2.6.

2.3 Redox supercapacitors

In the last few years, redox supercapacitors are becoming increasingly important, to overcome the energy-capacity trade-off characteristic of LIB and supercapacitors [33]. In this case, the supercapacitors incorporate functionalized redox groups at the electrode surface that can provide a fast redox answer, increasing the capacity by Faradaic processes, pseudocapacitance, while maintaining the fast charge–discharge response.

The use of SL of mass-selected ions for the preparation of well-defined electrode surfaces enables an unprecedented level of control, with uniform deposition of ions with specific composition, charge state, and kinetic energy. Fig. 2.5 shows the SL setup used for the functionalization of CNTs with polyoxometalates (POMs) [9], polyatomic ions characterized by their stability, multielectron redox activity [34], and strong interactions with electrodes [35]. Specifically, electrochemical supercapacitor devices were fabricated by immobilizing mass-selected $PMo_{12}O_{40}$ onto CNT-coated carbon fiber paper by a custom-designed SL instrument, described in detail in Refs. [36,37]. The ion SL instrument is equipped with an electrospray ionization (ESI) source, which produces charged

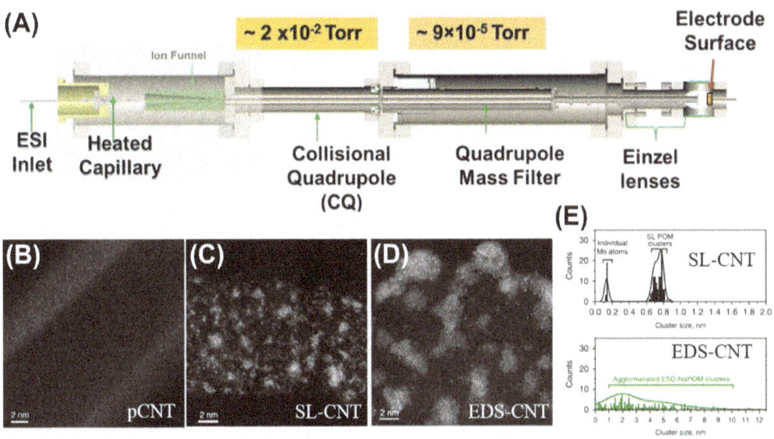

FIGURE 2.5 (A) Schematic diagram of the high-flux ion soft-landing instrument. (B–D). HAADF-STEM images of pCNT, SL-CNT, and ESD-CNT, respectively. (E) Histogram of cluster size distribution in SL-CNT and ESD-CNT. *CNT*, carbon nanotube; *ESD*, electrospray deposition; *HAADF-STEM*, high-angle annular dark-field scanning transmission electron microscopy; *SL*, soft landing. *Reproduced from V. Prabhakaran, et al., Rational design of efficient electrode–electrolyte interfaces for solid- state energy storage using ion soft landing. Nat. Commun. 7 (2016) 11399.*

droplets from $Na_3[PMo_{12}O_{40}]$ x H_2O solution in methanol. The charged droplets are transferred into the vacuum system by a heated stainless steel inlet and two electrodynamic ion funnels to produce the desolvated ions. The ions are then collimated in a radio frequency (RF)-only collision quadrupole and through a quadrupole mass filter (QMF), only the anions, $[PMo_{12}O_{40}]^{3-}$, are allowed to pass. After that, the beam is refocused with three lenses before SL deposition on the CNT electrode. To evaluate the advantages of using SL among other systems, other electrodes were tested as pristine CNT electrode, and CNT electrode with $Na_3[PMo_{12}O_{40}]$ deposited by electrospray deposition (ESD) technique. With ESD, there is no mass nor charge selection, and on CNTs, both the anions, $[PMo_{12}O_{40}]^{3-}$, and countercations, Na^+, are present.

The samples were characterized by high-angle annular dark-field scanning transmission electron microscopy (HAADF-STEM). Fig. 2.5B–D shows the HAADF-STEM images of pristine CNT electrode (pCNT) and the CNT electrodes incorporating POM by SL (SL-CNT) and by ESD (ESD-CNT). STEM images indicate that $PMo_{12}O_{40}$ ions are predominantly discrete without significant agglomeration and uniform cluster coverage when anions are deposited by SL, while important aggregation is observed in ESD deposition, probably due to the presence of countercations (Na^+). Fig. 2.5E shows a histogram of the cluster sizes for the samples SL-CNT and ESD-CNT. In the first case, the diameter of the clusters was measured to be 0.74 ± 0.04 nm, which matches the theoretical calculated size for $[PMo_{12}O_{4}0]^{3-}$ ions [34], and the presence of a minor fraction of individual Mo atoms at 0.15 nm. In contrast, the cluster size in ESD-CNT electrode is in the range of 3–10 nm with an average size of 3.66 ± 2.11 nm, indicating the high agglomeration present in the sample. These aggregates may decrease the overall interfacial surface area of the electrode and introduce additional contact resistance due to their low conductivity. These two effects would decrease the overall capacitance of the supercapacitor, as it is observed in the electrochemical response of the fabricated SCs.

The CVs of the supercapacitors prepared by symmetrical electrodes of pCNT, SL-CNT, and EDS-CNT are plotted in Fig. 2.6A–C. For the three electrodes, the electrochemical response of their SCs shows rectangular CV shape indicating that they exhibit an ideal capacity-like behavior. Even this rectangular shape is more remarkable in the presence of polyoxometalate (POM), suggesting faster transmission of ions at the electrode–electrolyte interface and lower related internal resistance during the polarity change. The total specific capacity versus scan rate for the three SCs is represented in Fig. 2.6D, demonstrating that for all the scan rates, SL-CNT shows the highest capacity.

The Faradaic contribution of the total specific capacitance, $C_{sp[PMo12]}$, can be calculated by the following equation:

$$C_{sp[PMo_{12}]} = \frac{(C_{sp,t} * W_{t,CNT+PMo_{12}}) - (C_{sp,}CNT * W_{CNT})}{W_{PMo_{12}}} \quad (2.6)$$

where $C_{sp,t}$ and $C_{sp,CNT}$ are the total specific capacitance (F/g) of modified CNT and pristine CNT electrodes, respectively; $W_{t,CNT+PMo12}$ is the sum of the weights (g) of loaded $[PMo_{12}O_{40}]$ clusters in each case and CNT material; which is the same in all the

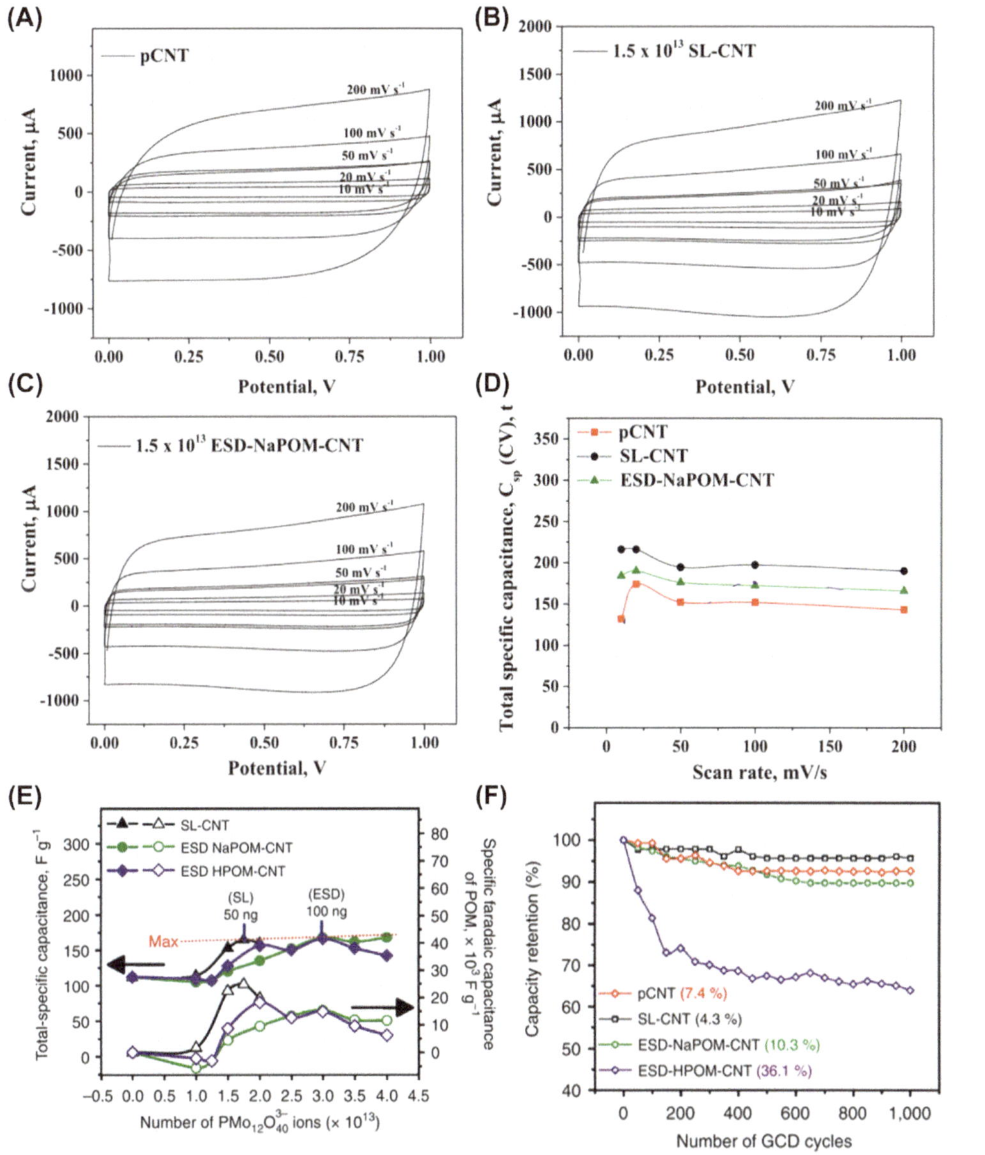

FIGURE 2.6 Representative electrochemical response of the different electrodes assembled in symmetrical supercapacitors (SCs) and using $EMIMBF_4$-based membrane used as a separator. CVs at different scan rates (10, 20, 50, 100, and 200 mV/s) of: (A) pCNT; (B) SL-CNT; (C) EDS-CNT. (D) Relation between specific capacitance and scan rates of the supercapacitors, calculated from the CVs represented from (A) to (C). (E) Specific capacitance and specific Faradaic capacitance (calculated from Eq. 2.6) in function of the loaded amount of $PMo_{12}O_{40}$ in the electrodes, obtained from galvanostatic charge–discharge measurements at 8 A/g. (F) Capacity retention as a function of the number of galvanostatic charge–discharge cycles at 8 A/g. *CNT*, carbon nanotube; *ESD*, electrospray deposition; *SL*, soft landing. *Reproduced from V. Prabhakaran, et al., Rational design of efficient electrode–electrolyte interfaces for solid- state energy storage using ion soft landing. Nat. Commun. 7 (2016) 11399.*

cases that is W_{CNT} (25 μg); W_{PMo12} is the mass of $[PMo_{12}O_{40}]$ alone in the modified electrodes. The values obtained for SL-CNT and EDS-CNT are shown in Fig. 2.6E. It is noticeable that the maximum total capacitance was obtained only with 1.75 × 1013 $[PMo_{12}O_4]^{3-}$ ions (50 ng POM on 25 μg of CNT material), demonstrating the high contribution of nanogram quantities of pure $PMo_{12}O_{40}$ to the total capacitance. Also, it is observed that the contribution in the specific Faradaic capacitance of SL-CNT is greater than that of ESD-CNT. The technique used for the deposition of $PMo_{12}O_{40}$ ions on CNT affects the electrochemical stability of the electrodes, as seen in Fig. 2.6F. After 1000 cycles, the capacitance of LS-CNT was decreased by 4.3%, followed by pCNT with a loss of 7.4% and ESD-CNT with 10.3%. The lower stability of these electrodes was attributed to side reactions that contribute to the degradation of the electrode surface, but the exact mechanism is not clear. Finally, the effect of the countercation when the deposited technique is ESD is observed in Fig. 2.6E and F, in which the total capacitance and the capacity retention of electrodes fabricated are decreased when the countercation Na^+ is changed by H^+ in the samples called, ESD-HPOM-CNT.

This study demonstrates that the high control of the deposition of electroactive species by SL, such as mass, charge, amount on an electrode surface, is key to optimize the pseudocapacitance response of redox SCs. This control opens up many possibilities in the design of new electrodes with improved energy storage maintaining fast charge–discharge response, thus in engineering new systems on the top-right of the Ragone plot [38].

2.4 Li-ion batteries

The energy storage in LIBs is based on the difference of chemical potential of lithium when it is integrated into two different electrodes by means of intercalation, alloying, or conversion reaction. Then, assimilating the electrode potential to the chemical potential of Li^+ in the electrode, μ, with respect to the chemical potential of the reference μ_{ref} (such as Li metal film), the voltage of the system could be written as [39]:

$$V = \frac{\mu - \mu_{ref}}{e} \tag{2.7}$$

where e is the positive elementary charge. The fundamental element that accounts the charging ability is the chemical capacitance, C_μ, that is defined as the ratio between the modification in electrode charge state produced by a change in the electrode potential:

$$C_\mu = eN\frac{dx}{dV} \tag{2.8}$$

where N and x are, respectively, the total concentration and the molar fraction of Li^+ ions. We note here that C_μ and C_{dl} are essentially different. The chemical capacitance measures the response to the variations in the chemical potential, while the double-layer capacitance is of dielectric nature, that is, it depends on the changes of the electrical field at the interface.

Taking this into mind, the electrode modifies its structure and electrochemical state with Li^+ ions. Then, the idea behind the works [11,12] is to introduce NP scaffolds into the active bulk material to increase the electrochemical response of the electrode, but would not be modified during the lithiation/delithiation. The two independent works selected the same active bulk material, Si, while use different metals for the NP scaffold, Ta and Cu.

Si was selected to be one of the most promising anode materials due to its high specific capacity, 3579 mAh/g, which is approximately 10 times higher than that of commercial graphite anode (372 mAh/g) [40]. Also, the low potential of discharge process ($\approx$0.5 V vs. Li/Li^+) is beneficial for high LIB voltage, deriving from the formation of solid electrolyte interphase (SEI) layer caused from the decomposition of the organic electrolyte [41]. The need of nanostructuring Si anodes comes from the drastic volume expansion/contraction of Si anodes during lithium insertion/extraction (320%) associated to its high capacity. This dramatic volume change causes large internal strains resulting in electrode fracture with the loss of active material and electrical conductance and the gradual thickening of the SEI layer due to the reexposed fresh Si surface. In consequence, Si anodes show fast capacity fade and low cyclability. The most used strategy to avoid Si anode drawbacks is Si nanostructuring to facilitate better accommodation of the strain during Li insertion/extraction in the Si anode, improving its cycle life, and there is extensive literature in this topic [40,42–45]. Also, nanostructuring reduces the path lengths for electric and ionic transport to overcome the low Si electrical conductivity and low Li^+ ion diffusion that leads to poor electrochemical reaction kinetics and limited rate capability.

Fig. 2.7A shows a scheme of the samples prepared by subsequent deposition of Si and metallic NPs. The electrode material is bottom-up assembled directly on the substrate in a high-vacuum chamber obtaining high-purity and binder-free films, offering reduced resistance. The depositions of Si and metallic NPs are performed by independent systems, which can work subsequently or at the same time. Then, the deposition conditions can be carefully chosen for each layer, selecting the thickness and number of layers. The SEM image in Fig. 2.7B shows the multilayer structure with two differentiated phases, obtained for the subsequent deposition of Cu NPs and amorphous Si (a-Si) by RF sputtering. The NP scaffolds are deposited by CBD described in detail in Refs. [12,46]. Both systems consist in a sputtering and a cluster growth chambers, but in the first system (Fig. 2.7C), there is a mass selection chamber by QMF, which allows a narrow size distribution of Ta NPs centered at 3 nm, while the deposition rate is around two order of magnitudes faster in the second system (Fig. 2.7D) to fabricate the Cu scaffold. The deposition conditions for the two systems are collected in Table 2.1.

The Si growth onto the NP scaffolds is amorphous, which plays a positive role in the improvement of structure stability since the added stress between crystalline Si and amorphous Li_xSi is avoided [47]. Furthermore, in both studies, XPS data show the absence of Si–Ta or Si– Cu bonding, indicating that Si chemistry is not modified by the presence of the studied metallic NPs.

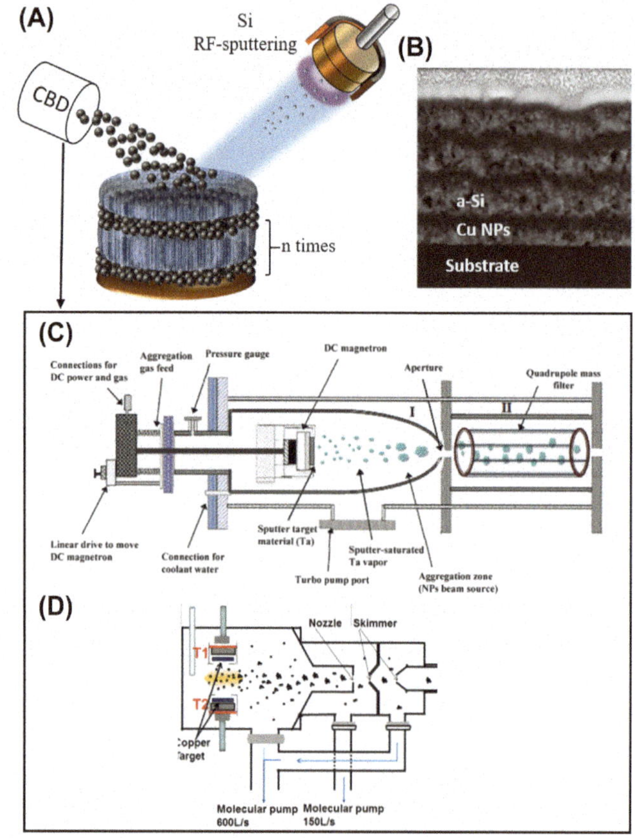

FIGURE 2.7 (A) Scheme of a sample prepared by sequential deposition of a layer of clusters by cluster beam deposition (CBD), which behave as NP scaffold, and a thin film of Si by RF sputtering. (B) Scanning transmission electron microscope (STEM) image of a four-layer composite of Cu NPs and amorphous Si (A-Si). (C) Schematic diagram of magnetron-sputter inert-gas condensation set-up used for growth Ta NP scaffold. NPs are formed in the aggregation zone (labeled-I), then size-selected using quadrupole mass filter (QMF) (labeled-II). (D) Schematic drawing of the plasma-gas-aggregation-type cluster beam deposition for growth of Cu NP scaffold. *(B) and (D) Reprinted with permission from L. Lin, et al., Copper-nanoparticle-induced porous Si/Cu composite films as an anode for lithium ion batteries. ACS Nano 11 (7) 2017 6893–6903. Copyright 2017 American Chemical Society. (C) Reproduced from V. Singh, et al., Assembly of tantalum porous films with graded oxidation profile from size-selected nanoparticles. J. Nanoparticle Res. 16 (5) 2014 2373.*

Table 2.1 Deposition conditions for the intercalated NP and Si layers composing the electrode films.

	Si layer thickness	Deposit ion time	Targets for NPs deposition	Work power	Ar gas flow	NP scaffold thickness	Deposition time
Ref [11]	40 nm ± 10 nm	1800 s	Ta d = 3 inch	400 W	350 sccm	10–20 nm	3600 s
Ref [12]	70 nm ± 10 nm	263 s	Cu d = 3 inch	400 W	20 scm	60–70 nm[a]	214 s[a]

[a]Parameters indicated for the multilayer sample.

The main characteristics reported for these Si composite anodes incorporating NP scaffolds are (1) increased stability of the electrode, (2) clamping effect, and (3) higher charge/discharge rate capability.

Fig. 2.8A collects the capacity retention of Si/Cu composite anodes in a semibattery after 100 cycles [12]. The solid black circles represent the anodes of a monolayer of Si growth on different Cu scaffolds of different time of deposition. From a flat substrate (0 s of Cu time deposition), the capacity retention is 66.6%. The capacity retention increases when there is Cu NP scaffold, rising with the time of deposition up to 86.8% after 171s of Cu deposition. Further Cu NPs deposition decreases the capacity retention providing a volcano shape. Then, there is an optimum morphology (time deposition) of the pre-deposited Cu-NP assembled film to increase capacity retention. When there are too few or too many Cu NPs in the scaffold, this could not work well as the growth direction template. The presence of Cu NP scaffold favors a porous construction of the composite film that can offer extra room to effectively accommodate the drastic volume expansion of Si material, which enhances the cyclability.

This is illustrated in Fig. 2.8B and C, which show two SEM images of Si anodes after 100 cycles without Cu NP scaffold and with Cu scaffold (171 s deposition), respectively.

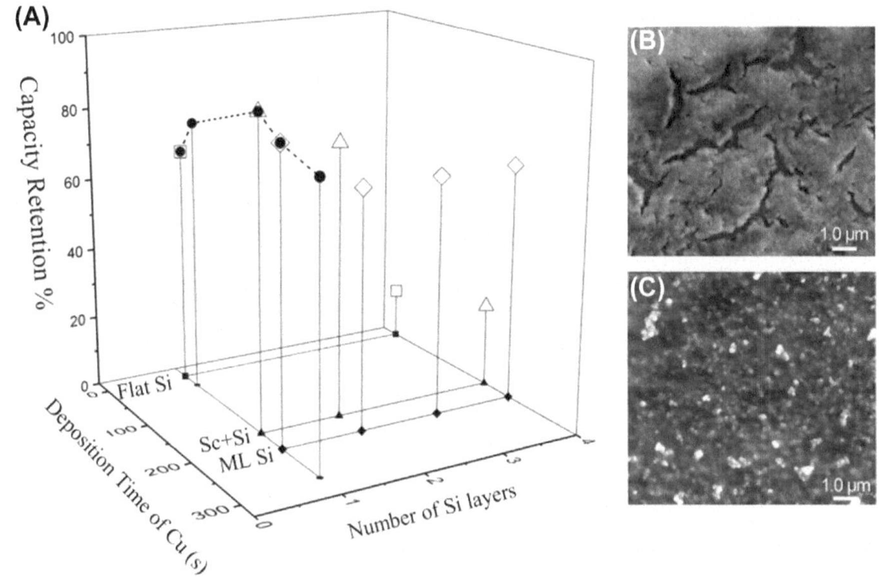

FIGURE 2.8 (A) Estimation of the capacity retention (%) after 100 cycles at cycling performances at 1 A/g of the following Si/Cu composite anodes: one monolayer of Si (263 s time deposition) onto substrates with different time of Cu NPs deposition from 0 s (flat Si) to 286 s represented in solid black circles (and joined by *dash line*), flat Si anode (without Cu NPs) of different time of deposition 263 (one layer) and 1052 s (four layers × 263 s) represented in open squares, Si of different thickness on one layer of Cu NP scaffold (Sc + Si) represented in *open triangles*, and multilayer structure (ML Si) in which Cu NP scaffolds and a-Si layers are deposited alternatively represented in *open diamonds*. SEM images after 100 charge–discharge cycles of: (B) flat Si and (C) single-layer Si/Cu composite anode. *(A–C) Data obtained from L. Lin, et al., Copper-nanoparticle-induced porous Si/Cu composite films as an anode for lithium ion batteries. ACS Nano 11 (7) 2017 6893–6903. Copyright 2017 American Chemical Society.*

In the flat Si, there exist many cracks on the film due to the large volume change during the charge/discharge processes. Opposite, in the Si/Cu composite film, the granular-like surface is well maintained after 100 cycles, with no obvious cracks.

The effect of increased cyclability with the presence of Cu NP scaffolds is more noticeable with increased Si time deposition. Y-axis of Fig. 2.8A shows the times that Si is sputtered during 263 s. Then, the empty squares show the capacity retention of the flat Si films. It is observed that the capacity retention drops off to 14.2% when the Si deposition is 1052 s (four times 263 s). This decrease is less accentuated when there is a first Cu NP scaffold (empty triangles); capacity retention is above 20% when the deposited Si is 1052 s. Nevertheless, the best improvement is obtained when a multilayer structure is built, in which Cu NP scaffolds and amorphous Si layers are alternately deposited (empty diamonds). With this configuration, the capacity retention is the same for two, three, and four layers of Si (263 s deposition), around 66%. Intercalating NP scaffolds and Si layers is a great strategy to increase the amount of Si material necessary to provide the required energy for real applications, avoiding the cracks reported in the literature for Si thin-film anodes with thickness above 100 nm [48].

For the multilayer structure of Si/Cu composite anodes, it was also observed that the initial specific capacity for the multilayer Si/Cu configuration is lower than one layer of Si, flat or on Cu NP scaffold. In that study, the authors comment that Cu NP scaffold may reduce the volume expansion of Si without providing further detail. Nevertheless, this effect is also observed in the Si/Ta multilayer anodes and related to the nanomechanical properties as we discuss below.

In the case of Si/Ta composite anodes (intercalated five layers of Ta NP scaffolds and Si films), it was found that the cyclability was also dependent on the conditions of electrode charging/discharging. Fig. 2.9A shows the specific capacity of Si/Ta composite anode during 200 cycles. When the charge–discharge was performed at 0.5 C, the capacity retention is 100%, but the specific capacity is limited to ca. 700 mAh/g. On the other hand, when the two first cycles are running at 0.2 C and after the cycling is at 0.5 C, the specific capacity at the first cycle is 2300 mAh/g, but the capacity retention decreases to 43% after 200 cycles. This behavior can be explained in terms of nanomechanical properties. Fig. 2.9D shows the stiffness mapping of Si/Ta composite and Si anodes and the probability density function. The Si/Ta composite anode shows pronounced heterogeneous stiffness that does not fit with the morphology features observed by AFM (Fig. 2.9B). The stiffness probability density function (Fig. 2.9C) of the Si anode forms a narrow distribution centered at 49 GPa in contrast to the wider one obtained for the Si/Ta composite, which ranges from 20 to 200 GPa with a main peak at 73 GPa and a shoulder at 49 GPa. Stiffness, k, is related to the displacement, δ, produced in a material under the action of a force, F:

$$K = \frac{F}{\delta} \tag{2.9}$$

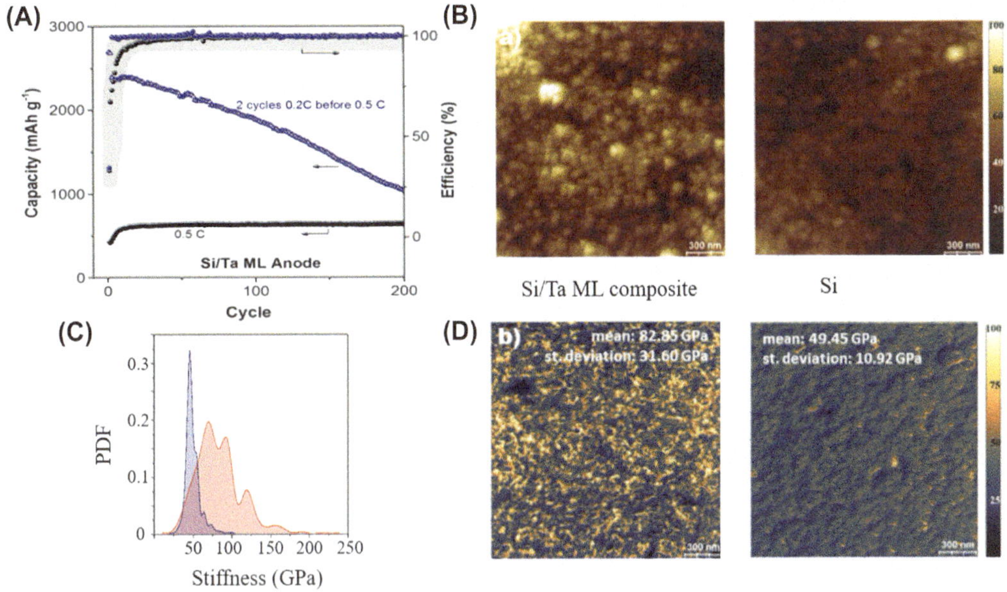

FIGURE 2.9 (A) Capacity and Coulombic efficiency of Si/Ta composite anode along 200 cycles, when it is cycled always at 0.5 C and when it is cycled at 0.5 C after two first cycles at 0.2 C. (B) Topographical atomic force microscope (AFM) images of Si/Ta composite (left) and Si (right) anodes. (C) Probability Density Function (PDF) of the stiffness mappings shown in (D) for Si/Ta composite (left) and Si (right) anodes measured by peak force quantitative nanomechanical (PF-QNM) mapping. *Reproduced from M. Haro, et al., Nanoscale Heterogeneity of Multilayered Si Anodes with Embedded Nanoparticle Scaffolds for Li-Ion Batteries. Adv. Sci. 4 (10) 2017.*

Therefore, for a same executed force inside two materials, the one with higher stiffness suffers less displacement than the other. In this case, the insertion of Li^+ ions produces strains in the Si anode which causes the volume expansion of the electrode. Then, the higher stiffness recorded for Si/Ta composite may limit the volume expansion of the electrode and then the capacity. At low charge/discharge rate, the insertion of Li^+ ions may force the rupture of the electrode, allowing high capacity but decreasing the cyclability. Taking all this in mind, Ta NP scaffolds seem to produce clamping effect similar to the introduction of columns [49]. This effect is not well studied, and better understanding could improve the capacity/cyclability relationship.

The last characteristic of this kind of anodes is a fast charge/discharge rate. Both Si/Cu and Si/Ta composites can be charged and discharged in few minutes. Si/Cu composite shows a capacity of ca. 2000 mAh/g at 20 A/g (Fig. 2.10A), while Si/Ta composite shows c. 1200 mAh/g at 10 C (33 A/g). In this last case, the authors show that flat Si anode cannot be charged/discharged at such a current density, but flat Si and Si/Ta composite anodes show similar charge/discharge plots when they are cycled at lower current density (Fig. 2.10B).

The characteristic Nyquist plot of LIB is shown in Fig. 2.10C, together with the corresponding ECMs [39]. The first arc at high frequency is ascribed to interfacial

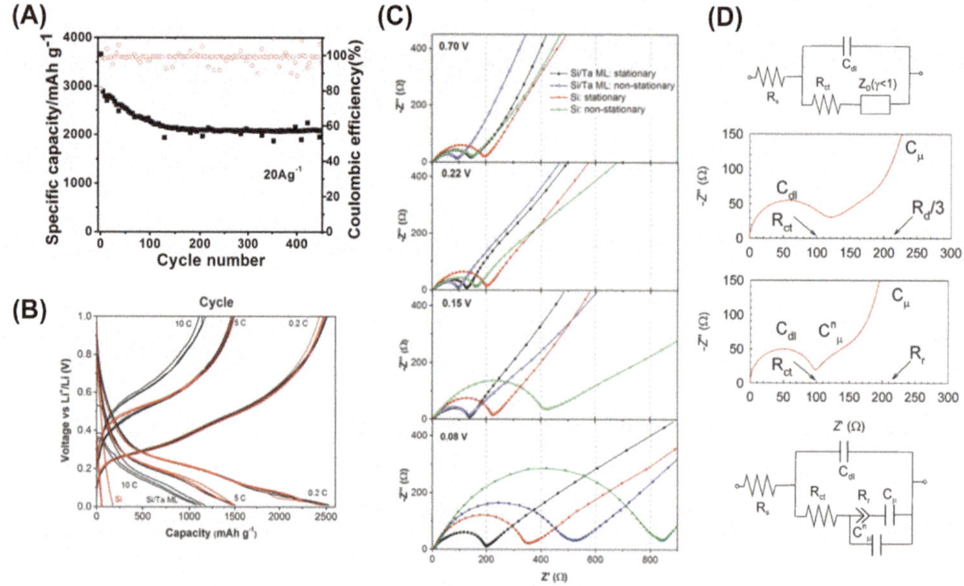

FIGURE 2.10 (A) Capacity and Coulombic efficiency of Si/Cu composite (one layer) anode along 450 cycles at 20 A/g. (B) Charge–discharge plots of Si/Ta composite and Si anodes at 0.2, 5, and 10. (C) For 0.2 and 5 C, the cycles shown are from two to four since they are more representative than the first cycle, while for 10 C the cycles depicted are from one to three to discuss the first response of Si anode. (C) Nyquist plot of Si/Ta ML and Si anodes from electrochemical impedance spectroscopy (EIS) measurements performed at stationary and nonstationary conditions. (D) Equivalent circuit model (ECM) and their corresponding impedance plot (Z″–Z′ complex plane) for Li-ion battery (LIB), considering two different mechanisms at low region frequency: top, anomalous spatially limited diffusion model ZD with disorder parameter $\gamma < 1$; bottom, reaction-limited model including the generalization of reaction resistance. *(C) Reproduced from M. Haro, et al., Nanoscale Heterogeneity of Multilayered Si Anodes with Embedded Nanoparticle Scaffolds for Li-Ion Batteries. Adv. Sci. 4 (10) 2017. (D) Reproduced from N. Vicente, M. Haro, G. Garcia-Belmonte, New approaches to the lithiation kinetics in reaction-limited battery electrodes through electrochemical impedance spectroscopy. Chem. Commun. (2018).*

phenomena, two parallel resistances and capacities related to the SEI layer and the CT resistance, R_{ct}, and the double-layer capacitance, C_{dl}. After a semiarc at intermediate frequency, the impedance increases indicating capacity behavior, which is the chemical capacitance, C_{μ}. The intermediate inclined semiarc is associated to the resistance accompanied the C_{μ}, and it has been extensively ascribed to a diffusion resistance, Rd, and represented by a Warburg element [50], but also it can be related to the lithiation reaction itself, R_r, such as alloying process in a Si anode [51]. In the study of Si/Ta anodes, it was observed that R_{ct} decreases notably when the Si is nanostructured with the Ta NP scaffolds in pseudoequilibrium and nonequilibrium conditions. This effect was explained in terms of fast CT due to rougher surface and faster Li^+ diffusion inside the electrode facilitating the new incorporation of Li^+ ions inside the electrode. The reduction on R_{ct} values was also observed in the study of Si/Cu anode, in which also was estimated the reduction in the diffusion resistance, R_d. Taking this into mind, it can be concluded that the incorporation of NP scaffolds provides a Si nanostructure that accelerates Li^+ diffusion in the anode and reduces the CT resistance. In consequence, the Si anodes with this structure show faster charge/discharge rate than Si anodes. In other words, the introduction of NP scaffolds is a good strategy to increase the power of Si anodes.

2.5 Li–O_2 batteries

Increasing the energy density storage in LIB technology requires advancing beyond Li^+ intercalation mechanisms, such as Li–O_2 cathodes, in which the energy is stored by the direct reaction between Li^+ ions and O_2 in a porous electrode that acts as an electric conductive substrate. Particularly, the chemical energy storage is based on the oxygen reduction reaction (ORR) during discharge and oxygen evolution reaction (OER) in the charge:

$$2Li^+ + O_2 + 2e^- \leftrightarrow Li_2O_2 \quad E^o = 2.96V \tag{2.10}$$

with theoretical specific energy of 3500 Wh/kg, which is one order of magnitude higher than current commercial LIB cathodes based on intercalation mechanism [52]. This change in the chemistry of energy storage means that the electrode develops a passive role in the redox process, but it is a scaffold to support the reaction and the electron transport media (10). Nevertheless, the reaction (10) shows high hysteresis with an important overpotential for the charge process with an accumulation of Li_2O_2 discharge product, electrically insulator, with the consequent battery death after few cycles by electrode clogging [53] or electrode surface passivation [54]. It is estimated that the capacity at cell death corresponds to a Li_2O_2 layer of 5–6 nm, thickness at which electrochemistry is expected to cease [55]. Exploration of Li–O_2 chemistry is at its initial stages, and fundamental understanding of the source of the high overpotential on charge is required, and the mechanisms of both ORR and OER reactions in nonaqueous electrolytes. To enhance reversibility that reduces charge overpotential and Li_2O_2 accumulation, the role of catalysts is critical, which may dictate the followed mechanisms for Li_2O_2 formation and its morphology. In this context, CBD is a great platform to create carefully controlled cathode materials with deposited metal clusters of exact size and composition for obtaining insight into the role of specific active sites in Li–O_2 batteries and understanding the role of catalysis.

Fig. 2.11 shows the CBD setup used to deposit size-selected Ag clusters for Li–O_2 cells, detailed in Ref. [56]. Shortly, the laser vaporization cluster source uses Nd:YAG laser operating at 4.5 kHz. The beam of Ag clusters passes through a biased skimmer, collimator, and focusing lenses to a quadrupole mass selector. The mass-selected Ag clusters are deflected into an ion lens setup placed in front of the sample where the NPs are deposited. In particular, the authors studied the effect of Ag clusters of size 3, 9, 15, and 18 atoms on conductive graphitized carbon (GC) electrodes [10]. Here we only discuss Ag clusters of size 3 and 15 atoms, while clusters of size 9 and 18 show similar behavior, intermediate between Ag_3 and Ag_{15}. These size-selective clusters at subnanometre scale can have dramatically different catalytic properties compared with larger NPs or bulk metal surfaces [56,57].

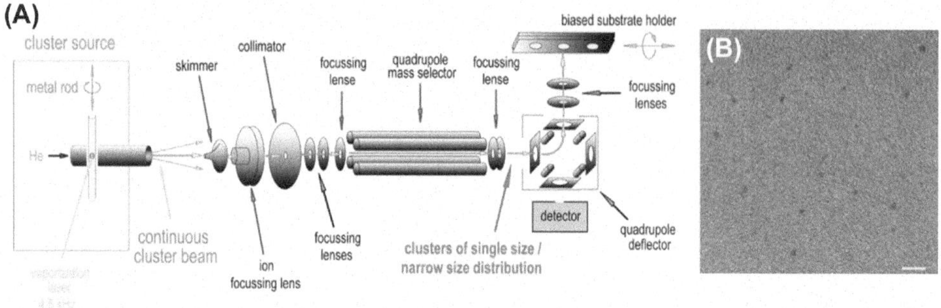

FIGURE 2.11 (A) Schematic diagram of the cluster beam deposition (CBD) setup for soft-landing and controlled surface coverage. (B) Low-magnification transmission electron microscope (TEM) image of Ag_{15} clusters on an amorphous carbon film showing uniform size and distribution of the cluster (scale bar, 5 nm). *(A) Reproduced from S. Vajda, et al., Subnanometre platinum clusters as highly active and selective catalysts for the oxidative dehydrogenation of propane. Nat. Mater. 8 (3) (2009) 213. (B) Reproduced from J. Lu, et al., Effect of the size-selective silver clusters on lithium peroxide morphology in lithium–oxygen batteries. Nat. Commun. (5) 2014 ncomms5895.*

To study the effect of subnanometre clusters on Li–O_2 reaction processes, it is necessary to eliminate defect sites in the carbon cathode from participating in the electrochemical processes. This is performed by atomic-layer deposition (ALD) of Al_2O_3 on the carbon with a thickness of 0.5 nm to guarantee that the surface is still electrically conductive. The clusters of different size were SL on the alumina under identical deposition and coverage conditions (12.35% atomic monolayer equivalent) to have the same total number of Ag atoms for direct comparison, despite the different cluster density (lower for the larger clusters). In these conditions, agglomeration at room temperature is prevented, as it is observed in the TEM image of Ag_{15} clusters in Fig. 2.11.

The results obtained for Ag_3 and Ag_{15} are summarized in Fig. 2.12. It is well recognized that the first step in the ORR is the formation of LiO_2 intermediate. The performed density functional theory (DFT) calculations indicate that the preferred mechanism is the reduction of O_2 on the surface of electrode into the superoxide anion (O_2^-), and after O_2^- desorption reaction with a Li^+ cation to form a solvated LiO_2 molecule. This is shown in Fig. 2.12A.

Following the formation of LiO_2, the authors proposed two possible mechanisms for the formation of Li_2O_2, which depends on the Ag cluster size. By DFT calculations determined the electronic structures of the Ag clusters on a hydroxylated alumina surface. The results are shown for Ag_3 and Ag_{15} clusters in Fig. 2.12B and C. The results indicate that there is a gap at the Fermi level for the Ag_3 but not for Ag_{15}. The large difference in density of states suggests that the Ag_{15} site will have more favorable electron transfer than Ag_3, resulting in the faster formation of LiO_2. Moreover, since the Ag_3 charge profile and capacity are similar to those of the alumina-coated carbon cathode without clusters, the ORR sites are not likely the clusters but rather other cathode surface sites. Then, a slow electron transfer rate for Ag_3 clusters would cause slow formation rate of solvated LiO_2 molecules. This would enable to LiO_2 react with a second superoxide

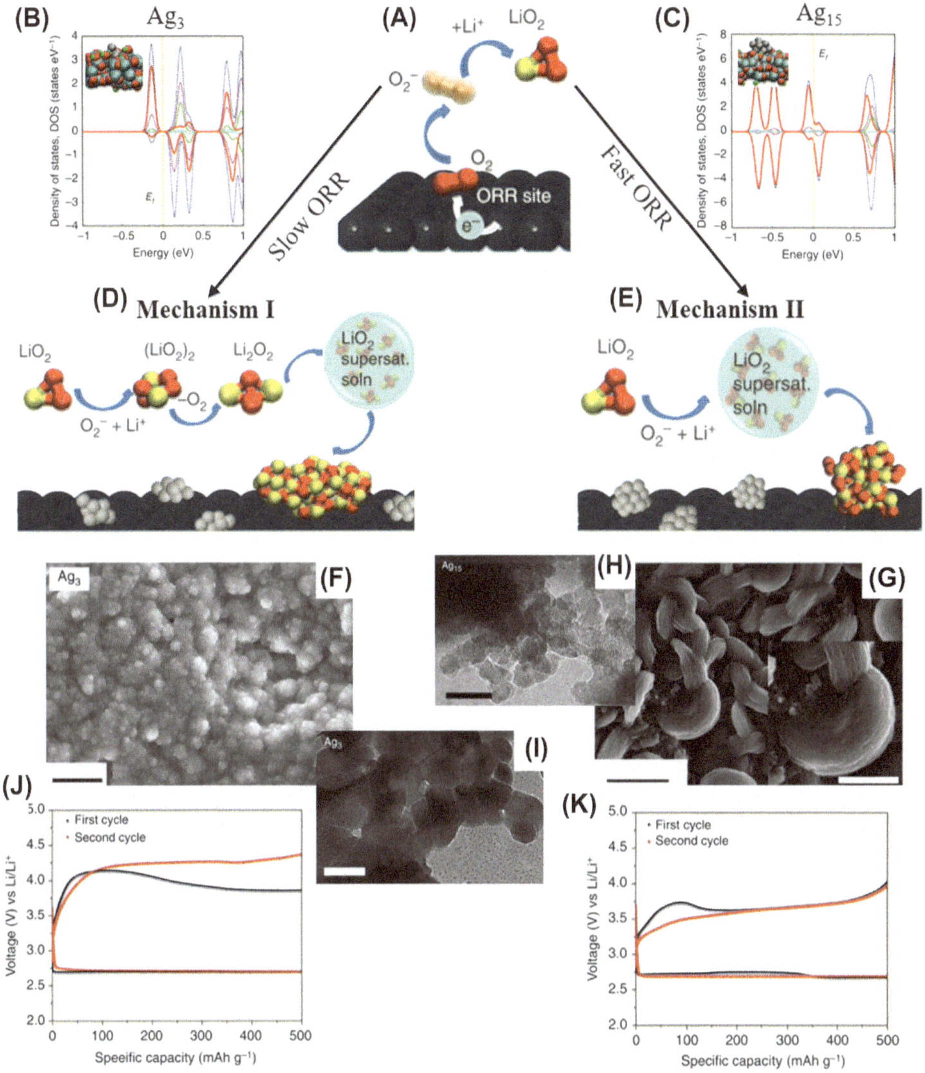

FIGURE 2.12 Illustration of the discharge reaction mechanism: (A) Initial oxygen reduction reaction (ORR) with an electron transfer to O_2 molecule at an active site followed by O_2 desorption and reaction with Li^+ to form solvated LiO_2. After this, two different paths can be followed depending on if ORR is slow (left side, as it happens for Ag_3 clusters) or fast (right side, as it happens for Ag_{15} clusters), which depends on the density of states of the clusters shown in (B) for Ag_3 and (C) for Ag_{15}. The two mechanisms are as follows. (D) Mechanism I: solvated LiO_2 addition incorporates O_2^- and Li^+ ions to form LiO_2 dimers that disproportionate into Li_2O_2 and O_2. Then, Li_2O_2 surface nucleation and growth happens when it is formed saturated Li_2O_2 solution. (E) Mechanism II: LiO_2 supersaturated solution is formed before dimerization. The LiO_2 supersaturated solution nucleates and grows on the surface before LiO_2 undergoes disproportionation. The two mechanisms result in different reaction production as scanning electron microscope (SEM) images show for the cathodes containing (F) Ag_3 and (G) Ag_{15} clusters (scale bar, 1 μm, scale bar of inset image, 500 nm). Transmission electron microscope (TEM) images of discharge products on (H) Ag_{15}-based cathode material (scale bar, 20 nm) and (I) Ag_3-based cathode material (scale bar, 200 nm). The two cycles of capacity-controlled voltages profiles of Li–O_2 batteries using the clusters (J) Ag_3 and (K) Ag_{15}. *Reproduced from J. Lu, et al., Effect of the size-selective silver clusters on lithium peroxide morphology in lithium–oxygen batteries. Nat. Commun. (5) 2014 ncomms5895.*

anion and a Li^+ cation resulting in a $(LiO_2)_2$ dimer, which has very short lifetime [58], and it will disproportionate to Li_2O_2 with loss of O_2, as Fig. 2.12D (Mechanism I) represents. Subsequently, Li_2O_2 reaches a supersaturated solution and nucleates and grows on surface electrode. On the other hand, a fast electron transfer (Ag_{15} clusters) will favor the formation of solvated LiO_2 molecules over dimer disproportionation in solution that results in supersaturated LiO_2 solution, which nucleates and grows on the surface at nucleation sites (Fig. 2.12E, Mechanism II).

The two mechanisms for the Li_2O_2 formation result in different morphologies of the discharge product. The SEM images (Fig. 2.12F and G) reveal that the discharge product resulting for the Ag_3 cluster–based cathode is film-like, while large toroid-like shape (reported in other $Li-O_2$ batteries [59–61]) are observed for Ag_{15} cluster. TEM images show the discharge product of Ag_3 (Fig. 2.12I) and Ag_{15} (Fig. 2.12H) clusters. TEM images show that Li_2O_2 product form plates (100 nm) in the presence of Ag_3, that agrees with the expected result of mechanism I and nanosized grains (10 nm) in the presence of Ag_{15}. These nanosized grains are embedded in the amorphous region with grain boundaries that can be either Li_2O_2 and/or remaining LiO_2, according to mechanism II.

The nature and morphology of the discharge product affect the overpotential of charge process [62]. The larger Li_2O_2 particles formed for Ag_3 clusters have lower electronic conductivity than the grain boundary network with embedded NPs obtained for Ag_{15}. As consequence, the overpotential of charge process is 0.7 V lower for Ag_{15} than Ag_3 (Fig. 2.12J and K) and Ag_{15} shows longer cyclability.

The authors point out that there are other mechanisms suggested in the literature [63–66], but the mechanisms proposed here are the ones that best fit with their study since they have the same electrolyte solution and charge/discharge density current.

This study of well-defined clusters on cathode surfaces provides evidence that the nature of the subnanometer surface species can strongly influence the morphology of the discharge products in $Li-O_2$ cells. These results imply that characterization of the surface structure of cathodes at an atomic/subnanometer level is a key factor in understanding the discharge and charge chemistries in $Li-O_2$ batteries and improving their performance by tailoring the cathode structure.

2.6 Perspectives and conclusions

After reviewing the work performed in energy storage systems incorporating electrodes with clusters deposited using CBD technique, we can return to Fig. 2.1 in which is indicated the works within this topic. We can see that clusters deposited by CBD can be used to functionalize electrodes, as redox units in redox SCs and as catalysts for $Li-O_2$ batteries. These clusters are deposited with high control of size, charge state, and percentage of surface coverage. This control allows to obtain redox SCs with enhanced specific capacity and longer cyclability than the values obtained when the ions are deposited by other techniques. And fundamental insights in catalysis mechanisms in $Li-O_2$ batteries are disclosed, thanks to the high control of the catalyst size. The observed effect was the discharge mechanism depends on the cluster (catalyst) size, determining the morphology of Li_2O_2 discharge product, which determines the overpotential of the charge process and the battery cyclability. CBD can be also used to synthesize new materials for SCs and LIBs. For the first group, we found numerous works using C clusters as building blocks for ns-C electrodes, in which the properties are determined by the interactions, size, and composition of the clusters in place of dealing with an "infinitely" extended solid. In the case of LIBs, clusters are also used as building blocks, but to tailor the structure of the top amorphous layer active in the storage of Li^+ ions. The presence of these scaffolds of clusters determines the mechanical and electrochemical properties of the electrodes in LIBs.

The advantages of using CBD for NPs deposition are similar in all the cases, independently of the role that clusters do in the energy storage mechanism. These are SL of the clusters that allows to maintain their individuality and properties in the thin film in the so-called "memory effect" [14], the not need of cluster surface passivation of the NPs due to the direct and UHV conditions [67], the not need of blinders in the electrode assembling (avoiding to reduce the conductivity and the active mass of the electrode), and high control on the cluster size and composition. These characteristics of CBD offer great advantage to integrate clusters and cluster layers in microelectronics compared to other synthesis methods of clusters, such as electrodeposition or wet chemistry. In this direction, SCs assemble with SCBD have been integrated into more complex systems, such as SC electrolyte gated transistors [7] or driving a low power temperature sensor [8].

To finalize, I am going to refer to the chapter "From cluster design to energy storage device engineering." Great advances have been performed in the design of clusters using CBD technique since the 1990s, with rapid progress in the cluster sources, manipulation in the gas phase, in-depth theoretical studies to elucidate the main cluster deposition mechanism, and characterization of nanostructured films, thanks to the development of electronic microscopies. This knowledge opens unique opportunities to engineer energy storage systems with improved properties than the current devices and integrated in multifunction-energy storage elements. To date, the electrochemical energy storage systems in which CBD was used are SCs, redox SCs, LIBs, and $Li{-}O_2$, but it can be used in many other systems, such as Na batteries or Li–S cathodes. In the same way, CBD was only utilized in the fabrication of electrodes, but there are many opportunities in the design of the electrolyte (solid or hybrid) or as an electrode protective layer. Furthermore, there are many possibilities in the integration of energy storage systems in multifunctional devices using CBD technique that should be further investigated. In one sentence, the great advances in the design of clusters performed by CBD technique opens new frontiers in engineering improved and multifunctional devices that have great potential to explore.

References

[1] L. Diederich, et al., Supercapacitors based on nanostructured carbon electrodes grown by cluster-beam deposition, Appl. Phys. Lett. 75 (17) (1999) 2662–2664.

[2] L.G. Bettini, et al., Electrochemical impedance spectroscopy on nanostructured carbon electrodes grown by supersonic cluster beam deposition, J. Nanoparticle Res. 15 (2) (2013) 1429.

[3] L.G. Bettini, et al., Planar thin film supercapacitor based on cluster-assembled nanostructured carbon and ionic liquid electrolyte, Carbon 59 (2013) 212–220.

[4] L.G. Bettini, et al., Nickel nanoparticles effect on the electrochemical energy storage properties of carbon nanocomposite films, Nanotechnology 25 (43) (2014) 435401.

[5] L. Bettini, et al., Flexible, ionic liquid-based micro-supercapacitor produced by supersonic cluster beam deposition, Electrochim. Acta 170 (2015) 57–62.

[6] L.G. Bettini, et al., Interfacial properties of a carbyne-rich nanostructured carbon thin film in ionic liquid, Nanotechnology 27 (11) (2016) 115403.

[7] F. Soavi, et al., Miniaturized supercapacitors: key materials and structures towards autonomous and sustainable devices and systems, J. Power Sources 326 (2016) 717–725.

[8] L.G. Bettini, et al., Supersonic cluster beam printing of carbon microsupercapacitors on paper, Flex. Print. Electron. 2 (2) (2017) 025002.

[9] V. Prabhakaran, et al., Rational design of efficient electrode–electrolyte interfaces for solid- state energy storage using ion soft landing, Nat. Commun. 7 (2016) 11399.

[10] J. Lu, et al., Effect of the size-selective silver clusters on lithium peroxide morphology in lithium–oxygen batteries, Nat. Commun. 5 (2014) ncomms5895.

[11] M. Haro, et al., Nanoscale heterogeneity of multilayered Si anodes with embedded nanoparticle scaffolds for Li-ion batteries, Adv. Sci. 4 (10) (2017).

[12] L. Lin, et al., Copper-nanoparticle-induced porous Si/Cu composite films as an anode for lithium ion batteries, ACS Nano 11 (7) (2017) 6893–6903.

[13] B.E. Conway, Transition from "supercapacitor" to "battery" behavior in electrochemical energy storage, J. Electrochem. Soc. 138 (6) (1991) 1539–1548.

[14] V. Paillard, et al., Diamondlike carbon films obtained by low energy cluster beam deposition: evidence of a memory effect of the properties of free carbon clusters, Phys. Rev. Lett. 71 (25) (1993) 4170.

[15] J. Yan, et al., Fast and reversible surface redox reaction of graphene–MnO_2 composites as supercapacitor electrodes, Carbon 48 (13) (2010) 3825–3833.

[16] P. Simon, Y. Gogotsi, Materials for electrochemical capacitors, Nat. Mater. 7 (11) (2008) 845.

[17] L.L. Zhang, X. Zhao, Carbon-based materials as supercapacitor electrodes, Chem. Soc. Rev. 38 (9) (2009) 2520–2531.

[18] E. Frackowiak, F. Beguin, Carbon materials for the electrochemical storage of energy in capacitors, Carbon 39 (6) (2001) 937–950.

[19] M. Inagaki, H. Konno, O. Tanaike, Carbon materials for electrochemical capacitors, J. Power Sources 195 (24) (2010) 7880–7903.

[20] A. Pandolfo, A. Hollenkamp, Carbon properties and their role in supercapacitors, J. Power Sources 157 (1) (2006) 11–27.

[21] C. Liu, et al., Toward superior capacitive energy storage: recent advances in pore engineering for dense electrodes, Adv. Mater. 30 (17) (2018) 1705713.

[22] F. Béguin, E. Frackowiak, Carbons for Electrochemical Energy Storage and Conversion Systems, CRC Press, 2009.

[23] P. Simon, T. Brousse, Supercapacitors Based on Carbon or Pseudocapacitive Materials, John Wiley & Sons, 2017.

[24] K. Wegner, et al., Cluster beam deposition: a tool for nanoscale science and technology, J. Phys. D: Appl. Phys. 39 (22) (2006). R439.

[25] F. Caruso, A. Bellacicca, P. Milani, High-throughput shadow mask printing of passive electrical components on paper by supersonic cluster beam deposition, Appl. Phys. Lett. 108 (16) (2016) 163501.

[26] E. Barborini, et al., Synthesis of carbon films with controlled nanostructure by separation of neutral clusters in supersonic beams, Chem. Phys. Lett. 300 (5–6) (1999) 633–638.

[27] G. Bongiorno, et al., Electronic properties and applications of cluster-assembled carbon films, J. Mater. Sci. Mater. Electron. 17 (6) (2006) 427–441.

[28] F. Borghi, et al., Nanoscale roughness and morphology affect the isoelectric point of titania surfaces, PloS One 8 (7) (2013) e68655.

[29] G. Rasines, et al., Electrochemical response of carbon aerogel electrodes in saline water, J. Electroanal. Chem. 671 (2012) 92–98.

[30] B.E. Conway, Electrochemical Supercapacitors: Scientific Fundamentals and Technological Applications, Springer Science & Business Media, 2013.

[31] J. Kang, et al., Development of an equivalent circuit model for electrochemical double layer capacitors (EDLCs) with distinct electrolytes, Electrochim. Acta 115 (2014) 587–598.

[32] N. Kurra, et al., MXene-on-paper coplanar microsupercapacitors, Adv. Energy Mater. 6 (24) (2016).

[33] F. Béguin, et al., Carbons and electrolytes for advanced supercapacitors, Adv. Mater. 26 (14) (2014) 2219–2251.

[34] H. Wang, et al., In operando X-ray absorption fine structure studies of polyoxometalate molecular cluster batteries: polyoxometalates as electron sponges, J. Am. Chem. Soc. 134 (10) (2012) 4918–4924.

[35] P.J. Kulesza, et al., Fabrication of network films of conducting polymer-linked polyoxometallate-stabilized carbon nanostructures, Electrochim. Acta 51 (11) (2006) 2373–2379.

[36] K.D.D. Gunaratne, et al., Design and performance of a high-flux electrospray ionization source for ion soft landing, Analyst 140 (9) (2015) 2957–2963.

[37] K.D.D. Gunaratne, et al., Controlling the charge state and redox properties of supported polyoxometalates via soft landing of mass-selected ions, J. Phys. Chem. C 118 (48) (2014) 27611–27622.

[38] T. Christen, M.W. Carlen, Theory of Ragone plots, J. Power Sources 91 (2) (2000) 210–216.

[39] N. Vicente, M. Haro, G. Garcia-Belmonte, New approaches to the lithiation kinetics in reaction-limited battery electrodes through electrochemical impedance spectroscopy, Chem. Commun. 54 (9) (2018) 1025–1040.

[40] H. Wu, Y. Cui, Designing nanostructured Si anodes for high energy lithium ion batteries, Nano Today 7 (5) (2012) 414–429.

[41] R.A. Huggins, Lithium alloy negative electrodes, J. Power Sources 81–82 (1999) 13–19.

[42] W.-J. Zhang, A review of the electrochemical performance of alloy anodes for lithium-ion batteries, J. Power Sources 196 (1) (2011) 13–24.

[43] D. Ma, Z. Cao, A. Hu, Si-based anode materials for Li-ion batteries: a mini review, Nano-Micro Lett. 6 (4) (2014) 347–358.

[44] X. Su, et al., Silicon-based nanomaterials for lithium-ion batteries: a review, Adv. Energy Mater. 4 (1) (2014).

[45] X. Zuo, et al., Silicon based lithium-ion battery anodes: a chronicle perspective review, Nano Energy 31 (2017) 113–143.

[46] V. Singh, et al., Assembly of tantalum porous films with graded oxidation profile from size- selected nanoparticles, J. Nanoparticle Res. 16 (5) (2014) 2373.

[47] M.T. McDowell, et al., 25th anniversary article: understanding the lithiation of silicon and other alloying anodes for lithium-ion batteries, Adv. Mater. 25 (36) (2013) 4966–4985.

[48] T. Moon, C. Kim, B. Park, Electrochemical performance of amorphous-silicon thin films for lithium rechargeable batteries, J. Power Sources 155 (2) (2006) 391–394.

[49] S.W. Lee, et al., Kinetics and fracture resistance of lithiated silicon nanostructure pairs controlled by their mechanical interaction, Nat. Commun. 6 (2015) 7533.

[50] J. Guo, et al., Cyclability study of silicon–carbon composite anodes for lithium-ion batteries using electrochemical impedance spectroscopy, Electrochim. Acta 56 (11) (2011) 3981–3987.

[51] M. Haro, et al., Germanium coating boosts lithium uptake in Si nanotube battery anodes, Phys. Chem. Chem. Phys. 16 (33) (2014) 17930–17935.

[52] P.G. Bruce, et al., Li–O_2 and Li–S batteries with high energy storage, Nat. Mater. 11 (1) (2012) 19.

[53] M. Haro, N. Vicente, G. Garcia-Belmonte, Oxygen reduction reaction promotes Li+ desorption from cathode surface in Li-O_2 batteries, Adv. Mater. Interfaces 2 (16) (2015).

[54] B. McCloskey, et al., Twin problems of interfacial carbonate formation in nonaqueous Li–O_2 batteries, J. Phys. Chem. Lett. 3 (8) (2012) 997–1001.

[55] A. Luntz, et al., Tunneling and polaron charge transport through Li_2O_2 in Li–O_2 batteries, J. Phys. Chem. Lett. 4 (20) (2013) 3494–3499.

[56] Y. Lei, et al., Increased silver activity for direct propylene epoxidation via subnanometer size effects, Science 328 (5975) (2010) 224–228.

[57] S. Vajda, et al., Subnanometre platinum clusters as highly active and selective catalysts for the oxidative dehydrogenation of propane, Nat. Mater. 8 (3) (2009) 213.

[58] V.S. Bryantsev, M. Blanco, F. Faglioni, Stability of lithium superoxide LiO_2 in the gas phase: computational study of dimerization and disproportionation reactions, J. Phys. Chem. 114 (31) (2010) 8165–8169.

[59] R.R. Mitchell, et al., Mechanisms of morphological evolution of Li_2O_2 particles during electrochemical growth, J. Phys. Chem. Lett. 4 (7) (2013) 1060–1064.

[60] B.D. Adams, et al., Current density dependence of peroxide formation in the Li–O_2 battery and its effect on charge, Energy Environ. Sci. 6 (6) (2013) 1772–1778.

[61] X. Gao, et al., Promoting solution phase discharge in Li–O_2 batteries containing weakly solvating electrolyte solutions, Nat. Mater. 15 (8) (2016) 882.

[62] B.M. Gallant, et al., Influence of Li_2O_2 morphology on oxygen reduction and evolution kinetics in Li–O_2 batteries, Energy Environ. Sci. 6 (8) (2013) 2518–2528.

[63] L. Johnson, et al., The role of LiO_2 solubility in O_2 reduction in aprotic solvents and its consequences for Li–O_2 batteries, Nat. Chem. 6 (12) (2014) 1091.

[64] C.O. Laoire, et al., Influence of nonaqueous solvents on the electrochemistry of oxygen in the rechargeable lithium– air battery, J. Phys. Chem. C 114 (19) (2010) 9178–9186.

[65] S.H. Oh, et al., Synthesis of a metallic mesoporous pyrochlore as a catalyst for lithium–O_2 batteries, Nat. Chem. 4 (12) (2012) 1004.

[66] X. Ren, et al., Oxygen reduction reaction catalyst on lithium/air battery discharge performance, J. Mater. Chem. 21 (27) (2011) 10118–10125.

[67] P. Moriarty, Nanostructured materials, Rep. Prog. Phys. 64 (3) (2001) 297.

3

Cluster-assembled devices for solar energy conversion

Marcel Di Vece

CIMAINA AND DEPARTMENT OF PHYSICS, UNIVERSITÀ DEGLI STUDI DI MILANO, MILANO, ITALY

Chapter outline

3.1 Introduction

Solar cell technology is important for many reasons, of which the ecological sustainability is the most important. The first commercial solar cells based on silicon wafer technology (Fig. 3.1) have been around for many decades and this technology has shown incremental progress. Currently, solar energy contributes to a bit more than 1% of the global energy demand (in some countries up to 7%), which clearly indicates that the price per kWh should be further reduced to become competitive (with conventional electricity sources) in more locations worldwide to obtain significantly higher market shares. Because the price of a solar cell is determined to a large extent by the module and installation costs, the efficiency of a solar cell determines its commercial success. This forms the basis of the ongoing solar cell research: the increase of photovoltaic efficiency. Research aims at higher efficiencies with cheap materials and using novel techniques [1,2].

Cluster Beam Deposition of Functional Nanomaterials and Devices. https://doi.org/10.1016/B978-0-08-102515-4.00003-9

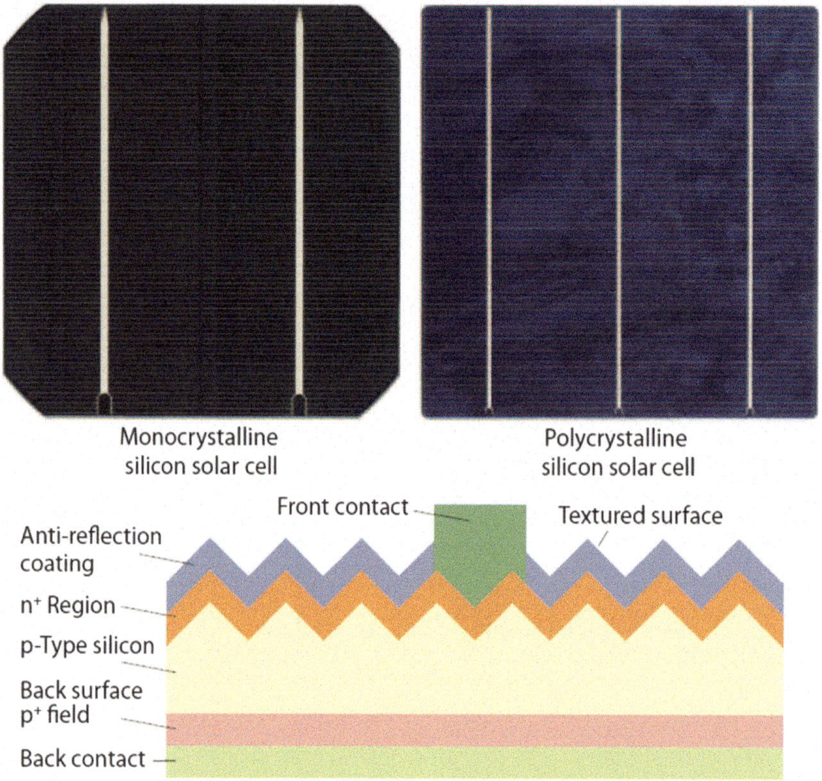

FIGURE 3.1 Typical mono- and polycrystalline silicon solar cells (*top*), and simplified cross-section of a commercial monocrystalline silicon solar cell (*bottom*). *Reprinted with permission of Saga T. (2010).*

The efficiency of a solar cell has fundamental limits as determined by Shockley and Queisser (SQ), who calculated a maximum efficiency using a single p-n junction of 32% for c-Si solar cells [3]. With modern approaches such as tandem cells, the theoretical efficiency is 45%. One of the most important processes to limit solar cell efficiency is the recombination of electrons and holes inside the active layer. Particularly for thick solar cells such as crystalline or multicrystalline Si wafer solar cells, this reduces the efficiency very strongly. The obvious advantage of lower cost and lower energy consumption during fabrication forms a strong motivation for fabricating thin-film solar cells. It is clear that by reducing the thickness of the active layer in a solar cell, the optical absorption likely decreases as well. Therefore, the main aim in current thin-film solar cell development is the improvement of absorption by the choice of composition and with advanced light management. Several materials are currently being explored for thin-film solar cell technology, of which the a-Si, CdTe, and CIGS are already commercially available [4,5], and the perovskites are knocking on the door.

Light management makes use of nanostructures which have intricate interactions with the incident light, which can be used to stimulate optical absorption in thin-film solar cells [6]. Both the composition of a solar cell and light management with nanostructures form attractive possibilities for using nanoparticles. Light management can be achieved by light scattering on, for example, metal particles, light in-coupling, and local field enhancement [7]. Additionally, light in the UV and in the infrared range is not harvested efficiently in solar cells. In the UV range, a considerable amount of energy is wasted by thermalization of electrons, which are excited far beyond the band gap. In the infrared range, the band gap is too large to be able to absorb the light. Therefore up- and down-converter materials are being investigated [8], which can be in the form of particles, to split the UV photons into two photons and combine the infrared photons to a single photon, all resulting in an energy package comparable to that of the semiconductor band gap. Nanoparticles are particularly interesting as quantum dots which are semiconductor particles with a radius smaller than the Bohr radius and which enable the tuning of the band gap by changing the size [9].

Most light management techniques currently being explored make use of lithography. Although this is very rewarding as a proof of principle, it is economically not viable. Therefore the use of nanoparticles forms an industrially interesting alternative because they can be deposited as a bottom-up method, making it fast and cheap [10]. Nanoparticles can be fabricated by chemical or physical techniques, each of which have their advantages and disadvantages. The chemical route to nanoparticle fabrication and deposition is generally well known, as are the complications of matching it with existing solar cell fabrication techniques. Although the physical deposition of nanoparticles, particularly with cluster sources, has been around for several decades, the use in solar cell research and development has barely begun. In this chapter the scientific and technological opportunities of the (physical) fabrication and implementation of nanoparticles (clusters) for photovoltaics are explored with two examples: Mie scattering and plasmonics.

3.2 Thin-film solar cells

Thin-film solar cells can be made of many possible materials, which are all currently the object of research, and the thickness ranges from several μm to hundreds of nm. An overview of the most investigated, used, and promising materials by Polman et al. [11] shows that the increase of efficiency due to research is the fastest for quantum dots, CZTS, CdTe, organic and perovskite solar cells. Except for Si (homojunction and heterojunction), GaAs, and GaInP, most compositions are at least 25% below their SQ limit, which implies that there is much room for improvement (Fig. 3.2).

The thin-film solar cell with a large market share is based on a-Si:H, which also forms an important research platform for light management studies. The advantages of a-Si:H-based solar cells are its conformal growth, which creates many perspectives in combination with nanoscale approaches, and the possibility to form flexible solar cells. Unfortunately the efficiency of the a-Si:H thin-film solar cell did not improve much

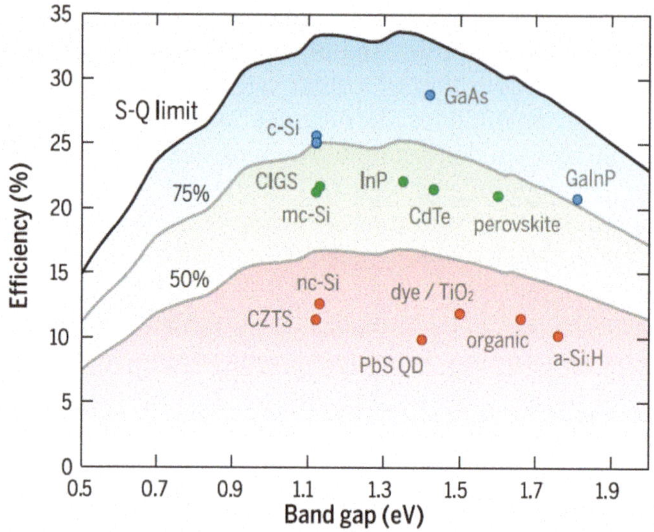

FIGURE 3.2 Theoretical Shockley-Queisser detailed-balance limit as a function of band gap (*black line*) and 75% and 50% of the limit (*gray lines*). The record efficiencies for different materials are plotted for the corresponding band gaps. *Reprinted with permission of A. Polman, M. Knight, E.C. Garnett, B. Ehrler, W.C. Sinke, Science 352 (2016).*

beyond 12% for many years. The main reason is the abundant presence of defects, which form recombination centers for the electrons and holes. Although CdTe and the chalcopyrite compound copper indium gallium diselenide (CIGS) are currently marketed as thin-film solar cells [12], more research is ongoing to increase their laboratory efficiency, to make them more useful.

Perovskites are currently being very intensively investigated, and impressive laboratory efficiencies of up to 25.2% have been reported (NREL efficiency chart 2019). Although the advantages of perovskites include ease of fabrication, strong solar absorption, and low nonradiative carrier recombination rates, their instability under ambient conditions and illumination due to the presence of organic compounds hinders commercial success. Replacement of the organic components is an important goal, which will bring this material to the market. The low resistance of present perovskite technology to moist air and water vapor, together with the inability to produce high-quality large surfaces, forms a serious bottleneck. Although perovskites can be used as a thin-film solar cell, they can also be used as a dye particle in, for example, a TiO_2 electrode.

3.3 Light management

In thick, conventional crystalline silicon solar cells, light management techniques such as antireflection texture and coatings have been employed successfully for a long time. Because such an antireflection texture has dimensions much larger than the thickness of thin-film solar cells, they are not suitable. Therefore new light-trapping techniques at the nanoscale, which are compatible with the thin-film solar cell, have to be found [13–15]. Here we describe two novel concept light management approaches: Mie Scattering with TiO_2 nanoparticles and plasmonics with gold nanoparticles, both combined with a-Si thin films.

3.3.1 Mie scattering

A new approach to more efficiently couple light into a thin-film solar cell is the use of Mie scatterers as shown in Fig. 3.3. In a Mie scatterer, usually a dielectric particle such as silicon [16] with the dimension of the wavelength of light (down to 10%), the scattering of light often has a stronger intensity in the forward direction, which is useful when a light absorbing material is placed beneath it. The forward scattering couples the light efficiently into the layers below it [17]. Since a Mie scatterer is made out of a dielectric material and not a metal with electric resistance acting on the plasmonic resonance, its optical losses are minimal. When the Mie scatterer is large, the captured light will resonate at particular wavelengths, depending on the interference conditions, which enhances its optical cross-section for these wavelengths. The circulating light in this Mie scatterer can be coupled into a solar cell beneath it by leaking, which makes more efficient and selective harvesting of light possible. Mie scattering can provide a broad band optical absorption with, for example, silicon nanocone arrays [18]. The Mie scatterer is not limited to solar cells but can also be employed in solar fuel generation by light [19,20,21]. The Mie scatterer composition can be chosen to obtain the desired optical properties as shown by using silicon carbide [22], silica [23], and germanium [24] particles.

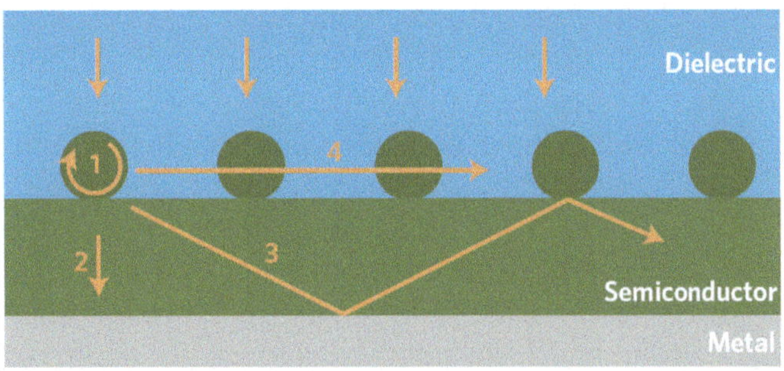

FIGURE 3.3 Schematic showing how various optical resonances supported by a model cell structure can be excited to enhance light absorption in the active semiconductor material (*green*). The numbers 1–4 label four distinct coupling mechanisms: 1: excitation of an optical (Mie) resonance of hexapolar symmetry in the NWs. This resonance is hybridized with a guided resonance in the underlying Si layer. 2: Fabry-Perot standing-wave resonance that results from the confinement of light between the reflecting top surface of the high-index Si film and the metallic back-reflector. 3: guided resonance of the Si layer. 4: excitation of diffracted modes. Such modes can be excited when incident light is redirected into the plane of the light-trapping layer, where they concentrate most of the field. These modes also extend into the underlying semiconductor layer where they can induce useful absorption. *Reprinted with permission of M.L. Brongersma, Y. Cui, S. Fan, Nat. Mater. 13 (2014) 451–460.*

In most of the current research of Mie scatterers for solar cells, top-down techniques such as lithography and laser printing [25] are used to create the desired nanostructures. As this is commercially not yet attractive, alternative methods (i.e., bottom-up) to fabricate Mie scatterers for solar cells need to be explored. Although wet chemical techniques are used to fabricate Mie scattering structures [26,27], more industrially compatible techniques are desirable. A recent work by Ref. [28] shows that TiO_2 Mie scatterers can be produced with a gas aggregation cluster source and combined with a-Si thin films. TiO_2 has also been used as a Mie scattering composite in dye-sensitized solar cells [29] and as a scatterer for solar cells [30]. Since thin a-Si films are used in solar cells, they function as a photovoltaic model system. TiO_2 Mie scatterers can easily be fabricated with a gas-phase aggregation cluster source [31,32], which is fast, cheap, vacuum-compatible, and provides ultraclean and well-defined particles. TiO_2 particles were also produced by a pulsed microplasma cluster source [31], which demonstrates its versatility.

3.3.1.1 *Mie scattering for solar cell example*

One new approach to more efficiently couple light into a solar cell is the use of Mie scatterers for thin-film solar cells, as these are not suitable for standard light trapping techniques [13,15]. In a Mie scatterer, with the dimension of the wavelength of light (down to 10%), the scattering of light often has a stronger intensity in the forward direction, which is useful when a light absorbing material is placed beneath it. The forward scattering enables the light to couple efficiently into the layers below it [17].

In this section the optical interaction of TiO_2 Mie scatterers, with a-Si thin film is investigated based on Giannakoudakis et al. [28]. TiO_2 has been used as Mie scattering composite in dye sensitized solar cells [29] and has a sufficient high refractive index. Thin a-Si films are used in solar cells and can also function as a photovoltaic model system. Since TiO_2 is inexpensive, it could lead to a reduction of the amount of a-Si in the solar cell. TiO_2 has a high refractive index (2.6 [33]) and has been investigated before as a scatterer for solar cells [30]. Here the TiO_2 Mie scatterers are fabricated with a gas phase aggregation cluster source [31,32], which is fast, cheap, vacuum-compatible, and provides ultraclean and well-defined particles. Additionally, for thin-film solar cells, conventional light management strategies don't work due to their large size [14], which is circumvented in this approach. Previously TiO_2 particles were produced by a pulsed microplasma cluster source, which demonstrates its versatility. After deposition of the TiO_2 particles on a-Si layers, the systems were studied by measuring the optical transmission and reflection, from which the optical absorption could be determined.

Strong forward scattering of the TiO_2 particles, depending on the film thickness and composition, allow the detailed study of the optical cross-section of such structures and the efficiency with which light is redirected. This work opens a very promising new research avenue by using cluster sources for energy-related research.

The TiO_2 nanoparticles were produced by a gas aggregation magnetron sputtering cluster source (NC200U-B, Oxford Applied Research Ltd.). The TiO_2 particle yield was 2.5 particles/μm^2/min. The TiO_2 particles were investigated by deposition on an a-C TEM

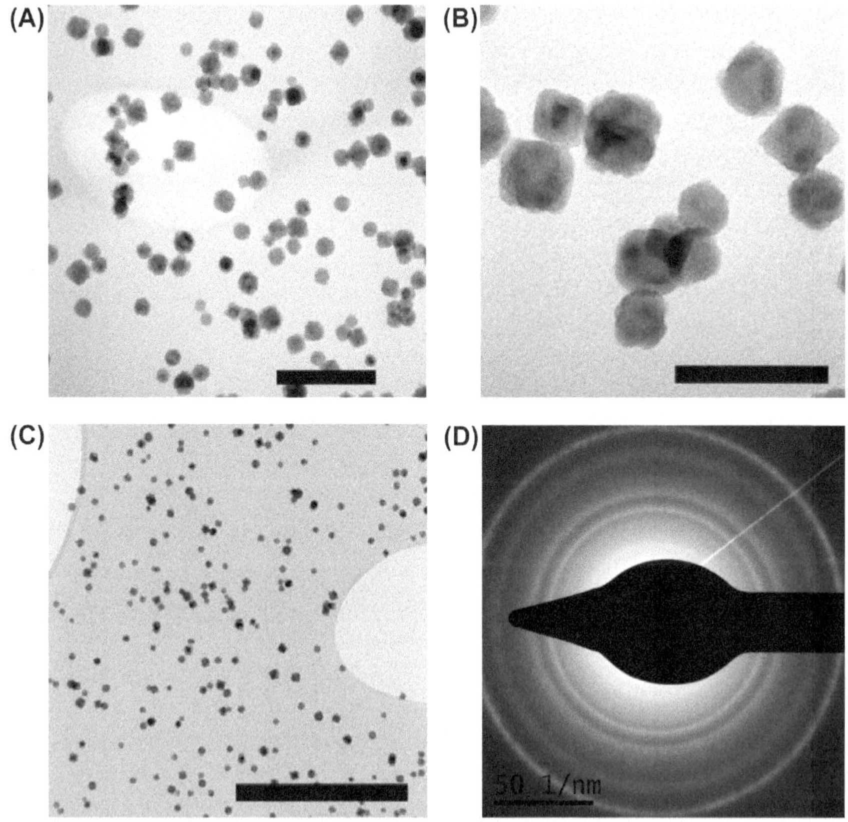

FIGURE 3.4 (A) and (B) TEM images of TiO_2 particles produced by the gas phase aggregation cluster source. The shape of the particles is predominantly cubic. In (C) the uniformity of the deposition is clear. (D) an electron diffraction pattern in which the clear rings confirm crystallinity. The scale bars are for (A) 100 nm, (B) 40 nm, and (C) 400 nm.

grid after which transmission electron microscopy (TEM) was performed as shown in Fig. 3.4A–C. Although the cubic shapes are predominant, other shapes such as hexagonal, rhomboid, "cauliflower" [34]-like particles, which are agglomerates of smaller particles, are also present. Each TiO_2 deposition produces a different ratio of these particles' shapes, leading to slightly different optical response. TEM electron diffraction analysis of the particles as shown in Fig. 3.4D, provided lattice distances of 2.57 Å, 2.26 Å, 1.71 Å, and 1.48 Å, which correspond well with a mixture of rutile and anatase structures. The samples selected for this work agreed best with the simulations and therefore likely consist mainly of cubic particles.

The a-Si thin films were obtained by magnetron sputtering of two nominally 30 and 70 nm thin films. The reasons for choosing 30 and 70 nm thick films was first of all to have the thickness as small as possible and still maintain reasonable optical absorption. The very thin a-Si film solar cell is commercially very interesting. Moreover, for this work it was necessary to not have maximum optical absorption in order to be able to measure optical absorption increases by the TiO_2 particles. Another complication of much thicker a-Si films is the presence of multiple interference effects which were avoided by choosing such a thin film. The TiO_2 particles were deposited on a-Si thin films with two different thicknesses. This was necessary because the interference of light in and on a thin film causes two effects: 1) destructive interference in reflection, leading to strongly reduced reflection at the resonance wavelength, which is used in products as antireflection coating, and 2) constructive interference in reflection, leading to a strong reflection (and reduced transmission) at the resonance wavelength. By selection of two a-Si thicknesses, these two modes were explored in combination with TiO_2 Mie scatterers. The thicknesses of 30 and 70 nm provides interference at 312 nm (3.97 eV) and 364 nm (3.41 eV), respectively. The total reflection, diffuse reflection, and transmission were measured from 250 to 1200 nm and as a reference a blank glass substrate was used. To obtain the optical response of the TiO_2 particles exclusively, the difference of a sample with and without TiO_2 particles was used.

The atomic force microscopy images of the TiO_2 particles on the a-Si thin films yielded clearly distinguishable particles with densities of 53 and 54 particles/μm^2 for the 30 and 70 nm-thin a-Si film, respectively (Fig. 3.5).

Finite-difference time-domain (FDTD) simulations in 3D were performed with commercial software (Lumerical Solutions Inc.) on a flat a-Si layer (30 and 70 nm) with cubic TiO_2 particles of three different sizes: 20, 40, and 70 nm. To simulate a fully periodic and infinitely uniform distribution of TiO_2 particles, we used Bloch boundary conditions in the direction of incident light, while in the perpendicular direction periodic

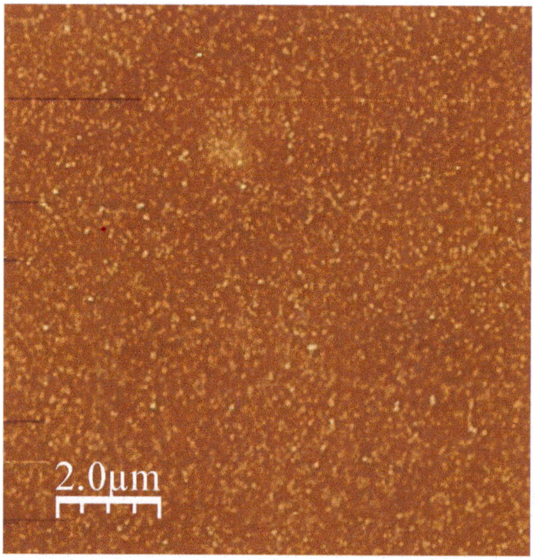

FIGURE 3.5 Atomic force microscopy (AFM) micrograph of 70 nm a-Si film with TiO_2 nanoparticles on top.

(anti)symmetric boundaries were used to reduce calculation time. The simulation area had a length and width of 513 and 560 nm.

In Fig. 3.6 the square TiO_2 nanoparticle on the a-Si thin film is schematically depicted for the condition of destructive (A) and constructive (B) interference for the reflected light. The light scattering of the TiO_2 particle is depicted as a dashed ellipse to indicate the large area of interaction. Since the scattering of light on a Mie particle results in a spectrally continuous (white) response [35], the scattered light therefore interferes with the incoming, reflected, and transmitted light on the thin film.

In Fig. 3.7A the optical absorptance of the 30 nm-thick a-Si thin film with TiO_2 nanoparticles after subtraction with the reference a-Si film is shown. With this method the effect of only the TiO_2 particles on the optical response remains. A clear difference of

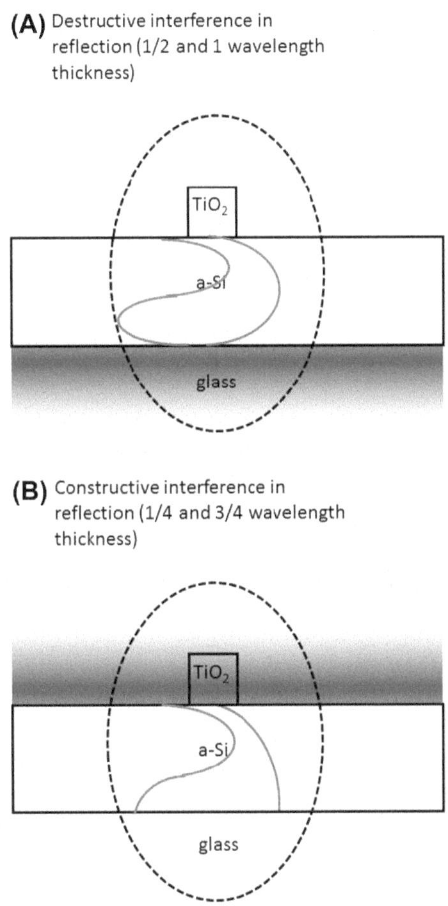

FIGURE 3.6 The two possible optical properties of a thin a-Si film: (A) with destructive interference for reflection providing optimal absorption and transmission and (B) constructive interference for reflection, providing reduced absorption and transmission. The TiO_2 particle is expected to interact strongly with the strong field from reflection in situation (B).

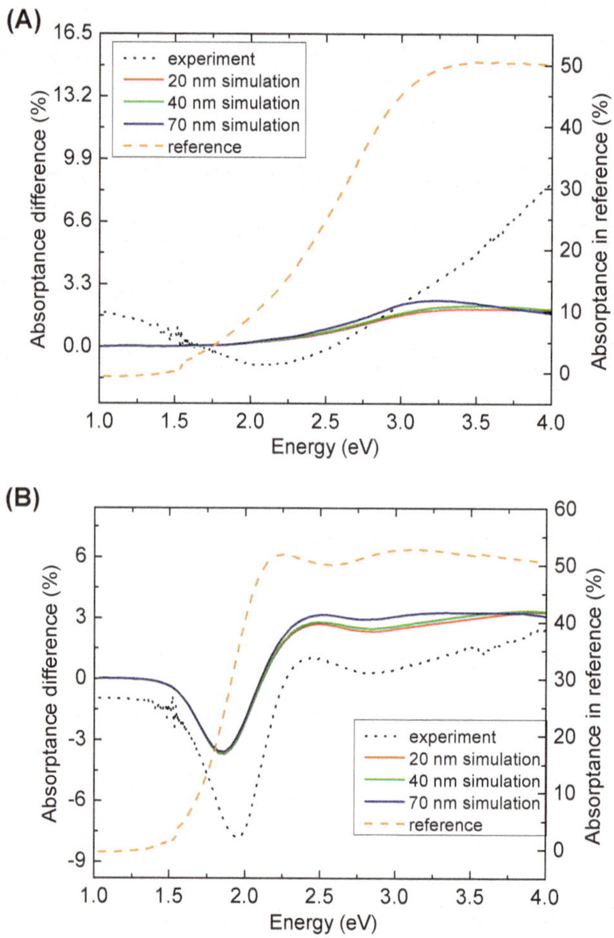

FIGURE 3.7 Experimental (*black*) and simulated (*red, green, and blue*) optical absorptance difference with an a-Si reference (*orange*) for (A) 30 nm (antireflection) and (B) 70 nm (antitransmission) a-Si film. The different TiO_2 particle sizes are indicated by color. The reference a-Si thin film always has the same thickness as the a-SI thin film with TiO_2 particles.

several percent between the reference thin film and the thin film with TiO_2 particles is observed. The spectral shape of the absorptance difference is significantly different from the reference absorptance spectrum, indicating an independent effect of the TiO_2 particle.

The increase of absorptance difference going from the blue to the UV is caused by the optical absorption of TiO_2 with a band gap around 3.1 eV [36,37]. Since the solar spectrum (AM1.5) contains very little intensity at such energies, this effect is not relevant for photovoltaics. The absorptance increase of a few percent between 1 and 1.5 eV and the reduced absorptance at about 2 eV is the result of light scattering on the TiO_2 particles. Considering the particle density of only 54 particles/μm^2 as obtained from AFM inspection, this is a strong response.

In the simulation of the different TiO_2 particle sizes, no significant absorptance difference is observed from 1 to 2 eV, possibly because the a-Si thin film does not absorbs light in that range. From about 2 eV until about 3 eV the absorptance difference increases in line with the experiment due to scattering. In contrast to the experiment, the absorptance difference remains constant from 3 eV till the UV. The absence of a slope for the absorptance difference from 3 eV to the UV for the simulations is likely the result of the presence of defects in the experimental TiO_2 particles, which are optically very active. The difference in magnitude of the absorptance difference between experiment and simulation is the result of different TiO_2 particle surface density on the a-Si layer.

In Fig. 3.7B the absorptance (difference) spectra for the 70 nm-thin a-Si film with and without TiO_2 particles are shown. Except for a dip to negative values around 1.9 eV both the shape of the experimental and simulated graphs are very similar to the reference a-Si absorptance curve. The positive values of the absorptance difference over a wide spectral range suggests that the main effect of the TiO_2 particles is to increase the overall absorptance due to scattering. Since the 70 nm-thin a-Si film has enhanced reflection, it may be argued that the scattering on the TiO_2 particles reduces the reflection, which leads to increased optical absorption.

The dip of the absorptance difference in both the simulation and experiment around 1.9 eV is the result of the reducing refractive index of a-Si. When the refractive index of a-Si is lower with respect to the refractive index of TiO_2, the scattering properties of the TiO_2 particles change and more light is scattered backward, away from the a-Si thin film. The increasing absorptance from about 3.1 eV is again the result of optical absorption in defect-rich TiO_2.

The results of the FDTD simulation for three particles of different size (20, 40, 70 nm) agree remarkably well with the experimental curves. The three different particle sizes show no discernible differences, which confirms that their Mie scattering properties, which are mainly size-independent, dominate the optical effects. It is precisely this size independence of the Mie scattering of the TiO_2 particles that enables such a good agreement between experiment and simulations.

The absorptance of the 30 and 70 nm-thick films is similar around 50% because the increased reflection in the 70 nm film reduces its optical absorption and the reduced reflection of the 30 nm film increases its optical absorption.

In Fig. 3.8 the E and H fields are shown in cross-sections of the 40 nm TiO_2 particle on the 30 and 70 nm a-Si thin films. The choice of size and thickness is arbitrary, since the different particle sizes and both a-Si thin films provide very similar results. From the pattern around the particles it is clear that the TiO_2 particles affect the electromagnetic field around them. The E-field is clearly distorted mainly outside the particle as can be seen from Fig. 3.8A and C. It is known that the E-field induces displacement currents on the dielectric surface of a particle, which results in a much stronger magnetic response [38] as can be seen from the high intensity H field inside the particle (Fig. 3.8B and D). These cross-sections demonstrate the significant electromagnetic interaction of the TiO_2 particles with the light field on an a-Si thin film.

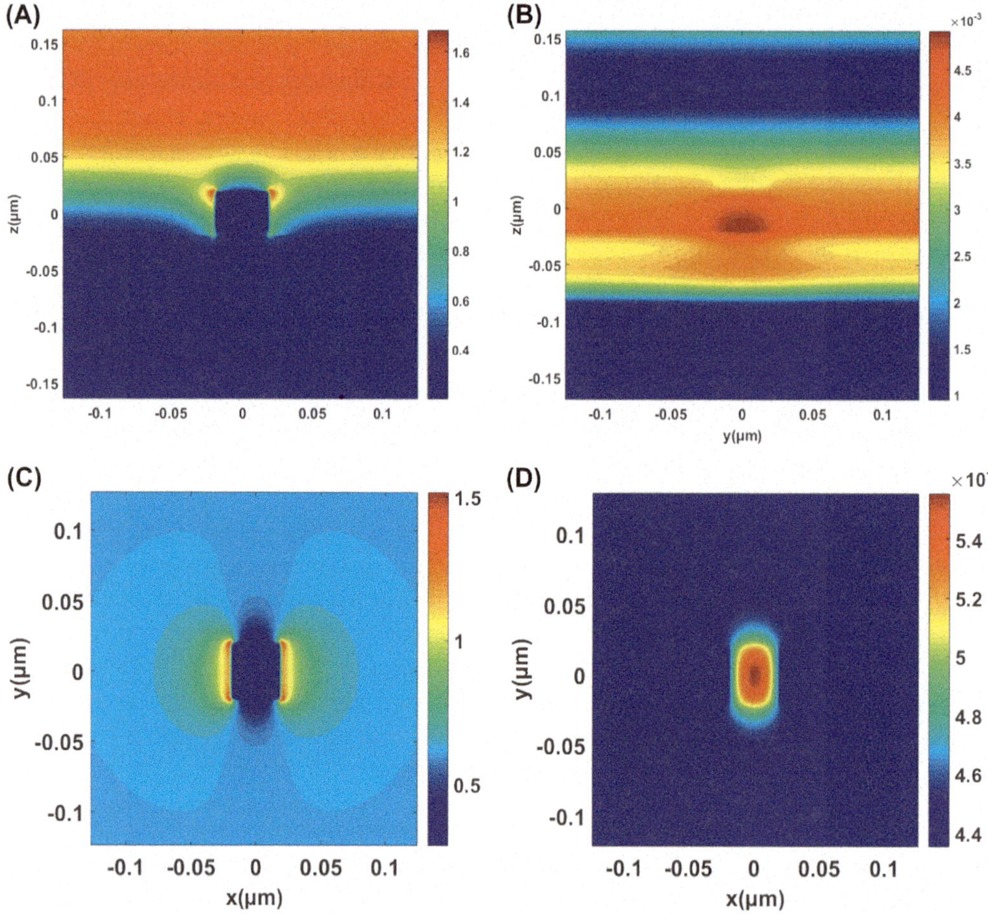

FIGURE 3.8 Time integrated (A) E-field (logarithmic scale) and (B) H-field at 541 nm (2.29 eV) of 40 nm TiO_2 particle on 70 nm-thick a-Si film (bottom part of graph). The high E-field intensity above the a-Si thin film confirms it antitransmission properties. The TiO_2 particle affects both the E and H fields strongly at a local level. The E (logarithmic scale) and H fields at the surface of the a-Si film underneath the 40 nm TiO_2 particle are shown in (C) and (D), respectively (450 nm).

Although the optical effects in both the experiment and simulation are in the order of a few percent, increasing the particle density on the a-Si thin film surface would significantly increase the optical interaction and subsequently optical absorption. Since this is only a proof of principle study, future research toward device should explore the various parameters in more detail.

3.3.1.2 Conclusions

The presence of optical broad band Mie scattering of TiO_2 particles in the fields of constructive or destructive interference on a thin a-Si film results in a complex interaction. For the thin antireflecting and enhanced reflection of a-Si films, both increased and decreased optical absorption has been observed for certain photon energy ranges. The TiO_2 particles absorb photons larger than the band gap of TiO_2, which was particularly strong in the experiment due to defects. The experiment and simulations agreed well on many spectral features in the different systems.

3.3.2 Plasmonics

Plasmonics, the electronic response of light interacting with metal nanostructures, is a promising novel concept in the quest for photovoltaic devices with high efficiencies and low-cost materials. The interest in plasmonics, which merges optics with electronics [39,40] at a very small scale, has increased considerably since about a decade. Five main effects are distinguished with the metal plasmon and optically active neighboring materials [41]: 1) Plasmonic field enhancement [7,42–44] (a near-field effect) increases the electromagnetic field close to the metal surface by orders of magnitude. This is the result of the (dipole) electric field around the metal nanoparticle in response to illumination. This is particularly interesting for thin-film solar cells, as light concentration has a significant effect on the overall absorption. 2) Scattering at/by the metal nanoparticle [45] increases the light path length in a solar cell and is therefore an important mechanism to increase efficiency. Incident light is scattered off the metal object into a distribution of optical modes within the semiconductor. 3) Coupling of plasmon resonance to waveguide modes in thin semiconductor slabs. 4) The coupling of an (semiconductor) emitter to plasmon modes also affects both the radiative and nonradiative decay rates. This phenomenon is based on the principle that the strength with which an emitter couples with an electromagnetic field depends on its environment [46–48]. 5) Energy transfer from the absorber/emitter to plasmon modes of a nearby nanoparticle or vice versa can occur by, for example, Förster Energy Transfer (FRET) [49]. This energy will be partially dissipated or coupled to radiation [50,51]. The balance between these effects depends strongly on the geometry, shape, size, and distance [52]. A schematic of some plasmonic interactions is shown in Fig. 3.9. However, inclusion of metal nanostructures in solar cells may also be responsible for an efficiency drop of the device performance, since the introduction of additional metallic structures into the semiconductor photoactive layer causes further Ohmic losses and may introduce significant structure defects, which are quite detrimental to the photocurrent by increasing the recombination rate [53,54].

One of the first experimental investigations of plasmonic enhancement for photovoltaics involved silver nanoparticles in an organic solar cell [55,56]. The first papers to explore the potential of localized surface plasmon resonance (LSPR) in solid-state photovoltaics were published in the late 1990s by Stuart and Hall [57]. Their device consisted of a thin silicon-on-insulator wafer, where the Si was 165 nm thick, and since the absorbing Si layer was separated from the Si substrate by a layer of SiO_2, it acts as a thin waveguide. Metal island formation was achieved by the deposition of a thin

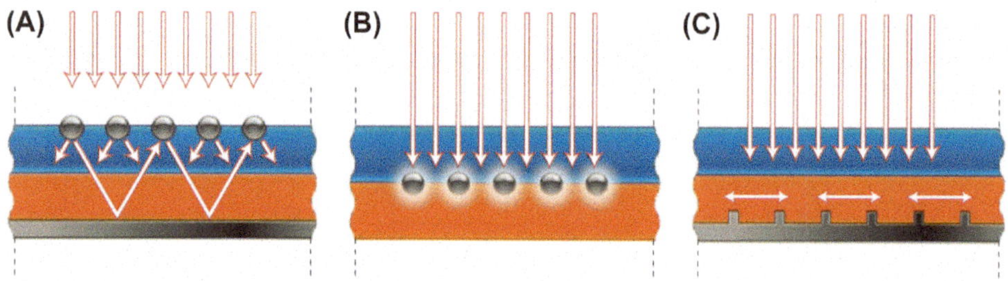

FIGURE 3.9 Plasmonic light-trapping geometries for thin-film solar cells. (A) Light trapping by scattering from metal nanoparticles at the surface of the solar cell. Light is preferentially scattered and trapped into the semiconductor thin film by multiple and high-angle scattering, causing an increase in the effective optical path length in the cell. (B) Light trapping by the excitation of localized surface plasmons in metal nanoparticles embedded in the semiconductor. The near-field of the excited particles causes the creation of electron–hole pairs in the semiconductor. (C) Light trapping by the excitation of surface plasmon polaritons at the metal/semiconductor interface. A corrugated metal back surface couples light to surface plasmon polariton or photonic modes that propagate in the plane of the semiconductor layer. *Reprinted with permission of H.A. Atwater, A. Polman, Nat. Mater. 9 (2010) 205–213.*

metallic film and annealing under N_2 so that the film coalesced into discrete islands. Subsequent studies show that the local field enhancement and scattering of plasmonic nanostructures increase the solar cell performance [7,39,42,43]. Additionally, work by Barnard et al. [58] shows the enhancing effect of silver nanostructures on the photocurrent in silicon. Underneath a metal nanoantenna, the field intensity is enhanced by a factor of 2.8 while locally the field enhancement can be significantly higher (up to 20) in "hot spots" under the corners of the antenna. A theoretical study of plasmonic nanoparticles for solar cells shows that an absorption increase of 10%–15% is more than feasible [59].

To date, most research is focused on noble metals [60,61] which have strong resonances in the visible part of the spectrum and have the advantage of being inert. However, the high price of noble metals makes them unsuitable for commercial solar cells. Although aluminum has its bulk plasmon resonance in the far UV range at around 150 nm, its plasmon resonance can be shifted, making it a cheap alternative. The plasmon resonance wavelength of aluminum nanoparticles can be shifted toward the visible range by selection of size, shape, configuration, and dielectric environment [62–68]. The electron collision frequency at a plasma frequency in the visible range in aluminum is in between that of gold and silver [69], which make Ohmic losses acceptable. A theoretical study by Akimov et al. [59] has demonstrated that aluminum particles increase the optical absorption in a-Si:H over a wide wavelength range better than silver particles.

Nanoparticle plasmonics: although previous research on using plasmonics for solar cells included both nanoparticles and nanostructures formed by lithography as shown in Fig. 3.10, much work remains to be done as commercial applications are still not realized. Although lithography has been very rewarding as a proof of principle method, this top-down technique is economically not yet interesting. However, the roll-to-roll nano

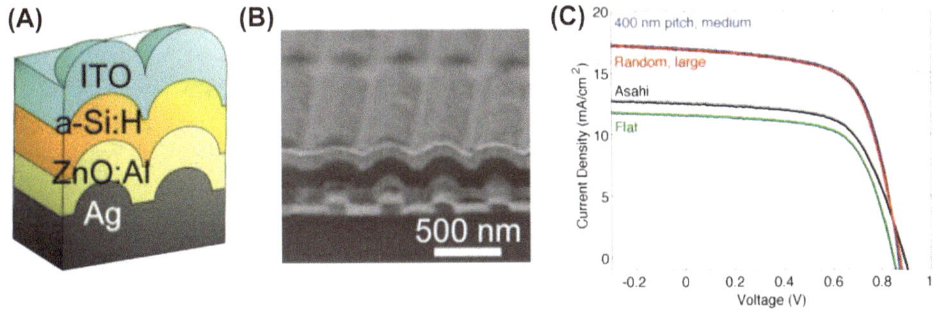

FIGURE 3.10 Current voltage characteristics of nanopatterned cells. The cells are conformally deposited over the patterned substrate, as shown schematically in (A) and in SEM cross-section in (B). The maximum particle diameter in the backpattern was chosen so that nanostructures in the ITO top layer would touch without overlap. (C) Current density voltage measurements for the most efficient cells on the substrate with 90 nm intrinsic layers. The optimized periodic pattern and the pseudorandom pattern have efficiencies of 9.6% and 9.4%, respectively. *Reprinted with permission of Ferry V.E. et al. (2011).*

imprint lithography technique has been used in such studies and is industrially more interesting [70,71]. The use of (plasmonic) nanoparticles from a bottom-up approach, such as with a gas aggregation cluster source, as demonstrated by Faraone et al. [72], has many advantages such as speed and low cost, which brings these novel concepts within industrial reach.

3.3.2.1 Plasmonics for solar cell example

It is of paramount importance to tailor the size, density, shape, and location of the nanoparticles in the design of an appropriate solar cell architecture in order to provide both an efficient light trapping scheme and a reduction of the energy losses allowing an enhancement over the entire spectrum of the photovoltaic performance [53,67]. Although Ohmic losses in the metallic nanoparticles are considered an obstacle for thick solar cells [73], for very thin- film solar cells the benefits may outweigh these parasitic losses, since their optical absorption can benefit from the smallest boost [74]. In particular, in thin a-Si photoactive layers, the light trapping scheme should be able to achieve a large absorption cross-section near the band-edge regions (600–800 nm) of the absorption spectrum where the absorption coefficients typically become small and the efficiency is low. Therefore a better understanding of the optical properties of a-Si nanocomposite containing metallic nanoparticles is important and may lead the further optimization of light-trapping schemes being effective across the whole useful solar spectrum. This is particularly of importance to thin-film solar cells, which have a considerable market share and offer interesting prospects [75,76].

In this section the possibility to tune the light absorption inside thin a-Si films deposited on glass substrates and incorporating a random array of gold nanoparticles from the visible toward the NIR as a function of the film-thickness is discussed based on Faraone et al. [72], Fig. 3.11. The nanoparticles were fabricated by a gas phase aggregation magnetron sputter cluster source which produces ultraclean nanoparticles of any composition at a fast rate [77]. Recently such nanocluster sources have been used to explore light management for photovoltaics [28,34,68], providing a novel and economically interesting approach.

The effective dielectric environment causes a redshift of the gold nanoparticles plasmonic resonance inside a-Si with increasing thickness. These experimental findings are explained with finite difference time domain (FDTD) simulations, by which also the role played by the near field coupling between gold nanoparticles is elucidated. The intensity and the broadening of the plasmonic resonance as a function of the a-Si layer thickness and gold nanoparticles deposition time is discussed as well.

Gold nanoparticles were deposited with a commercial gas aggregation nanocluster source based on magnetron sputtering (NC200U-B Oxford Applied Research Ltd.) [78,79] on glass substrates. The deposited nanoparticles were subsequently covered with a conformally grown a-Si thin film, also by magnetron-sputtering from a polycrystalline silicon target. The thickness of the deposited a-Si films was controlled by varying the thin-film deposition time: 100, 200, 300, and 400 s. A schematic cross-sectional representation of the samples in the present work is given in Fig. 3.11. An experimental scanning electron microscopy (SEM) cross-section of a similar sample confirming the conformal growth can be found in Raza et al. [80]. FDTD simulations were performed on systems consisting of two spherical gold nanoparticles supported on a glass substrate and embedded in a-Si films with different thicknesses. The mean size of the gold nanoparticles (24 nm) determined from the TEM micrographs was used to model the diameter of the simulated spherical nanoparticles, whereas the mean interparticle distance obtained from a statistical analysis performed on the AFM images was used as the distance between the two simulated spherical gold particles. The simulation of only two interacting gold particles was chosen to encompass all the present properties and interactions to describe the system.

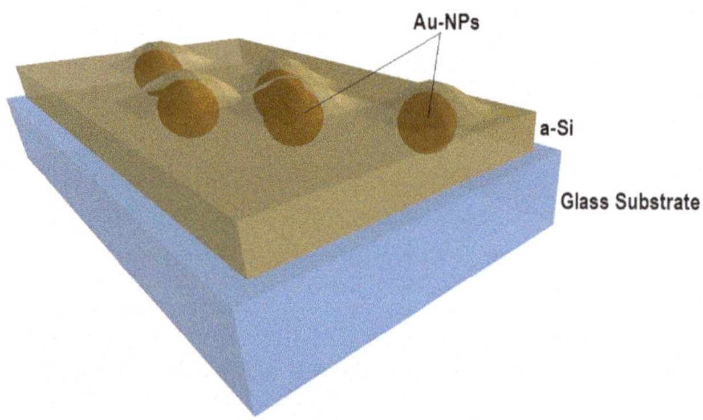

FIGURE 3.11 Schematic depiction of a cross-section of a sample characterized in the present work.

The optical properties of a nanocomposite depend both on the material and morphological properties of its structural units (cluster or nanoparticles) and of its surfaces [35]. The geometrical structure as well as the mean size of the deposited gold nanoparticles are determined primarily by the parameters chosen during the deposition. Different studies on the fabrication of nanoparticles by the gas aggregation process have shown that the cluster size distribution can be accurately described by a lognormal distribution [81,82]. This is confirmed by statistical analysis of the TEM micrographs of the isolated gold nanoparticles (Fig. 3.12A and B), which have a mean diameter of 24 ± 1.5 nm. The deposition of nanoparticles with gas aggregation techniques produces both small Au-clusters and large nanoparticles having typically icosahedral or dodecahedral geometries [83] in agreement with the here-observed hexagonal shapes.

Since the typical energy of the particles ejected from a gas-aggregation cluster source resides in the low energy impact (soft landing) regime, <0.1 eV/atom, the NPs deposited on the glass substrates are expected to remain almost intact forming a film consisting of crystallites with dimensions of the original clusters [84–86]. The height value from AFM in Fig. 3.13A and B gives a good estimate for the size of the deposited gold nanoparticle and agrees well with the mean size derived from the TEM micrographs. Therefore the gold nanoparticles are not flattened upon impact and their size remains unaltered even after embedding them in a conformally grown a-Si layer. The lateral size in Fig. 3.13B is larger than the nanoparticle height because of the AFM-tip-convolution. The AFM image in Fig. 3.13A shows also that the topology of the surface is characterized by the presence of single and randomly scattered nanoparticles, as well as of large groups of close particles separated from each other by particle-free zones. This peculiar topology may be caused by the interplay between the surface roughness of the glass substrate, which affects gold nanoparticle mobility and the presence of charged gold nanoparticles [79], which are attracted or repelled, depending on their charge state. The deposition time of 2 min is, however, still sufficiently low that the formation of coagulation aggregates of gold nanoparticles is less probable and the resulting particle distribution on the surface is 2D isotropic and random.

The experimental optical absorbance spectra of gold nanoparticles deposited on glass substrates and embedded in conformally grown a-Si films with different thicknesses are shown in Fig. 3.14A. The measured plasmon resonance wavelength of the glass supported "bare" gold nanoparticles is red-shifted by about 22 nm compared to the simulated plasmonic resonance wavelength. This is an effect caused by 1) the difference between the Palik dielectric function of gold and the actual dielectric function and 2) the size distribution of the gold particles leading to a slightly different plasmon resonance energy for each NP-size.

The red-shift of the plasmonic resonance with increasing thickness of the embedding a-Si film is clearly visible for the 6 and 12 nm thickness. The plasmonic peak shifts from the visible at 535 nm for the supported gold nanoparticles in air toward 555 nm for the

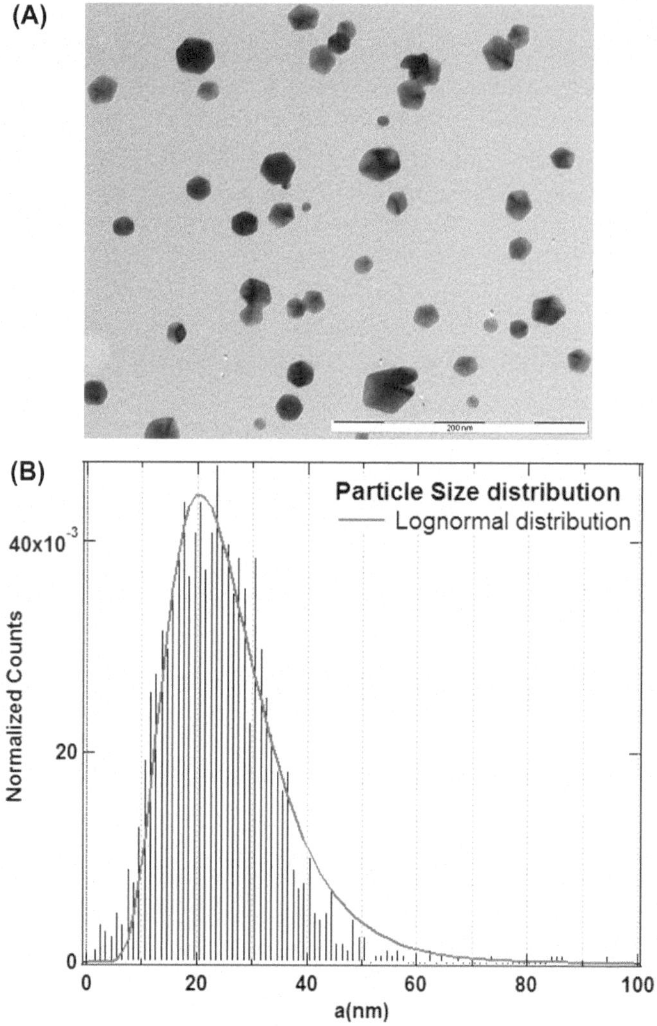

FIGURE 3.12 (A) TEM micrograph of gold nanoparticles deposited on a holey carbon TEM grid made with an aggregation distance of 60 mm and deposition time of 20 s. The scale bar is 200 nm. (B) The experimental single-particle size distribution obtained from a statistical analysis performed on the size of 1700 gold nanoparticles measured from the TEM micrographs and a best fit with normalized lognormal distribution with mean $\ln \bar{a} = 3.81$ and standard deviation $\ln \sigma_a = 0.41$ (*red solid line*).

gold nanoparticles in a 6 nm a-Si film and then to the NIR, at 616 nm, for gold nanoparticles in a 12 nm a-Si film. In the samples with thicker a-Si films (17 and 23 nm), the plasmonic resonance disappeared. It is likely that the plasmon resonance is much more shifted to the red, outside the observation range, due to the formation of the silicide, which has a larger refractive index. The formation of silicide is stimulated by the presence of the plasma, which heats up the sample. For the thicker samples the heating takes longer, resulting in more silicide being formed.

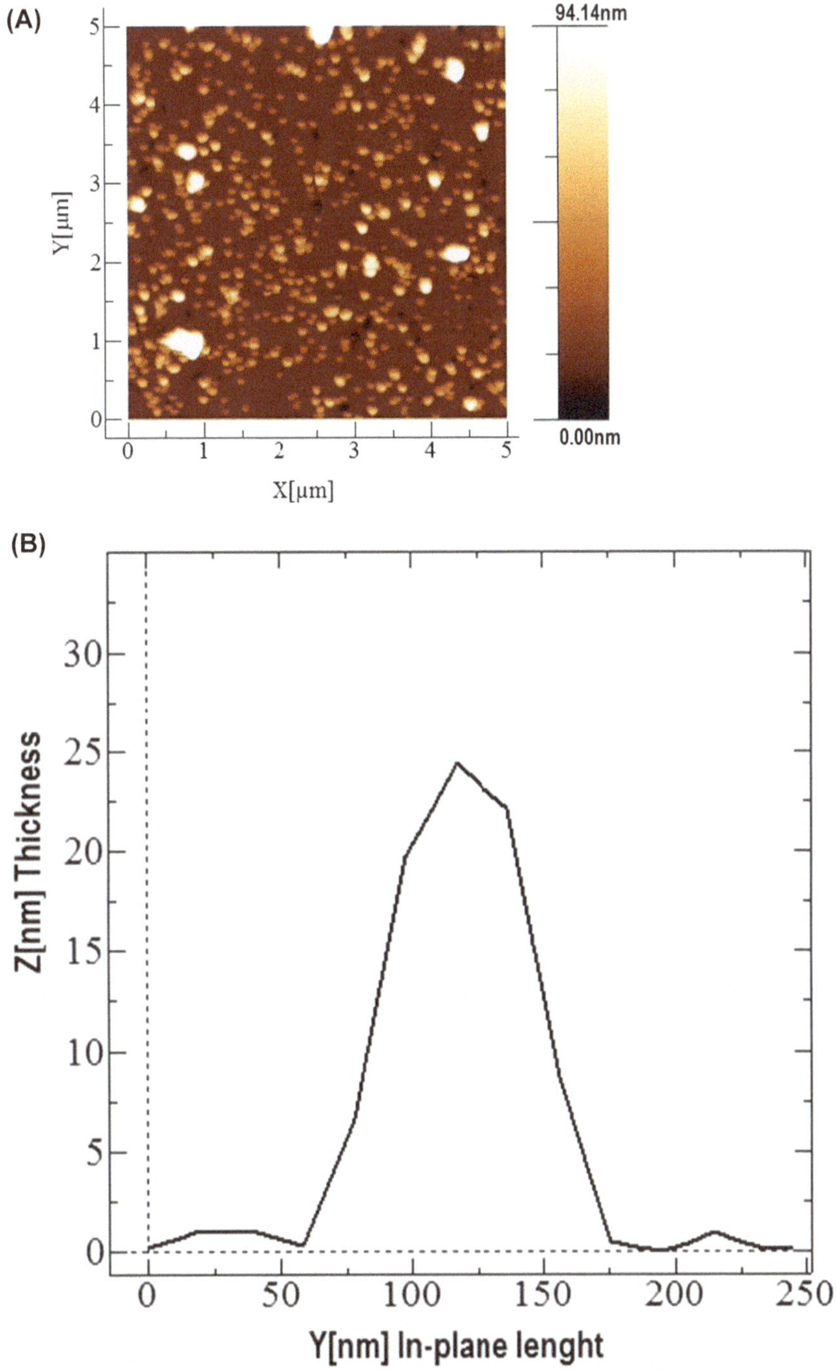

FIGURE 3.13 (A) AFM image gold nanoparticle on glass and embedded in a 12 nm a-Si film. The density of gold nanoparticles is 32 μm^{-2} with a surface coverage of 22.1%. (B) Measured profile of a 25 nm-thick feature on the sample surface corresponding to the a-Si layer conformally grown around a gold nanoparticle.

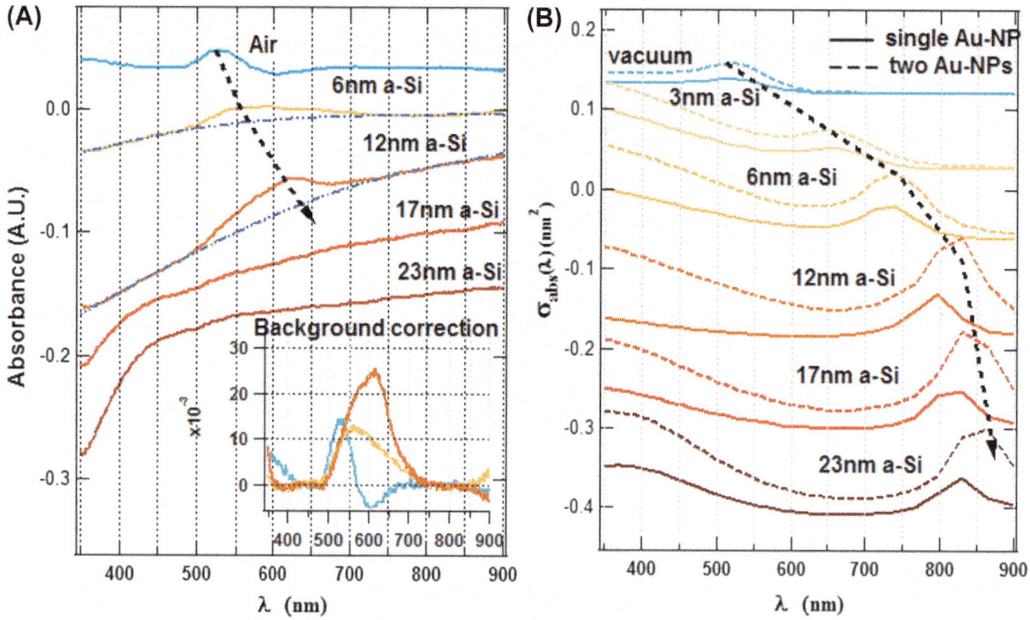

FIGURE 3.14 (A) Experimental optical absorption spectra of gold nanoparticles embedded in a-Si films with different thickness. **Inset:** Results of the background subtraction performed on the 6 and 12 nm samples. (B) Comparison between the FDTD simulations for the total absorption cross-section of a single (24 nm) spherical gold nanoparticle (*solid lines*) and two spherical (same diameter) gold nanoparticles with a center to center distance of 69 nm (*dashed lines*) embedded in a-Si films with different thickness. The simulated curves have been magnified of a factor 1.8×10^{13} for a better comparison with the experimental curves in (A). For a better presentation both the experimental (A) and the simulated curves (B) have arbitrary offsets along the y-axis. The *dashed arrows* in both the figures indicate the direction of the plasmonic resonance shift.

The results of the FDTD simulations for the total absorption cross-section are reported in Fig. 3.14B. The simulations confirm that the red-shift of the plasmonic resonance occurs with increasing thickness of the embedding a-Si layer shifting the position of the peak to the IR. Moreover, the red-shift is stronger for an interacting gold nanoparticle pair as compared with a single isolated gold nanoparticle. The plasmonic resonance wavelength of the gold nanoparticle pair attains a constant value at about 830 nm when the particles are embedded in a-Si films thicker than 17 nm. In an effective medium theory, the red-shift of the plasmonic resonance is explained as a consequence of the change in the effective refractive index of the medium surrounding the nanoparticles [35,87]. By increasing the thickness of the a-Si layer, the effective refractive index changes and increases in value until it reaches the value of bulk a-Si. If the thickness of the embedding a-Si layer is kept below 17 nm, it is possible to tune the plasmonic response of the nanoparticles and, therefore, the absorption of the NIR-radiation inside the a-Si film. Furthermore, from the FDTD simulations it is clear that when the a-Si film has a thickness greater than 12 nm, the plasmonic resonance of the interacting gold nanoparticle pair is more red-shifted than a single isolated gold nanoparticle. The individual plasmonic resonances of two nanoparticles separated by a nanoscale dielectric gap hybridize forming a bonding dipole plasmon mode (BDP) which has lower energy (longer wavelengths) than the plasmonic mode of a single isolated nanoparticle [88–91]. When the gold-nanoparticle pair is embedded in a-Si layers having a thickness more than 12 nm, the near field hybridization between the plasmonic modes of the isolated nanoparticles is more effective.

In Fig. 3.15 the position of the experimental and simulated plasmonic resonance wavelength is shown as a function of the a-Si thickness. The simulated results were well fitted with an exponential function, while the experimental could be fitted with a second degree polynomial. There is a significant discrepancy between the experimental and simulated plasmon resonance wavelength for the 6 and 12 nm and their extrapolation to larger a-Si thickness. Although it is possible that the formation of an interfacial double-layer of charge (Schottky layer) [92,93], due to a difference in work function, increases the free electron density on the gold nanoparticle, which leads to a plasmon resonance shift toward the blue, this charge transfer is too small to cause a detectable effect. It is, therefore, more likely that a difference between the experimental gold particles and a-Si layers and their FDTD model results in the observed discrepancy.

The vanishing of the plasmon resonance for a-Si thicknesses of 17 and 24 nm cannot be explained from a classical point of view as the FDTD simulations contain all the required parameters and nevertheless result in clear gold plasmon peaks. Both the optical absorption of gold (plasmonic and inter band) and a-Si are accounted for in the used dielectric functions. Considering the large size of the gold nanoparticles (24 nm), quantum effects are also unlikely. The interface effects, which are not included in the FDTD simulations, should be comparable with that of the thinner a-Si films and can therefore also not explain the disappearance of the gold plasmon resonance. The formation of a silicide [94,95] cannot be excluded and could influence the plasmon resonance strength and position.

In order to study the intensity and the full width half maximum (FWHM) of the plasmonic resonances of gold nanoparticles embedded in the a-Si film, a background subtraction was performed on the absorption spectra of the 6 and 12 nm samples. The "background" was modeled performing a spline-interpolation fit on two wavelength-ranges between 450–500 nm and 750–850 nm. The results of the background correction are shown in the inset of Fig. 3.14A where they are compared also with the plasmonic peak of "bare" (i.e., in air) gold nanoparticles supported on glass. This comparison shows that the embedding of an a-Si film modifies, besides the position, the FWHM and the intensity of the plasmonic resonance.

The intensity of the plasmon resonance is strongly increased in the sample with a 12 nm a-Si film with respect to the intensity of the resonance in the sample with a 6 nm a-Si film. Furthermore, in the latter the intensity of the plasmon resonance is comparable with the intensity of the plasmon peak of bare gold nanoparticles. Since the resonance in the thinner 6 nm a-Si film is closer to the threshold for the onset of

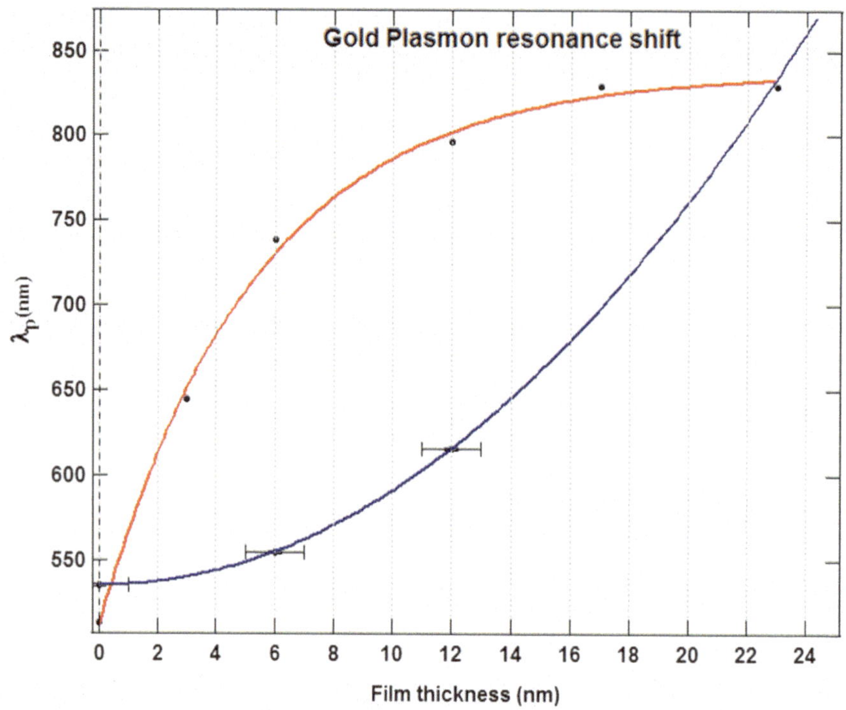

FIGURE 3.15 Simulated (*red*) and experimentally (*blue*) measured position of the plasmonic resonance wavelength versus the thickness of the embedding a-Si films. The *red solid line* results from a curve-fitting with an exponential law: $A + Be^{-C \cdot d}$, where $A = (836.94 \pm 6.93)$ nm, $B = (-325.26 \pm 9.31)$, $C = (0.186 \pm 0.014)$ nm. The *blue solid line* results from modeling the thickness dependence of the measured positions of the plasmonic resonance with a second degree polynomial $Ax^2 + Bx + C$, with: $A = 0.570$ nm^{-2}, $B = -0.083$ nm, $C = 535$ nm.

interband transitions in gold ($\hbar\omega^{\text{inter}} = 2.3$ eV, $\lambda^{\text{inter}} = 540$ nm), the intensity of the plasmonic peak is reduced by the "background extinction" caused by the 5d-6sp band transitions [35,87]. However, the resonance in the sample with the 12 nm a-Si film is significantly red-shifted, away from the interband transitions threshold and therefore the reduction in its intensity caused by the effects due to the interband transitions becomes weaker.

The FWHM of the plasmon resonance is generally determined by "inhomogeneous" and "homogenous" broadening effect [35]. The "inhomogeneous" broadening is typically caused by the nonuniform topology and size distribution of the deposited NPs. The "homogenous" broadening is caused by the material and interface properties. The interface properties of the nanoparticle affect the broadening of the plasmonic resonance, especially if the particle is embedded in a dielectric matrix, which creates many surface defect states. The FWHM of the plasmonic resonances in the samples with a 6 nm and a 12 nm a-Si film are identical (130 nm) and both broader than the FWHM (71 nm) of the bare gold nanoparticles on glass (see inset Fig. 3.14A). Since the gold nanoparticles have been deposited keeping the same conditions in all the samples, this shows that 1) the "inhomogeneous broadening" due to the topology and size of the particles, contributes in an identical way, not depending on the thickness of these a-Si film and 2) the "homogeneous broadening", instead, is responsible of an extended broadening by the presence of an additional damping mechanism at the gold nanoparticle/a-Si interface related to the nature of the electronic states between the a-Si and gold interface [93,96,97]. For nanoparticle composites on different substrates and surrounding mediums this is commonly known as *chemical interface damping* [35,96]. Since this interface effect is also independent on the a-Si thickness, it is ruled out as explanation for the disappearance of the gold plasmon for thicker a-Si films. It is possible that the a-Si growth over the gold nanoparticles may have different modes and results in different densities, defect states, and degrees of crystallinity, which may affect its dielectric properties. Particularly, the formation of a silicide will have a strong effect. The thicker the a-Si layer the more energy is available to stimulate the formation of silicide. This could shift the gold nanoparticle plasmon resonance more toward the red, perhaps beyond our detection capability.

In Fig. 3.16 the optical absorption, as obtained from the FDTD, simulations in the gold nanoparticles and the surrounding a-Si is shown, which clearly demonstrates the significantly increased optical absorption in a-Si, caused by the plasmonic properties of the gold nanoparticles. In Fig. 3.16A and B the optical absorption in a-Si and the gold nanoparticles shows that only a fraction of the light is "parasitically" absorbed by the gold particle, while a clear absorption increase in a-Si around 850 nm is measured. This optical absorption increase in the a-Si layer, which can be up to a factor of 2, is the result of plasmonic field enhancement and scattering of the gold nanoparticle. In Fig. 3.16C the optical absorption is shown spatially over the entire a-Si film and gold nanoparticle. It is clear that the gold nanoparticle induces increased optical absorption far away from its location due to local field enhancement and scattering. Moreover, it is clear that, although the gold nanoparticle causes small parasitic optical absorption, the beneficial optical absorption for a-Si is considerable in the wavelength range which needs improvements.

3.3.2.2 Conclusions

The control of the red-shift of the plasmon resonance by embedding a random array of glass supported gold nanoparticles in different thicknesses of a-Si thin films forms a route to precisely tune the optical absorption properties. This could be exploited in the design of efficient light-trapping techniques in a-Si thin-film solar cells, as well as in the design of NIR-tunable optical coatings. Although the experimental red-shift of the plasmonic resonance is qualitatively reproduced by FDTD simulations, a mismatch of the plasmonic resonance wavelength is explained by differences between model and experimental parameters. Moreover, for the thicker a-Si layers the gold plasmon vanishes, which could not be explained by classical arguments as the FDTD simulations show a clear gold plasmon resonance. The formation of a silicide could explain the

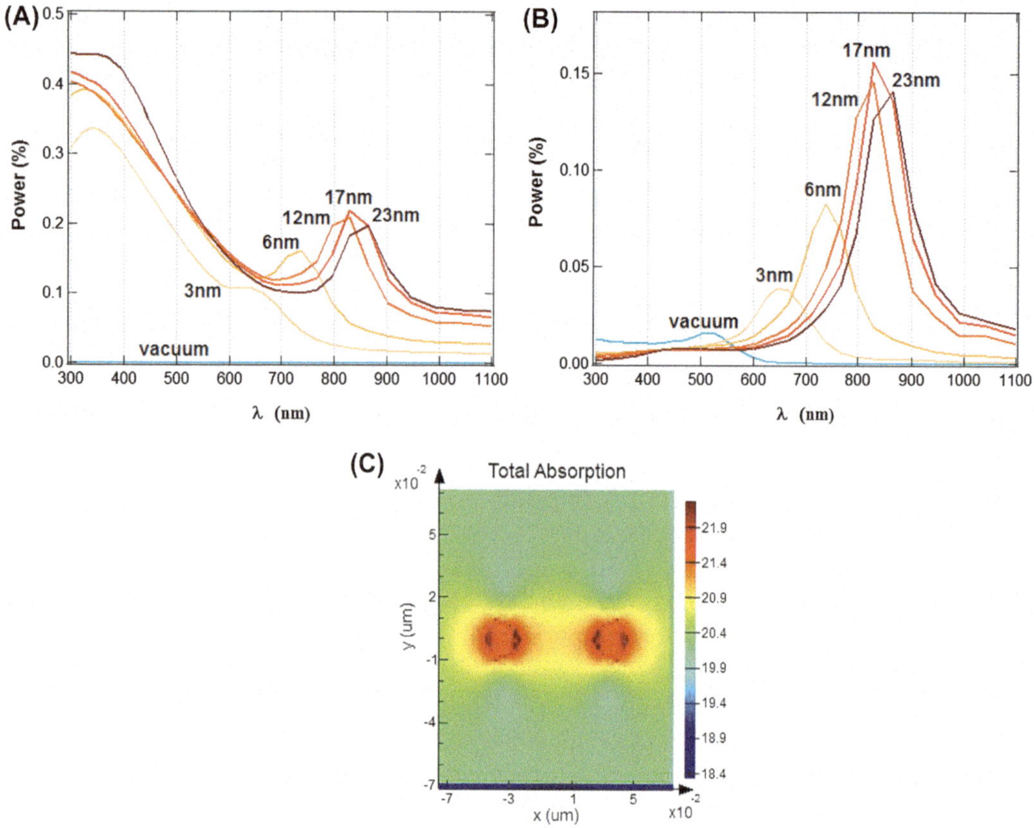

FIGURE 3.16 Optical absorption as function of wavelength calculated by the FDTD simulations of (A) a-Si and (B) gold nanoparticles. The optical absorption in the plane of the 12 nm a-Si film on glass is shown in (C). The extra optical absorption (*yellow*) of about an order of magnitude in the a-Si layer, extending far away from the gold nanoparticles strongly demonstrates the beneficial effect of the gold nanoparticles.

plasmon vanishing. Comparison of the experimental and simulated plasmon resonance broadening confirm that the nanoscale interfaces between gold and a-Si explain the measured widths of the plasmonic resonances. The FDTD simulations show that considerable optical gain can be obtained by the presence of the gold nanoparticles due to local field enhancement and scattering. This work provides new methods and insights, which are highly relevant for ongoing research to use metal nanostructures in light harvesting systems.

3.4 General conclusions

Using nanoparticles for photovoltaics is a promising bottom-up approach for various aspects of the thin-film solar cell. Nanoparticles can be used as a light management structure in or on thin-film solar cells. Plasmonic particles can scatter light or the local field enhancement can increase absorption. Dielectric Mie scatterers can also scatter light or can function as light antennas to subsequently couple light into the active solar cell layer. The gas aggregation cluster source is a new fabrication technique for nanoparticles covering the compositional and light management parts in thin-film solar cells. Because thin-film solar cells require only one or several monolayers of particles, this is feasible. By merging two historically independent research fields, a bright future for nanoparticles and solar cells is on the horizon. In this chapter two state-of-the-art examples of using TiO_2 Mie scatterers and plasmonic gold nanoparticles on/in a-Si layers for solar cell purposes have been discussed. The TiO_2 and gold nanoparticles are deposited in a matter of minutes with a laboratory cluster source, which could be improved to seconds when upscaling and technological improvements are applied. The TiO_2 and gold nanoparticles indeed increase the optical absorption, which can be tuned to a wide range of wavelengths. This recent work with respect to using nanoparticles from gas aggregation cluster sources to improve the efficiency of solar cells brings new opportunities for research and technology for both fields.

References

[1] M.A. Green, Third Generation Photovoltaics: Advanced Solar Energy Conversion, Springer, Berlin, 2003.

[2] G. Conibeer, Mater. Today 10 (2007) 42–50.

[3] W. Shockley, H.J. Queisser, J. Appl. Phys. 32 (1961) 510–519.

[4] T.D. Lee, A.U. Ebong, Renew. Sustain. Energy Rev. 70 (2017) 1286–1297.

[5] P. Mints, in: 2010 35th IEEE Photovoltaic Specialists Conference, 2010, pp. 002400–002404.

[6] E.T. Yu, J. van de Lagemaat, MRS Bull. 36 (2011) 424–428.

[7] H.A. Atwater, A. Polman, Nat. Mater. 9 (2010) 205–213.

[8] X. Huang, S. Han, W. Huang, X. Liu, Chem. Soc. Rev. 42 (2013) 173–201.

[9] G.E. Jabbour, D. Doderer, Nat. Photonics 4 (2010) 604–605.

[10] M. Di Vece, Using nanoparticles as a bottom-up approach to increase solar cell efficiency. KONA Powder and Particle Journal 36 (2019) 72–87.

[11] A. Polman, M. Knight, E.C. Garnett, B. Ehrler, W.C. Sinke, Science 352 (2016).

[12] R.W. Miles, G. Zoppi, I. Forbes, Mater. Today 10 (2007) 20–27.

[13] M.L. Brongersma, Y. Cui, S. Fan, Nat. Mater. 13 (2014) 451–460.

[14] S. Chattopadhyay, Y.F. Huang, Y.J. Jen, A. Ganguly, K.H. Chen, L.C. Chen, Mater. Sci. Eng. R Rep. 69 (2010) 1–35.

[15] A.I. Kuznetsov, A.E. Miroshnichenko, M.L. Brongersma, Y.S. Kivshar, B. Luk'yanchuk, Science 354 (2016).

[16] P. Spinelli, A. Polman, IEEE J. Photovoltaics 4 (2014) 554–559.

[17] P. Spinelli, M.A. Verschuuren, A. Polman, Nat. Commun. 3 (2012) 692.

[18] Z.Y. Wang, R.J. Zhang, S.Y. Wang, M. Lu, X. Chen, Y.X. Zheng, L.Y. Chen, Z. Ye, C.Z. Wang, K.M. Ho, Sci. Rep. 5 (2015) 7810.

[19] Y.Y. Cheng, B. Fückel, R.W. MacQueen, T. Khoury, R.G.R.C. Clady, T.F. Schulze, N.J. Ekins-Daukes, M.J. Crossley, B. Stannowski, K. Lips, T.W. Schmidt, Energy Environ. Sci. 5 (2012).

[20] S.J. Kim, I. Thomann, J. Park, J.-H. Kang, A.P. Vasudev, M.L. Brongersma, Nano Lett. 14 (2014) 1446–1452.

[21] H. Xu, X. Chen, S. Ouyang, T. Kako, J. Ye, J. Phys. Chem. C 116 (2012) 3833–3839.

[22] J.A. Schuller, R. Zia, T. Taubner, M.L. Brongersma, Phys. Rev. Lett. 99 (2007) 107401.

[23] S. Nunomura, A. Minowa, H. Sai, M. Kondo, Appl. Phys. Lett. 97 (2010) 063507.

[24] L. Cao, J.-S. Park, P. Fan, B. Clemens, M.L. Brongersma, Nano Lett. 10 (2010) 1229–1233.

[25] U. Zywietz, A.B. Evlyukhin, C. Reinhardt, B.N. Chichkov, Nat. Commun. 5 (2014) 3402.

[26] A. Araújo, A. Pimentel, M.J. Oliveira, M.J. Mendes, R. Franco, E. Fortunato, H. Águas, R. Martins, Flexible Printed Electron. 2 (2017) 014001.

[27] M. Karg, T.A.F. König, M. Retsch, C. Stelling, P.M. Reichstein, T. Honold, M. Thelakkat, A. Fery, Mater. Today 18 (2015) 185–205.

[28] G. Giannakoudakis, M. Di Vece, Eur. Phys. J. D 71 (2017) 101.

[29] S. Son, S.H. Hwang, C. Kim, J.Y. Yun, J. Jang, ACS Appl. Mater. Interfaces 5 (2013) 4815–4820.

[30] S.H. Kang, J.-Y. Kim, H.S. Kim, H.-D. Koh, J.-S. Lee, Y.-E. Sung, J. Photochem. Photobiol. A Chem. 200 (2008) 294–300.

[31] E. Barborini, I.N. Kholmanov, A.M. Conti, P. Piseri, S. Vinati, P. Milani, C. Ducati, Eur. Phys. J. D 24 (2003) 277–282.

[32] O. Polonskyi, T. Peter, A. Mohammad Ahadi, A. Hinz, T. Strunskus, V. Zaporojtchenko, H. Biederman, F. Faupel, Appl. Phys. Lett. 103 (2013) 033118.

[33] J.R. DeVore, J. Opt. Soc. Am. 41 (1951) 416–419.

[34] W. Tang, J.J. Eilers, M.A. van Huis, D. Wang, R.E.I. Schropp, M. Di Vece, J. Phys. Chem. C 119 (2015) 11042–11047.

[35] U. Kreibig, M. Vollmer, Optical Properties of Metal Clusters, Springer, Berlin; New York, 1995.

[36] K. Madhusudan Reddy, S.V. Manorama, A. Ramachandra Reddy, Mater. Chem. Phys. 78 (2003) 239–245.

[37] X. Chen, S.S. Mao, Chem. Rev. 107 (2007) 2891–2959.

[38] J. van de Groep, A. Polman, Opt. Express 21 (2013) 26285–26302.

[39] S.A. Maier, M.L. Brongersma, P.G. Kik, S. Meltzer, A.A.G. Requicha, H.A. Atwater, Adv. Mater. 13 (2001) 1501–1505.

[40] E. Ozbay, Science 311 (2006) 189–193.

[41] S.A. Maier, Plasmonics Fundamentals and Applications, Springer, 2007.

[42] R.A. Pala, J. White, E. Barnard, J. Liu, M.L. Brongersma, Adv. Mater. 21 (2009) 3504–3509.

[43] V.E. Ferry, J.N. Munday, H.A. Atwater, Adv. Mater. 22 (2010) 4794–4808.

[44] S. Pillai, K.R. Catchpole, T. Trupke, M.A. Green, J. Appl. Phys. 101 (2007) 093105.

[45] C.F. Bohren, D.R. Huffman, Absorption and Scattering of Light by Small Particles, Wiley-VCH, Weinheim, 2004.

[46] E. Purcell, Phys. Rev. 69 (1946) 681.

[47] L. Novotny, B. Hecht, Principles of Nano-Optics, Cambridge University Press, 2006.

[48] P. Anger, P. Bharadwaj, L. Novotny, Phys. Rev. Lett. 96 (2006) 113002.

[49] A.O. Govorov, W. Zhang, T. Skeini, H. Richardson, J. Lee, N.A. Kotov, Nanoscale Res. Lett. 1 (2006) 84–90.

[50] A.R. Clapp, I.L. Medintz, H. Mattoussi, ChemPhysChem 7 (2006) 47–57.

[51] X. Michalet, F.F. Pinaud, L.A. Bentolila, J.M. Tsay, S. Doose, J.J. Li, G. Sundaresan, A.M. Wu, S.S. Gambhir, S. Weiss, Science 307 (2005) 538–544.

[52] A.J. Haes, S. Zou, G.C. Schatz, R.P. Van Duyne, J. Phys. Chem. B 108 (2004) 109–116.

[53] A. Ghahremani, A.E. Fathy, Energy Sci. Eng. 4 (2016) 334–343.

[54] R. Santbergen, R. Liang, M. Zeman, IEEE (2010) 000748–000753.

[55] O. Stenzel, A. Stendal, K. Voigtsberger, C. von Borczyskowski, Sol. Energy Mater. Sol. Cells 37 (1995) 337–348.

[56] M. Westphalen, U. Kreibig, J. Rostalski, H. Lüth, D. Meissner, Sol. Energy Mater. Sol. Cells 61 (2000) 97–105.

[57] H.R. Stuart, D.G. Hall, Appl. Phys. Lett. 69 (1996) 2327–2329.

[58] E.S. Barnard, R.A. Pala, M.L. Brongersma, Nat. Nanotechnol. 6 (2011) 588–593.

[59] Y.A. Akimov, W.S. Koh, K. Ostrikov, Opt. Express 17 (2009) 10195–10205.

[60] C. Hägglund, M. Zäch, B. Kasemo, Appl. Phys. Lett. 92 (2008) 013113.

[61] K. Nakayama, K. Tanabe, H.A. Atwater, Appl. Phys. Lett. 93 (2008) 121904.

[62] Y. Ekinci, H.H. Solak, J.F. Löffler, J. Appl. Phys. 104 (2008) 083107.

[63] M.W. Knight, L. Liu, Y. Wang, L. Brown, S. Mukherjee, N.S. King, H.O. Everitt, P. Nordlander, N.J. Halas, Nano Lett. 12 (2012) 6000–6004.

[64] C. Langhammer, M. Schwind, B. Kasemo, I. Zorić, Nano Lett. 8 (2008) 1461–1471.

[65] J. Olson, A. Manjavacas, L. Liu, W.-S. Chang, B. Foerster, N.S. King, M.W. Knight, P. Nordlander, N.J. Halas, S. Link, Proc. Natl. Acad. Sci. U. S. A 111 (2014) 14348–14353.

[66] S.J. Tan, L. Zhang, D. Zhu, X.M. Goh, Y.M. Wang, K. Kumar, C.-W. Qiu, J.K.W. Yang, Nano Lett. 14 (2014) 4023–4029.

[67] T.L. Temple, D.M. Bagnall, J. Appl. Phys. 109 (2011) 084343.

[68] T. van der Vliet, M. Di Vece, Thin Solid Films 603 (2016) 404–407.

[69] P. Tassin, T. Koschny, M. Kafesaki, C.M. Soukoulis, Nat. Photonics 6 (2012) 259–264.

[70] V.E. Ferry, M.A. Verschuuren, H.B.T. Li, E. Verhagen, R.J. Walters, R.E.I. Schropp, H.A. Atwater, A. Polman, Opt. Express 18 (2010) A237–A245.

[71] N. Kooy, K. Mohamed, L.T. Pin, O.S. Guan, Nanoscale Res. Lett. 9 (2014) 320.

[72] G. Faraone, R. Modi, S. Marom, A. Podestà, M. Di Vece, Opt. Mater. 75 (2018) 204–210.

[73] P. Spinelli, A. Polman, Opt. Express 20 (2012) A641–A654.

[74] A.C. Lesina, G. Paternoster, F. Mattedi, L. Ferrario, P. Berini, L. Ramunno, A. Paris, A. Vaccari, L. Calliari, Plasmonics 10 (2015) 1525–1536.

[75] T.M. Razykov, C.S. Ferekides, D. Morel, E. Stefanakos, H.S. Ullal, H.M. Upadhyaya, Sol. Energy 85 (2011) 1580–1608.

[76] A. Feltrin, T. Meguro, E. Van Assche, T. Suezaki, M. Ichikawa, T. Kuchiyama, D. Adachi, O. Inaki, K. Yoshikawa, G. Koizumi, H. Uzu, H. Ueda, T. Uto, T. Fujimoto, T. Irie, H. Hayakawa, N. Nakanishi, M. Yoshimi, K. Yamamoto, Sol. Energy Mater. Sol. Cells 119 (2013) 219–227.

[77] V.N. Popok, I. Barke, E.E.B. Campbell, K.-H. Meiwes-Broer, Surf. Sci. Rep. 66 (2011) 347–377.

[78] H. Haberland, J. Vac. Sci. Technol. 10 (1992) 3266.

[79] H. Haberland, J. Vac. Sci. Technol. 12 (1994) 2925.

[80] S. Raza, S. Kadkhodazadeh, T. Christensen, M. Di Vece, M. Wubs, N.A. Mortensen, N. Stenger, Nature Communications **6** (2015) 8788.

[81] J. Söderlund, L.B. Kiss, G.A. Niklasson, C.G. Granqvist, Phys. Rev. Lett. 80 (1998) 2386–2388.

[82] C.G. Granqvist, J. Appl. Phys. 47 (1976) 2200.

[83] R.L. Johnston, Atomic and Molecular Clusters, Taylor & Francis, London ; New York, 2002.

[84] P. Milani, S. Iannotta, Cluster Beam Synthesis of Nanostructured Materials, Springer Berlin Heidelberg, Berlin, Heidelberg, 1999.

[85] F. Borghi, E. Sogne, C. Lenardi, A. Podestà, M. Merlini, C. Ducati, P. Milani, J. Appl. Phys. 120 (2016) 055302.

[86] A. Podestà, F. Borghi, M. Indrieri, S. Bovio, C. Piazzoni, P. Milani, J. Appl. Phys. 118 (2015) 234309.

[87] M. Quinten, Optical Properties of Nanoparticle Systems: Mie and Beyond, Wiley-VCH, Weinheim, Germany, 2011.

[88] P. Nordlander, C. Oubre, E. Prodan, K. Li, M.I. Stockman, Nano Lett. 4 (2004) 899–903.

[89] E. Hao, G.C. Schatz, J. Chem. Phys. 120 (2004) 357–366.

[90] J.M. McMahon, S.K. Gray, G.C. Schatz, Phys. Rev. B 83 (2011).

[91] Y. Huang, L. Ma, M. Hou, J. Li, Z. Xie, Z. Zhang, Sci. Rep. 6 (2016).

[92] R.C. Plumb, J. Phys. 25 (1964) 69–73.

[93] A. Hilger, T. Von Hofe, U. Kreibig, Nova Acta Leopold. 92 (2005) 9–19.

[94] L. Braicovich, C.M. Garner, P.R. Skeath, C.Y. Su, P.W. Chye, I. Lindau, W.E. Spicer, Phys. Rev. B 20 (1979) 5131–5141.

[95] S. Vandré, E. Narducci, A. Goldoni, C. Lenardi, L. Calliari, L. Casalis, L. Gregoratti, J. Kovač, M. Kiskinova, M. Sancrotti, Surf. Sci. 377 (1997) 266–270.

[96] H. Hövel, S. Fritz, A. Hilger, U. Kreibig, M. Vollmer, Phys. Rev. B 48 (1993) 18178–18188.

[97] A. Hilger, N. Cüppers, M. Tenfelde, U. Kreibig, Eur. Phys. J. D. 10 (2000) 115.

4

Cluster nanoportals for the hydrogenation of underlying nanofilms

Panagiotis Grammatikopoulos[1], Sushant Kumar[2]

[1]OKINAWA INSTITUTE OF SCIENCE AND TECHNOLOGY (OIST) GRADUATE UNIVERSITY, OKINAWA, JAPAN; [2]DEPARTMENT OF CHEMICAL AND BIOCHEMICAL ENGINEERING, INDIAN INSTITUTE OF TECHNOLOGY PATNA, PATNA, INDIA

Chapter outline

4.1 Introduction

It is a well-established fact that since the industrial revolution, energy demands around the world have been multiplied significantly (by a factor of more than 30). Today, the industrialized world relies mainly on fossil fuels (and secondarily on nuclear fission) to satisfy these demands, with alternative sources such as wind, geothermal, or solar power contributing but not very substantially yet. As all future energy projections assume

Cluster Beam Deposition of Functional Nanomaterials and Devices. https://doi.org/10.1016/B978-0-08-102515-4.00004-0

further and steeper increase in global energy demands, little doubt remains that new energy sources should be developed over the coming decades. With global warming being one of the key issues in current worldwide energy policies, it is essential that these sources are sustainable: "living on the Earth's income rather than eroding its capital" and "keeping the consumption of renewable natural resources within the limits of their replenishment" [1].

But global warming is not the only large-scale challenge humanity faces today. Shortages of food and fresh water supplies are of great importance, especially in the developing world, posing continuous threats of potential turmoil. In fact, all these problems can be directly or indirectly ascribed to that of energy supply. Furthermore, growing industrialization of nonindustrial nations has recently been added to this picture, offsetting the energy gain from conservation efforts in the industrialized world and making the need for drastic solutions even more imperative [2]. From an international politics point of view, petroleum is associated with a region of political instability. But also from a national-scale perspective, it is commonly accepted that any healthy society needs a multitude of energy sources: economic and political stability demands that no nation is dependent on just a single energy source [3].

For all these reasons, substantial research and investment is being focused on the development and implementation of novel, sustainable sources of energy, both large-scale (mainly for stationary applications) and small-scale (mainly for mobile applications). To this day, the most promising option for the former seems to be solar energy, and, potentially in the future, nuclear fusion [3]; for the latter, hydrogen poses as the most desirable candidate [4,5].

For a successful leap from our current hydrocarbon economy to the often-cited hydrogen economy, it is necessary to develop safe, reliable, and efficient methods of hydrogen production and storage, which is the point of interest of this chapter. In this context, efficient storage translates into fast and easy storage and delivery, and at a high compression rate. Formerly, liquid hydrogen tanks were proposed as the storage medium, as they offer almost double capacity by volume than pressurized hydrogen. However, even that was deemed insufficient as it is considerably lower than the US Department of Energy requirements; moreover, the process requires substantial amounts of energy for compressing hydrogen into tanks [6]. Currently, the most promising method appears to be by storing hydrogen atomically inside a metallic matrix, especially of lightweight metals, thus forming a metal hydride, as it avails much higher capacities and is much safer than, for instance, compressed hydrogen gas [7].

The advantages and disadvantages of using bulk and nanoscale metal hydrides will be discussed next. So far, the most promising material option for solid-state hydrogen storage has been Mg, for various reasons; this will be the focus of Section 4.3. A novel architecture with nanoparticles acting as nanoportals for the hydrogenation of Mg films will be the subject of Section 4.4 and the main thesis of this chapter, which will be concluded with a short summary and future outlook in Section 4.5.

4.2 Bulk vs. nano for hydrogen storage

4.2.1 Advantages and disadvantages of downsizing hydrogen storage materials

Following are the basic steps involved during metal-to-metal hydride transformation (schematically illustrated in Fig. 4.1) [6]:

- Dissociation/recombination of hydrogen molecules at the surface.
- Penetration of hydrogen atoms in subsurface layers.
- Diffusion of hydrogen atoms in the α-phase (generally fast), followed by diffusion in the β-phase (upon continuous hydrogenation). β-phase diffuses fast in interstitial hydrides, but is rather slower in ionic and covalent hydrides.
- Nucleation of the α/β phases.
- Propagation of the hydride layer through forward movement of the metal/metal hydride interface.

Through extensive experimental and theoretical work, it was clearly understood that each of the aforementioned steps and the overall hydrogen uptake capacity of any material are vastly dependent on their microstructure and, more specifically, on the actual size of the structural features [6]. For a nanostructured material, the factors that concur to enhance the kinetics of hydrogen absorption are:

- **Surface area effect**: Owing to downsizing, metals and metal hydrides can offer larger surface areas over which hydrogen molecules can effectively dissociate into atoms, accelerating the first step of metal-to-metal hydride transformation (if the surface is catalytically active) [8]. Further, the surfaces/interfaces can also act as sites for heterogeneous nucleation of the new phase [9,10].
- **Grain size effect**: The reduction in grain size maximizes the density of grain boundaries in solid metal hydrides (e.g., in nanofilms [11] or nanocrystals [12,13]).This reduces the length over which hydrogen has to diffuse in the system in order to complete the transformation. Consequently, this results in faster hydrogen diffusion into the host matrices and, hence, the overall absorption

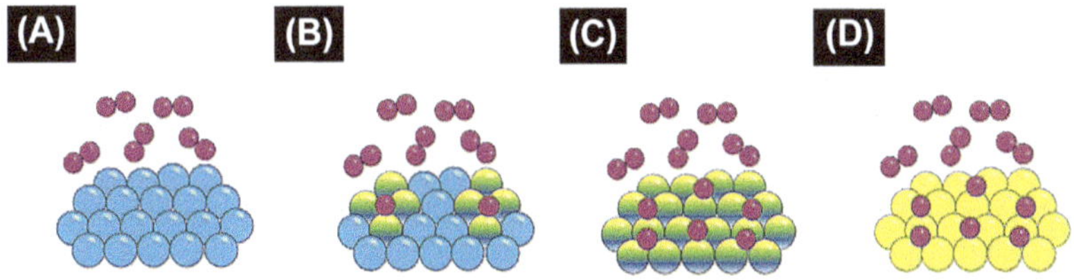

FIGURE 4.1 Schematic illustration of (A) exposure of metal to hydrogen atmosphere, (B) dissociation of hydrogen molecules on the metal surface and penetration of hydrogen atoms to subsurface layers, leading to the formation of the α-phase, (C) further dissociation of hydrogen molecule and subsequent diffusion into the sublayers forming the β-phase, and (D) forward movement of the metal/metal hydride interface that can lead to complete metal-to-metal hydride transformation.

kinetics can be significantly improved [14]. Moreover, grain boundaries may also act as short-circuit diffusion pathways for hydrogen, which can further speed up the transport of hydrogen at the interface or inside the crystal. Additionally, small-sized materials have relatively more defects such as dislocations or vacancies compared with the bulk; this also helps in absorbing hydrogen at a more rapid rate. Finally, the reduced size of metal hydride domains facilitates the desorption step. Such a phenomenon is attributed to surface energies and materials properties at the nanoscale, which assist in reducing the hydrogen release temperature.
- **Mechanical stability**: The absorption and/or desorption of hydrogen involves substantial changes in volume (expansion and contraction) that lead to significant mechanical deformation [4,14,15]. During this step, particles undergo cyclic plastic deformation, which results in fracture of the structure, and generates internal imperfections; in their own turn, these also have strong influence over hydrogen diffusion and kinetics [16]. To accommodate such changes, better plastic stability of metals is desired, such as that demonstrated by nanoparticulated structures, to accommodate large dislocations. Hence, nanostructured materials can be beneficial for hydrogen storage purposes [17].

Because of these factors, numerous nanomaterials such as carbon nanotubes, magnesium, complex hydrides, boron nanotubes, alanates, polymer nanocomposites, and metal organic frameworks have shown improved kinetic results for both absorption and desorption of hydrogen [18,19].

However, in spite of the aforementioned advantages, hydrogen storage at the nanoscale has its own limitations:

- **Oxidation challenge**: When metals are reduced to nanoscale sizes, they are subject to quick oxidation upon air exposure, which significantly impedes their hydrogen uptake [20–22]. To avoid such situations, the surface of nanoparticulated materials can be modified using a gas-selective organic polymer or elemental metal [23,24]. However, the nonuniform distribution of these layers over nanoparticle surfaces (due to limitations in control during synthesis) consequently leads to loss of hydrogen storage capacity [25].
- **Sintering challenge**: Microstructural stability of nanoparticles is crucial for practical applications, but it can be affected by coalescence/sintering, following the growth of nanodomains during hydrogen absorption or desorption [26,27]. Additionally, for optimal hydrogen absorption or desorption, synthesis of nanoparticles as uniform as possible (in shape, structure, and size) is vital; such homogeneity may be compromised by random sintering phenomena.
- **Fabrication/instability challenge**: Theoretical calculations and related experimental work suggests that any significant change in thermodynamic properties can only occur to very small particles, of just a few nanometers in diameter [28]. However, even if the synthesis of such small particles in quantities substantial enough to accommodate appreciable amounts of hydrogen was not already problematic enough, their instability after a few absorption/desorption cycles poses an even more challenging task in this field.

4.2.2 Hydride thermodynamics: from bulk to nanoscale

As expected, thermodynamics at the nanoscale is different from that of bulk matter. For example, consider the reversible hydrogen absorption/desorption reaction:

$$M(s)+H_2(g)\leftrightarrow MH_2(s) \tag{4.1}$$

where M is a metal. The thermodynamic properties for this reaction can be given by the Van't Hoff relation:

$$\ln P_{H2}=\frac{\Delta H^\circ}{RT}-\frac{\Delta S^\circ}{R} \tag{4.2}$$

where P_{H2} is the equilibrium pressure of hydrogen; ΔH^o and ΔS^o are the equilibrium enthalpy and entropy, respectively; R is the gas constant, and T is the absolute temperature.

At the nanoscale, the surface-to-volume ratio is high, thus increasing the number of potential adsorption sites, the binding energies are weak compared to the bulk, and the surface free energy has a strong dependence on radius (r), i.e., the size of nanocluster. In order to include the effects of above factors, a modified Van't Hoff relation is proposed for nanoscale materials [14,29]:

$$\ln P_{H2}=\frac{\Delta H^\circ(r)}{RT}+\frac{3\overline{\gamma V_M}}{rRT}-\frac{\Delta S^\circ(r)}{R} \tag{4.3}$$

In this modified relation (Eq. 4.3), there is an additional term due to the surface energy. In particular, $\frac{3\overline{V_M}}{r}$ represents the average surface area. The contribution from surface energy for hydride formation can be mathematically given by $\frac{3\overline{V_M}}{r}*\overline{\gamma}$ (note that the smaller the radius, the higher the surface energy becomes). Here, the term $\overline{\gamma}$ denotes surface free energies of the metal hydride and the metal particle with an additional energy term (E_{ads}), as shown in (Eq. 4.4)

$$\overline{\gamma}=\left[\gamma_{MH2}\left(\frac{\overline{V_{MH2}}}{V_M}\right)^{\frac{2}{3}}-\gamma_M\right]+E_{ads} \tag{4.4}$$

The additional term E_{ads} accounts for the binding of hydrogen at the surface of both the metal hydride and the metal. The combination of various terms in (Eq. 4.4) may cause shifting of the equilibrium toward the left or the right side, depending on the local environment, hydrogen pressure, and the metal properties. Since the interaction of the hydrogenated with the nonhydrogenated part of the system may change with the hydrogen pressure or the local environment, E_{ads} may increase or decrease the excess energy difference between the forward or reverse reaction. Also, please note that if $\gamma_{MH2} > \gamma_M$, decreasing the size of the nanoparticles will lower the stability of the hydride.

The other two thermodynamic parameters (ΔH^o and ΔS^o) are also radius-dependent, and can shift the equilibrium according to Eq. (4.3). The dependence of the modified

Van't Hoff equation on nanocluster size provides a way to optimize nanostructured metal hydrides. However, to date, information about the surface energy term for metal/metal hydride is scarce, and ways to quantify E_{ads} are not readily available.

Therefore, downsizing particles to nanoscale sizes may lead to a positive or negative thermodynamic change (stabilization/destabilization). This can be understood from the viewpoint of excess surface energy. Hence, if the nanosize metal hydride particles have higher surface energy than that of nonhydrogenated metal particles, desorption will occur at lower temperatures [14,29–31]. Nevertheless, more work is required to understand the effect of various contributions to the stability of a metal particle and its hydride.

4.3 Hydrogen storage in magnesium

In this chapter, we present Mg as a model system to understand hydrogen sorption behavior at the nanoscale. In addition to its attractive physical and chemical properties, Mg is also cheap and readily available. Different synthetic routes to form Mg nanofilms and/or nanoparticles, often mixed with transition metals such as Pd, Ti, Ta, etc. to promote hydrogen storage, are described in detail.

4.3.1 Bulk Mg as a hydrogen storage material

The reaction of hydrogen with pure magnesium (hexagonal, $P6_3/mmc$) yields magnesium hydride, MgH_2 (tetragonal, $P4_2/mnm$). Magnesium hydride, in particular, is a promising candidate because of its high gravimetric capacity (7.6 wt.% H_2) [32]. However, the slow absorption and desorption kinetics hinder its practical usage. Moreover, considerable amount of hydrogen uptake is only possible at high temperature (350°C–400°C) and high pressure (>30bar) conditions, which is certainly undesirable for large-scale applications. The sluggish absorption rate of hydrogen in Mg can be attributed to (i) surface adsorbed species and impurity gases (such as CO_2, H_2O, C_xH_y), (ii) the low diffusion constant of hydrogen into existing MgH_2, and (iii) the tendency of the Mg surface for early oxidation.

The former factor can deprive Mg of active sites for catalytic dissociation of hydrogen molecules by covering the nucleation sites for hydride formation. The second factor restricts the fast diffusion of hydrogen atoms in existing MgH_2 and, hence, can affect the overall hydrogen uptake capacity substantially [25]. Here, the growth of the MgH_2 phase occurs via a sluggish motion of Mg-MgH_2 interfacial boundary, which is primarily dependent on the diffusion coefficient value of hydrogen in MgH_2 ($D_H^{MgH_2} = 1.1 \times 10^{-20} m^2 s^{-1}$ at 33°C); this value is prohibitively low for fast kinetics to be achieved [33]. It is observed that hydrogenation of Mg can stop abruptly when the thickness of the hydride layer exceeds 30–50 μm [34]. Finally, the latter factor of early surface oxide formation is suppressing the hydrogenation of metallic Mg, since hydrogen molecules cannot effectively dissociate on the surface oxide layer due to the high activation energy barrier (2.34–2.94 eV) [35–37]. Therefore, to initiate absorption the oxide layer is often ruptured by an activation process: the activation process is a step where oxidized Mg undergoes heating and cooling cycling in a vacuum or hydrogen atmosphere [38,39] to enable the pristine Mg layer to surface; a novel alternative approach is that of using nanoparticles as nanoportals for the hydrogen to be introduced to Mg through discontinuities in the oxide layer, as will be explained in Section 4.4.

Stampfer et al. [40] and Kennelley et al. [41] pioneered the study of hydrogenation/dehydrogenation behavior of Mg and determined the thermodynamic properties such as enthalpy (−70 to −75 kJ/mol) and entropy (−127.3 to −130.6 J/molK). Later, Vigeholm et al. performed thorough investigations for commercially pure Mg [42–44]. In their work, the highest absorption rate was observed at temperatures around 340°C–350°C (Fig. 4.2A). However, at higher temperatures, the hydrogen uptake value decreased. They also concluded that the activation step is not necessarily required, or it should be performed only for a few cycles under normal absorption conditions. Further, they also investigated the desorption curve and showed that desorption kinetics increases with temperature (Fig. 4.2B). Owing to its attractive hydrogen storage properties, Mg has been continuously studied thereafter.

Keeping the aforementioned issues in mind, efforts were made to decrease the sorption enthalpies of MgH_2, lower the desorption temperature, and enhance the absorption/desorption kinetics. Hence, Mg-based systems have been modified by alloying with other transition metals and their oxides, by changing reaction pathways (e.g., with the addition of Si, Al, or Ge, and the utilization of catalysts) [45,46], and by nanostructuring (i.e., in thin films, nanowires, or nanoparticles). Excellent reviews are available regarding the former approaches, which provide a detailed overview of hydrogen sorption characteristics of such modified systems [9,39,47–51]; the latter approach, that of nanostructuring, will be discussed next.

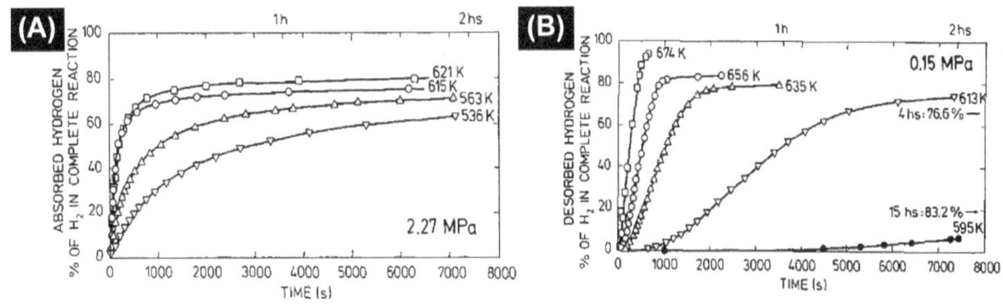

FIGURE 4.2 (A) Absorption and (B) desorption curve for hydrogen in pure magnesium studied by Vigeholm et al. *Reproduced with permission from B. Vigeholm, J. Kjøller, B. Larsen, A. Schrøder Pedersen, Hydrogen sorption performance of pure magnesium during continued cycling, Int. J. Hydrogen Energy 8(10) (1983) 809; Copyright 1983 Elsevier.*

4.3.2 Nanosize Mg for hydrogen storage

Nanosize Mg is regarded as a good choice for hydrogen storage because of its high surface-to-volume ratio, short diffusion length, and enhanced surface reactivity [52]. Nanosize Mg is also expected to have better thermodynamic properties; for instance, De Jong et al. found that desorption enthalpy for MgH_2 nanoclusters (0.9 nm in diameter) is 63 kJ/mol H_2, and the desorption temperature is just 473 K [28]. On the basis of such theoretical calculations, it is generally accepted that due to the weakened bonding of H with Mg, thermodynamic properties can only be modified for very small Mg particles, below 3 nm in diameter [28]. It should be noted, however, that with the decrease in particle size, the importance of surface oxidation is also accentuated, which can concomitantly impede the absorption/desorption kinetics, as discussed above. Clearly, there are competing factors that can eventually determine the percentage reversibility or the overall hydrogen uptake capacity of Mg nanoparticles. Hence, a method, either physical or chemical which can be used to readily synthesize Mg nanoparticles in a controlled way is still missing.

Broadly, synthesis techniques for Mg nanoparticles can be categorized as (i) top-down (e.g., mechanical milling [27,53]) or (ii) bottom-up (e.g., electrochemical deposition [54–56]). Unfortunately, none of these techniques has provided the desired reversible hydrogen storage capacity at mild condition so far. The main reason may be the lack of uniformity in size for Mg nanoparticles, and the rapid formation of a thin, randomly distributed surface oxide layer. Numerous reviews are available focusing on the details of hydrogen sorption properties of Mg nanoparticles and their alloys produced by ball milling [8,47,57]. As known, mechanical milling/reactive milling (under argon/hydrogen atmosphere) has been extensively used for the reduction of Mg particle sizes and the subsequent examination of their hydrogen uptake capacity. However, the milling method is often limited by continuous fracturing, agglomeration, and a cold-welding process. The particles obtained using this method are nonuniform in size and are usually present in metastable forms which tend to agglomerate upon cycling [58]. The latter route, that of bottom-up synthesis, has also been widely used for the fabrication of Mg nanoparticles. The resulting nanoparticles have shown neither an appreciable absorption/desorption uptake capacity, nor reversibility at mild condition; however, studies based on this synthesis method have provided insights to the fundamental understanding of sorption behavior. Finally, it should be mentioned that there also exist various alternative techniques less frequently used to fabricate Mg nanoparticles such as sonoelectrochemistry [59], cryochemistry [60], or deposition from high-temperature molten salts [61], whose hydrogen sorption behavior was subsequently explored at different sets of temperatures and pressures.

Mg nanoparticles can also be prepared through the reduction of organomagnesium, where physical properties can be controlled by varying the electron-carrier type (most often naphthalene), as well as the concentration of reactants; this popular method is widely known as the "electroless method," developed by Rieke [62,63]. Liu et al. obtained

Mg nanoparticles 8 nm in diameter by electroless reduction of di-n-butylmagnesium with lithium in the presence of naphthalene as an electron carrier. This study showed a significant decrease in thermodynamic property values for particle sizes below 25 nm, reaching 63.5 kJ/mol H_2 and 118.4 J/(K mol H_2) for enthalpy and entropy, respectively, for diameters of 15 nm [64]. Similarly, Liu et al. reported a single-step method for the synthesis of alkali metal-coated Mg nanoparticles (where, M = Li, Na, K) in the size range of 7–52 nm, and their respective hydride/perovskite phase under hydrogen atmosphere [65]. Among the studied metals and their respective hydrides, MgK showed low values in their thermodynamic properties (enthalpy: -63.2 ± 2.8 kJ/mol H_2, entropy: -117.8 ± 4.6 J/(K mol H_2)) compared to bulk Mg. Moreover, better desorption kinetics was also noticed for the MgK system, due to the superior catalytic properties of the K metal [65]. In another study, Mg nanofibers 0.4–4 μm in length and 40 nm in width were synthesized using the electroless reduction method. These nanofibers readily absorbed hydrogen at 300°C (>90% of hydrogen absorption in 20 min) and released hydrogen at 350°C (>90% of hydrogen was desorbed in 20 min) [66].

Gas-phase deposition is another, promising bottom-up approach to synthesize ultrafine nanosize Mg and its alloys with adequate control on their average size and purity [22,67]. In principle, this tactic involves two basic steps: (i) atomization of the material by some physical or chemical method (e.g., evaporation, sputtering), and (ii) rapid but controlled condensation into clusters and nanoparticles [68]. In this technique, ultrafine particles (10–100 nm in diameter) can be synthesized within a low-pressure inert-gas atmosphere, the main advantage lying in the fact that their sizes can be controlled by varying the gas pressure [69]. Using the gas-phase deposition method, Li et al. showed enhanced hydrogen storage properties in Mg nanowires, with a desorption enthalpy of 65.3 kJ/mol H_2 for 30–50 nm nanowires, which is significantly lower than that of bulk MgH_2 (75 kJ/mol H_2), but this result was consistent only for few cycles [70]. The work also demonstrates the dependency of kinetics and thermodynamics on the thickness of Mg nanowires.

In another study [71], Pasquini et al. synthesized Pd decorated-Mg/MgO core/shell nanoparticles (sizes ranging from ~100 nm to ~1 μm) for hydrogen storage purposes [72] (Fig. 4.3A–C) In this work, complete metal to metal-hydride transformation was achieved in about 10 min at 333 K and 100 Pa H_2 pressure only. Moreover, it was also noted that hydride forms via a one-dimensional growth process (fitted with the Johnson-Mel-Avrami (JMA) model) in small nanoparticles (~35 nm), controlled by diffusion through the hydride layer. On the other hand, the H-desorption step involved interface-controlled three-dimensional growth of nuclei at constant rate. The corresponding desorption curves illustrate that desorption of maximum absorbed H-fraction decreases with increasing pressure, as well as the number of cycles [72].

In a similar study of the same group, it is concluded that larger Mg nanoparticles might be preferable because of their possibility for mass production while keeping oxide contamination at low levels. However, in small nanoparticles the highest gravimetric capacity is limited to 4.5 wt.% H due to the high MgO weight fraction [73] (Fig. 4.3D–E).

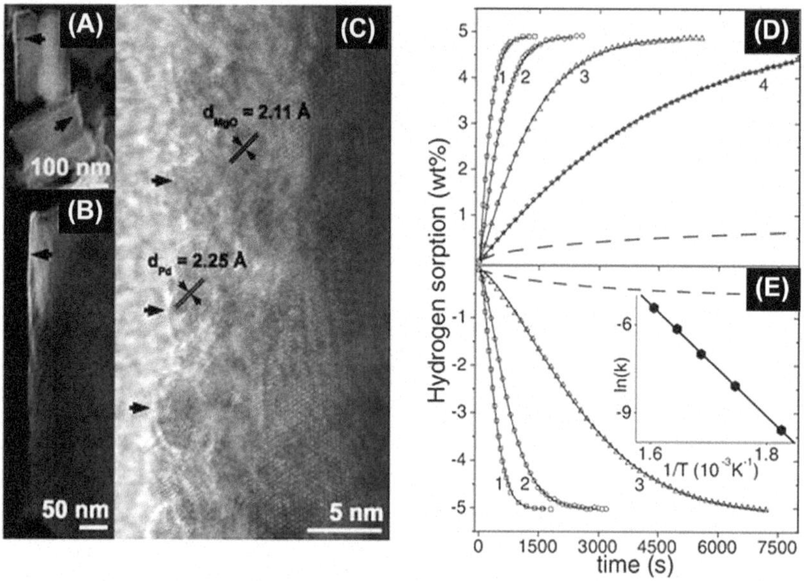

FIGURE 4.3 (A–B) STEM images of Pd-decorated Mg nanoparticles and (C) HR-TEM image of region indicated in (B) by the black arrow. (D–E) Hydrogen adsorption and desorption kinetics, respectively, for the same material, fitted with the Johnson-Mel-Avrami (JMA) model (the inset represents an Arrhenius plot of the rate constant, *K*, for desorption at different temperatures, i.e., 550 K–625 K). *Reproduced with permission from E. Callini, L. Pasquini, E. Piscopiello, A. Montone, M. Vittori Antisari, E. Bonetti, Hydrogen sorption in Pd-decorated Mg-MgO core-shell nanoparticles, Appl. Phys. Lett. 94 (2009) 221905; Copyright 2009 American Institute of Physics.*

The calculated activation energies (obtained from an Arrhenius plot, see inset of Fig. 4.3E) for absorption and desorption are 145 ± 5 kJ/mol H and 155 ± 5 kJ/mol H, respectively.

Interestingly, the gas-phase condensation method was also employed to study the interesting role of MgO surface layers over Mg nanoparticles. In one such work, Kooi et al. synthesized Mg nanoparticles (10–80 nm) and investigated the as-formed nanoparticles using high-resolution transmission electron microscopy (HR-TEM). According to this study, the hexagonal prism shape was dominant, terminated by Mg {0002} and Mg {100} facets. Nonetheless, slight variations in shape were noticed and attributed to the presence of a MgO shell (~3 nm in thickness) at the periphery. The voids between the Mg and MgO shells were due to the faster outward diffusion of Mg ions compared to the inward diffusion of oxygen. This occurs to compensate for unequal material flow due to the different atomic diffusivities, and therefore exemplifies the well-known Kirkendall effect [22,74,75]. To understand clearly the major role of the thin MgO surface layer (2–3 nm) over the complete hydrogenation of Mg nanoparticles (core-shell structure), transmission electron microscopy (TEM) has been widely used recently; efforts are mainly directed toward understanding the time dependence of the growth of the surface oxide layer, and to the effect of the interaction between nanoparticles and an underlying substrate that can alter the thermodynamics of sorption [76,77].

4.3.3 Nanoconfinement of Mg nanoparticles

Most of the aforementioned methods provide nonuniform sizes of Mg nanoparticles that remain prone to oxidation. To avoid rapid oxidation, surface contamination, and grain growth of Mg nanoparticles, efforts have been made to nanoconfine these materials [78,79]. Moreover, as mentioned earlier, hydride nanoparticles can agglomerate during the absorption or desorption cycle, which can lower the overall hydrogen uptake amount. Hence, to avoid aggregation of the nanoparticles, it has been suggested to embed hydrogen storage materials into micro- and meso-porous structures of scaffolds. Several nanoporous scaffold-confined hydrogen systems have been studied [80–83], revealing that, in general, two methods are usually employed: (i) melt infiltration, and (ii) impregnation. Here, melt/solution of hydrides permeates through the scaffold surface and fills all the available pores, which eventually prohibit the aggregation of adjacent nanoparticles. Using the melt infiltration method, De Jongh et al. reported first the preparation of Mg nanoparticles nanoconfined into nanoporous carbon [84]. The obtained size of the Mg crystallites (~2–5 nm) could be tailored according to the pore size of the nanoporous carbon; no bulk Mg could be detected in the samples with loading <15 wt.% of carbon. Along the same line, several reports exhibit the use of di-n-butylmagnesium precursor (i.e., $(C_4H_9)_2Mg$) for the preparation of MgH_2 according to the reaction: $(C_4H_9)_2Mg + 2H_2 \rightarrow MgH_2 + 2C_4H_{10}$ [85,86]. Interestingly, the hydrogen desorption rate could also be tailored by the pore size distribution of the scaffold material. For instance, smaller pores lead to a faster desorption rate, which could be attributed to the decreased size of nanoconfined MgH_2 [21]. Recently, Huang et al. synthesized MgH_2/graphene nanosheet nanocomposites using the thermal decomposition reaction between $(C_4H_9)_2Mg$ and Mg [87]. The resulting structure of MgH_2 nanoparticles uniformly grown inside a graphene nanosheet matrix illustrated a reversible capacity of about 4.5 wt.% for 6 cycles of hydriding/dehydriding at 300°C with a small decrease in kinetic performance.

Recently, Mg nanoparticles (~3 nm) nanoconfined in a LiCl matrix exhibited a decrease in enthalpy (11.9 kJ/mol H_2) and entropy (19.1 J/K/mol H_2) compared with bulk Mg (enthalpy: 75 kJ/mol H_2, entropy: 133.4 J/K/mol H_2), but showed very slow absorption kinetics even at 300°C [83]. In another study, Jeon et al. synthesized Mg nanocrystals confined in a gas-selective polymer (poly-methyl-methacrylate, PMMA) that enabled high hydrogen storage of up to 6 wt.% Mg and 4 wt.% for the composite and rapid kinetics (i.e., loading in <30 min at 473.15 K) [25] (Fig. 4.4A–C). Such enhanced hydrogen sorption behavior was attributed to the encapsulating property of the PMMA polymer, which curtails the exposure of Mg nanoparticles to O_2 and H_2O atmosphere. Further, this work also demonstrated the theoretical modeling of the experimental results by the JMA model, and found that the hydrogenation of Mg nanoclusters proceeds through a one-dimensional growth mechanism.

However, the operating condition (200°C, 35 bar) for the hydrogen absorption step was far from the set US DOE target. Makridis et al. used a laser ablation technique to

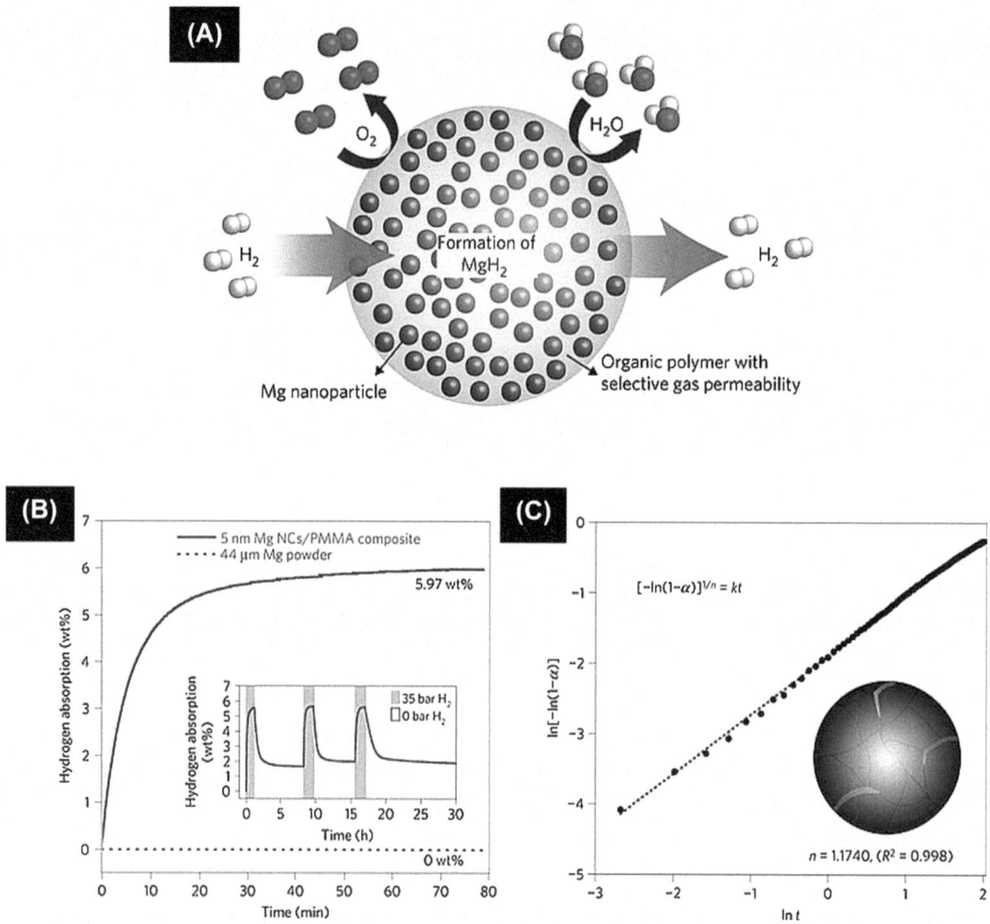

FIGURE 4.4 (A) Pictorial representation of a Mg nanoparticle encapsulated by a selective gas-permeable polymer (composite of Mg nanoclusters in PMMA), synthesized using lithium naphthalide and bis (cyclopentadienyl)-magnesium. (B) Hydrogen absorption data for the Mg nanocluster/PMMA composite at 200°C and 35 bar. (C) Obtained kinetics data fitted with the JMA model and further growth mechanism of MgH_2 inside the Mg nanocluster/PMMA composite. *Reproduced with permission from K.J. Jeon, H.R. Moon, A.M. Ruminski, B. Jiang, C. Kisielowski, R. Bardhan, J.J. Urban, Air-stable magnesium nanocomposites provide rapid and high-capacity hydrogen storage without using heavy-metal catalysts, Nature Mater. 10 (2011) 286; Copyright 2011 Springer.*

create Mg nanoparticles (5 nm in diameter) dispersed in PMMA, and study their hydrogen sorption behavior. Their study showed an uptake of hydrogen (<20 min at 250°C) with a high capacity (6 wt.% in Mg, 5.5 wt.% overall) [88]. Along the same line, Mg nanoparticles (of average crystallite size of 19.3 nm) were confined in carbon aerogels (nanoconfinement was performed through hydrogenation of infiltrated dibutyl-magnesium) and showed an absorption capacity of 2.35 wt.% at 698 K [89]. The apparent activation energy for hydrogen absorption was also determined as 29.4 kJ/mol H_2. Interestingly, Mg nanoparticles were also encapsulated in reduced graphene oxide, which acts as a hydrogen gas selective matrix [90–92], and density functional theory calculations were performed on the nanocomposite, revealing improved mechanical and chemical stability of the chosen nanocomposite [93]. To complement the theoretical calculation, Cho et al. used Mg doped with Ni wrapped by reduced graphene oxide to store hydrogen. Their results illustrate absorption of 6.3 wt.% H_2 (200°C, 15 bar, 30 min), and complete desorption of absorbed hydrogen at 300°C and 0 bar in 10 min [94]. Another study by Lim et al. fabricated Mg nanocrystals within a porous metal-organic framework, which could store hydrogen by both physisorption and chemisorption [95]. However, the reported overall uptake capacity for the confined system was quite low (0.71 wt.% at 473 K under 30 bar). In this regard, several review articles were recently published to exhibit the hydrogen storage behavior of nanoconfined metal hydrides [96–101].

In summary, it is observed that nanoconfinement of Mg nanoparticles leads to improved kinetics, thermodynamic destabilization, and cycling stability, because aggregation and growth of Mg nanoparticles is greatly hindered. However, the storage capacities of the nanoconfined systems are generally low because of the limited loadings of the Mg nanoparticles inside confining materials. Moreover, this method also inherently creates a massive "dead weight" (of inactive scaffolds), which lowers the hydrogen uptake amount. Finally, such an approach is also tedious; therefore, it was deemed as of inadequate overall efficiency [102–104].

4.4 Combined approach: nanoportals for hydrogenation of Mg nanofilms

4.4.1 General introduction

An alternative approach that combines some of the advantages of both nanoparticulated and coarser scale systems is that of nanoportals for the hydrogenation of thin films. On the one hand, the actual storage medium is still a 2D film (or even 3D, if thick enough), which means it does not have problems such as those of restricted volume or "dead weight" discussed above. On the other hand, the catalyst dissociating molecular hydrogen is still in nanoparticle form, taking advantage of the large surface-to-volume ratio.

Nanoportals discussed here are a specific nanoarchitecture that consists of native oxide–terminated Mg films decorated with hydrogen-permeable nanoparticles. The oxide layer forms a barrier for the hydrogenation of the Mg film; however, the presence of nanoparticles in random positions on the film's surface inhibits oxide growth, thus remaining an Achilles heel in the film's hydrogen-proof armor and enabling hydrogen to enter through the nanoparticle/metallic-Mg interface. The nanoportals reported in literature so far contain Pd nanoparticles, since they have been reported as highly active in dissociating H_2 molecules, and since they can be easily hydrogenated into Pd-hydrides, thus enabling the transport of hydrogen atoms through their volume all the way down to the interface with the Mg film [105–107].

The unique architecture of the nanoportals allows for both spatial and temporal decoupling of the nucleation and growth processes, since hydride nucleation occurs only under the Pd nanoparticles, and growth ensues inside the whole Mg film. As a result, it can be considered a valuable tool for studying hydrogenation under various controllable experimental conditions, thus providing in-depth insights into the underlying reaction mechanisms. Moreover, spatial decoupling between different nanoportals also enables the use of Scanning Electron Microscopy (SEM), atomic force microscopy (AFM), or optical transmission techniques [108], since the Pd nanoparticles do not obscure the observation of the phase transformation inside the underlying Mg film.

4.4.2 Nanoportal fabrication by cluster beam deposition

Cluster Beam Deposition (CBD) characterizes a class of bottom-up, gas-phase synthesis methods for producing multifunctional clusters and nanoparticles from individual atoms or molecules [109]. It is well suited to become the standard approach for the fabrication of nanocluster-based devices and functional nanomaterials. The synthesis process starts by the nucleation of subnanometric clusters in an inert-gas atmosphere followed by cluster coagulation, coalescence, and growth. Such methods allow for flexible growth of multifunctional hybrid nanoparticles, often in metastable morphologies owing to the fast kinetics and nonequilibrium process they entail. After formation, clusters can be deposited on suitable substrates, or integrated into functional devices [110,111].

The nanoportals which are the focus of this section were all fabricated by a specific CBD method, namely magnetron-sputtering inert-gas condensation [105–107]. Over the past decades, this fabrication method has been increasingly utilized for the synthesis of single and multicomponent metal and semiconductor nanoparticles [112–119]; both theoretical and technological advances facilitated, on the one hand, a better understanding of the main principles of particle formation [120–122], and, on the other hand, higher cluster yields and better control over the cluster physical and chemical ordering [112,123–130].

A high-vacuum magnetron-sputtering inert-gas condensation apparatus such as those used for the fabrication of the nanoportal architectures of interest in this section is illustrated in Fig. 4.5. It consists of three chambers (marked as A–C in the illustration), the Aggregation Chamber (A), the Mass Filter (B), and the Deposition Chamber (C). The chronological order of the deposition process is marked by numerals 1 and 2 in the figure: first, Mg nanofilms are grown by RF-sputtering on various supports (namely, silicon or sapphire substrates, or TEM grids for subsequent characterization); the rationale behind these choices will be discussed in the following section. Afterward, Pd nanoparticles are grown by DC-magnetron sputtering and deposited onto the Mg films. In this setup, a sputter target is placed on a DC magnetron gun, inside a water-cooled

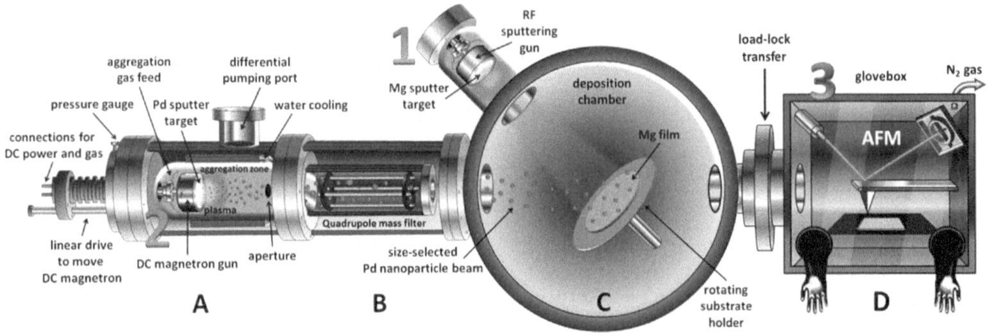

FIGURE 4.5 Schematic illustration of the experimental setup for the fabrication of the nanoportal architecture based on the deposition of size-selected Pd nanoparticles on an underlying Mg film, consisting of the Aggregation Chamber (A), the Mass Filter (B), and the Deposition Chamber (C). An inert-gas glovebox containing an atomic force microscope is also connected via load-lock (D). The sequential deposition processes of the Mg film by RF-sputtering and the Pd nanoparticles by magnetron-sputtering inert-gas condensation, as well as the subsequent topography characterization, are designated by Arabic numerals 1, 2, and 3, respectively. *Adapted with permission from S. Kumar, T. Pavloudis, V. Singh, H. Nguyen, S. Steinhauer, C. Pursell, B. Clemens, J. Kioseoglou, P. Grammatikopoulos, M. Sowwan, Hydrogen flux through size selected Pd nanoparticles into underlying Mg nanofilms, Adv. Energy Mater. 8 (2018) 1701326; Copyright 2018 Wiley-VCH.*

sputtering reactor, while inert gas, typically Ar, is introduced into the reactor and is heated into ionized plasma [68]. The Ar^+ ions of the plasma are directed toward the target due to the presence of the magnetic field of the magnetron gun, and, as they bombard its surface, target atoms are sputtered off [131]. These highly energetic atoms collide with Ar atoms and, in the process, slow down; simultaneously, they also collide with one another, forming nuclei for cluster formation [132–135]. Eventually, the size of these nascent clusters increases with more Ar atoms landing on them, or by their coalescing with each other [136–139]. A pressure differential between the Aggregation Chamber (at high pressure) and the Deposition Chamber (at low pressure), along with the inert-gas drift velocity, drives the nanoparticles toward their deposition onto the support [112,140].

The inherent flexibility of the sputtering method enables the possibility to adjust various deposition parameters (such as power, U/V ratio, inert-gas flow rate, and aggregation zone length) and thus produce a relatively monodisperse nanoparticle beam with good control over shapes and sizes. The use of a quadrupole mass filter (QMF, B in the illustration) enables further postgrowth size selection. Also, it is important to note that the two-step nature of the fabrication process allows for an optional intermediate step of air-exposure, after that of film growth but before nanoparticle deposition, thus enabling the formation of a full Mg oxide layer on the surface of the films. This way, both Pd/Mg and Pd/MgO interfaces can be formed in a well-controlled manner.

After substrate landing, nanoportal-loaded films can be load-lock transferred to an inert-gas (N_2) glovebox (D in Fig. 4.5), where their surface morphology is characterized by AFM (step 3 in the illustration). The obtained images can be useful in evaluating the degree of metal-to-hydride conversion, since the hydride is significantly larger in volume than metallic Mg.

4.4.3 Hydrogenation of Mg nanofilms via Pd nanoportals

A qualitative description of the working principle of the nanoportal architecture is summarized in Fig. 4.6: Pd nanoparticles overlaid on the surface of the Mg film locally inhibit its oxidation [106], thus forming sporadic localized metal-to-metal interfaces; simultaneously, upon hydrogen exposure, they adsorb and dissociate hydrogen molecules (incident A in Fig. 4.6). Subsequently, individual hydrogen atoms diffuse inside the Pd nanoparticles (through interstitial octahedral sites, see Refs. [141–143]), and reach the interface with the support. If the Mg film was already oxidized before being decorated by the Pd nanoparticle (inactive nanoportal, incident B), hydrogen diffusion is prohibited; if not, the Pd/Mg interface allows hydrogen penetration into the Mg film and, eventually, the nucleation of a MgH_2 domain (active nanoportal, incident C). Further growth of this domain is associated with volumetric expansion, as a result of the ongoing Mg-to-MgH_2 transformation. This expansion also leads to the formation of protrusions, which are readily discernible by AFM, thus allowing direct experimental observation of hydride growth. Protrusion formation may have two competing effects on the

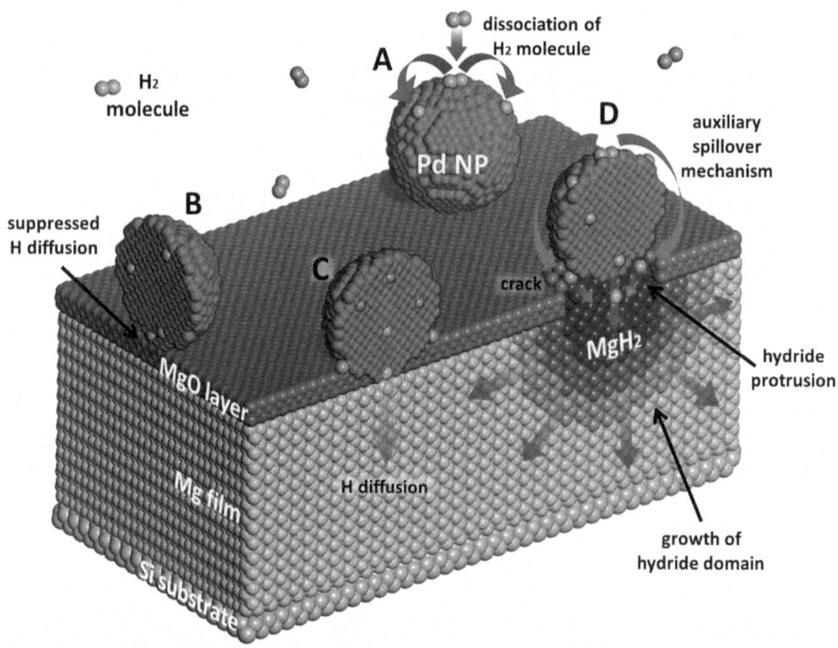

FIGURE 4.6 Schematic illustration of the working principle behind hydrogenation of a Mg film via the nanoportal architecture. *Reprinted with permission from S. Kumar, T. Pavloudis, V. Singh, H. Nguyen, S. Steinhauer, C. Pursell, B. Clemens, J. Kioseoglou, P. Grammatikopoulos, M. Sowwan, Hydrogen flux through size selected Pd nanoparticles into underlying Mg nanofilms, Adv. Energy Mater. 8 (2018) 1701326. Copyright 2018 Wiley-VCH.*

hydrogenation process: (i) on the one hand, it may rupture the surface of the surrounding oxide layer, thus allowing for an auxiliary spillover mechanism through the cracks to assist the hydrogenation of the Mg film (incident D) [144], while (ii), on the other hand, it may assist encroachment of the oxide layer under the nanoportal, with the potential result of eventually closing it down.

The main assumption behind the potential functionality of the nanoportal structure, i.e., that the oxide layer prevents the hydrogenation of the underlying film whereas the metal-to-metal interface does not, was made based on experimental observations [35–39]. However, it was recently proven by analytical ab initio atomistic simulations [107]. First, it was shown that upon landing the Pd nanoparticles are energetically expected to undergo a crystallographic phase transformation from their initial face-centered cubic structure (space group: $Fm\bar{3}m$ 225) to a hexagonal close-packed structure (space group: $P6_3/mmc$ 194), as dictated by the structure of the underlying Mg film. Then, the energy barriers were calculated—and compared—for hydrogen to diffuse through the interface between the nanoparticle and the (i) metallic film, (ii) native oxide layer, and (iii) a previously formed MgH_2 domain, as summarized in Fig. 4.7A. Note that the barriers shown here are after full hydrogenation of the Pd nanoparticle; however, similar trends were demonstrated for nonhydrogenated Pd clusters [107]. Clearly, the barrier of the PdH/Mg interface is low (0.49 eV), enabling hydrogen penetration through the interface. The corresponding diffusion path of a hydrogen atom through the Pd/Mg interface is shown in Fig. 4.7B. In contrast, the energy barrier for the single hydrogen atom to diffuse through the PdH/MgO interface is much higher, reaching up to 2.21 eV. As shown in Fig. 4.7C, the hydrogen atom is strongly attracted by the oxygen atoms of the native oxide, which are pulled closer to it and, thus, impede its motion. Hydrogen atom diffusion from the nanoparticle into Mg hydride is somewhat easier (with a penetration barrier of 1.63 eV), but still in full agreement with the experimental observation that hydrogenation is significantly slowed down by already existing hydride domains. The corresponding meandering diffusion path of a hydrogen atom is shown in Fig. 4.7D.

Thanks to the flexibility of the fabrication method in adjusting parameters of the deposition, various contributions could be readily compared and evaluated with respect to their effects and their importance on the hydrogenation process. Next, we will go through these effects; we have grouped them in two sets, for a more systematic approach. Parameters describing the hydrogenation itself and its timing are summarized in Fig. 4.8, whereas parameters depending on choices during the deposition are grouped in Fig. 4.9.

First, we examine the growth of the hydride domains as a function of hydrogenation time, based on AFM scans revealing the development of protrusions of increasing heights. Fig. 4.8A and B show AFM topography maps of a 25 nm-thick Mg film decorated by Pd nanoparticles 5 nm in diameter with 3.6% coverage, after hydrogenation under 1 bar H_2 pressure at 298 K for 10 and 180 min, respectively. Noticeably, hydride formation has only just begun after 10 min (~2% fractional hydride coverage), but has approached saturation at 180 min (~48% fractional hydride coverage). This is more

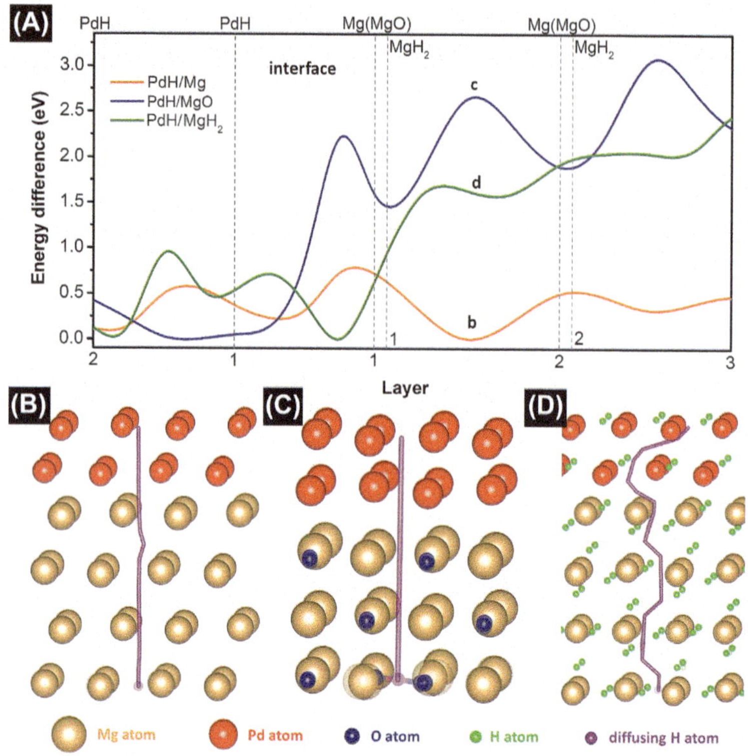

FIGURE 4.7 (A) Energy barriers for hydrogen diffusion perpendicular to the interface. Assuming full hydrogenation of the overlaid Pd NPs, a comparison between the barriers reveals that hydrogenation may be prohibited through the native oxide, and significantly suppressed through previously formed MgH_2 domains. (B–C–D). Corresponding diffusion paths of hydrogen atoms through Pd/Mg, Pd/MgO, and PdH/MgH_2 interfaces, respectively. *Adapted with permission from S. Kumar, T. Pavloudis, V. Singh, H. Nguyen, S. Steinhauer, C. Pursell, B. Clemens, J. Kioseoglou, P. Grammatikopoulos, M. Sowwan, Hydrogen flux through size selected Pd nanoparticles into underlying Mg nanofilms, Adv. Energy Mater. 8 (2018) 1701326; Copyright 2018 Wiley-VCH.*

clearly indicated in Fig. 4.8C, showing the time dependence of fractional hydride coverage; one can observe three different stages of hydride formation, indicated by Latin numerals. First, a slow induction period is evident (marked with "*i*" in the figure), associated with the formation of small, roughly spherical hydride domains directly below active nanoportals. Since their hydrogen content is still relatively low, practically no protrusion can be measured by AFM. As the reaction progresses, it significantly speeds up (marked with "*ii*" in the figure) due to the contribution of an increasing number of nucleation sites, thus indicating the correlation between the growth rate and the number of available active nanoportals. Owing to their random positions, hydride domains often merge, leading to the observed irregular shaped hydride regions. The process slows down after 60 min and eventually stops (marked with "*iii*" in the figure), despite the fact that part of the film still remains in metallic form. Therefore, it can be deduced that not all the Pd nanoparticles are active nanoportals, contributing to the nucleation of MgH_2,

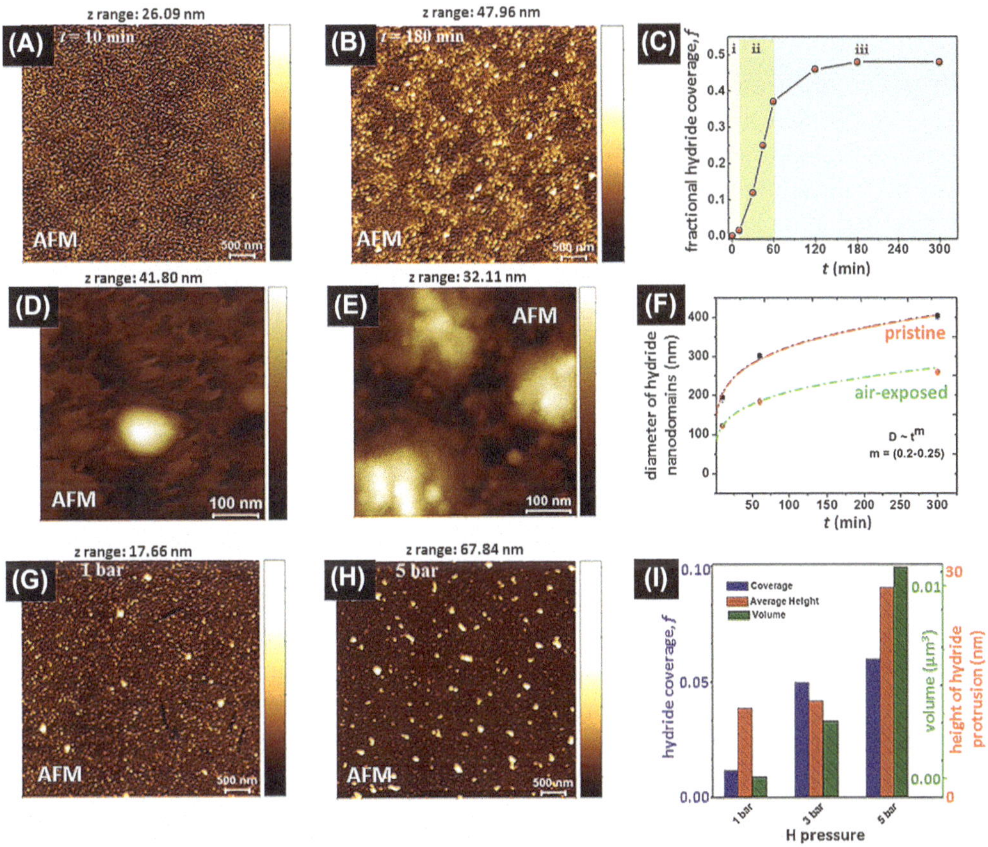

FIGURE 4.8 (A–B, D–E, G–H) AFM images comparing the growth of hydride domains under identical conditions (per pair). (A–B) Comparison after 10 and 180 min, respectively, and (C) time dependence of fractional hydride coverage; three different stages are marked by Latin numerals and indicate saturation despite residual metallic Mg regions in the film. (D–E) Comparison of pristine vs. air-exposed samples after 60 min of hydrogenation, and (F) time dependence of corresponding fractional hydride coverages; clearly, more nanoportals are active in the pristine sample. (G–H) Comparison of samples hydrogenated under different hydrogen pressures, namely 1 and 5 bar, respectively, and (I) hydride coverage, volume, and protrusion heights as a function of hydrogen pressure; portals which can be considered inactive under low pressure can be activated under higher pressures. *Adapted with permission from S. Kumar, T. Pavloudis, V. Singh, H. Nguyen, S. Steinhauer, C. Pursell, B. Clemens, J. Kioseoglou, P. Grammatikopoulos, M. Sowwan, Hydrogen flux through size selected Pd nanoparticles into underlying Mg nanofilms, Adv. Energy Mater. 8 (2018) 1701326; Copyright 2018 Wiley-VCH, and S. Kumar, V. Singh, C. Cassidy, C. Pursell, C. Nivargi, B. Clemens, M. Sowwan, Hydrogenation of Mg nanofilms catalyzed by size-selected Pd nanoparticles: observation of localized MgH_2 nanodomains J. Catalysis 337 (2016) 14; Copyright 2016 Elsevier.*

and also that there are impenetrable barriers within the film through which the hydrogen cannot diffuse.

To clarify the difference between active and inactive nanoportals, Fig. 4.8D–F compares the hydride growth rates of a pristine and air-exposed Mg film under identical conditions (100 nm Mg films, decorated by Pd nanoparticles 5 nm in diameter and

~20% number density). By pristine, we mean that the nanoparticles were deposited prior to exposure to air, whereas in the air-exposed sample the nanoparticles were deposited after the sample surface was allowed to oxidize fully. This led to a remarkable decrease in the number of active nanoportals, as fewer nanoparticles were in direct contact with metallic Mg; in fact, none would be expected, but the nonnegligible surface roughness of the film meant that there were still sites where the native oxide layer was not substantial enough to prevent the formation of nanoportals. As expected, this had a strong effect in the hydride formation rate, as shown in Fig. 4.8F. However, it is clear that the growth rates for both samples are qualitatively alike, showing similar behavior as that observed in Fig. 4.8C. Both follow a correlation of D ~ t^m (where m ~ 0.25), approximating 3D growth (where m ~ 0.33). 2D or 3D growth modes for hydride growth depend on the thickness of the film; actually, for thinner films 2D growth has been observed [105] and explained [107]. The deviation from the expected exponent value (i.e., 0.33) can be ascribed to the native oxide partially undercutting the Pd nanoparticles, as discussed above, thus reducing the area of each individual nanoportal.

However, it should be stressed that the distinction between active and inactive nanoportals is by no means a clear one. The nanoportal functionality is hardly an inherent property of the nanoparticles themselves, or of their interface with the film, but it very much depends on the hydrogenation conditions themselves. For example, one can activate more nanoportals simply by increasing the applied hydrogen pressure, as shown in Fig. 4.8G and H, where all parameters remained constant save for the applied hydrogen pressure (5 nm Pd nanoparticles at 3.6% coverage hydrogenated for 10 min at 298 K at hydrogen pressures of 1 and 5 bar, respectively). Clearly, more nucleation sites can be observed under pressure of 5 bar, with the corresponding increase in overall hydrogen coverage, protrusion heights, and overall volume (Fig. 4.8I). In other words, one should be careful not to mistake the schematic (and, as such, qualitative) description of the process shown in Fig. 4.6 for the real, much more complex, experimental configuration. Observations from electron energy loss spectroscopy (EELS), X-ray photoelectron spectroscopy (XPS) measurements attest the coexistence of MgH_2, MgO, and metallic Mg under active nanoportals, indicating that the Pd nanoparticles may be in contact with both/either metallic Mg and/or the native oxide, and that their nanoportal effectiveness depends on whether the MgO can interfere with hydrogen diffusion or not [106].

However, the actual area of the Pd/Mg interface of individual nanoportals can, indeed, have an important effect on the hydrogenation reaction, as indicated in Fig. 4.9A–C, where nanoparticles of different sizes (2.5 and 5 nm in diameter, respectively for Fig. 4.9A and B show remarkably different hydrogenation efficiency under identical conditions (1 bar hydrogen for 60 min at 298 K on 25 nm-thick film decorated with Pd nanoparticles of similar number densities). Although it is expected that smaller size nanoparticles should be more efficient catalysts for the dissociation of molecular hydrogen, it seems that, at least within the current nanoparticle size limits, this factor is

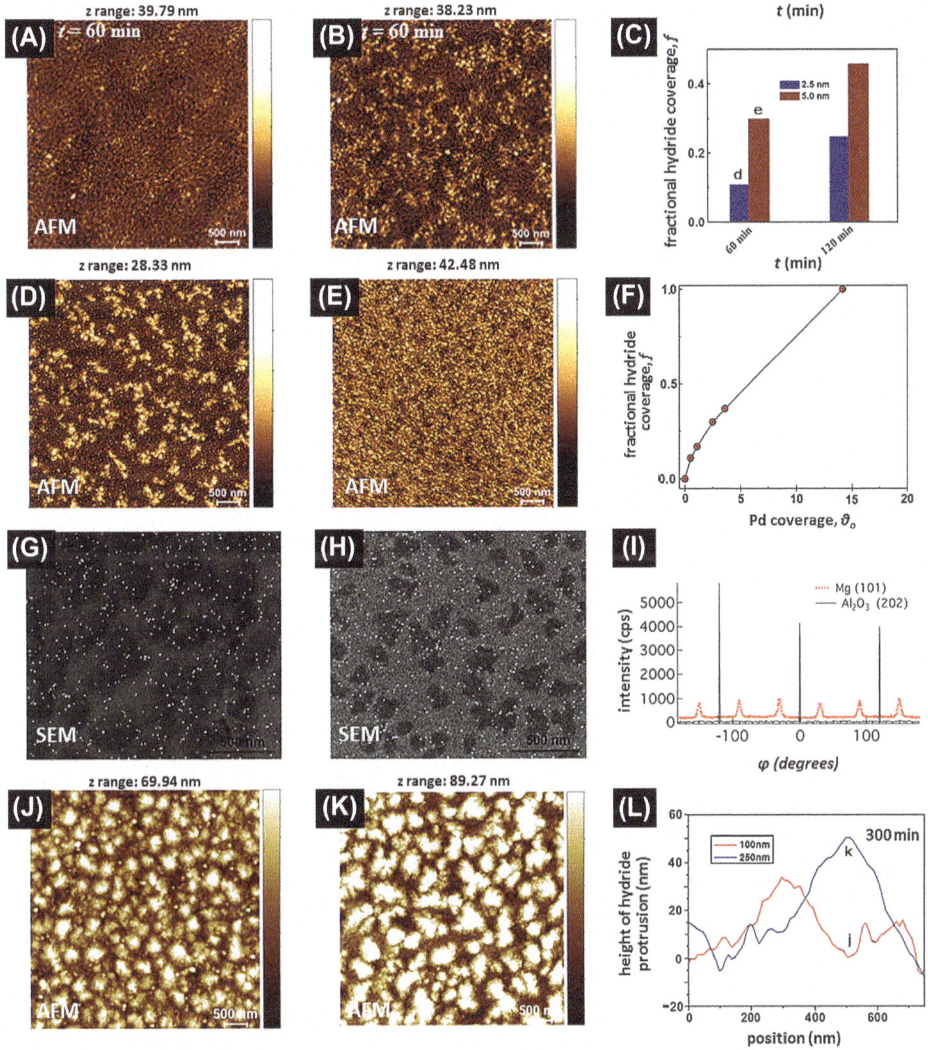

FIGURE 4.9 (A–B, D–E, G–H, J–K) AFM or SEM images comparing the growth of hydride domains under identical conditions (per pair). (A–B) Comparison of hydrogenated Mg films decorated by Pd nanoparticles 2.5 and 5 nm, respectively, and (C) nanoparticle size effect on fractional hydride coverage indicating that higher nanoportal area enhances the hydrogenation. (D–E) Comparison of hydrogenated Mg films decorated by Pd nanoparticles 5 nm in diameter but at different coverages, namely 2.5% and 14.1%, and (F) fractional hydride coverage as a function of Pd nanoparticle coverage. (G–H) Comparison of epitaxial and nonepitaxial Mg films with similar degree of hydrogenation, where the grain boundaries of the latter suppress hydrogen diffusion and, as such, are outlined by the hydride domains. (I) X-ray diffraction φ scans of the epitaxial film shown in (G), indicating the epitaxial in-plane alignment. (J–K) Comparison of hydrogenated Mg films of different thicknesses (100 and 250 nm, respectively), and (L) corresponding protrusion heights indicating that the hydride pockets of the thinner film reaches full capacitance, whereas those of the thicker film do not. *Adapted with permission from S. Kumar, T. Pavloudis, V. Singh, H. Nguyen, S. Steinhauer, C. Pursell, B. Clemens, J. Kioseoglou, P. Grammatikopoulos, M. Sowwan, Hydrogen flux through size selected Pd nanoparticles into underlying Mg nanofilms, Adv. Energy Mater. 8 (2018) 1701326; Copyright 2018 Wiley-VCH, and S. Kumar, V. Singh, C. Cassidy, C. Pursell, C. Nivargi, B. Clemens, M. Sowwan, Hydrogenation of Mg nanofilms catalyzed by size-selected Pd nanoparticles: observation of localized MgH_2 nanodomains, J. Catalysis 337 (2016) 14. Copyright 2016 Elsevier, and C.-J. Chung, C. Nivargi, B. Clemens, Nanometer-scale hydrogen "portals" for the control of magnesium hydride formation, Phys. Chem. Chem. Phys. 17 (2015) 28977; Copyright 2015 Royal Society of Chemistry.*

far less important than that of hydrogen atom diffusion through the nanoportal, which is, therefore, the rate-limiting step of the process.

From this discussion it can also be expected that the total interfacial area (i.e., the sum of all interfacial areas of individual active nanoportals) can determine the overall hydrogenation behavior of the sample. The surface coverage of Pd nanoparticles controls the density of nucleation sites, and, thus, the global degree of hydrogenation. This is indicated in Fig. 4.9D–F, showing AFM scans under identical conditions as above for nanoparticles 5 nm in diameter of two samples with different coverage (2.5% and 14.1%, respectively), and the dependence of hydride coverage on nanoparticle coverage. Interestingly, the size of individual local hydride domains is similar at any given reaction time [106].

Not only the nanoparticle deposition but also the Mg film growth can be variable, leading to different hydrogenation results. In Fig. 4.9G–I, for example, it is shown how a different choice of support on which the Mg film is deposited can play a decisive role in the resultant morphology of the hydrogenated film. As always, hydrogenation conditions were identical for both samples (hydrogenation for 8.5 min under 700 torr H_2 at room temperature). In Fig. 4.9G an SEM image of a hydrogenated film grown on sapphire is shown, where the original Mg film had a high degree of both out-of-plane and in-plane alignment with the support; therefore, even though the film was polycrystalline, it was still devoid of high-angle grain boundaries. In contrast, the hydrogenated film of Fig. 4.9H was grown on an oxidized Si substrate, where no epitaxial in-plane alignment led to the formation of high-angle grain boundaries, due to the grains having random orientation with respect to the surface normal. As a result, in the former case the hydride domains are regular, nearly circular in shape, indicating a hydrogenation propagation front of roughly equal speed along all crystallographic directions, which is not hindered by any obstacles. In the latter case, the domains are more irregular, with sometimes a dendritic-like form. Apparently, the hydride regions follow the grain structure of the parent Mg film, which, in its own turn, is dictated by the underlying support. More importantly, the grains forming pockets that coincide with the grain boundaries signify that these boundaries are not permeable to hydrogen diffusion, posing physical barriers, and explaining the initial observation of $<100\%$ convergence of the film to the hydride phase, despite reaching saturation (see Fig. 4.8C).

Of course, the saturation point of individual hydride pockets can be offset by varying the thickness of the Mg film, as shown in Fig. 4.9J–K, after two films 100 and 250 nm in thickness, respectively, were hydrogenated under identical conditions for 5 h (1 bar H_2 pressure, at 298 K). Volumetric expansion due to full hydrogenation into MgH_2 is of the order of ~32% [145]; therefore, the 100 nm film did reach the saturation stage (as indicated by a protrusion of ~34 nm in Fig. 4.9L), whereas no hydride domains of the 250 nm film reached this stage. This is so because for the domains to start protruding from the initial film surface, they have to grow in the vertical direction; this happens mostly when growth along the other directions (i.e., lateral and in-depth) are prohibited by impenetrable grain boundaries. The larger grain size of the thicker film means that its pockets are much more difficult to fill before protrusions are produced. Naturally, even after a pocket has reached its hydride capacitance, further hydrogenation of the overall film can occur, presumably at a slower rate, until all pockets containing active nanoportals are also saturated.

Since the unique architecture of nanoportals allows growing and characterizing a configuration of distinct hydride domains, it also offers the possibility to calculate the apparent diffusion coefficient of hydrogen in a Mg film, which would otherwise be very difficult to confirm experimentally. By *apparent* we mean that the obtained model encompasses both the hydride phase formation within a single, perfectly crystalline domain, and its propagation throughout an imperfect, multigrain film. With this in mind and based on Fick's second law, MgH_2 formation can be represented by the diffusion of hydrogen of a concentration C, which can be described by a parabolic partial differential equation with the Laplace operator in cylindrical coordinates:

$$\frac{\partial C}{\partial t} = D\left(\frac{\partial^2 C}{\partial r^2} + \frac{1}{r}\frac{\partial C}{\partial r} + \frac{\partial^2 C}{\partial z^2}\right)$$

whose boundary conditions are: $C = C_0$, $z = 0$ and $|r| \leq 150$,

$$\frac{\partial C}{\partial z} = 0, \quad z = 0 \quad \text{and} \quad 150 < |r| \leq 250$$

$$C = 0, \quad -250 \leq z < 0 \quad \text{and} \quad |r| = 250$$

$$C = 0, \quad z = -250 \quad \text{and} \quad |r| \leq 250$$

$C_0 = 1.45 \times 10^{-21}$ g/nm^3 is the saturated concentration of MgH_2, whereas the initial concentration is zero. This equation was said to reveal an apparent diffusion coefficient of $D_H^{tot} \approx 8 \times 10^{-18}$ m^2/s, comparable, at least in magnitude, with literature results for other hydrogenated Mg film configurations [146]. Numerical values were fitted based on TEM analysis of a focused ion beam (FIB)-prepared sample, as shown in Fig. 4.10.

4.5 Future outlook and conclusion

A search for a suitable hydrogen storage material has been extensively performed in the past decade. The ability to reversibly store (i.e., absorb and desorb) hydrogen with faster kinetics (at mild condition) still forms a challenge for the establishment of the much sought-after hydrogen economy. Solid-stage storage, in the form of light-metal hydrides, offers the best compression rates and safety during hydrogen compression and storage available so far, with bulk Mg or Mg films appearing as the most suitable candidate to date. However, there are certain hindrances that need to be overcome for their widespread utilization, the slow diffusion rate being, arguably, the most important one.

The nanoportal structure presented in this chapter, consisting of nanoparticles deposited on Mg films by some CBD technique, is an excellent tool for the in-depth investigation of the hydrogenation mechanism, both qualitative and quantitative. It avails the spatial and temporal decoupling of the hydrogenation through individual

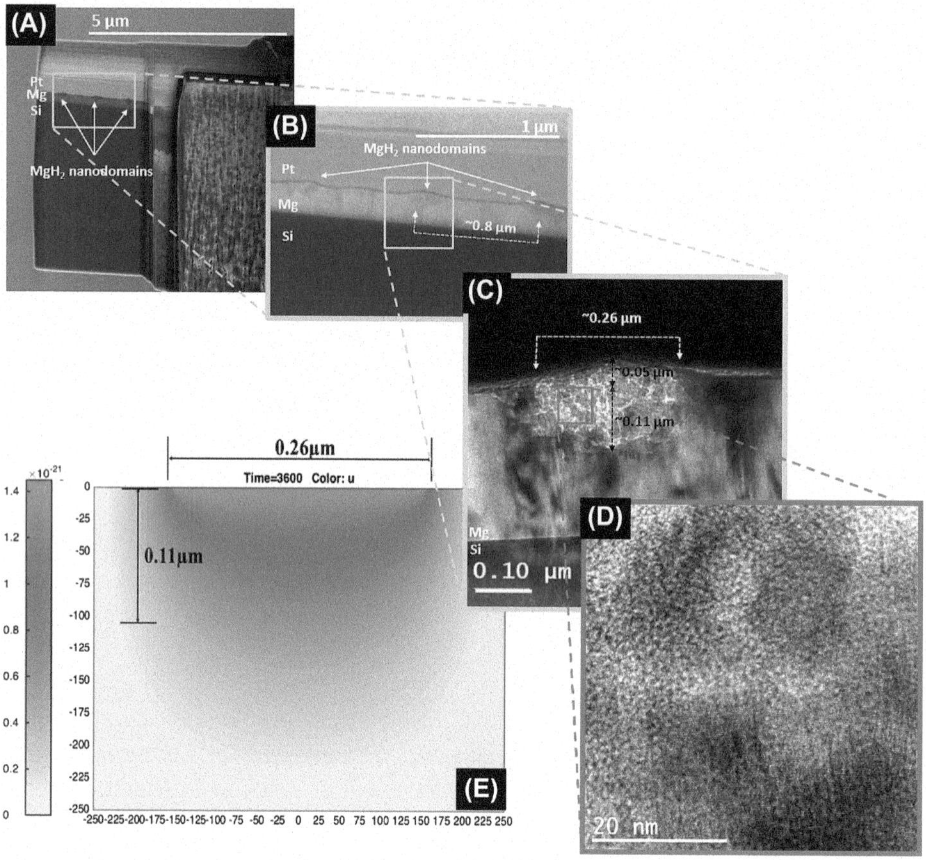

FIGURE 4.10 (A–D) Various magnification levels of sample used for the fitting of diffusion model for the calculation of hydrogen apparent diffusion coefficient in a polycrystalline Mg film 250 nm thick. (A) Preparation of TEM lamella using FIB. (B) SEM image showing spatial separation between hydride domains. (C) TEM micrograph of a single hydride domain. (D) HR-TEM zoom-in inside the hydride domain. (E) Finite-element diffusion model fitted on experimental data, estimating $D_H^{tot} \approx 8 \times 10^{-18}$ m^2/s. *Adapted with permission from S. Kumar, T. Pavloudis, V. Singh, H. Nguyen, S. Steinhauer, C. Pursell, B. Clemens, J. Kioseoglou, P. Grammatikopoulos, M. Sowwan, Hydrogen flux through size selected Pd nanoparticles into underlying Mg nanofilms, Adv. Energy Mater. 8 (2018) 1701326; Copyright 2018 Wiley-VCH.*

nanoparticles; as such, it enables the underlying principles of the hydrogen penetration (or not) inside metallic Mg, Mg oxide, or Mg hydrides and the calculation of the apparent diffusion coefficient of hydrogen in the Mg films. Potentially, the nanoportals (possibly embedded into rolled sheets of thin film) may offer solutions for desorption, which is another hindrance factor. Specifically, at a large scale, the thermodynamics of charging/discharging of metal hydrides need to be carefully engineered for efficient use of involved energy. For instance, use of onboard heat exchanger can save a lot of waste heat during charging of metals. Moreover, more compositions may be investigated, owing to the flexibility and good control of the CBD method.

Further, it would be interesting to note that the usage of nanoportals may be expanded to other areas of research, where single-nanoparticle performance is of importance, such as, for example, catalysis or electrochemistry [147].

References

[1] J. Houghton, Global Warming: The Complete Briefing, Cambridge University Press, 1997.

[2] G. Marbán, T. Valdés-Solís, Towards the hydrogen economy? Int. J. Hydrogen Energy 32 (12) (2007) 1625–1637.

[3] R.A. Gross, Fusion Energy, John Wiley & Sons, 1984.

[4] L. Schlapbach, A. Zuttel, Hydrogen-storage materials for mobile applications, Nature 414 (2001) 353–358.

[5] A. Zuttel, A. Borgschule, L. Schlapbach, Hydrogen as a Future Energy Carrier, Wiley-VCH Verlag GmbH & Co. KGaA, 2008.

[6] R.A. Varin, T. Czujko, Z.S. Wronski, Nanomaterials for Solid State Hydrogen Storage, Springer, USA, 2009.

[7] www.eere.energy.gov/hydrogenandfuelcells.

[8] M.U. Niemann, S.S. Srinivasan, A.R. Phani, A. Kumar, D.Y. Goswami, E.K. Stefanakos, Nanomaterials for hydrogen storage applications: a review, J Nanomater 2008 (2008) 1–9, https://doi.org/10.1155/2008/950967. Article ID 950967.

[9] J.-C. Crivello, B. Dam, R.V. Denys, M. Dornheim, D.M. Grant, J. Huot, T.R. Jensen, P. de Jongh, M. Latroche, C. Milanese, D. Milčius, G.S. Walker, C.J. Webb, C.Z.V.A. Yartys, Review of magnesium hydride-based materials: development and optimization, Appl. Phys. Mater. Sci. Process 122 (2016) 97.

[10] Y. Sun, C. Shen, W. Liu, D.-W. Wang, K.-F. Aguey-Zinsou, Tailoring magnesium based materials for hydrogen storage through synthesis: current state of the art, Energy Storage Mater. 10 (2018) 168–198.

[11] A. Baldi, M. Gonzalez-Silveira, V. Palmisano, B. Dam, R. Griessen, Destabilization of the Mg-H system through elastic constraints, Phys. Rev. Lett. 102 (22) (2009) 226102.

[12] B. Zahiri, B.S. Amirkhiz, D. Mitlin, Hydrogen storage cycling of MgH_2 thin film nanocomposites catalyzed by bimetallic Cr Ti, Appl. Phys. Lett. 97 (2010) 083106.

[13] T.K. Nielsen, U. Bösenberg, R. Gosalawit, M. Dornheim, Y. Cerenius, F. Besenbacher, T.R. Jensen, A reversible nanoconfined chemical reaction, ACS Nano 4 (7) (2010) 3903–3908.

[14] A. Zaluska, L. Zaluski, J.O. Ström-Olsen, Structure, catalysis and atomic reactions on the nano-scale: a systematic approach to metal hydrides for hydrogen storage, J. Appl. Phys. 72 (2) (2001) 157–165.

[15] C. Zlotea, J. Lu, Y. Andersson, Formation of one-dimensional MgH_2 nano-structures by hydrogen induced disproportionation, J. Alloy. Comp. 426 (1–2) (2006) 357–362.

[16] S. Singh, S.W.H. Eijt, M.W. Zandbergen, W.J. Legerstee, V.L. Svetchnikov, Nanoscale structure and the hydrogenation of Pd-capped magnesium thin films prepared by plasma sputter and pulsed laser deposition, J. Alloy. Comp. 441 (1–2) (2007) 344–351.

[17] Q. Yu, L. Qi, R.K. Mishra, X. Zeng, A.M. Minor, Size-dependent mechanical properties of Mg nanoparticles used for hydrogen storage, Appl. Phys. Lett. 106 (26) (2015) 261903.

[18] G.F. Froudakis, Hydrogen storage in nanotubes & nanostructures, Mater. Today 14 (7–8) (2011) 324–328.

[19] S. Orimo, Y. Nakamori, J.R. Eliseo, A. Zuttel, C.M. Jensen, Complex hydrides for hydrogen storage, Chem. Rev. 107 (10) (2007) 4111–4132.

[20] A.M. Ruminski, R. Bardhan, A. Brand, S. Aloni, J.J. Urban, Synergistic enhancement of hydrogen storage and air stability *via* Mg nanocrystal–polymer interfacial interactions, Energy Environ. Sci. 6 (2013) 3267–3271.

[21] T.K. Nielsen, K. Manickam, M. Hirscher, F. Besenbacher, T.R. Jensen, Confinement of MgH_2 nanoclusters within nanoporous aerogel scaffold materials, ACS Nano 3 (11) (2009) 3521–3528.

[22] G. Krishnan, R.F. Negrea, C. Ghica, G.H. ten Brink, B.J. Kooi, G. Palasantzas, Synthesis and exceptional thermal stability of Mg-based bimetallic nanoparticles during hydrogenation, Nanoscale 6 (20) (2014) 11963–11970.

[23] E.S. Cho, A.M. Ruminski, S. Aloni, Y.-S. Liu, J. Guo, J.J. Urban, Graphene oxide/metal nanocrystal multilaminates as the atomic limit for safe and selective hydrogen storage, Nat. Commun. 7 (2016) 10804.

[24] W. Liu, E.J. Setijadi, K.F. Aguey-Zinsou, Tuning the thermodynamic properties of MgH_2 at the nanoscale via a catalyst or destabilizing element coating strategy, J. Phys. Chem. C 118 (48) (2014) 27781–27792.

[25] K.J. Jeon, H.R. Moon, A.M. Ruminski, B. Jiang, C. Kisielowski, R. Bardhan, J.J. Urban, Air-stable magnesium nanocomposites provide rapid and high-capacity hydrogen storage without using heavy-metal catalysts, Nat. Mater. 10 (4) (2011) 286–290.

[26] M.A. Lillo-Ródenas, Z.X. Guo, K.F. Aguey-Zinsou, D. Cazorla-Amorós, A. Linares-Solano, Effects of different carbon materials on MgH_2 decomposition, Carbon 46 (1) (2008) 126–137.

[27] K.F. Aguey-Zinsou, J.R. Ares-Fernández, Hydrogen in magnesium: new perspectives toward functional stores, Energy Environ. Sci. 3 (5) (2010) 526–543.

[28] R.W.P. Wagemans, J.H. van Lenthe, P.E. de Jongh, A. Jos van Dillen, K.P. de Jong, Hydrogen storage in magnesium Clusters: quantum chemical study, J. Am. Chem. Soc. 127 (47) (2005) 16675–16680.

[29] K.C. Kim, B. Dai, J.K. Johnson, D.S. Sholl, Assessing nanoparticle size effects on metal hydride thermodynamics using the Wulff construction, Nanotechnology 20 (20) (2009), 204001, 7 pp.

[30] P.E. de Jongh, P. Adelhelm, Nanosizing and nanoconfinement: new strategies towards meeting hydrogen storage goals, Chemsuschem 3 (12) (2010) 1332–1348.

[31] E. Callini, K.F. Aguey-Zinsou, R. Ahuja, J.R. Ares, S. Bals, N. Biliskov, S. Chakraborty, G. Charalambopoulou, A.L. Chaudhary, F. Cuevas, B. Dam, P.E. de Jongh, M. Dornheim, Y. Filinchuk, J.G. Novakovic, M. Hirscher, T.R. Jensen, P.B. Jensen, N. Novakovic, Q. Lai, F. Leardini, D.M. Gattia, L. Pasquini, T. Steriotis, S. Turner, T. Vegge, A. Zuttel, A. Montone, Nanostructured materials for solid-state hydrogen storage: a review of the achievement of COST Action MP1103, Int. J. Hydrogen Energy 41 (32) (2016) 14404–14428.

[32] M. Dornheim, S. Doppiu, G. Barkhordarian, U. Boesenberg, T. Klassen, O. Gutfleisch, R. Bormann, Hydrogen storage in magnesium-based hydrides and hydride composites, Scr. Mater. 56 (10) (2007) 841–846.

[33] P. Spatz, H.A. Aebischer, A. Krozer, L. Schlapbach, The diffusion of H in Mg and the nucleation and growth of MgH_2 in thin films, Z. Phys. Chem. 181 (1993) 393.

[34] B. Vigeholm, J. Kjøller, B. Larsen, A.S. Pedersen, Formation and decomposition of magnesium hydride, J. Less common met. 89 (1) (1983) 135–144.

[35] T. Ito, A. Kawanami, K. Toi, T. Shirakawa, T. Tokuda, Hydrogen adsorption on ultraviolet-irradiated magnesium oxide, J. Phys. Chem. 92 (13) (1988) 3910–3914.

[36] H. Kobayashi, M. Yamaguchi, T. Ito, Ab initio MO study on adsorption of a hydrogen molecule onto magnesium oxide (100) surface, J. Phys. Chem. 94 (18) (1990) 7206–7213.

[37] G. Wu, J. Zhang, Y. Wu, Q. Li, K. Chou, X. Bao, Adsorption and dissociation of hydrogen on MgO surface: a first-principles study, J. Alloy. Comp. 480 (2) (2009) 788–793.

[38] A. Zaluska, L. Zaluski, J.O. Ström-Olsen, Nanocrystalline magnesium for hydrogen storage, J. Alloy. Comp. 288 (1–2) (1999) 217–225.

[39] I.P. Jain, C. Lal, A. Jain, Hydrogen storage in Mg: a most promising material, Int. J. Hydrogen Energy 35 (10) (2010) 5133–5144.

[40] J.F. Stampfer Jr., C.E. Holley Jr., J.F. Suttle, The magnesium-hydrogen system, J. Am. Chem. Soc. 82 (14) (1960) 3504–3508.

[41] J.A. Kennelley, J.W. Varwig, H.W. Myers, Magnesium-hydrogen relationships, J. Am. Chem. Soc. 64 (1960) 703–704.

[42] B. Vigeholm, J. Kjøller, B. Larsen, Magnesium for hydrogen storage, J. Less common met. 74 (2) (1980) 341–350.

[43] A.S. Pedersen, J. Kjøller, B. Larsen, B. Vigeholm, Magnesium for hydrogen storage, Int. J. Hydrogen Energy 8 (3) (1983) 205–211.

[44] B. Vigeholm, J. Kjøller, B. Larsen, A. Schrøder Pedersen, Hydrogen sorption performance of pure magnesium during continued cycling, Int. J. Hydrogen Energy 8 (10) (1983) 809–817.

[45] J.J. Vajo, F. Mertens, C.C. Ahn, R.C. Bowman, B. Fultz, Altering hydrogen storage properties by hydride destabilization through alloy Formation: LiH and MgH_2 destabilized with Si, J. Phys. Chem. 108 (37) (2004) 13977–13983.

[46] F.E. Pinkerton, M.S. Meyer, G.P. Meisner, M.P. Balogh, J.J. Vajo, Phase boundaries and reversibility of $LiBH_4/MgH_2$ hydrogen storage material, J. Phys. Chem. 111 (35) (2007) 12881–12885.

[47] B. Sakintuna, F. Lamari-Darkrim, M. Hirscher, Metal hydride materials for solid hydrogen storage: a review, Int. J. Hydrogen Energy 32 (9) (2007) 1121–1140.

[48] C.J. Webb, A review of catalyst-enhanced magnesium hydride as a hydrogen storage material, J. Phys. Chem. Solids 84 (2015) 96–106.

[49] Y. Wang, Y. Wang, Recent advances in additive-enhanced magnesium hydride for hydrogen storage, Pro. Nat. Sci. Mater. 27 (1) (2017) 41–49.

[50] L. Wang, J. Jaing, A. Ma, Y. Li, D. Song, A critical review of Mg-based hydrogen storage materials processed by equal channel angular pressing, Metals 7 (9) (2017) 324.

[51] M. Zhu, Y. Lu, L. Ouyang, H. Wang, Thermodynamic tuning of Mg-based hydrogen storage alloys: a review, Materials 6 (10) (2013) 4654–4674.

[52] J. Li, B. Li, H. Shao, W. Li, H. Lin, Catalysis and downsizing in Mg-based hydrogen storage materials, Catalysts 8 (2) (2018) 89.

[53] J.R. Ares, K.-F. Aguey-Zinsou, T. Klassen, R. Bormann, Influence of impurities on the milling process of MgH_2, J. Alloy. Comp. 434–435 (2007) 729–733.

[54] N. Amir, Y. Vestfrid, O. Chusid, Y. Gofer, D. Aurbach, Progress in nonaqueous magnesium electrochemistry, J. Power Sources 174 (2) (2007) 1234–1240.

[55] M. Matsui, Study on electrochemically deposited Mg metal, J. Power Sources 196 (16) (2011) 7048–7055.

[56] L. Viyannalage, V. Lee, R.V. Dennis, D. Kapoor, C.D. Haines, S. Banerjee, From Grignard's reagents to well-defined Mg nanostructures: distinctive electrochemical and solution reduction routes, Chem. Commun. 48 (42) (2012) 5169–5171.

[57] H. Shao, L. He, H. Lin, H.-W. Li, Progress and trends in magnesium-based materials for energy-storage research: a review, Energy Technol. 6 (3) (2018) 445–458.

[58] P.-A. Huhn, M. Dornheim, T. Klassen, R. Bormann, Thermal stability of nanocrystalline magnesium for hydrogen storage, J. Alloy. Comp. 404–406 (2005) 499–502.

[59] I. Haas, A. Gedanken, Synthesis of metallic magnesium nanoparticles by sonoelectrochemistry, Chem. Commun. 0 (15) (2008) 1795–1797.

[60] G.B. Sergeev, Cryochemistry of metal nanoparticles, J. Nano Res. 5 (5–6) (2003) 529–537.

[61] A. Kisza, J. Kaźmierczak, B. Børresen, G.M. Haarberg, R. Tunold, Kinetics and mechanism of the magnesium electrode reaction in molten magnesium chloride, J. Appl. Electrochem. 25 (10) (1995) 940–946.

[62] R.D. Rieke, Preparation of organometallic compounds from highly reactive metal powders, Science 246 (4935) (1989) 1260–1264.

[63] M.R. Song, M. Chen, Z.-J. Zhang, Preparation and characterization of Mg nanoparticles, Mater. Char. 59 (5) (2008) 514–518.

[64] W. Liu, K.F. Aguey-Zinsou, Size effects and hydrogen storage properties of Mg nanoparticles synthesised by an electroless reduction method, J. Mater. Chem. 2 (25) (2014) 9718–9726.

[65] W. Liu, K.F. Aguey-Zinsou, Hydrogen storage properties of in-situ stabilised magnesium nanoparticles generated by electroless reduction with alkali metals, Int. J. Hydrogen Energy 40 (47) (2015) 16948–16960.

[66] Y. Sun, K.F. Aguey-Zinsou, Synthesis of magnesium nanofibers by electroless reduction and their hydrogen interaction properties, Part. Part. Syst. Charact. 34 (2017), 1600276, 8 pp.

[67] V.A. Vons, A. Anastasopol, W.J. Legerstee, F.M. Mulder, S.W.H. Eijt, A. Schmidt-Ott, Low-temperature hydrogen desorption and the structural properties of spark discharge generated Mg nanoparticles, Acta Mater. 59 (8) (2011) 3070–3080.

[68] P. Grammatikopoulos, S. Steinhauer, J. Vernieres, V. Singh, M. Sowwan, Nanoparticle design by gas-phase synthesis, Adv. Phys. X 1 (1) (2016) 81–100.

[69] C.C. Koch (Ed.), Nanostructured Materials, Elsevier, London, 2007.

[70] W. Li, C. Li, H. Ma, J. Chen, Magnesium Nanowires: enhanced kinetics for hydrogen absorption and desorption, J. Am. Chem. Soc. 129 (21) (2007) 6710–6711.

[71] L. Pasquini, E. Callini, E. Piscopiello, A. Montone, M. Vittori Antisari, E. Bonetti, Metal-hydride transformation kinetics in Mg nanoparticles, Appl. Phys. Lett. 94 (4) (2009) 041918.

[72] L. Pasquini, M. Brighi, A. Montone, M.V. Antisari, B. Dam, V. Palmisano, E. Bonetti, Magnesium nanoparticles for hydrogen storage: structure, kinetics and thermodynamics, IOP Conf. Ser. Mater. Sci. Eng. 38 (2012), 012001, 8pp.

[73] E. Callini, L. Pasquini, E. Piscopiello, A. Montone, M. Vittori Antisari, E. Bonetti, Hydrogen sorption in Pd-decorated Mg–MgO core-shell nanoparticles, Appl. Phys. Lett. 94 (22) (2009) 221905.

[74] B.J. Kooi, G. Palasantzas, J.T.M. De Hosson, Gas-phase synthesis of magnesium nanoparticles: a high-resolution transmission electron microscopy study, Appl. Phys. Lett. 89 (16) (2006) 161914.

[75] J.-W. Li, C. Zhao, C. Feng, Z. Zhou, G. Yu, Impact of interface manipulation of oxide on electrical transport properties and low-frequency noise in MgO/NiFe/MgO heterojunctions, AIP Adv. 5 (8) (2015) 087139.

[76] F. Venturi, M. Calizzi, S. Bals, T. Perkisas, L. Pasquini, Self-assembly of gas-phase synthesized magnesium nanoparticles on room temperature substrates, Mater. Res. Express 2 (2015) 015007.

[77] V. Singh, C. Cassidy, F. Abild-Pedersen, J.H. Kim, K. Aranishi, S. Kumar, C. Lal, C. Gspan, W. Grogger, M. Sowwan, Engineering high-performance Pd core–MgO porous shell nanocatalysts via heterogeneous gas-phase synthesis, Nanoscale 7 (32) (2015) 13387–13392.

[78] J.J. Vajo, Influence of nano-confinement on the thermodynamics and dehydrogenation kinetics of metal hydrides, Curr. Opin. Solid State Mater. Sci. 15 (2) (2011) 52–61.

[79] C. Zlotea, F. Cuevas, J. Andrieux, C.M. Ghimbeu, E. Leroy, E. Leonel, S. Sengmany, C. Vix-Guterl, R. Gadiou, T. Martens, M. Latroche, Tunable synthesis of (Mg–Ni)-based hydrides nanoconfined in templated carbon studied by in situ synchrotron diffraction, Nano Energy 2 (1) (2013) 12–20.

[80] A. Gutowska, L. Li, Y. Shin, C.M. Wang, X.S. Li, J.C. Linehan, R.S. Smith, B.D. Kay, B. Schmid, W. Shaw, M. Gutowski, T. Autrey, Nanoscaffold mediates hydrogen release and the reactivity of ammonia borane, Angew. Chem. Int. Ed. 44 (23) (2005) 3578–3582.

[81] A.F. Gross, J.J. Vajo, S.L. Van Atta, G.L. Olson, Enhanced hydrogen storage kinetics of $LiBH_4$ in nanoporous carbon scaffolds, J. Phys. Chem. C 112 (14) (2008) 5651–5657.

[82] M. Konarova, A. Tanksale, J.N. Beltramini, G.Q. Lu, Effects of nano-confinement on the hydrogen desorption properties of MgH_2, Nano Energy 2 (1) (2013) 98–104.

[83] Z. Zhao-Karger, J. Hu, A. Roth, D. Wang, C. Kübel, W. Lohstrh, M. Fichtner, Altered thermodynamic and kinetic properties of MgH_2 infiltrated in microporous scaffold, Chem. Commun. 46 (44) (2010) 8353–8355.

[84] P.E. de Jongh, R.W.P. Wagemans, T.M. Eggenhuisen, B.S. Dauvillier, P.B. Radstake, J.D. Meeldijk, J. W. Geus, K.P. de Jong, The preparation of carbon-supported magnesium nanoparticles using melt infiltration, Chem. Mater. 19 (24) (2007) 6052–6057.

[85] S. Zhang, A.F. Gross, S.L. Van Atta, M. Lopez, P. Liu, C.C. Ahn, J.J. Vajo, C.M. Jensen, The synthesis and hydrogen storage properties of a MgH_2 incorporated carbon aerogel scaffold, Nanotechnology 20 (2009) 204027–204032.

[86] M. Konarova, A. Tanksale, J.N. Beltramini, G.Q. Lu, Porous MgH_2/C composite with fast hydrogen storage kinetics, Int. J. Hydrogen Energy 37 (10) (2012) 8370–8388.

[87] Y. Huang, G. Xia, J. Chen, B. Zhang, Q. Li, X. Yu, One-step uniform growth of magnesium hydride nanoparticles on graphene, Pro. Nat. Sci. Mater. 27 (1) (2017) 81–87.

[88] S.S. Makridis, E.I. Gkanas, G. Panagakos, E.S. Kikkinides, A.K. Stubos, P. Wagener, S. Barcikowski, Polymer-stable magnesium nanocomposites prepared by laser ablation for efficient hydrogen storage, Int. J. Hydrogen Energy 38 (26) (2013) 11530–11535.

[89] Y. Liu, J. Zou, X. Zeng, X. Wu, H. Tian, W. Ding, J. Wang, A. Walter, Study on hydrogen storage properties of Mg nanoparticles confined in carbon aerogels, Int. J. Hydrogen Energy 38 (13) (2013) 5302–5308.

[90] H. Suda, K. Haraya, Gas permeation through micropores of carbon molecular sieve membranes derived from kapton polyimide, J. Phys. Chem. B 101 (20) (1997) 3988–3994.

[91] H.W. Kim, H.W. Yoon, S.M. Yoon, B.M. Yoo, B.K. Ahn, Y.H. Cho, H.J. Shin, H. Yang, U. Paik, S. Kwon, J.Y. Choi, H.B. Park, Selective gas transport through few-layered graphene and graphene oxide membranes, Science 342 (6154) (2013) 91–95.

[92] S.P. Surwade, S.N. Smirnov, I.V. Vlassiouk, R.R. Unocic, G.M. Veith, S. Dai, S.M. Mahurin, Water desalination using nanoporous single-layer graphene, Nat. Nanotechnol. 10 (2015) 459–464.

[93] L.F. Wan, Y.S. Liu, E.S. Cho, J.D. Forster, S. Jeong, H.T. Wang, J.J. Urban, J. Guo, D. Prendergast, Atomically thin interfacial suboxide key to hydrogen storage performance enhancements of magnesium nanoparticles encapsulated in reduced graphene oxide, Nano Lett. 17 (9) (2017) 5540–5545.

[94] E.S. Cho, A.M. Ruminski, Y.S. Liu, P.T. Shea, S.Y. Kang, E.W. Zaia, J.Y. Park, Y.D. Chang, J.M. Yuk, X. Zhou, T.W. Heo, J. Guo, B.C. Wood, J.J. Urban, Hierarchically controlled inside-out doping of Mg nanocomposites for moderate temperature hydrogen storage, Adv. Funct. Mater. 27 (47) (2017) 1704316.

[95] D.-W. Lim, J.W. Yoon, K.Y. Ryu, M.P. Suh, Magnesium nanocrystals embedded in a metal–organic framework: hybrid hydrogen storage with synergistic effect on physi- and chemisorption, Angew. Chem. Int. Ed. 51 (39) (2012) 9814–9817.

[96] P.E. de Jongh, M. Allendorf, J.J. Vajo, C. Zlotea, Nanoconfined light metal hydrides for reversible hydrogen storage, MRS Bull. 38 (6) (2013) 488–494.

[97] C. Zlotea, M. Latroche, Role of nanoconfinement on hydrogen sorption properties of metal nanoparticles hybrids, Colloids Surf A Physicochem. Eng. Asp 439 (2013) 117–130.

[98] X. Yu, Z. Tang, D. Sun, L. Ouyang, M. Zhu, Recent advances and remaining challenges of nanostructured materials for hydrogen storage applications, Prog. Mater. Sci. 88 (2017) 1–48.

[99] J. Ren, N.M. Musyoka, H.W. Langmi, M. Mathe, S. Liao, Current research trends and perspectives on materials-based hydrogen storage solutions: a critical review, Int. J. Hydrogen Energy 42 (1) (2017) 289–311.

[100] T. Sadhasivam, H.T. Kim, S. Jung, S.H. Roh, J.H. Park, H.Y. Jung, Dimensional effects of nanostructured Mg/MgH_2 for hydrogen storage applications: a review, Renew. Sustain. Energy Rev. 72 (2017) 523–534.

[101] P.E. de Jongh, Keeping out the oxygen, Nat. Mater. 10 (4) (2011) 265–266.

[102] G. Xia, D. Li, X. Chen, Y. Tan, Z. Tang, Z. Guo, H. Liu, Z. Liu, X. Yu, Carbon-coated Li_3N nanofibers for advanced hydrogen storage, Adv. Mater. 25 (43) (2013) 6238–6244.

[103] J. Shao, X. Xiao, X. Fan, X. Huang, B. Zhao, S. Li, H. Ge, Q. Wang, L. Chen, Enhanced hydrogen storage capacity and reversibility of $LiBH_4$ nanoconfined in the densified zeolite-templated carbon with high mechanical stability, Nano Energy 15 (2015) 244–255.

[104] M.L. Christian, K.F. Aguey-Zinsou, Core–shell strategy leading to high reversible hydrogen storage capacity for $NaBH_4$, ACS Nano 6 (9) (2012) 7739–7751.

[105] C.-J. Chung, C. Nivargi, B. Clemens, Nanometer-scale hydrogen 'portals' for the control of magnesium hydride formation, Phys. Chem. Chem. Phys. 17 (43) (2015) 28977–28984.

[106] S. Kumar, V. Singh, C. Cassidy, C. Pursell, C. Nivargi, B. Clemens, M. Sowwan, Hydrogenation of Mg nanofilms catalyzed by size-selected Pd nanoparticles: observation of localized MgH_2 nanodomains, J. Catal. 337 (2016) 14–25.

[107] S. Kumar, T. Pavloudis, V. Singh, H. Nguyen, S. Steinhauer, C. Pursell, B. Clemens, J. Kioseoglou, P. Grammatikopoulos, M. Sowwan, Hydrogen flux through size selected Pd nanoparticles into underlying Mg nanofilms, Adv. Energ. Mater. 8 (4) (2018) 1701326.

[108] R. Gremeaud, C.P. Broedersz, D.M. Borsa, A. Borgschulte, P. Mauron, H. Schreuders, J.H. Rector, B. Dam, R. Griessen, Hydrogenography: an optical combinatorial method to find new light-weight hydrogen-storage materials, Adv. Mater. 19 (19) (2007) 2813–2817.

[109] Y. Huttel (Ed.), Gas-phase Synthesis of Nanoparticles, Wiley-VCH, Weinheim, 2017.

[110] M. Haro, V. Singh, S. Steinhauer, E. Toulkeridou, P. Grammatikopoulos, M. Sowwan, Nanoscale heterogeneity of multi-layered Si anodes with embedded nanoparticle scaffolds for Li-ion batteries, Adv. Sci. 4 (2017), 1700180, 10pp.

[111] S. Steinhauer, J. Vernieres, J. Krainer, A. Köck, P. Grammatikopoulos, M. Sowwan, In situ chemoresistive sensing in the environmental TEM: probing functional devices and their nanoscale morphology, Nanoscale 9 (22) (2017) 7380–7384.

[112] Y.-H. Xu, J.-P. Wang, Direct gas-phase synthesis of heterostructured nanoparticles through phase separation and surface segregation, Adv. Mater. 20 (5) (2008) 994–999.

[113] G. Krishnan, M.A. Verheijen, G.H. Ten Brink, G. Palasantzas, B.J. Kooi, Tuning structural motifs and alloying of bulk immiscible Mo-Cu bimetallic nanoparticles by gas-phase synthesis, Nanoscale 5 (12) (2013) 5375–5383.

[114] D. Llamosa, M. Ruano, L. Martínez, A. Mayoral, E. Roman, M. García-Hernández, Y. Huttel, The ultimate step towards a tailored engineering of core-shell and core-shell-shell nanoparticles, Nanoscale 6 (22) (2014) 13483–13486.

[115] P. Grammatikopoulos, J. Kioseoglou, A. Galea, J. Vernieres, M. Benelmekki, R.E. Diaz, M. Sowwan, Kinetic trapping through coalescence and the formation of patterned Ag-Cu nanoparticles, Nanoscale 8 (18) (2016) 9780–9790.

[116] J. Vernieres, M. Benelmekki, J.-H. Kim, P. Grammatikopoulos, J.-F. Bobo, R.E. Diaz, M. Sowwan, Single-step gas phase synthesis of iron aluminide nanoparticles with soft magnetic properties, Apl. Mater. 2 (2014) 116105.

[117] M. Bohra, P. Grammatikopoulos, R.E. Diaz, V. Singh, J. Zhao, J.-F. Bobo, A. Kuronen, F. Djurabekova, K. Nordlund, M. Sowwan, Surface Segregation in chromium-doped NiCr alloy nanoparticles and its effect on their magnetic behavior, Chem. Mater. 27 (9) (2015) 3216–3225.

[118] M. Benelmekki, M. Bohra, J.-H. Kim, R.E. Díaz, J. Vernieres, P. Grammatikopoulos, M. Sowwan, A facile single-step synthesis of ternary multicore magneto-plasmonic nanoparticles, Nanoscale 6 (2014) 3532–3535.

[119] M. Benelmekki, J. Vernieres, J.-H. Kim, R.E. Diaz, P. Grammatikopoulos, M. Sowwan, On the formation of ternary metallic-dielectric multicore-shell nanoparticles, Mater. Chem. Phys. 151 (2015) 275–281.

[120] P.V. Kashtanov, B.M. Smirnov, R. Hippler, Magnetron plasma and nanotechnology, Phys. Uspekhi 50 (5) (2007) 455–488.

[121] B.M. Smirnov, Processes involving clusters and small particles in a buffer gas, Phys. Uspekhi 54 (7) (2011) 691–721.

[122] P. Brault, E.C. Neyts, Molecular dynamics simulations of supported metal nanocatalyst formation by plasma sputtering, Catal. Today 256 (1) (2015) 3–12.

[123] M. Imanaka, T. Katayama, Y. Ohshiro, S. Watanabe, H. Arai, T. Nakagawa, Nanocluster ion source by plasma-gas aggregation, Rev. Sci. Instrum. 75 (5) (2004) 1907–1909.

[124] G.N. Iles, S.H. Baker, S.C. Thornton, C. Binns, Enhanced capability in a gas aggregation source for magnetic nanoparticles, J. Appl. Phys. 105 (2009) 024306.

[125] T. Momin, A. Bhowmick, A new magnetron based gas aggregation source of metal nanoclusters coupled to a double time-of-flight mass spectrometer system, Rev. Sci. Instrum. 81 (2010) 075110.

[126] S. Stoyanov, Y. Huang, Y. Zhang, V. Skumryev, G.C. Hadjipanayis, D. Weller, Fabrication of ordered FePt nanoparticles with a cluster gun, J. Appl. Phys. 93 (10) (2003) 7190–7192.

[127] M.K. Kennedy, F.E. Kruis, H. Fissan, B.R. Mehta, S. Stappert, G. Dumpich, Tailored nanoparticle films from monosized tin oxide nanocrystals: particle synthesis, film formation, and size-dependent gas-sensing properties, J. Appl. Phys. 93 (1) (2003) 551–560.

[128] I.M. Goldby, B.V. Issendorff, L. Kuipers, R.E. Palmer, Gas condensation source for production and deposition of size-selected metal clusters, Rev. Sci. Instrum. 68 (9) (1997) 3327–3334.

[129] A. Majumdar, D. Köpp, M. Ganeva, D. Datta, S. Bhattacharyya, R. Hippler, Development of metal nanocluster ion source based on dc magnetron plasma sputtering at room temperature, Rev. Sci. Instrum. 80 (9) (2009) 095103.

[130] D.M. Foster, R. Ferrando, R.E. Palmer, Experimental determination of the energy difference between competing isomers of deposited, size-selected gold nanoclusters, Nat. Commun. 9 (2018), 1323, 9pp.

[131] J. Vernieres, S. Steinhauer, J. Zhao, A. Chapelle, P. Menini, N. Dufour, R.E. Diaz, K. Nordlund, F. Djurabekova, P. Grammatikopoulos, M. Sowwan, Gas phase synthesis of multifunctional Fe-based nanocubes, Adv. Funct. Mater. 27 (2017), 1605328, 12pp.

[132] M. Bohra, P. Grammatikopoulos, V. Singh, J. Zhao, E. Toulkeridou, S. Steinhauer, J. Kioseoglou, J.-F. Bobo, K. Nordlund, F. Djurabekova, M. Sowwan, Tuning the onset of ferromagnetism in heterogeneous bimetallic nanoparticles by gas phase doping, Phys. Rev. Mater. 1 (2017), 066001, 12pp.

[133] J. Zhao, E. Baibuz, J. Vernieres, P. Grammatikopoulos, V. Jansson, M. Nagel, S. Steinhauer, M. Sowwan, A. Kuronen, K. Nordlund, F. Djurabekova, Formation mechanism of Fe nanocubes by magnetron sputtering inert gas condensation, ACS Nano 10 (4) (2016) 4684–4694.

[134] J. Zhao, V. Singh, P. Grammatikopoulos, C. Cassidy, K. Aranishi, M. Sowwan, K. Nordlund, F. Djurabekova, Crystallization of silicon nanoclusters with inert gas temperature control, Phys. Rev. B 91 (2015), 035419, 12pp.

[135] V. Singh, C. Cassidy, P. Grammatikopoulos, F. Djurabekova, K. Nordlund, M. Sowwan, Heterogeneous gas-phase synthesis and molecular dynamics modeling of janus and core-satellite Si-Ag nanoparticles, J. Phys. Chem. C 118 (25) (2014) 13869–13875.

[136] P. Grammatikopoulos, C. Cassidy, V. Singh, M. Benelmekki, M. Sowwan, Coalescence behaviour of amorphous and crystalline tantalum nanoparticles: a molecular dynamics study, J. Mater. Sci. 49 (2014) 3890–3897.

[137] P. Grammatikopoulos, C. Cassidy, V. Singh, M. Sowwan, Coalescence-induced crystallisation wave in Pd nanoparticles, Sci. Rep. 4 (2014), 5779, 9pp.

[138] P. Grammatikopoulos, E. Toulkeridou, K. Nordlund, M. Sowwan, Simple analytical model of nanocluster coalescence for porous thin film design, Model. Simul. Mater. Sci. Eng. 23 (2015), 015008, 15pp.

[139] V. Singh, P. Grammatikopoulos, C. Cassidy, M. Benelmekki, M. Bohra, Z. Hawash, K.W. Baughman, M. Sowwan, Assembly of tantalum porous films with graded oxidation profile from size-selected nanoparticles, J. Nano Res. 16 (2014), 2373, 10pp.

[140] C. Cassidy, V. Singh, P. Grammatikopoulos, F. Djurabekova, K. Nordlund, M. Sowwan, Inoculation of silicon nanoparticles with silver atoms, Sci. Rep. 3 (2013), 3083, 7pp.

[141] K. Sköld, G. Nelin, Neutron study of the diffusion of hydrogen in palladium, Solid State Commun. 4 (6) (1996) 303–306.

[142] K. Sköld, G. Nelin, Diffusion of hydrogen in the α-phase of Pd-H studied by small energy transfer neutron scattering, J. Phys. Chem. Solids 28 (12) (1967) 2369–2380.

[143] S. Hong, T.S. Rahman, Adsorption and diffusion of hydrogen on Pd(211) and Pd(111): results from first-principles electronic structure calculations, Phys. Rev. B 75 (2007), 155405, 10 pp.

[144] B.D. Adams, A. Chen, The role of palladium in a hydrogen economy, Mat. Today 14 (6) (2011) 282–289.

[145] C.-J. Chung, S. Lee, J.R. Groves, E.N. Brower, R. Sinclair, B.M. Clemens, Interfacial alloy hydride destabilization in Mg/Pd thin films, Phys. Rev. Lett. 108 (2012), 106102, 4pp.

[146] H.T. Uchida, S. Wagner, M. Hamm, J. Kürschner, R. Kirchheim, B. Hjörvarsson, A. Pundt, Absorption kinetics and hydride formation in magnesium films: effect of driving force revisited, Acta Mater. 85 (2015) 279–289.

[147] A. Datta, A.J. Porkovich, P. Kumar, G. Nikoulis, J. Kioseoglou, T. Sasaki, S. Steinhauer, P. Grammatikopoulos, M. Sowwan, Single nanoparticle activities in ensemble: a study on Pd cluster nanoportals for electrochemical oxygen evolution reaction, J. Phys. Chem. C 123(43) (2019) 26124–26135.

5 Applications of polymer films with gas-phase aggregated nanoparticles

Ondřej Kylián[1], Vladimir N. Popok[2]

[1]DEPARTMENT OF MACROMOLECULAR PHYSICS, FACULTY OF MATHEMATICS AND PHYSICS, CHARLES UNIVERSITY, PRAGUE, CZECH REPUBLIC; [2]DEPARTMENT OF MATERIALS AND PRODUCTION, AALBORG UNIVERSITY, AALBORG, DENMARK

Chapter outline

5.1 Introduction

Nanoparticles (NPs) of different types are widely used in research and industry, which utilize their unique physical and chemical properties for a variety of applications. In practice, NPs are either used in solutions or as deposited/embedded on/into solid matrixes. Among hosting media, polymers are very attractive because they are plastic, relatively light, easily processed or formed, and of low-cost [1–3].

One of the new functionalities which can be added by embedding metal NPs into dielectric polymer media is related to electric conductance. An example for the resistivity change in the case of increasing volume fraction of gold NPs in polymeric matrix is presented in Fig. 5.1. By increasing the metal filling factor one can tune the electron transport mechanisms from a tunneling or variable range hopping (VRH) between

Cluster Beam Deposition of Functional Nanomaterials and Devices. https://doi.org/10.1016/B978-0-08-102515-4.00005-2

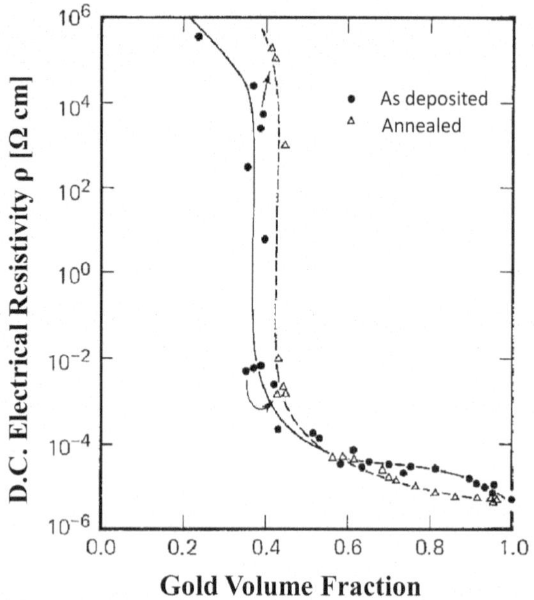

FIGURE 5.1 Variation of resistivity as a function of gold volume fraction in Au/C:F nanocomposites. *Adapted with permission from E. Kay, Synthesis and properties of metal clusters in polymeric matrices, Z. Phys. D Atoms, Mol. Clust. 3 (1986) 251–262.*

individual NPs, which is a low conductance regime, toward a high conductance level under the conditions when the NPs start percolating and forming a quasi-continuous nanoparticular layer, i.e., stipulating an insulator-to-metal transition (IMT) in the composite material [4–9]. This is an efficient way for the production of elastomeric electrodes with required parameters and different types of paper-based electronic components, which will be discussed in more detail below [10–13]. On the other hand, one can utilize plasticity of polymers having fixed surface coverage or volume filling factor of NPs. In this case, the resistance can be tuned in an easy way by stretching or contracting the composite films [10]. Polymer films with metal NPs have also been suggested as simple strain gauges, where the very small applied mechanical forces can be measured by change of conductance [14]. Similar effect, i.e., rapid change of the nanocomposite conductivity induced by small alteration in the metal filling factor, may also be observed during the swelling of certain polymeric matrixes upon exposure to the organic vapor molecules or during their illumination by light with a proper wavelength. This enabled production of fast and reversible gas sensors [15] or photoswitchable devices [16]. Controlling the percolation level of NPs, furthermore, allows forming switching devices enabling to "remember" the conductance level. These devices are named *memristors* and can be used to build neuromorphic architectures mimicking configuration of neuron synapses [17,18]. Metal NPs embedded into a dielectric matrix may also act as deep traps for the injected charges. This effect was utilized to prevent formation of bulk space charge in polymers used for cable insulation in high voltage direct current applications [19] or to suppress dielectric charging in radio frequency (RF) microelectromechanical switches [20].

Another area of significant practical interest is utilization of plasmonic properties of metal NPs [22–27]. This phenomenon gives rise to characteristic extinction and strong local field, which are widely used for optical sensing of gases, environmental species, and various bio-objects [28–32]. Use of metallic nanostructures as plasmonic sources has become very important for surface enhanced Raman scattering (SERS) or surface enhanced fluorescence (SEF) applications [28,33–37] allowing detection of ultralow concentration of analytes down to a single molecule recognition [38,39]. Depositing or implanting NPs on/in polymers adds a desired functionality related to plasticity of the host medium. For example, polymer composite films with embedded NPs showed stable plasmonic tuning under cyclic strain conditions [40]. Utilizing the localized surface plasmon resonance (LSPR) allows designing the metamaterials based on polymer media with either perfect absorption or full transparency for visible light [41,42]. The latter can be combined with conductivity of the embedded metal NPs, thus making this polymer-based composite to be a potential candidate for substitution of indium tin oxide. Recently, a strong enhancement in imaging of biomolecules using matrix assisted laser desorption/ionization (MALDI) has been reported by silver and gold NPs embedded into thin near-surface layers of biopolymer samples [43,44].

Interest to optical properties of polymers with NPs is not only based on LSPR. It is found that adding semiconductor particles like Si, ZnS, and iron sulfides into polymer materials allows tuning the refractive index. This can be used to form passive optical elements such as filters and reflectors [45,46]. Embedding of ZnO and TiO_2 NPs with very low filling factor (a fraction of at. %) into organic films demonstrates good photo-protective properties against UV light while keeping them transparent in the visible spectral interval [47]. Mixing the semiconductor quantum dots into polymers allows production of flexible materials with photoluminescent properties [48–50]. The films made of semiconductor NPs also have a potential to be used for solar cell technologies. This topic is separately addressed in Chapter 3 of this book.

Polymer coatings filled by composite NPs containing ferromagnetic materials demonstrate excellent properties in absorbing electromagnetic waves of GHz frequencies, which is of high interest for microwave technologies [45]. Controlling a filling factor of NPs of ferromagnetic materials (Fe, Co, and Ni) in polymer media provides a route for synthesis of composites with magnetic properties, which can be tuned between superparamagnetic and ferromagnetic ones [51–55].

Adding of NPs to polymers is reported for synthesis of membranes used in filtration and separation processes. Such composites greatly promote synergistic effects positively affecting a number of important parameters: permeability, hydrophilicity, conductivity, mechanical strength, thermal stability, as well as the antiviral and antibacterial properties [56,57]. Bactericidal properties of polymer films with deposited or embedded metal and metal oxide NPs can also be used in food packaging or in medicine, thus opening one more large application area [58–61].

However, thorough understanding of the physics and chemistry of organic composites with added particles is still needed. From the practical point of view, fabrication of polymer composites with NPs requires specific methodological and technological approaches in order to obtain the materials with required functionalities and properties. A short overview on main approaches for particle preparation, deposition, and embedment into host organic matrixes will be presented below with focus on the cluster beam technologies providing a number of practical advantages in assembling of nanomaterials.

5.2 Synthesis routes of polymers with nanoparticles

Polymer composites with NPs can be prepared using chemical and physical methods. Since the main purpose of this chapter is to describe polymer composites with NPs obtained by gas-phase aggregation technique, the other methods are only briefly overviewed in this section. More details can be found in the provided references.

Chemical approaches can be divided into ex situ and in situ [1,62]. In the first case, the particles are synthesized ex situ using one of chemical techniques. The popular ones are through reduction of metal ions from their salt solution or decomposition of metal complexes [63,64]. Then, these NPs are dispersed in a polymer solution. To avoid aggregation, surfaces of NPs can be functionalized or stabilized by organic coatings. Under the ex situ approaches there are possibilities for size and shape control, which are important options for optical and bio-related applications [65]. In the in situ case, NPs are directly synthesized in an organic matrix. One of the widely used approaches is the sol-gel method. Both metal and metal oxide NP formation can be performed in parallel with the polymer synthesis [45].

One of the well-known physical approaches for formation of NPs in polymers is ion implantation [51,52,66–68]. At high fluences, concentration of the implanted atoms can overcome the solubility limit in a polymer that causes nucleation of NPs. With further fluence increase one can even obtain a granular film in the near-surface layer [69] (see schematic picture for nanocluster formation in Fig. 5.2 [70]). Main advantages of this method are that practically any chemical element can be implanted into any type of substrate and by varying the implantation energy one can control the depth at which the particles are formed. However, this method also has certain drawbacks. One of them is a statistical distribution of the implanted impurity around so-called mean projected range of ions that leads to the formation of NPs within a wide size range both in depth and in the plane parallel to the surface. Another problem, which is very pronounced in polymers, is a structural damage caused by energetic ions which also leads to compositional evolution of organic materials. The projectiles can cause scission of polymer chains that facilitates cross-linking [71,72]. Rupture of chemical bonds leads to release of volatile compounds [73,74]. It can be, for example, H_2, CH_4, C_2H_2, CO, CO_2, and other molecules depending on polymer composition. This process is called *degassing* and causes matrix carbonization, which becomes very significant under high fluences needed for NP

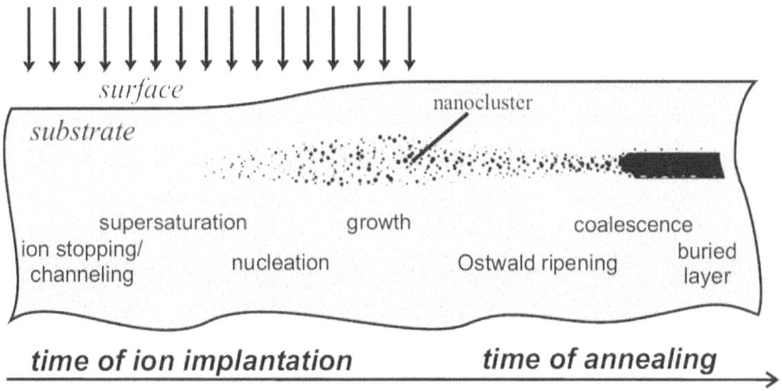

FIGURE 5.2 Schematic picture of nanoparticle nucleation and growth under ion implantation. *Adapted with permission from K.H. Heining, T. Müller, B. Schmidt, M. Strobel, W. Möller, Interfaces under ion irradiation: growth and taming of nanostructures, Appl. Phys. A 77 (2003) 17–25.*

formation [68,73–75]. Atmospheric oxygen can penetrate into the implanted layer through the ion tracks and decorate the radiation damaged areas by making chemical bonds with radicals, for example, carbonyl groups [76,77]. Thus, the structure and composition of polymers implanted with high fluences are dramatically changed and the material can lose some functional properties required for practical applications.

Among the other physical methods, vapor phase deposition is the most common approach. A material for NP formation can be vaporized by thermal heating, electron beam irradiation, and different sputtering techniques [3,78–80]. The atomic vapor deposited on a surface can self-aggregate forming NPs, especially in case of metals. If the NPs are needed to be embedded into the film, thermal treatment above the polymer glass transition temperature T_g can be used facilitating particle immersion beneath the surface [81–84]. Alternatively, the deposition can be carried out into polymer melt [85]. However, in both cases the NPs become embedded into a relatively thin layer. To obtain more homogeneous spreading of NPs in the bulk, the polymer can be codeposited, for example, by the evaporation of monomers and polycondensation on the substrate, thermal cracking, and repolymerization.

An interesting alternative to conventional polymers is the use of so-called plasma polymers. These macromolecular solids are formed when an organic precursor or gas passes through plasma, where it is activated and fragmented. Formed active species subsequently undergo (either in a plasma bulk or on a substrate) radical polymerization that gives rise to a material with irregular and highly branched structure [86–89]. The starting material is mostly supplied into the plasma zone in the form of a gas or vapor (plasma enhanced chemical vapor deposition, PE-CVD). However, the solid-state polymers may be used as well—in this case they are either thermally evaporated (plasma assisted vacuum thermal decomposition, PA-VTD [90,91]) or used as targets in RF magnetron sputtering [92–94]. The plasma polymerization was in the last decades proven to be not only highly versatile in terms of physicochemical or bioadhesive properties of produced materials that can be controlled by the operational parameters (e.g., used precursor, delivered power, pressure, system geometry), but also well-suited for the fabrication of metal/plasma polymer nanocomposites. The latter is achieved by different strategies utilizing plasma polymerization of metal organic compounds [95], simultaneous plasma polymerization and sputtering of a metallic target [96,97], sputtering from the composite target or from two independent magnetron sources [98], or more recently by plasma polymerization at atmospheric pressure using a precursor mixed with metallic salts (e.g., silver nitrate [99,100]). The main disadvantages of the above mentioned vapor phase deposition techniques are related to self-aggregation of NPs leading to different sizes and shapes, as well as to impossibility to independently control the matrix properties on the one hand and size and filling factor of metal inclusions on the other hand. This drawback has been recently overcome by direct introduction of prefabricated metal NPs into the feed gas [101,102].

For many applications, control of NP parameters as well as deposition or embedding conditions into polymer matrix is a critical issue. These requirements can be achieved by the cluster beam method, where the particles are formed in vacuum by gas-phase aggregation, then collimated into a beam, and deposited or implanted with well-controlled surface coverage or volume filling factor [103–107]. When describing cluster beam approach, we will use terms "nanoparticle" and "cluster" as synonyms; thus, the reader should not be confused. Main advantages of the cluster beam method can be summarized as the following (see also schematic illustrations in Figs. 5.3 and 5.4):

- **(i)** a good control of cluster composition (due to formation from ultrapure targets) as well as production of not only homoatomic but also alloy and compound clusters with tunable structure (Janus-, dumbbelllike, and core@shell [108–119]);
- **(ii)** selection of cluster sizes (masses) either by aerodynamic lens systems or by means of electromagnetic mass filters [120,121];
- **(iii)** tuning of cluster kinetic energy to adjust the mechanisms of cluster-surface interaction from soft-landing to implantation [122];
- **(iv)** patterning of NP films by a stencil mask [120] or formation of areas with gradient in particle density [123].

Cluster beam technique is already known since a few decades. However, with the development of nanotechnologies in recent years it has been used more and more as a versatile tool providing a number of unique capabilities for materials synthesis and manipulation on the nanoscale. It is out of scope of this chapter to overview these works, but the reader can refer to a few review papers and books as well as Refs. [120,122,124–127]. Since a few years ago, cluster beams have been applied for the synthesis of metal-polymer composites [128–130] and demonstrated great capabilities in the formation of materials with advanced properties. These studies will be overviewed in more detail in the sections below.

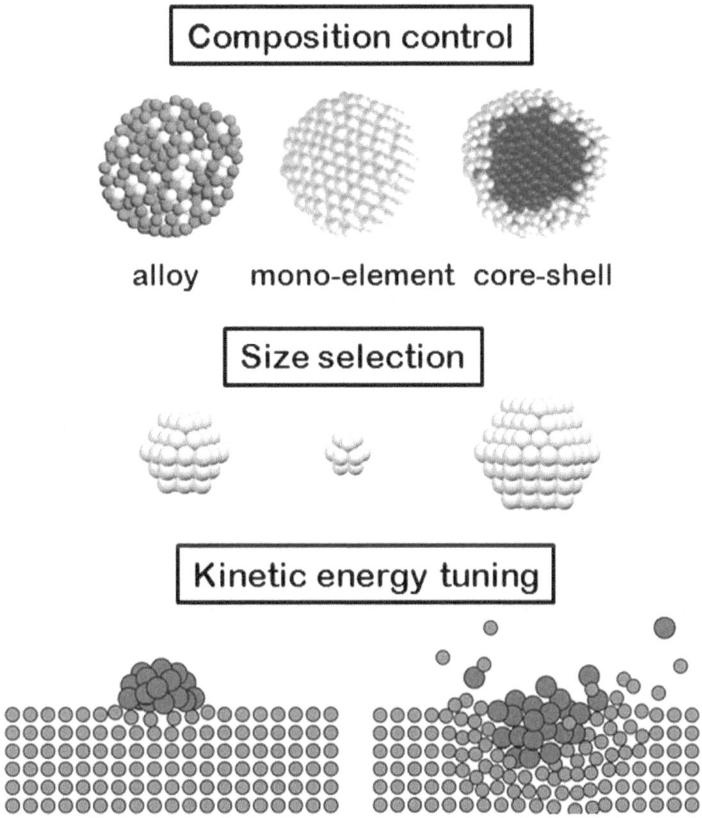

FIGURE 5.3 Schematic picture illustrating control of cluster composition, size, and interaction energy allowed by cluster beam technique.

5.3 Formation of polymer composites using cluster beam technique

One of the simplest ways to obtain a polymer covered by NPs is to soft-land, i.e., deposit them at low kinetic energies from a cluster beam [131]. By controlling the deposition time one can obtain a required coverage. However, adhesion of NPs to polymer surface in this case is typically low, and if the composites are used in the procedures involving wet chemistry or aqueous solutions, the particles can easily be washed out. One of the possibilities to improve adhesion is partly embed the NPs into the supporting polymer medium. This can be realized by a simple thermal treatment facilitating immersion. The driving force for the embedding is a difference in surface tension (surface energy) γ between the particle material and common polymers. Latter have relatively low values of γ, which is typically on the order of a few tens mJ/m^2, while γ of bulk metals, their oxides, and semiconductors is one to two orders of magnitude higher [132,133,134]. It is still under discussion how the surface energy changes on the nanoscale. Some of the

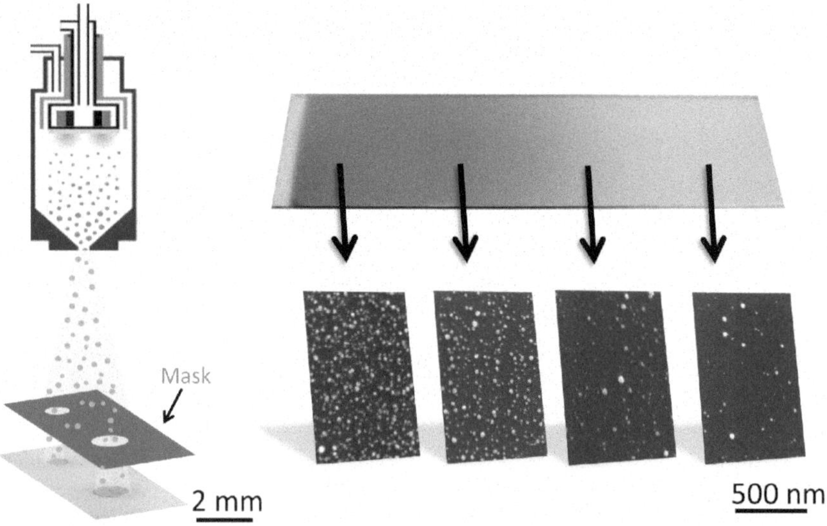

FIGURE 5.4 Cluster beam deposition of patterned (*left*) or gradient (*right*) coatings.

experiments show that γ of metal NPs can be much greater compared to the bulk metals reaching values of a few J/m^2 [85,135]. Thus, it increases the difference with polymers even more. The above-mentioned thermal treatment is needed to increase flexibility of polymer chains and provide necessary room for NP indentation. In approximation of a rigid particle immersing into a soft material, one can develop an inequality condition for the embedding [136]

$$\gamma_{NP} > \gamma_{pol} + \gamma_{NP/pol}, \tag{5.1}$$

where γ_{NP}, γ_{pol}, and $\gamma_{NP/pol}$ are the tensions at the interfaces of NP/air, polymer/air, and NP/polymer, respectively. The immersion is continued until the equilibrium condition is reached, which can be found by balancing the surface tension components projected on the plane tangential to the particle surface

$$\gamma_{NP} = \gamma_{NP/pol} + \gamma_{pol} \cos \varphi, \tag{5.2}$$

where φ is the contact angle as illustrated in Fig. 5.5. For a spherical particle, by considering values of γ for different interfaces in Eq. (5.2) and isolating $\cos\varphi$ one can make a simple calculation to estimate the particle indentation

$$z = r(1 + \cos \varphi). \tag{5.3}$$

One should be aware that the presented equation is an oversimplification of contact phenomena at the rigid particle/soft substrate interface. Real cases, especially on the nanoscale, require quantification of the forces related to the substrate deformation, for example, as presented in Ref. [137]. It is also shown elsewhere [85] that the immersion can undergo through the formation of wetting layer (as shown in Fig. 5.6A), which creates a capillary pressure on the NP.

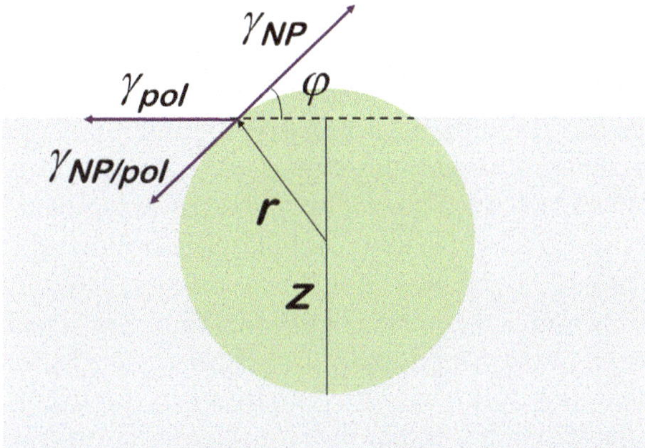

FIGURE 5.5 Schematic image of a rigid nanoparticle embedded into soft material. See text for details.

Embedding of NPs depending on thermal heating of polymers is studied experimentally. In Ref. [138] it was found by X-ray scattering that the NP immersion (gold in this case) starts at temperatures below T_g of the polymer (polystyrene, PS) due to increased mobility of the PS segments at the surface. Furthermore, the embedding depends on the cluster surface coverage because a short-range metal-polymer interaction can exceed the van der Waals forces and reduce the chain mobility, thus, requiring higher temperatures for the immersion.

Another efficient technique that monitors embedding of NP is atomic force microscopy (AFM). It is suggested in Ref. [139] that a thin near-surface layer of polymer is characterized by higher molecular mobility and lower glass transition temperature compared to the bulk that facilitates the indentation to the depth of 3–4 nm already at temperatures below T_g. Later experiments on embedding of gold NPs and relaxation of holes on the polymer surface confirm this suggestion [140,141]. In Ref. [142], it is found

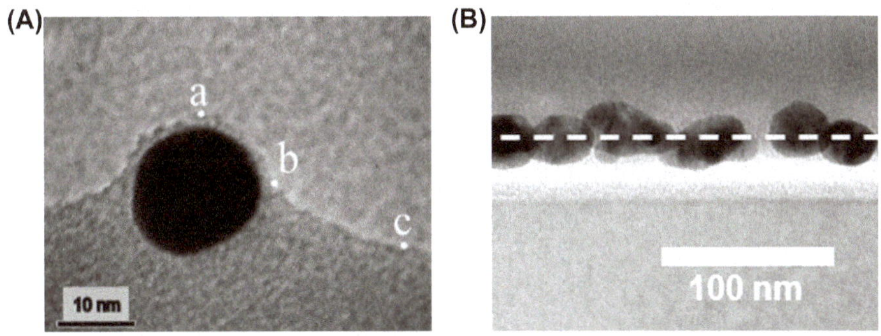

FIGURE 5.6 HRTEM images of (A) gold NP embedded into polystyrene melt with wetting layer covering the particle and (B) silver NPs embedded into solid poly(methyl methacrylate) (PMMA) layer (bright stripe) using thermal annealing. Dashed line just provides visual guidance for the PMMA surface. *Image (A) is reprinted with permission from R.D. Deshmukh, R.J. Composto, Direct observation of nanoparticle embedding into the surface of a polymer melt, Langmuir 23 (2007) 13169–13173.*

that the thickness of the mobile layer near the glass transition point can be around 6 and 14 nm for PMMA and PS films, respectively. The dependences presenting the NP height changes at different temperatures are shown in Fig. 5.7. They demonstrate relatively fast NP embedding: above the T_g the particles immerse for about half of diameter in 10 min.

The results on immersion of particles deposited using cluster beam technique are presented in Refs. [84,143,144]. Postdeposition annealing at temperatures above T_g leads to embedding of Ag and Cu NPs into spin coated polymer films as found by measuring the particle height using AFM. Cross-sectional transmission electron microscopy (TEM) proves that the particles immerse but do not become flattened (see Fig. 5.6B). The study of size-selected particles shows very rapid (on the timescale of a few minutes) embedding (see Fig. 5.8). This very fast, even in comparison to the results shown in Fig. 5.7, dynamics could be related to the fact that the size-selected clusters immerse in the same manner, while in case of a size dispersion, the embedding can go fast for small particles while slower for larger ones. Unfortunately, there is no thorough study on size-dependence of particle embedding into polymers presented in literature. It is worth noting that if the spin coated polymer is not annealed, i.e., the film is of low density and can contain leftovers of solvent, the as-deposited NPs become partly embedded already at room temperature [84].

Clusters can be directly embedded into a near-surface layer of polymer by increasing their kinetic energy, thus, changing the cluster-surface interaction regime from soft-landing to implantation. Using a supersonic beam of palladium clusters (velocity of 1000 m/s, mean impact energy of about 0.5 eV/atom), it was possible to implant them into a few tens nm thick layers of PMMA [128]. This energy is below a typical cohesive

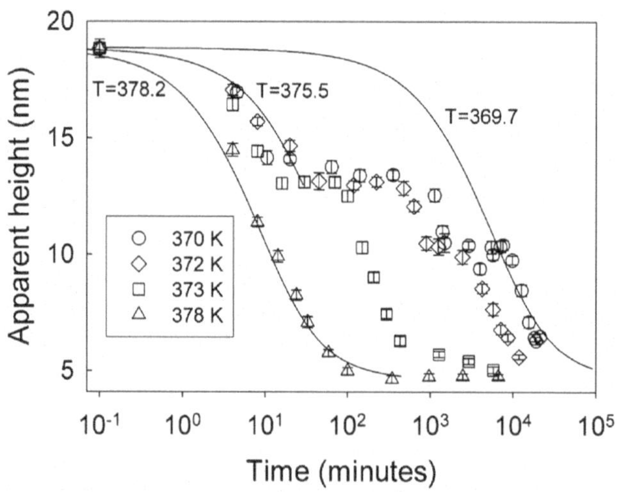

FIGURE 5.7 Height of 20 nm in diameter gold NP on PS film (100 nm thick) versus annealing time at different temperatures. 378 K is above T_g, 373 K is close to T_g while the other temperatures are below, thus demonstrating different dynamics of embedding. *Reprinted with permission from D. Qi, M. Ilton, J.A. Forrest, Measuring surface and bulk relaxation in glassy polymers, Eur. Phys. J. E 34 (2011) 56.*

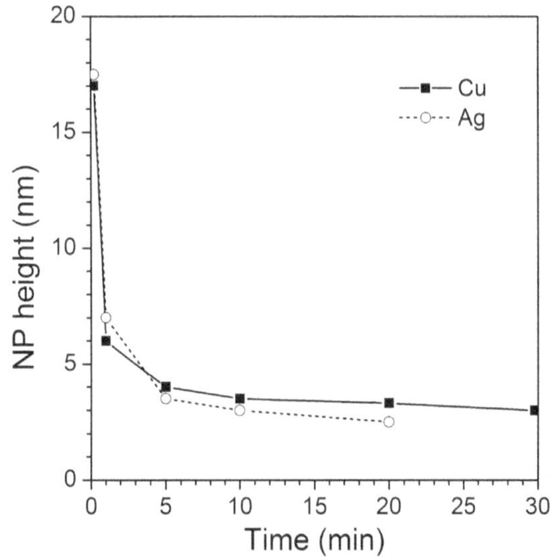

FIGURE 5.8 Time evolution of height for size-selected (ca. 17 nm in diameter) Ag and Cu NPs soft-landed on 50 nm thick PMMA films and annealed at 125°C in ambient atmosphere.

energy of metal atoms in an NP; thus, the particle will survive on impact. But taking into account that a cluster of 2–3 nm in diameter consists of few hundreds of atoms, a very high energy is delivered locally to the surface on a short timescale (few tens fs) leading to the thermalization effect. According to the calculations presented in Ref. [145], a 3 nm-radius gold cluster (about 6000 atoms) with a energy of 2.0 eV/atom causes a local heating along the track up to 600 K in polydimethylsiloxane (PDMS). The induced temperature wave is quickly dumped in the plane direction, but under a high-fluence cluster deposition (i.e., track density is high and time intervals between successive impacts is short) the overall temperature of the near-surface layer can increase to about 50° that is accompanied by the decrease in density by about 5% [146]. These thermally induced phenomena greatly facilitate metal cluster embedding into polymers. If the polymer substrate is preheated up to 95°C, the penetration depth is increased from approximately 50 nm (room temperature) to 70 nm. Thus, the cluster impact induces local thermal and pressure spike due to the multiple collision effect [147–149] causing embedding, which can be further promoted by additional heating. The cluster penetration in depth is momentum-driven and can be facilitated by so-called "clearing the way" effect in which the front atoms of the cluster push away the target atoms in collisions decreasing the stopping power (loss of energy per unit mass) for the rest of the atoms of the cluster [150,151].

It is worth mentioning that NPs of different species may have different geometrical shapes, thus affecting electronic and optical properties of produced nanocomposites. Small gas-aggregated metal clusters can represent different geometries, which can be driven by strong chemical bonds (covalent or ionic), or provide high packing density of atoms with relatively weak bonds, for example, different polyhedrons: icosahedrons have only hexagonal facets while others like truncated decahedrons or cuboctahedrons have square facets [152]. With increasing number of atoms in a cluster, there is a tendency to change the structure toward that of bulk matter. It typically occurs in the size range between approximately 500–600 and 1000–1500 atoms. It is found in many studies that clusters of silver, copper, gold, platinum, and indium larger than 4–5 nm in diameter represent the geometries close to spheres and have fcc structure [24,153–157]. However, recent experiments with silver clusters showed presence of 3D structures like icosahedrons, decahedrons, etc., even if the particles comprise billions of atoms [158]. Al and Ti NPs produced by magnetron sputtering sources show morphologies with hexagonal and rectangular facets or even cubic (see Fig. 5.9) [27,159]. Cubic-shaped clusters are found for iron as well as alloys of iron-cobalt and core-shell iron/cobalt-gold/silver structures [160,161]. Si NPs produced in plasma discharges from silane are also predominately faceted with cubic-shaped crystals [162]. After embedding into polymers, the particles typically preserve these shapes. That is why it is important to consider this point for practical applications.

Another possibility for the formation of organic composites with embedded NPs that is gaining increasing attention is combination of gas aggregation sources with plasma polymerization. Two distinct strategies are suggested. The first one is based on the sequential deposition of layers of plasma polymer either by PE-CVD or RF magnetron sputtering and metal NPs produced by cluster beam systems as shown in Fig. 5.10 that results in sandwiched or multilayered structures [61,154,163]. The great advantage of this approach is excellent control over the properties (chemical composition, thickness, density, and size of NPs) of each individual layer that in turn makes it possible to design

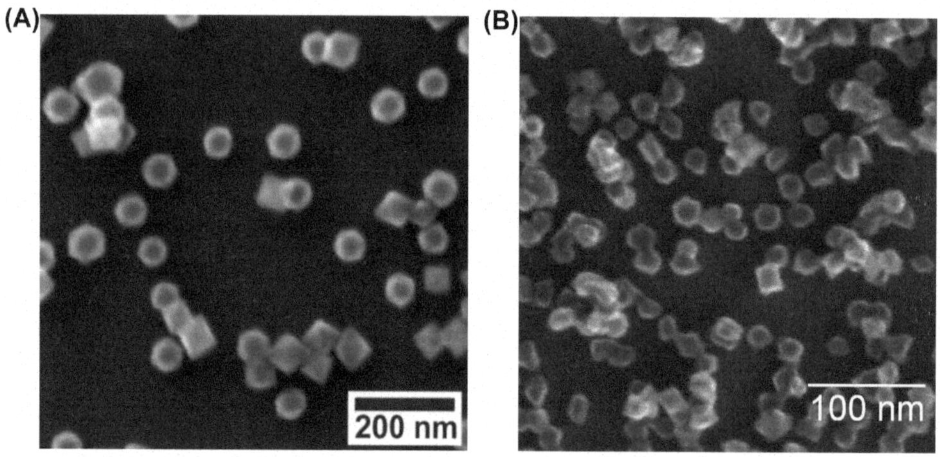

FIGURE 5.9 Scanning electron microscopy (SEM) images of clusters produced by magnetron sputtering and gas aggregation: (A) Al NPs and (B) Ti NPs on Si substrates. *Panel (A) is reprinted with permission from O. Polonskyi, O. Kylián, M. Drábik, J. Kousal, P. Solař, A. Artemenko, J. Čechvala, A. Choukourov, D. Slavínská, H. Biederman, Deposition of Al nanoparticles and their nanocomposites using a gas aggregation cluster source, J. Mater. Sci. 49 (2013) 3352–3360.*

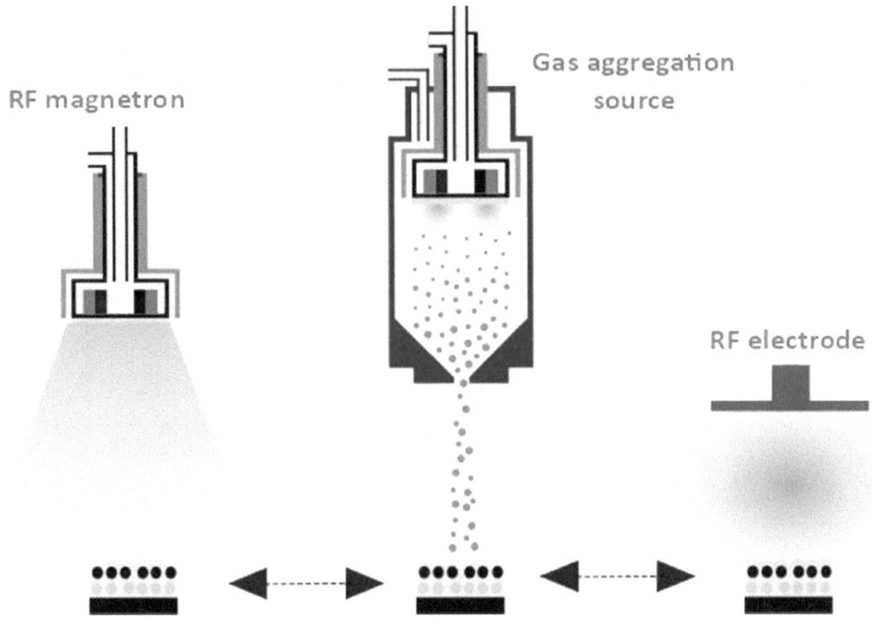

FIGURE 5.10 Sequential deposition of sandwiched or multilayered nanocomposites using gas aggregation source combined with either magnetron sputtering or PE-CVD of plasma polymer layers.

coatings with required enhanced functionality. An example of this is fabrication of dual LSPR peak materials based on layers of Ag and Cu NPs separated by the plasma polymer thin interlayers (see Fig. 5.11).

The second proposed strategy is simultaneous or quasisimultaneous deposition of plasma polymers and metal NPs. Various arrangements are tested that utilize, for instance, deposition on static (Fig. 5.12A) [27] or rotating substrates (Fig. 5.12B) [164,165]. In addition, in order to produce hard matrixes, the substrate to be coated may be positioned directly on an RF electrode used for the PE-CVD (see Fig. 5.13). In this case, the DC self-negative bias developed on the powered RF electrode accelerates the ions impinging on the substrate surface during the deposition, which results in enhanced densification and hardening of the growing film [166]. This approach is recently used to produce hard Ag/a-C:H and Cu/a-C:H nanocomposites with metal NP filling factor controlled by either the cluster flux or duty cycle of RF power applied to the RF electrode [167,168].

Although the metal/plasma polymers are traditionally deposited in the form of thin coatings, recent results also indicate a possibility to produce heterogeneous metal/plasma polymer particles with different structures. The first option that is tested for fabrication of such particles is based on the introduction of a precursor into the aggregation chamber of the cluster source equipped with an RF magnetron carrying metal target [169]. It is demonstrated that the competitive growth of metal NPs and plasma polymerization process results in the formation of highly heterogeneous particles. This

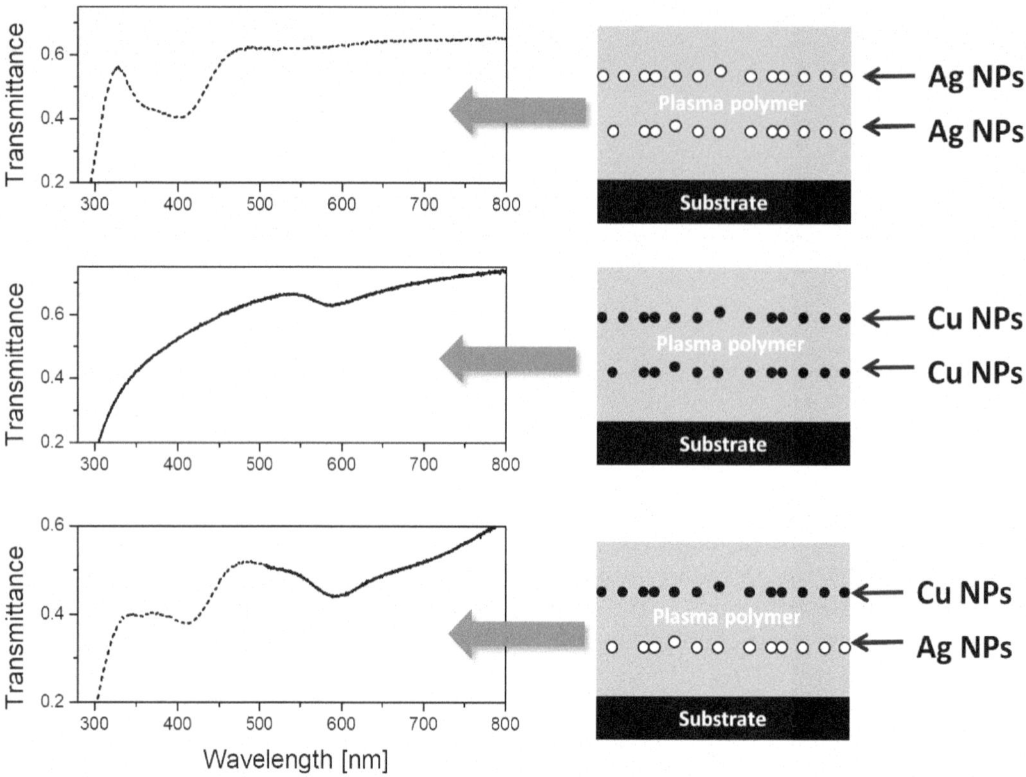

FIGURE 5.11 UV/Vis spectra of Ag (top), Cu (middle), and Ag/Cu (bottom) nanocomposites. Matrix material is C:H plasma polymer, thickness of C:H interlayers is 10 nm.

phenomenon is ascribed to the differences between the cohesive forces of the noble metal atoms, on the one hand, and interaction energies of metals and organics, on the other hand, which causes a phase segregation of metals and polymers. In addition, due to the subsequent coalescence of individual small metal/plasma polymer NPs the resulting features have a multi-core@shell structure, in which several small metal cores are embedded in a plasma polymer matrix. The size of metal cores and thickness of plasma polymer envelope are furthermore found to be adjustable by the concentration of added precursor, whereas the chemical composition of plasma polymer can be varied by selection of a working gas mixture.

The second strategy enabling production of metal/plasma polymer core@shell NPs is in-flight deposition of polymer or plasma polymer shell onto NPs. This may be achieved either by means of polymer evaporation source (see Fig. 5.14A) [170,171] or an auxiliary RF plasma positioned in between the gas aggregation source and the substrate (see Fig. 5.14B) [172,173]. In contrast to the previous case, the growth of NPs and matrix deposition are completely decoupled in this configuration; hence, the properties of cores and shells may be independently varied. As can be seen in Fig. 5.14C, the presence of a thin amorphous plasma polymer shell (typically several nm thick) does not alter the

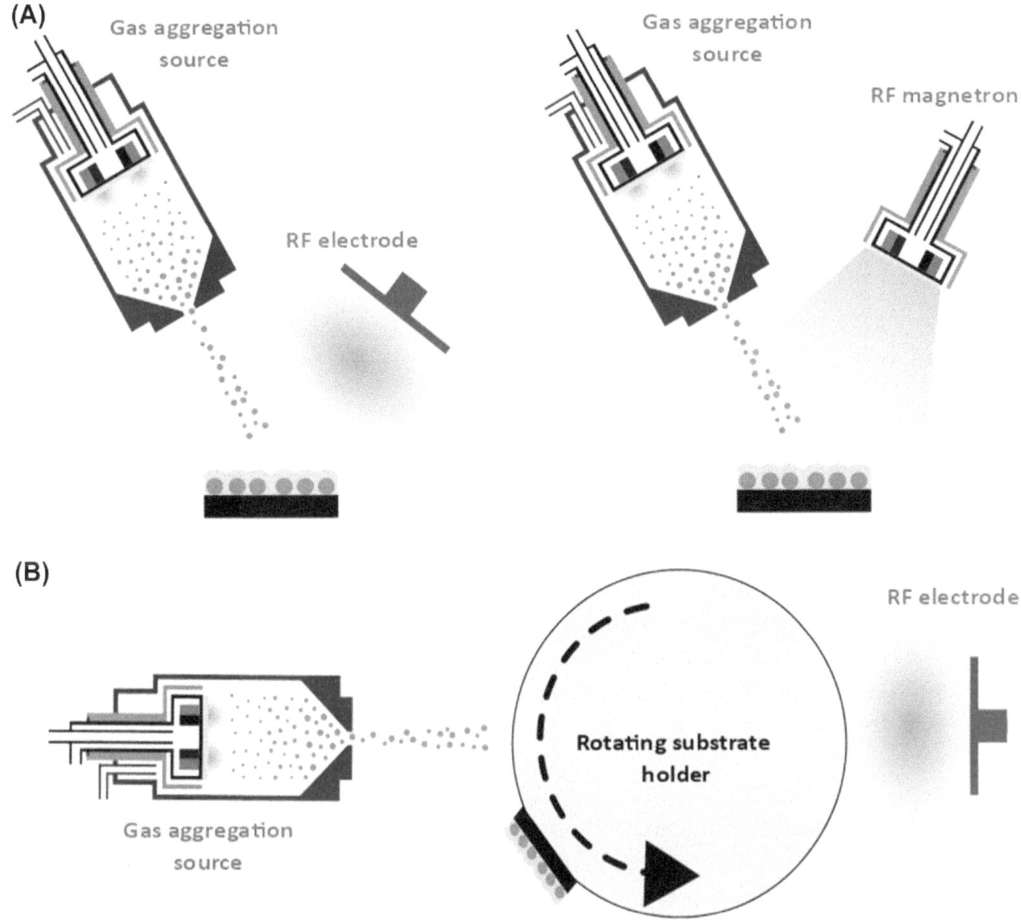

FIGURE 5.12 Fabrication of metal/plasma polymer nanocomposites by (A) simultaneous deposition onto a static substrate and (B) by deposition on a rotating substrate.

crystallinity of the metal core. Such an approach is highly valuable as the shell acts as a protective dielectric layer or may be utilized for subsequent immobilization of biomolecules needed for biomedical and biosensing applications.

5.4 Applications of polymers with nanoparticles

5.4.1 Materials with advanced electrical properties

Covering or filling polymer films with NPs cause significant change of physical and chemical properties of the resulting composites. One of them, which is of practical interest, is evolution of conductance. As shown in Fig. 5.1, there is a gradual decrease in resistance with increase of volume metal fraction followed by a stepwise change for several orders of magnitude corresponding to the percolation transition and formation

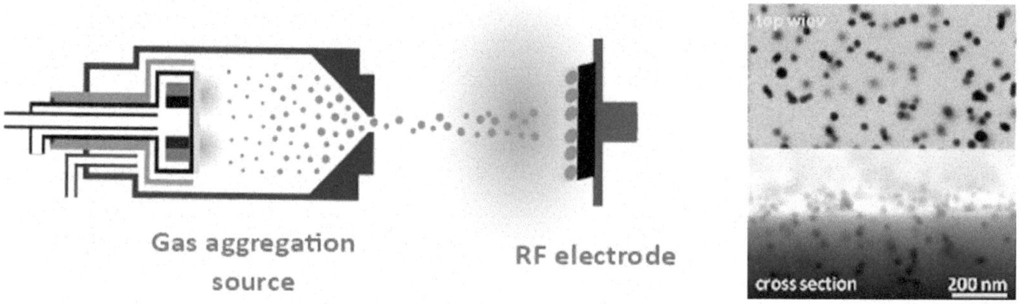

FIGURE 5.13 Schematic drawing of the experimental setup used for the fabrication of hard metal/plasma polymer nanocomposites and example of top and side view TEM images of Ag/a-C:H coatings. *Reprinted with permission from M. Vaidulych, J. Hanuš, T. Steinhartová, O. Kylián, A. Choukourov, J. Beranová, I. Khalakhan, H. Biederman, Deposition of Ag/a-C:H nanocomposite films with Ag surface enrichment, Plasma Process. Polym. 14 (2017) 1600256.*

of paths through metal NPs providing metallic type of conductance [9,21,174,175]. Further adding of metal NPs can lead to IMT, which can be found out by measuring temperature dependence of resistance and calculation of local activation energy [5,176,177]. Thus, by controlling the surface coverage or volume filling factor of metal NPs one can tune the conductance to a desired value, which is a very important capability for the formation of electrical contacts and electronic components. An idea to form elastomeric electrodes is realized, for example, by implanting gold clusters with supersonic velocities into near-surface layers of PMMA and PDMS to the concentrations above the percolation threshold [10]. The resistance of these composite films can be tuned by mechanical stretching. After repeating the load/unload strain cycles up to 50,000 times, the resistance difference between the stretched and normal conditions is gradually decreased that is related to reorganization of NPs in the medium into a percolating network less affected by the strain but the conductivity properties generally do not degrade. This is a big advantage compared to traditional thin metal films evaporated on polymers in which repeating strains lead to cracking and permanent brakeage of metal layer continuity, thus leading to loss of conductivity.

Recently, the supersonic gold cluster beam deposition was used for the fabrication of a stretchable keyboard made of PDMS connected to a touch sensor, which allows operating an LED matrix connected to an electrode pattern printed on paper also by cluster beam technique (see Fig. 5.15) [178]. Certainly, this is not a circuit on microscale, but it demonstrates good sustainability against deformations and proves a principle for functional device fabrication using very cheap materials. Actually, development of paper electronics has recently received an impulse due to possibility of low-cost solutions, high speed of device fabrication, and easy way for recycling after use [179]. Printing of electronic components on paper by cluster beams has been demonstrated for the formation of passive electronic components (resistors and capacitors) giving certain advantages (use of different cluster species, control of surface coverage, good adhesion, etc.) compared to ordinary evaporation or ink jet printing [12,180]. More complex devices

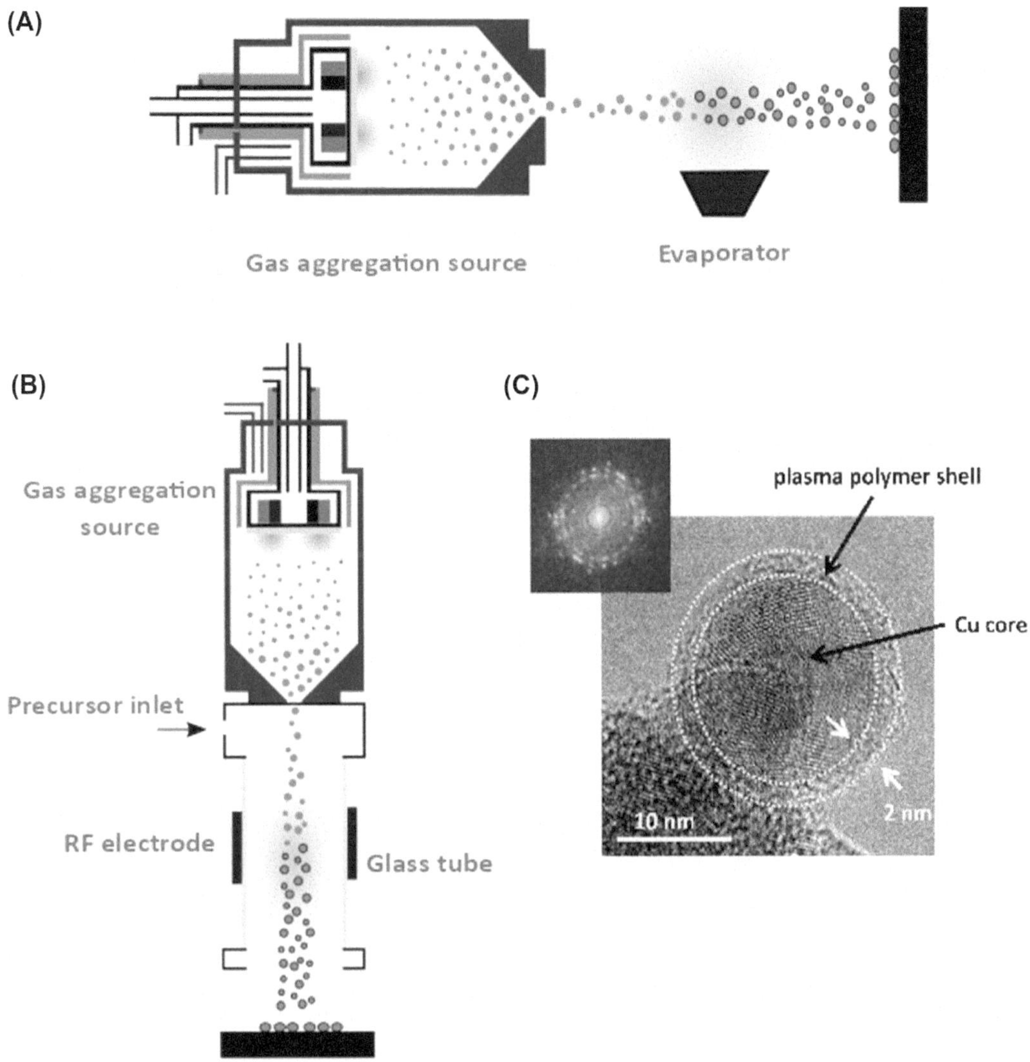

FIGURE 5.14 Schematic representation of system used for in-flight coating of NPs produced by gas aggregation source using (A) evaporator or (B) PE-CVD. (C) shows TEM image of Cu core coated by C:H plasma polymer with inset presenting a diffraction pattern obtained by fast Fourier transform. *Panel (C) is reprinted with permission from O. Kylián, A. Kuzminova, M. Vaydulych, M. Cieslar, I. Khalakhan, J. Hanuš, A. Choukourov, D. Slavínská, H. Biederman, Core@shell Cu/hydrocarbon plasma polymer nanoparticles prepared by gas aggregation cluster source followed by in-flight plasma polymer coating, Plasma Process. Polym. 15 (2018) 1700109.*

(antennas, generators, sensors, etc.) can be built on paper contributing to a new direction of so-called origami and kirigami electronics adding possibility to fold, unfold, and crease for storage of devices in small spaces [181,182].

The metal cluster deposition with controllable coverage on polymer films is also proposed for strain sensing [183]. A strain applied to such composite film changes the

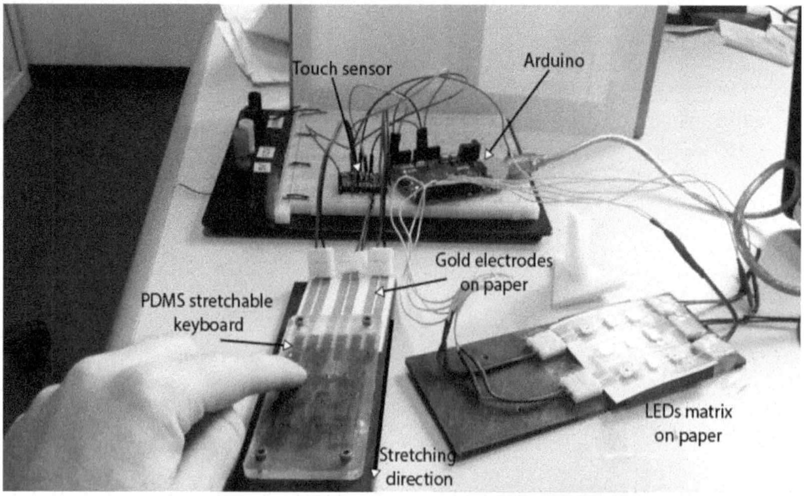

FIGURE 5.15 A circuit comprising LED matrix operated by touching a stretchable keyboard. PDMS keyboard, electrodes on paper, and matrix for mounting LEDs are fabricated by gold cluster beam deposition. *Reprinted from A. Bellacicca, P. Milani, Supersonic cluster beam fabrication of a stretchable keyboard with multilayer electrical interconnects integrated with paper-based flexible devices, ArXive 1706.08586.*

interparticle distance and, thus, the conductance (see Fig. 5.16). These devices are found to be extremely sensitive to very small relative strain ε of up to $\approx 0.3\%$ showing a gauge factor of 20 compared to 2, which is a typical value for metal-foil-based sensors. One

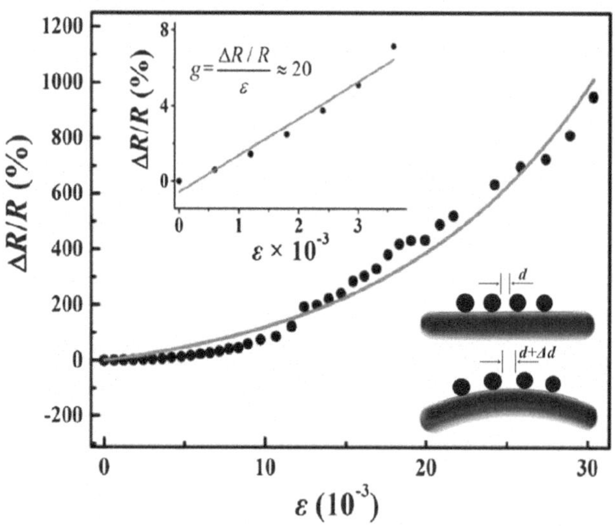

FIGURE 5.16 Relative change of resistance as a function of applied strain. Insert in left top corner shows the dependence for very small ε, insert in right bottom corner presents schematic picture of interparticle distance change affecting the resistance under applied strain. *Reprinted with permission from M. Zheng, W. Li, M. Xu, N. Xu, P. Chen, M. Han, B. Xie, Strain sensors based on chromium nanoparticle arrays, Nanoscale 6 (2014) 3930–3933.*

more advantage of the metal NP/polymer sensors is an easy way to scale down to be feasible for MEMS and NEMS applications.

Another area of cluster beam application to polymers can be formation of actuators. About 100 nm-thick electrodes produced by gold cluster implantation into a cross-linked ion conductive polymer enables the manufacturing of elastic soft actuators [184]. They show sensitivity to the applied electric field starting at 0.1 V and exhibit high-performance operation in response to electrical signals at low voltages as well as good durability in frequency response (up to 76,000 cycles at 2 V and 1 Hz). This type of actuators could be a possible technological solution in the production of components for biomimetic robots, soft sensors, and wearable biomedical devices.

Change of conductance in cluster-assembled films on interaction with volatile species is also proposed for gas sensing. Advantages of the cluster beam method are found to be in controlling particle size, surface coverage, and coalescence. Tuning of these parameters allows efficient detection of gaseous species in small concentration, which is shown by sensing of hydrogen and ammonia using deposited tin oxide and palladium clusters [185–188]. Other important practical parameters, which can be controlled in wider range compared to traditional gas sensors, are temperature and pressure. In Ref. [189], it has been shown that cluster beam technique can be developed toward parallel fabrication of sensor batches using nanostructured transition metal oxide films on microhotplate platforms. Functionality of such batches is tested by a prototype system based on WO_3 NPs for detection of ethanol and NO_2. Similar technology can be developed for other NP-based assembles relevant for gas sensing, namely, TiO_2, Fe_2O_3, MoO_3, ZnO, etc. The above-mentioned sensors are formed on different types of substrates but the approach can be transferred to polymers if needed, adding to the devices such attractive properties as plasticity and flexibility. Another advantage of NP sensors based on polymers is found to be in fast response to volatile compounds as demonstrated by gold NP layers embedded in different types of organic matrices [15].

Percolating systems of conductive particles can be used for fabrication of memristors. Recently, stable networks made of metal clusters with the conductance reflecting the configuration of synapses were realized [17,18]. The conductance can be modulated by electrical stimuli, and the synaptic structure can be reconfigured by voltage applied at different repetition rates, thus enabling the utilization of these atomic-switch networks in short- to long-term memory conversion, mimicking biological neurons. This research provides a promising route in fabrication of devices with neuromorphic properties, and polymer matrixes can be good candidates as a substrate material.

5.4.2 Composites with attractive optical properties

Another branch of applications of polymers with NPs is related to unique optical properties of nanoscale objects. We would like to start with examples utilizing LSPR [23]. NPs of gold, silver, and copper are well-known candidates for plasmonic applications. Depositing or embedding them into polymer films allow formation of functional materials to be used in waveguiding, sensing, and as flexible optical components (filters, reflectors, etc.). Using gas aggregation technique one can choose appropriate metals or alloys, tune NP size, control surface coverage or volume filling factor, as well as form NP patterns or areas with gradient in particle density. This excellent flexibility in control of parameters provides great capabilities for the formation of optical systems with required characteristics [109,110,116]. One example is shown in Fig. 5.17 for the size-selected Ag clusters (13.4 nm in diameter) deposited on PMMA film. Postdeposition thermal annealing causes partial embedding of the NPs leading to small red shift and significant increase of the plasmon band intensity [84]. Similar tendencies are found for copper NPs [143]. These composites can be used as optical transducers for sensing of bioobjects providing an advantage of good adhesion (because of partial embedding) of NPs to the substrate and, thus, appropriate stability in processes involving wet chemistry [32]. Particle sizes can be varied to adopt them to dimensions of biological objects. Plasmon band intensity can be controlled by surface coverage [130]. In terms of detection capabilities, silver NPs produced by gas aggregation demonstrate long-term stability of plasmonic properties in ambient conditions (see Fig. 5.18) [155], which is a big practical advantage in comparison to the silver NPs made by chemical means. It is worth mentioning that often the spectra of size-selected deposited clusters show two distinct plasmon bands (see Figs. 5.17 and 5.18). One of them is assigned to individual particles while the other one (at longer wavelength) to the formation of small agglomerates (for example, dimers, due to stochastic process of deposition) leading to longitudinal excitation mode. However, after the cluster immersion into polymers the second band

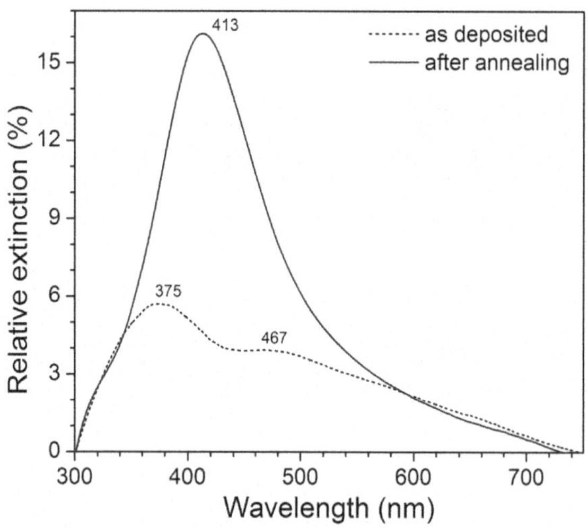

FIGURE 5.17 Normalized extinction spectra of PMMA films with as-deposited size-selected Ag clusters (13.4 nm in diameter) and after thermal annealing at 125°C for 10 min facilitating the cluster embedding into the polymer. Numbers in the panel show wavelengths of the intensity maxima. *Reprinted with permission from M. Hanif, R.R. Juluri, M. Chirumamilla, V.N. Popok, Poly (methyl methacrylate) composites with size-selected silver nanoparticles fabricated using cluster beam technique, J. Polym. Sci. B Polym. Phys. 54 (2016) 1152–1159.*

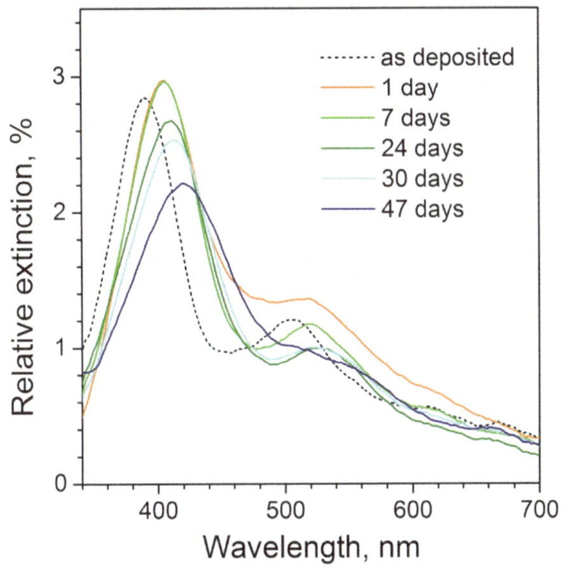

FIGURE 5.18 Evolution of normalized extinction spectra of size-selected Ag clusters (12.5 nm in diameter) on quartz with time under ambient conditions storage showing long-term stability of plasmonic properties.

disappears or becomes very weak (see Fig. 5.17), which probably indicates that the NPs penetrate into the polymer film individually, thus, breaking the agglomerates.

In addition to traditional metal species, such as gold and silver, providing optical responses in the visible interval of spectrum, Al NPs are important for plasmonic applications requiring resonance band in UV spectral range. For example, 60 nm in size Al NPs embedded into C:H film show a pronounced plasmon band with maximum at $\lambda \approx 307$ nm [27]. Varying the filling factor of NPs in the polymer from 1.6% to 9.3% causes significant rise of the resonance intensity (see Fig. 5.19). By changing the size from 60 to 30 nm, the band can be shifted to $\lambda \approx 255$ nm.

For sensing and imaging of organic compounds, SERS is often used. In many applications, specific ordering (arraying) of NPs for local field enhancement is required [34,190,191]. Recently, a combination of electron beam lithography (EBL) with cluster deposition has been suggested as a route to form plasmonic structures of desired configuration [144]. The procedure is schematically shown in Fig. 5.20A. A double-layer structure with polymethylglutarimide (PMGI) at the bottom and PMMA at the top is fabricated by spin coating on a substrate. A pattern is prepared by electron beam writing in PMMA and developed using a solvent. Then, metal clusters are deposited across the entire surface. Partial embedding of NPs into polymers is stimulated by thermal annealing, ensuring that they are not washed out at next step. Thereafter, PMMA liftoff is performed in appropriate solvent leaving on the surface only the clusters in the prepared pattern (embedded into PMGI). As a proof of concept, linear stripes of NPs are obtained (see Fig. 5.20(b)). More complex 2D patterning is possible using conventional EBL approach.

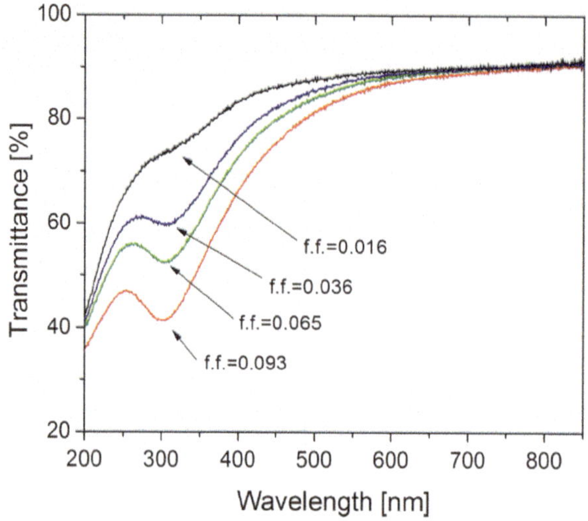

FIGURE 5.19 Transmission spectra of polymer films with different filling factors (f.f.) of Al NPs (mean diameter of 60 nm). *Reprinted with permission from O. Polonskyi, O. Kylián, M. Drábik, J. Kousal, P. Solař, A. Artemenko, J. Čechvala, A. Choukourov, D. Slavínská, H. Biederman, Deposition of Al nanoparticles and their nanocomposites using a gas aggregation cluster source, J. Mater. Sci. 49 (2013) 3352–3360.*

Plasticity of polymer films with metal NPs opens a way to form media with tunable plasmonic properties under applied strain. Such systems are demonstrated by gold cluster implantation into near-surface layer of PDMS [40]. It is shown that the plasmon band position and intensity change if the film is stretched with different degrees of elongation [28]. The stretching leads to the red shift of the band maximum which is explained by compression of the film in the direction of the incoming light (normal to the surface). Such compression increases the filling factor, thus decreasing interparticle distances and enhancing particle-particle interaction. An advantage of the cluster beam technique in fabrication of such plasmonic systems is in precise control of NP filling factor and in achieving stability in tuning of band position after a number of cycles, thus making the polymer nanocomposites to be promising media for the formation of tunable optical filters or components of stretchable optical devices.

Stretchable polymer nanocomposites produced by means of cluster beams can also be used as adaptive optical elements such as lenses, mirrors, gratings, etc. For example, fabrication of stretchable reflective gratings based on elastomer composites obtained by implantation of silver clusters into PDMS is reported in Ref. [192]. An advantage of the films filled by NPs compared to conventional continuous silver layers on polymer surface is in superior stability against thousands of mechanical cycles. Typically, silver layers undergo cracking while in the case of NP-filled films this problem is minor and does not affect optical quality of the grating. The stretching or compression changes the distances between the NPs and, of cause, after thousands of cycles the NP network is rearranged but this does not break the reflectivity of the entire layer.

Formation of compound clusters and their embedding into organic films is shown to be of interest for photovoltaics. Materials with such properties are obtained by magnetron sputtering of copper in presence of reactive hydrogen sulfide (H_2S) yielding $Cu_{2-x}S$ NPs [193]. The copper chemical state of the formed compounds is very important

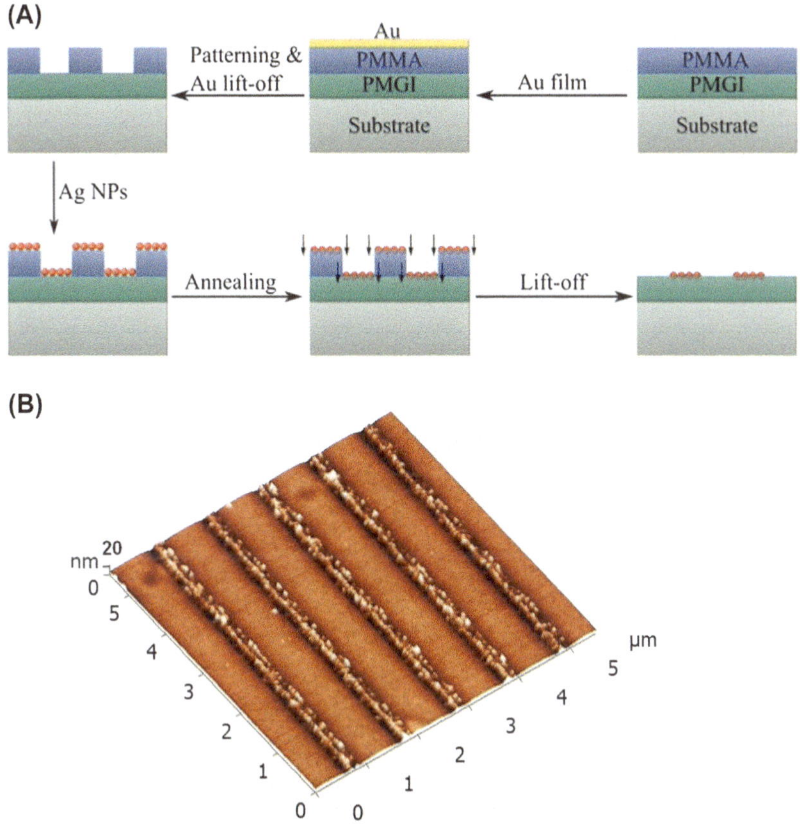

FIGURE 5.20 (A) Schematic illustration of procedure for formation of cluster arrays (see text for details) and (B) AFM image of 200 nm wide stripes by size-selected Cu clusters obtained using this procedure. *Reprinted with permission from F.A. Ceynowa, M. Chirumamilla, V.A. Zenin, V.N. Popok, Arrays of size-selected metal nanoparticles formed by cluster ion beam technique, MRS Adv. 3 (2019) 2771–2776.*

parameter for light absorption. It is found that NPs with Cu(I) are better suited for light harvesting compared to the high-band gap Cu(II) sulfide. The clusters are deposited both simultaneously and in a stepwise manner with pentacene and quaterthiophene as the matrix controlling the thickness of the composite films and filling factor of NPs. These organic films are not observed to impact the NP morphology or chemical structure, thus making the polymer nanocomposites attractive for hybrid photovoltaic devices.

5.4.3 Nanostructured coatings

Next family of materials, for which the cluster beam deposition is believed to play a significant role in the near future, is the nanostructured coatings. The cluster beams in this case act as a source of building blocks (precursors) for the growth of highly disordered nanostructured films composed of NPs that maintain their individuality during the assembling process.

The main characteristics of nanostructured coatings are their root-mean-square roughness w defined as the standard deviation of surface heights, i.e., a parameter that describes how different the surface is from perfect smoothness in vertical direction, and correlation length ξ that describes the statistical independence of two points on surface, being thus a measure of horizontal width of the surface. The evolution of w and ξ describes the kinetic roughening of the surface in both vertical and horizontal directions and is peculiar for different growth modes of thin films [194,195].

Assuming the constant flux of incoming clusters, i.e., constant deposition rate, following scaling laws have to be satisfied

$$w(t) \sim t^{\beta}, \tag{5.4}$$

$$\xi(t) \sim t^{1/z}, \tag{5.5}$$

where t is the deposition time (or more correctly the film thickness), β is the growth exponent, and z is the dynamic exponent. Finally, for the fixed deposition time/thickness and for the lateral dimensions of the system L smaller than the correlation length ξ, the roughness w scales as

$$w(t) \sim L^{\alpha}, \tag{5.6}$$

where α denotes the so-called roughness exponent. Knowledge of above defined scaling exponents, mutually linked by relation $z = \alpha/\beta$ [194], allows identification of the growth mechanism of cluster films. According to extensive work of the P. Milani's group on TiO_2 and ZrO_2 cluster beam deposition in soft-landing regime [196], the growth exponent β is independent of the material and size of used clusters; the size determines only the absolute value of the coating roughness, but it has no impact on the growth mode of the films (see Fig. 5.21). Furthermore, the value of the growth exponent β is found to be very

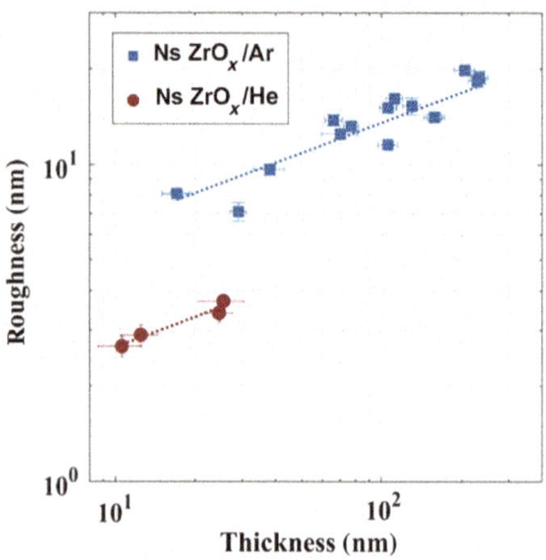

FIGURE 5.21 Scaling of the surface roughness with film thickness for ZrO_2 NPs. NPs prepared in argon (denoted as ZrO_x/Ar) have mean size of 7.3 ± 4.1 nm, NPs prepared in helium (denoted as ZrO_x/He) are 1.9 ± 1.0 nm in size. *Reprinted with permission from F. Borghi, A. Podestà, C. Piazzoni, P. Milani, Growth mechanism of cluster-assembled surfaces: from submonolayer to thin-film regime, Phys. Rev. Appl. 9 (2018) 044016. Copyright (2018) by the American Physical Society.*

close to the one characteristic to ballistic deposition (for which $\beta = 0.33$ [194]), i.e., the growth regime, in which the incoming clusters stick on the growing interface and do not significantly diffuse. Although the growth regime may slightly differ from the pure ballistic one, obtained results are of great importance as they show that by controlling the thickness of deposited nanoparticle films, which is directly proportional to the deposition time, it is possible to precisely control the roughness of resulting materials. Such a simple and highly reproducible procedure, thus, paves the way for the production of nanostructured coatings with well-defined nanoroughness needed in various highly demanding applications. Two examples of situations, in which roughness plays the key role and in which the cluster beam deposition was already tested, are wettability and interaction of surfaces with biomolecules or cells.

In the case of wettability, it is well-known that the roughness enhances the hydrophilic or hydrophobic character of the surface. This may be in most of the cases explained by two wetting models (more detailed description may be found for instance in Refs. [197,198]) that suppose either an enlarged area of contact between the surface and liquid droplet with increasing surface roughness (homogeneous wetting [199]) or presence of air pockets in the protrusions beneath the liquid droplet that limits the solid surface-liquid contact area (heterogeneous wetting regime [200]). Due to the strong impact of the surface topography on wetting, the possibility to control precisely the surface roughness is a prerequisite for the design of materials with desired wetting properties. The cluster beam deposition seems to be a very promising option to meet this requirement. The possibility to use self-assembled nanoparticulate films for regulating surface wettability is already reported by different authors. For instance, the deposition of TiO_2 [201] or $Ge_2Sb_2Te_5$ [202] NP-assembled films allows, in combination with their thermal posttreatment used either for the removal of organic contaminants and recovering of OH groups bonded to Ti atoms or for crystallization of $Ge_2Sb_2Te_5$ NPs, to vary the water contact angles in the range from highly hydrophilic up to 140 degrees. Cu cluster films prepared by means of a gas aggregation source are found to have water contact angles close to 150 degrees [203]. Cu/plasma polymer core@shell NPs coatings can exhibit, depending on the wetting properties of plasma polymer shell, either super-hydrophilic or super-hydrophobic character [172].

Also concerning the possibility to use self-assembled NP films to govern the adsorption of biomolecules or to tailor the growth and cell differentiation, the reported results are highly promising. For instance, it is shown that the adsorption of various proteins on NP films follows the Langmuir isotherm model and that the surface nanoscale morphology strongly influences the amount of incubated biomolecules. Interestingly, more than linear increase of the saturation uptake with increasing number of disposable adsorption sites is observed indicating that the nanometric pores generate the conditions for protein nucleation inside the pores [204]. In other studies, it is demonstrated that the spatial organization of the surface roughness of cluster-assembled films strongly influences also mechanotransductive signaling of cells that is in turn capable to alter the neuronal cells differentiation [205–207], or to assure long-term preservation of human pancreatic β-cells differentiation and function in vitro [208].

So far, mostly the bare cluster-assembled films were considered. However, such films are in many cases not well anchored on the substrate and are prone to damages. In order to overcome this issue but still maintain the possibility to control the roughness of produced materials, simple strategy may be followed that is based on overcoating the NP films by a thin layer of same or another material. This solves not only the problem connected with often observed insufficient fixation of NPs on the substrate material but, as the chemical structure of the overcoat layer may be selected depending on a particular application, it also broadens the flexibility in terms of functional materials that can be produced. Analogous to the bare cluster-assembled films, such approach is already reported to be suitable for production of coatings that govern cell adhesion (e.g., Ti or C:H nanoparticles coated with TiO_X films [129,209], see Fig. 5.22) or surfaces with well-defined wettability. Regarding the latter, probably the first attempt in this direction is reported in Ref. [210] using Ti clusters overcoated either with hydrophilic C:H:N:O plasma polymer prepared by magnetron sputtering of Nylon 6,6 target or with plasma polymerized hydrophobic n-hexane (both 40 nm thick). It is found that the increasing number of Ti NPs in the base layer leads to the enhancement of the surface roughness of the resulting coatings that in turn causes change of their wettability according to the Wenezel model [199]: a decrease of water contact angle from 42 degrees down to

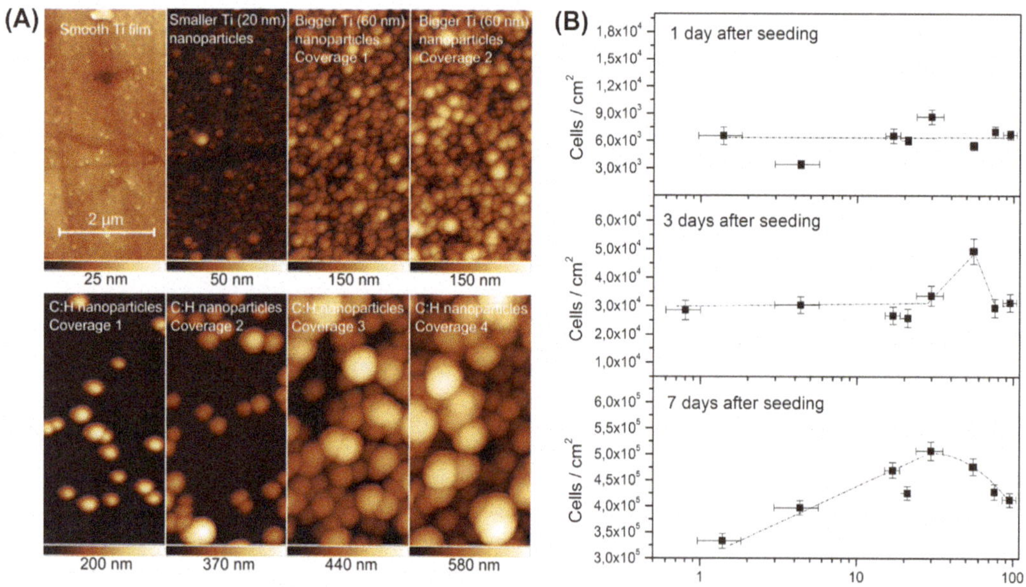

FIGURE 5.22 (A) AFM images of different Ti or C:H NP films overcoated with TiO_X film and (B) dependences of number of MG-63 cells on roughness of such coatings measured on day 1, 3, and 7 after seeding. *Reprinted with permission from P. Solař, O. Kylián, A. Marek, M. Vandrovcová, L. Bačáková, J. Hanuš, J. Vyskočil, D. Slavínská, H. Biederman, Particles induced surface nanoroughness of titanium surface and its influence on adhesion of osteoblast-like MG-63 cells, Appl. Surf. Sci. 324 (2015) 99–105.*

approximately 16 degrees is observed on hydrophilic C:H:N:O coated films while an increase of water contact angle from about 90 degrees up to 130 degrees is found when hydrophobic n-hexane is used to coat Ti NPs (see Fig. 5.23). Since then, numerous studies reveal that by proper selection of the size and amount of used NPs as well as by the thickness and chemical structure of the overcoat layer the full range of wettability from super-hydrophilic (i.e., water contact angle lower than 10 degrees) up to purely super-hydrophobic (i.e., water contact angle higher than 160 degrees) can be obtained [123,211–213].

Furthermore, the possibility to deposit clusters with markedly different sizes makes it even possible to fabricate coatings with multiscale roughness. Such materials may be prepared either using different gas aggregation sources or varying the operational conditions (e.g., power or aggregation length) in a gas aggregation source. Examples of SEM images of such structures, which are produced by combination of small Cu clusters and big C:H particles overcoated with C:H layer [214] or HMDSO-like particles subsequently coated by plasma polymerized HMDSO [215], are presented in Fig. 5.24A and B, respectively. The bimodal character of surface roughness mimics the hierarchical structures observed in nature (e.g., lotus leaf or petal rose [198,216,217]), and it is responsible for different wetting behavior that comprises super-hydrophobic but sticky surfaces (Fig. 5.24C) or robust super-hydrophobic and slippery coatings (Fig. 5.24D).

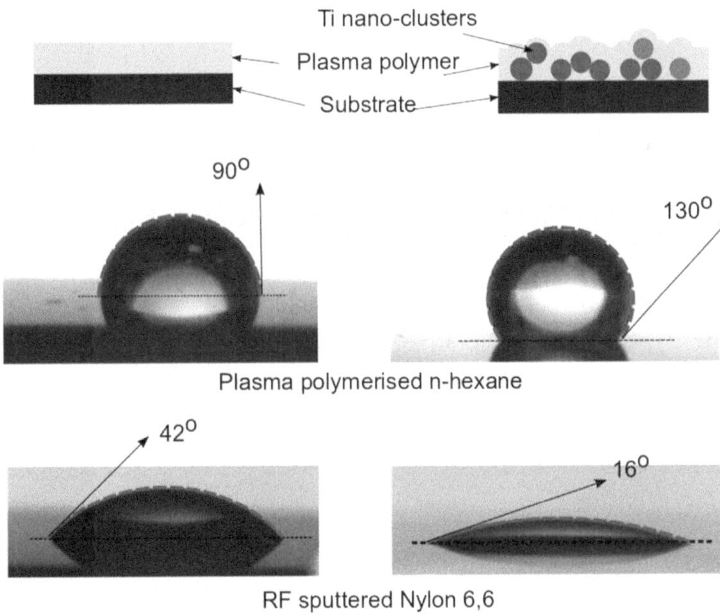

FIGURE 5.23 Water droplet deposited onto smooth plasma polymerized n-hexane and RF sputtered Nylon 6,6 (*left*) and water droplets on Ti NPs overcoated with plasma polymerized n-hexane and magnetron sputtered Nylon 6,6 (*right*). *Reprinted with permission from O. Kylián, O. Polonskyi, J. Kratochvíl, A. Artemenko, A. Choukourov, M. Drábik, P. Solař, D. Slavínská, H. Biederman, Control of wettability of plasma polymers by application of Ti nano-clusters, Plasma Process. Polym. 9 (2012) 180–187.*

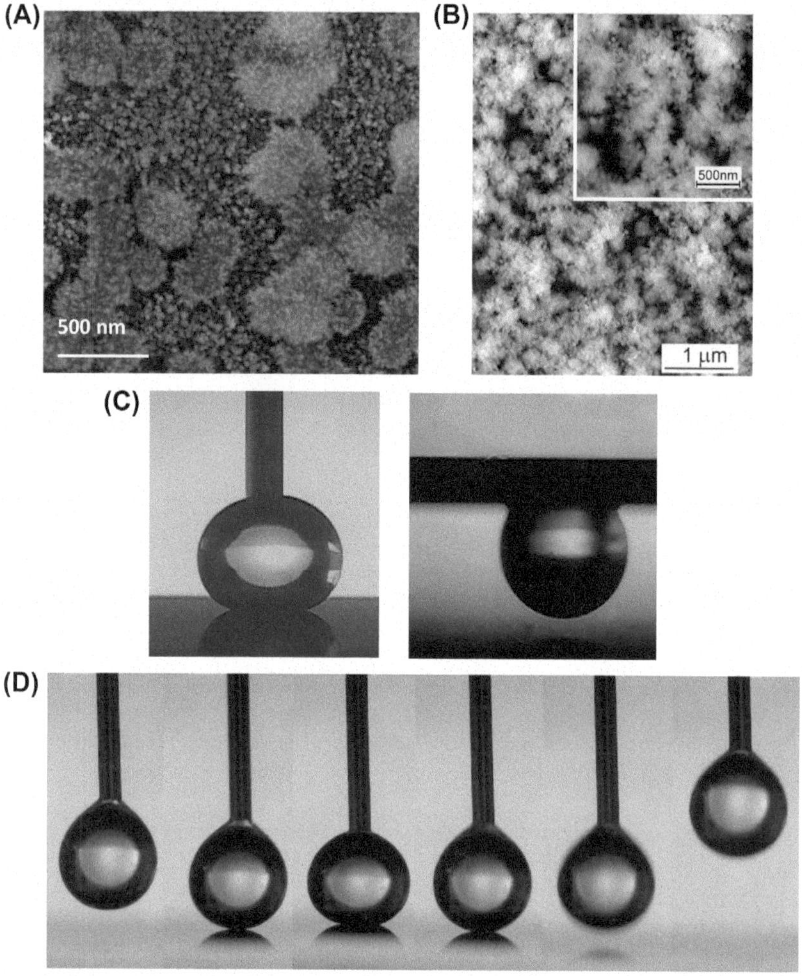

FIGURE 5.24 (A) SEM image of dual-scale roughness film based on small Cu and big C:H particles coated by C:H plasma polymer. (B) SEM image of hierarchical surface prepared by overcoating differently sized HMDSO-like particles by plasma polymerized HMDSO film. (C) Example of water droplet behavior on super-hydrophobic and sticky surface. (D) Sequence of images for water droplet approaching to and retracting from a super-hydrophobic and slippery hierarchical surface. *Images are republished with permission from M. Petr, J. Hanuš, O. Kylián, J. Kratochvíl, P. Solař, D. Slavínská, H. Biederman, Superhydrophobic fluorine-free hierarchical coatings produced by vacuum-based method, Mater. Lett. 167 (2016) 30–33; A. Shelemin, D. Nikitin, A. Choukourov, O. Kylián, J. Kousal, I. Khalakhan, I. Melnichuk, D. Slavínská, H. Biederman, Preparation of biomimetic nano-structured films with multi-scale roughness, J. Phys. D Appl. Phys. 49 (2016) 254001.*

However, it is worth mentioning that in spite of the fact that numerous studies revealed the procedure of a substrate seeding with NPs followed by deposition of an overcoat film to be suitable for applications, the control of surface properties (i.e., roughness, correlation length, and porosity) is not so straightforward as in the case of assembling of bare clusters (ballisticlike mode of the film growth). Depending on the properties of both the NP film and overcoat material, as well as on the flux and direction of incoming species for the overcoat layer formation, different growth regimes may occur that in turn leads to the formation of nanostructured materials with significantly diverse architectures. For instance, in the case of Ti NPs coated with plasma polymerized n-hexane, two consecutive regimes are identified [218]. In the first one, growing plasma polymer fills the voids in between the Ti NPs leading to constant roughness and correlation length. As soon as the filling is completed, the second regime starts, in which further deposition of plasma polymer results in preferentially filling of the valleys between the surface asperities. This causes gradual smoothening of the surface that may be seen from the negative value of the growth exponent as well as from the broadening of the topographical features in the lateral direction that is reflected as increase of the correlation length (see Fig. 5.25). The filling of voids in the early stage of polymer deposition is also observed for the case of Ag NPs coated by C:F plasma polymer

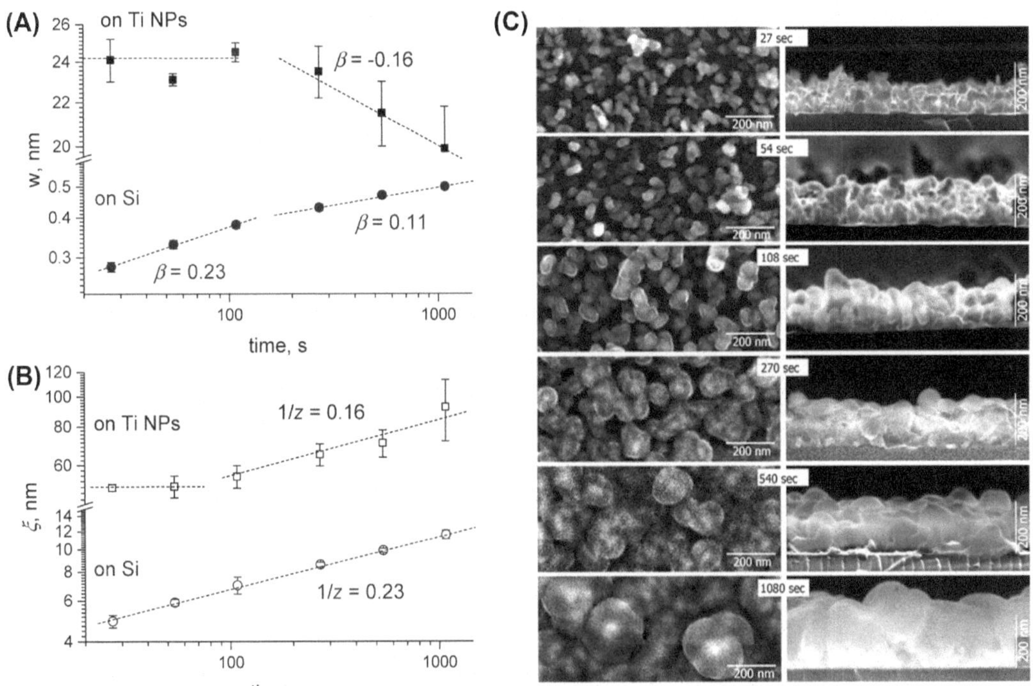

FIGURE 5.25 (A) The interface width, *w*, and (B) the correlation length, ξ, of n-hexane plasma polymers deposited on the Ti NPs (squares) and on flat Si (circles) for different periods of time. (C) SEM top view (*left column*) and cross-section (*right column*) images of Ti NPs overcoated with C:H plasma polymer for different periods of time. *Reprinted with permission from A. Choukourov, I. Melnichuk, A. Shelemin, P. Solař, J. Hanuš, D. Slavínská, H. Biederman, Plasma polymerization on mesoporous surfaces: n-hexane on titanium nanoparticles, J. Phys. Chem. C 119 (2015) 28906–28916.*

produced by RF magnetron sputtering [219]. However, in this case the plasma polymer is even found to penetrate beneath the NPs lifting them up. As the result of Ag NPs migration, no change of surface roughness is observed not only at the initial stage, but also at the later phase, i.e., after the complete filling of interparticle voids. The growth exponent remains to be zero and only lateral growth is observed that is characterized by increasing correlation length.

Significantly different situation may be seen when the substrates are tilted with respect to the flux direction of incoming species and when the diffusion of these species on a growing film is limited. This may lead, especially at higher tilting angles, to the formation of inclined nanocolumnar structures [220,221]. As can be seen in Fig. 5.26, where SEM images of structures formed by silver deposition on the top of C:H plasma polymer nanoparticles at 70 degrees tilt angle are presented, the number of grown nanocolumns is proportional to the number of NPs seeded onto the substrate. Such behavior is connected with the existence of "shadow" regions created by the NPs that are not directly accessible by incoming material. The possibility to grow such nanostructured films from different materials with controllable spacing between individual columns is a highly interesting option for production of matrixes with well-defined porosity that are needed for photovoltaic applications, catalysis, and fuel cells.

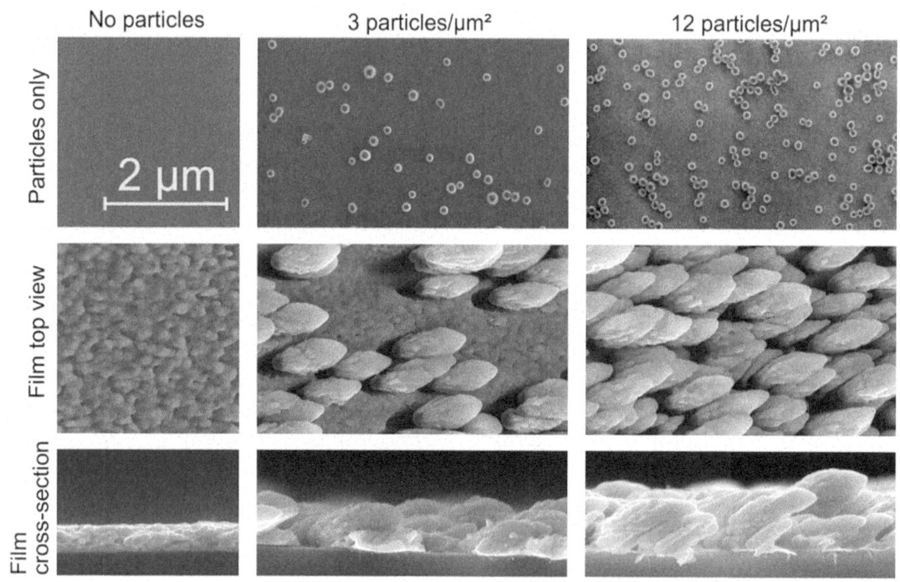

FIGURE 5.26 Silver films deposited under 70 degrees angle over different amounts of C:H plasma polymer particles. Top: as deposited C:H plasma polymer particles, middle: surface of the columnar silver film deposited under 70 degrees angle, bottom: cross-sectional view of the columnar silver film. *Reprinted with permission from P. Solař, O. Kylián, M. Petr, J. Hanuš, A. Choukourov, D. Slavínská, H. Biederman, Glancing angle deposition of silver promoted by pre-deposited nanoparticles, Plasma Process. Polym. 12 (2015) 486–492.*

5.4.4 Antibacterial applications of polymer nanocomposites

Nanomaterials prepared by means of gas aggregation sources have great application potential not only in electronics, plasmonics, sensing, and formation of biocompatible films, but they are also recently considered as promising antibacterial materials. This topic is separately addressed in Chapter 11. Therefore, we mention only a couple of interesting experiments.

It is well-known that certain metal (e.g., Ag, Cu) and metal-oxide (e.g., CuO, ZnO, TiO_2) NPs exhibit high bactericidal effect that is due to their ability to release bactericidal metallic ions, catalytic or photocatalytic characteristics, and their potency to block membrane transport channels and disturb electrochemical gradients [60]. However, due to the undesirable potential side effects that are connected with the release of the NPs into the human body or environment, the NPs have to be fixed on a substrate material. The first possibility is to embed soft-landed NPs into surface layer of polymer by means of thermal heating. This approach is recently applied for Cu and Ti/TiO_x NPs deposited onto thin PS films [159]. It is found that the controllable thermal heating provides partial embedding, i.e., improves the fixation of NPs: such produced samples withstand multiple washing cycles without any visible changes in their morphology. Both Cu and Ti/TiO_x containing samples show antibacterial action. Furthermore, in the case of Ti/TiO_x NP samples, which require UV illumination to trigger the antibacterial action, good temporal stability of antibacterial performance and possibility of multiple uses is reported.

The second strategy to fix NPs on a substrate is based on the production of layered structures, in which layers of metallic NPs are sandwiched in between the layers of matrix material deposited, for instance, by low-pressure plasma-based methods. Matrix material not only fixes the NPs on the surface and avoids their wash-out, but also acts as a barrier between the metal NPs and surrounding environment that governs both the rate of oxidative dissolution of metal NPs and the out-diffusion of formed metal ions, i.e., the principal bactericidal agents in this case, from the nanocomposite coatings [222]. According to the studies performed with Ag and Cu NPs embedded in HMDSO, SiO_2, or C:F matrixes, the antibacterial action of produced nanocomposites may be finely tuned by the amount of NPs, matrix properties (e.g., wettability, swelling, compactness), number of matrix/NPs/matrix layers or thickness of the topmost overcoat layer [61,163,223]. This is highly important as the amount of released metal ions may be easily dosed in the way that assures appropriate bactericidal performance without any undesirable cytotoxic effect.

Finally, the procedure enabling simultaneous deposition of metallic NPs and matrix material is recently established [167,168]. When a hard a-C:H matrix is used, mechanically stable Ag/a-C:H films with antibacterial character are obtained [168]. Furthermore, it is shown that it is possible to significantly enhance the antibacterial action of produced Ag/a-C:H coatings by their plasma etching. This effect is ascribed to partial removal of a carbonaceous layer from the outermost surface of produced nanocomposites that facilitate interaction of Ag NPs with water, as well as simplifies the release of formed Ag^+ ions [168].

5.5 Conclusions

The chapter presents an overview on formation of polymer composites with metal and compound metal nanoparticles using gas aggregation (cluster beam) technique. This approach provides a number of advantages in control of particle parameter and overall properties of composite films compared to other chemical and physical routes of synthesis. In particular, cluster beam method allows for (i) good control of cluster composition and tunability in obtaining different structural and geometric configurations, (ii) size selection, (iii) adjusting the dynamics of cluster-surface interaction, and (iv) formation of patterned structures or coverages with gradients in particle density. The chapter also provides with fundamental aspects on nanoparticle interaction with polymer surfaces showing that the driving force for embedding of particles into the polymer bulk is the large difference in surface tension between metals and organic materials. Practical aspects of nanoparticle embedding into polymers are discussed giving examples on different approaches for the formation of composites. Namely, one can utilize (i) soft-landing of clusters followed by thermal annealing by which the degree of NP immersion into polymer can be controlled, (ii) supersonic beams with sufficient kinetic energy for the direct implantation of clusters into thin near-surface layers, and (iii) co-deposition of NPs and plasma polymers allowing to grow composite materials. The rest of the chapter overviews various applications of polymer matrixes with NPs produced by gas aggregation mostly focusing on the electrical and optical characteristics. Good control in particle parameters and flexibility of the polymer host provide excellent capabilities in tuning the composite properties making them very attractive for use as advanced components for optics and electronics, memristors and transducers for sensors. Additionally, the formation of nanostructured coatings is discussed, where the main parameters under tuning are roughness and correlation length allowing efficient control of wettability and interaction with bio-objects. Examples of two different strategies on how to use metal and metal-oxide NPs on/in polymer films as antibacterial coatings are also presented. Analysis of all presented research results allows to conclude about great potential of cluster beam approach combined with advantages of polymers as host materials on the way toward industrial applications of elastic, low-cost, and low-weight electronic, electromechanical, and optical components, flexible plasmonic media, transducers for sensing, and antibacterial coatings to be used in food packaging or as components for medical instruments.

References

[1] F. Faupel, V. Zaporojtchenko, T. Strunskus, M. Elbahri, Metal-polymer nanocomposites for functional applications, Adv. Eng. Mater. 12 (2010) 1177–1190.

[2] J. Prakash, J.C. Pivin, H.C. Swart, Noble metal nanoparticles embedding into polymeric materials: from fundamentals to applications, Adv. Colloid Interface Sci. 226 (2015) 187–202.

[3] V. Torrisi, F. Ruffino, Metal-polymer nanocomposites: (Co-)Evaporation/(Co)sputtering approaches and electrical properties, Coatings 5 (2015) 378–424.

[4] M.C. Salvadori, F.S. Teixeira, L.G. Sgubin, M. Cattani, I.G. Brown, Surface modification by metal ion implantation forming metallic nanoparticles in an insulating matrix, Appl. Surf. Sci. 310 (2014) 158–163.

[5] V.N. Popok, M.G. Lukashevich, S.M. Lukashevich, R.I. Khaibullin, V.V. Bazarov, Charge carrier transport in polyimide with Co nanoparticles formed by ion implantation, Surf. Sci. 566–568 (2004) 327–331.

[6] M. Niklaus, H.R. Shea, Electrical conductivity and Young's modulus of flexible nanocomposites made by metal-ion implantation of polydimethylsiloxane: the relationship between nanostructure and macroscopic properties, Acta Mater. 59 (2011) 830–840.

[7] C. Tian-Xiang, Y. Shu-Du, H. Wei, F. Tao, L. Lin, Z. Sheng-Qiang, Charge transport and magneto-transport properties of polyimide irradiated by 80 keV Co ions, Chin. Phys. Lett. 26 (2009) 087201.

[8] M. Tie, A.-A. Dhirani, Conductance of molecularly linked gold nanoparticle films across an insulator-to-metal transition: from hopping to strong Coulomb electron-electron interactions and correlations, Phys. Rev. B 91 (2015) 155131.

[9] E. Barborini, G. Corbelli, G. Bertolli, P. Repetto, M. Leccardi, S. Vinati, P. Milani, The influence of nanoscale morphology on the resistivity of cluster-assembled nanostructured metallic thin films, New J. Phys. 12 (2010) 073001.

[10] G. Corbelli, C. Ghisleri, M. Marelli, P. Milani, L. Ravagnan, Highly deformable nanostructured elastomeric electrodes with improving conductivity upon cyclical stretching, Adv. Mater. 23 (2011) 4504–4508.

[11] F. Greco, A. Bellacicca, M. Gemmi, V. Cappello, V. Mattoli, P. Milani, Conducting shrinkable nanocomposite based on Au-nanoparticle implanted plastic sheet: tunable thermally induced surface wrinkling, ACS Appl. Mater. Interfaces 7 (2015) 7060–7065.

[12] F. Caruso, A. Bellacicca, P. Milani, High-throughput shadow mask printing of passive electrical components on paper by supersonic cluster beam deposition, Appl. Phys. Lett. 108 (2016) 163501.

[13] Z. Yi, L.G. Bettini, G. Tomasello, P. Kumar, P. Piseri, I. Valitova, P. Milani, F. Soavi, F. Cicoira, Flexible conducting polymer transistors with supercapacitor function, J. Polym. Sci. B Polym. Phys. 55 (2017) 96–103.

[14] G. Di Girolamo, M. Massaro, E. Piscopiello, L. Tapfer, Metal ion implantation in inert polymers for strain gauge applications, Nucl. Instrum. Methods Phys. Res. B 268 (2010) 2878–2882.

[15] C. Hanisch, N. Ni, A. Kulkarni, V. Zaporojtchenko, T. Strunskus, F. Faupel, Fast electrical response to volatile organic compounds of 2D Au nanoparticle layers embedded into polymers, J. Mater. Sci. 46 (2011) 438–445.

[16] C. Pakula, V. Zaporojtchenko, T. Strunskus, D. Zargarani, R. Herges, F. Faupel, Reversible light-controlled conductance switching of azobenzene-based metal/polymer nanocomposites, Nanotechnology 21 (2010) 465201.

[17] S.K. Bose, J.B. Mallinson, R.M. Gazoni, S.A. Brown, Stable self-assembled atomic-switch networks for neuromorphic applications, IEEE Trans. Electron Devices 64 (2017) 5194–5201.

[18] C. Minnai, A. Bellacicca, S.A. Brown, P. Milani, Facile fabrication of complex networks of memristive devices, Sci. Rep. 7 (2017) 7955.

[19] L. Milliere, K. Makasheva, C. Laurent, B. Despax, G. Teyssedre, Efficient barrier for charge injection in polyethylene by silver nanoparticles/plasma polymer stack, Appl. Phys. Lett. 105 (2014) 122908.

[20] K. Makasheva, C. Villeneuve-Faure, C. Bonafos, C. Laurent, A. Pugliara, B. Despax, L. Boudou, G. Teyssedre, Dielectric engineering of nanostructured layers to control the transport of injected charges in thin dielectrics, IEEE Trans. Nanotechnol. 15 (2016) 839–848.

[21] E. Kay, Synthesis and properties of metal clusters in polymeric matrices, Z. Physik D Atoms, Mol. Clust. 3 (1986) 251–262.

[22] M. Quinten, Optical Properties of Nanoparticle Systems, Wiley-VCH, Weinheim, 2011.

[23] U. Kreibig, M. Vollmer, Optics of nanosized metals, in: R.E. Hummel, P. Wissmann (Eds.), Handbook of Optical Properties, Optics of Small Particles, Interfaces, and Surfaces, vol. 2, CRC Press, 1997, pp. 145–190.

[24] G. Maggioni, A. Vomiero, S. Carturan, C. Scian, G. Mattei, M. Bazzan, C. de Julian Fernandez, P. Mazzoldi, A. Quaranta, G. Della Mea, Structure and optical properties of Au-polyimide nanocomposite films prepared by ion implantation, Appl. Phys. Lett. 85 (2004) 5712–5714.

[25] V.N. Popok, V.I. Nuzhdin, V.F. Valeev, A.L. Stepanov, Copper nanoparticles synthesized in polymers by ion implantation: surface morphology and optical properties of the nanocomposites, J. Mater. Res. 30 (2015) 86–92.

[26] V.N. Popok, M. Hanif, A. Mackova, R. Miksova, Structure and plasmonic properties of thin PMMA layers with ion-synthesized Ag nanoparticles, J. Polym. Sci. B Polym. Phys. 53 (2015) 664–672.

[27] O. Polonskyi, O. Kylián, M. Drábik, J. Kousal, P. Solař, A. Artemenko, J. Čechvala, A. Choukourov, D. Slavínská, H. Biederman, Deposition of Al nanoparticles and their nanocomposites using a gas aggregation cluster source, J. Mater. Sci. 49 (2013) 3352–3360.

[28] K.A. Willets, R.P. Van Duyne, Localised surface plasmon resonance spectroscopy and sensing, Annu. Rev. Phys. Chem. 58 (2007) 267–297.

[29] E. Petryayeva, U.J. Krull, Localized surface plasmon resonance: nanostructures, bioassays and biosensing - a review, Anal. Chim. Acta 706 (2011) 8–24.

[30] S. Shrivastava, N. Jadon, R. Jain, Next-generation polymer nanocomposite-based electrochemical sensors and biosensors: a review, Trends Anal. Chem. 82 (2016) 55–67.

[31] B. Van Dorst, J. Mehta, K. Bekaert, E. Rouah-Martin, W. De Coen, P. Dubruel, R. Blust, J. Robbens, Recent advances in recognition elements of food and environmental biosensors: a review, Biosens. Bioelectron. 26 (2010) 1178–1194.

[32] M. Hanif, R.R. Juluri, P. Fojan, V.N. Popok, Polymer films with size-selected silver nanoparticles as plasmon resonance-based transducers for protein sensing, Biointerface Res. Appl. Chem. 6 (2016) 1564–1568.

[33] S. Fateixa, H.I.S. Nogueira, T. Trindade, Hybrid nanostructures for SERS: materials development and chemical detection, Phys. Chem. Chem. Phys. 17 (2015) 21046–21071.

[34] N.A. Brazhe, A.B. Evlyukhin, E.A. Goodilin, et al., Probing cytochrome c in living mitochondria with surface-enhanced Raman spectroscopy, Sci. Rep. 5 (2015) 13793.

[35] M. Šubr, M. Petr, O. Kylián, J. Kratochvíl, M. Procházka, Large-scale Ag nanoislands stabilized by a magnetron-sputtered polytetrafluoroethylene film as substrates for highly sensitive and reproducible surface-enhanced Raman scattering (SERS), J. Mater. Chem. C 3 (2015) 11478–11485.

[36] J.R. Lakowicz, C.D. Geddes, I. Gryczynski, J. Malicka, Z. Gryczynski, K. Aslan, J. Lukomska, E. Matveeva, J. Zhang, R. Badugu, J. Huang, Advances in surface-enhanced fluorescence, J. Fluoresc. 14 (2004) 425–441.

[37] E. Fort, S. Grésillon, Surface enhanced fluorescence, J. Phys. D Appl. Phys. 41 (2008) 13001.

[38] K. Kneipp, H. Kneipp, H.G. Bohr, Single-molecule SERS spectroscopy, in: K. Kneipp, M. Moskovits, H. Kneipp (Eds.), Surface-Enhanced Raman Scattering – Physics and Applications, Springer-Verlag, Berlin, 2006, pp. 261–278.

[39] D. Radziuk, H. Moehwald, Prospects for plasmonic hot spots in single molecule SERS towards the chemical imaging of live cells, Phys. Chem. Chem. Phys. 17 (2015) 21072–21093.

[40] C. Minnai, P. Milani, Metal-polymer nanocomposite with stable plasmonic tuning under cyclic strain conditions, Appl. Phys. Lett. 107 (2015) 073106.

[41] M.K. Hedayati, M. Javaherirahim, B. Mozooni, R. Abdelaziz, A. Tavassolizadeh, V.S.K. Chakravadhanula, V. Zaporojtchenko, T. Strunkus, F. Faupel, M. Elbahri, An omnidirectional transparent conducting-metal-based plasmonic nanocomposite, Adv. Mater. 23 (2011) 5410–5414.

[42] M. Elbahri, M.K. Hedayati, V.S.K. Chakravadhanula, M. Jamali, T. Strunkus, V. Zaporojtchenko, F. Faupel, Design of a perfect black absorber at visible frequencies using plasmonic metamaterials, Adv. Mater. 23 (2011) 1993–1997.

[43] J.A. Schultz, T.F. Egan, E.K. Lewis, S. Ulrich, K.L. Waters, Nanoparticulate Assisted Nanoscale Molecular Imaging by Mass Spectrometry, US patent 8,614,416 B2, December 24, 2013.

[44] S.N. Jackson, K. Baldwin, L. Muller, V.M. Womack, J.A. Schultz, C. Balaban, A.S. Woods, Imaging of lipids in rat heart by MALDI-MS with silver nanoparticles, Anal. Bioanal. Chem. 406 (2014) 1377–1386.

[45] S. Li, M.M. Lin, M.S. Toprak, D.K. Kim, M. Muhammed, Nanocomposites of polymer and inorganic nanoparticles for optical and magnetic applications, Nano Rev. 1 (2010) 5214.

[46] C. Lu, C. Guan, Y. Liu, Y. Cheng, B. Yang, PbS/Polymer nanocomposite optical materials with high refractive index, Chem. Mater. 17 (2005) 2448–2454.

[47] D. Sun, N. Miyatake, H.-J. Sue, Transparent PMMA/ZnO nanocomposite films based on colloidal ZnO quantum dots, Nanotechnology 18 (2007) 215606.

[48] J. Lee, V.C. Sundar, J.R. Heine, M.G. Bawendi, K.F. Jensen, Full color emission from II–VI semiconductor quantum dot–polymer composites, Adv. Mater. 12 (2000) 1102–1105.

[49] H. Zhang, Z. Cui, Y. Wang, K. Zhang, X. Ji, C. Lu, B. Yang, M. Gao, From water-soluble CdTe nanocrystals to fluorescent nanocrystal–polymer transparent composites using polymerizable surfactants, Adv. Mater. 15 (2003) 777–780.

[50] M.J. Ventura, M. Gu, Engineering spontaneous emission in a quantum-dot-doped polymer nanocomposite with three-dimensional photonic crystals, Adv. Mater. 20 (2008) 1329–1332.

[51] R.I. Khaibullin, B.Z. Rameev, C. Okay, A.L. Stepanov, V.A. Zhikharev, I.B. Khaibullin, L.R. Tagirov, B. Aktas, Ion beam synthesis of magnetic nanoparticles in polymers, in: B. Aktas, L. Tagirov, F. Mikailov (Eds.), Nanostructured Magnetic Materials and Their Applications, NATO Science Series: II Mathematics, Physics and Chemistry, vol. 143, Kluwer, Dordrecht, 2004, pp. 33–54.

[52] V.N. Popok, Polymer films with ion-synthesized cobalt and iron nanoparticles: conductance and magnetism, Rev. Adv. Mater. Sci. 36 (2014) 1–12.

[53] A. Munoz-Bonilla, J. Sanchez-Marcos, P. Herrasti, Magnetic nanoparticles-based conducting polymer nanocomposites, in: V. Kumar, S. Kalia, H.C. Swart (Eds.), Conducting Polymer Hybrids, Springer International, 2017, pp. 45–80.

[54] J. Koo, H. Kim, K.-Y. Kim, Y.R. Jang, J.-S. Lee, S.W. Yoon, B.J. Suh, T. Yu, J. Bang, K. Yoon, G. Yuan, S. K. Satija, Controlling the magnetic properties of polymer-iron oxide nanoparticle composite thin films via spatial particle orientation, RSC Adv. 6 (2016) 55842–55847.

[55] C. Okay, B.Z. Rameev, R.I. Khaibullin, M. Okutan, F. Yildiz, V.N. Popok, B. Aktas, Ferromagnetic resonance study of iron implanted PET foils, Phys. Status Solid 203 (2006) 1525–1532.

[56] J. Kim, B. Van der Bruggen, The use of nanoparticles in polymeric and ceramic membrane structures: review of manufacturing procedures and performance improvement for water treatment, Environ. Pollut. 158 (2010) 2335–2349.

[57] L.Y. Ng, A.W. Mohammad, C.P. Leo, N. Hilal, Polymeric membranes incorporated with metal/metal oxide nanoparticles: a comprehensive review, Desalination 308 (2013) 15–33.

[58] S. Noimark, C.W. Dunnill, M. Wilson, I.P. Parkin, The role of surfaces in catheter-associated infections, Chem. Soc. Rev. 38 (2009) 3435–3448.

[59] C. Silvestre, D. Duraccio, S. Cimmino, Food packaging based on polymer nanomaterials, Prog. Polym. Sci. 36 (2011) 1766–1782.

[60] H. Palza, Antimicrobial polymers with metal nanoparticles, Int. J. Mol. Sci. 16 (2015) 2099–2116.

[61] A. Kuzminova, J. Beranová, O. Polonskyi, A. Shelemin, O. Kylián, A. Choukourov, D. Slavínská, H. Biederman, Antibacterial nanocomposite coatings produced by means of gas aggregation source of silver nanoparticles, Surf. Coat. Technol. 294 (2016) 225–230.

[62] G. Whittell, I. Manners, Metallopolymers: new multifunctional materials, Adv. Mater. 19 (2007) 3439–3468.

[63] L. Tröger, H. Hünnefeld, S. Nunes, M. Oehring, D. Fritsch, Structural characterization of catalytically active metal nanoclusters in poly(amide imide) films with high metal loading, J. Phys. Chem. B 101 (1997) 1279–1291.

[64] S. Li, J. Qin, A. Fornara, M. Toprak, M. Muhammed, D.K. Kim, Synthesis and magnetic properties of bulk transparent PMMA/Fe-oxide nanocomposite, Nanotechnology 20 (2009) 185607.

[65] M. Bhamidipati, L. Fabris, Multiparametric assessment of gold nanoparticle cytotoxicity in cancerous and healthy cells: the role of size, shape, and surface chemistry, Bioconjug. Chem. 28 (2017) 449–460.

[66] V.V. Bazarov, V.Yu. Petukhov, V.A. Zhikharev, I.B. Khaibullin, Conductivity of the granular metal films obtained by high dose ion implantation into PMMA, MRS Symp. Proc. 338 (1995) 417–421.

[67] T. Kobayashi, T. Iwata, Y. Doi, M. Iwaki, Metallic nanoparticles formed in polyimide by ion implantation, Nucl. Instrum. Methods Phys. Res. B 175–177 (2001) 548–553.

[68] V.N. Popok, Ion implantation of polymers: formation of nanoparticulate materials, Rev. Adv. Mater. Sci. 30 (2012) 1–26.

[69] R.I. Khaibullin, V.N. Popok, V.V. Bazarov, E.P. Zheglov, B.Z. Rameev, C. Okay, L.R. Tagirov, B. Aktas, Ion synthesis of iron granular films in polyimide, Nucl. Instrum. Methods Phys. Res. B 191 (2002) 810–814.

[70] K.H. Heining, T. Müller, B. Schmidt, M. Strobel, W. Möller, Interfaces under ion irradiation: growth and taming of nanostructures, Appl. Phys. A 77 (2003) 17–25.

[71] V.N. Popok, I.I. Azarko, V.B. Odzhaev, A. Toth, R.I. Khaibullin, High fluence ion beam modification of polymer surfaces: EPR and XPS study, Nucl. Instrum. Methods Phys. Res. B 178 (2001) 305–310.

[72] A. Kondyurin, M. Bilek, Ion Beam Treatment of Polymers, Elsevier, Amsterdam, 2008.

[73] V.N. Popok, Compositional and structural alterations of polymers under low-to-medium-energy ion implantation, in: Ch.P. Norris (Ed.), Surface Science Research, Nova Science Publishers, New York, 2005, pp. 147–193.

[74] D.V. Sviridov, Chemical aspects of implantation of high-energy ions into polymeric materials, Russ. Chem. Rev. 71 (2002) 315–327.

[75] V. Popok, I. Karpovich, V. Odzhaev, D. Sviridov, Structure evolution of implanted polymers: buried conductive layers formation, Nucl. Instrum. Methods Phys. Res. B 148 (1999) 1106–1110.

[76] V. Hnatowicz, J. Kvitek, V. Švorčik, V. Rybka, Oxidation of polyethylene implanted with as ions to different extents, Eur. Polym. J. 29 (1993) 1255–1258.

[77] V.B. Odzhaev, I.I. Azarko, I.A. Karpovich, I.P. Kozlov, V.N. Popok, D.V. Sviridov, V. Hnatowicz, O.N. Jankovskij, V. Rybka, V. Svorcik, The properties of polyethylene and polyamide implanted with B ions to high doses, Mater. Lett. 23 (1995) 163–166.

[78] P. Moriarty, Nanostructured materials, Rep. Prog. Phys. 64 (2001) 297–381.

[79] V. Zaporojtchenko, T. Strunskus, K. Behnke, C.v. Bechtolsheim, A. Thran, F. Faupel, Formation of metal–polymer interfaces by metal evaporation: influence of deposition parameters and defects, Microelectron. Eng. 50 (2000) 465–471.

[80] J. Röder, H.-U. Krebs, Tuning the microstructure of pulsed laser deposited polymer–metal nanocomposites, Appl. Phys. A 85 (2006) 15–20.

[81] K. Akamatsu, S. Deki, Characterization and optical properties of gold nanoparticles dispersed in nylon 11 thin films, J. Mater. Chem. 7 (1997) 1773.

[82] J. Siegel, P. Jurik, Z. Kolska, V. Svorcik, Annealing of silver nanolayers sputtered on polytetrafluoroethylene, Surf. Interface Anal. 45 (2013) 1063–1066.

[83] F. Ruffino, V. Torrisi, G. Marletta, M.G. Grimaldi, Effects of the embedding kinetics on the surface nano-morphology of nano-grained Au and Ag films on PS and PMMA layers annealed above the glass transition temperature, Appl. Phys. A 107 (2012) 669–683.

[84] M. Hanif, R.R. Juluri, M. Chirumamilla, V.N. Popok, Poly (methyl methacrylate) composites with size-selected silver nanoparticles fabricated using cluster beam technique, J. Polym. Sci. B Polym. Phys. 54 (2016) 1152–1159.

[85] R.D. Deshmukh, R.J. Composto, Direct observation of nanoparticle embedding into the surface of a polymer melt, Langmuir 23 (2007) 13169–13173.

[86] F. Faupel, V. Zaporojtchenko, H. Greve, U. Schürmann, C. Hanisch, V.S.K. Chakravadhanula, A. Kulkarni, A. Gerber, E. Quandt, R. Podschun, Deposition of nanpocomposites by plasmas, Contrib. Plasma Phys. 47 (2007) 537–544.

[87] H. Biederman (Ed.), Plasma Polymer Films, Imperial College Press, Singapore, 2004.

[88] H. Yasuda, Plasma Polymerization, Academic Press, Orlando, 1985.

[89] R. d'Agostino, Plasma Deposition, Treatment, and Etching of Polymers, Academic Press, 1990.

[90] A. Choukourov, I. Gordeev, O. Polonskyi, A. Artemenko, L. Hanyková, I. Krakovský, O. Kylián, D. Slavínská, H. Biederman, Poly(ethylene oxide)-like plasma polymers produced by plasma-assisted vacuum evaporation, Plasma Process. Polym. 7 (2010) 445–458.

[91] A. Choukourov, I. Gordeev, D. Arzhakov, A. Artemenko, J. Kousal, O. Kylián, D. Slavínská, H. Biederman, Does cross-link density of PEO-like plasma polymers influence their resistance to adsorption of fibrinogen? Plasma Process. Polym. 9 (2012) 48–58.

[92] D.T. Morrison, T. Robertson, R.F. sputtering of plastics, Thin Solid Films 15 (1973) 87–101.

[93] H. Biederman, S.M. Ojha, L. Holland, The properties of fluorocarbon films prepared by r.f. sputtering and plasma polymerization in inert and active gas, Thin Solid Films 41 (1977) 329–339.

[94] H. Biederman, RF sputtering of polymers and its potential application, Vacuum 59 (2000) 594–599.

[95] B.V. Tkachuk, N.Y. Marusii, E.P. Laurs, Formation of thin organotin polymeric flims under action of glow-discharge and investigation of their properties, Vysokomol. Soedin. A15 (1973) 2046–2051.

[96] E. Kay, A. Dilks, U. Hetzler, Incorporation of metals into fluoropolymer films synthesized by plasma techniques, J. Macromol. Sci. Part A - Chem. 12 (1978) 1393–1398.

[97] G.L. Harding, S. Craig, Magnetron sputtered metal carbide solar selective absorbing surfaces, J. Vac. Sci. Technol. 16 (1979) 857–862.

[98] H. Biederman, L. Holland, Metal doped fluorocarbon polymer films prepared by plasma polymerization using an RF planar magnetron target, Nucl. Instrum. Methods Phys. Res. 212 (1983) 497–503.

[99] R. Zimmermann, A. Pfuch, K. Horn, J. Weisser, A. Heft, M. Röder, R. Linke, M. Schnabelrauch, A. Schimanski, An approach to create silver containing antibacterial coatings by use of atmospheric pressure plasma chemical vapour deposition (APCVD) and combustion chemical vapour deposition (CCVD) in an economic way, Plasma Process. Polym. 8 (2011) 295–304.

[100] O. Beier, A. Pfuch, K. Horn, J. Weisser, M. Schnabelrauch, A. Schimanski, Low temperature deposition of antibacterially active silicon oxide layers containing silver nanoparticles, prepared by atmospheric pressure plasma chemical vapor deposition, Plasma Process. Polym. 10 (2013) 77–87.

[101] X. Deng, A. Nikiforov, D. Vujosevic, V. Vuksanovic, B. Mugoša, U. Cvelbar, N. De Geyter, R. Morent, C. Leys, Antibacterial activity of nano-silver non-woven fabric prepared by atmospheric pressure plasma deposition, Mater. Lett. 149 (2015) 95–99.

[102] A. Liguori, E. Traldi, E. Toccaceli, R. Laurita, A. Pollicino, M.L. Focarete, V. Colombo, M. Gherardi, Co-deposition of plasma-polymerized polyacrylic acid and silver nanoparticles for the production of nanocomposite coatings using a non-equilibrium atmospheric pressure plasma jet, Plasma Process. Polym. 13 (2016) 623–632.

[103] M.M. Kappes, S. Leutwyler, Molecular beams of clusters, in: G. Scoles (Ed.), Atomic and Molecular Beam Methods, vol. 1, Oxford University Press, New York, 1988, pp. 380–415.

[104] W.A. de Heer, The physics of simple metal clusters: experimental aspects and simple models, Rev. Mod. Phys. 65 (1993) 611–676.

[105] H. Haberland, Experimental methods, in: H. Haberland (Ed.), Clusters of Atoms and Molecules, Springer-Verlag, Berlin, 1994, pp. 205–252.

[106] O.F. Hagena, Cluster ion sources, Rev. Sci. Instrum. 64 (1992) 2374–2379.

[107] P. Milani, S. Iannotta, Cluster Beam Synthesis of Nanostructured Materials, Springer, Berlin, 1999.

[108] A. Shelemin, O. Kylián, J. Hanuš, A. Choukourov, I. Melnichuk, A. Serov, D. Slavínská, H. Biederman, Preparation of metal oxide nanoparticles by gas aggregation cluster source, Vacuum 120 (2015) 162–169.

[109] O. Polonskyi, A.M. Ahadi, T. Peter, K. Fujioka, J.W. Abraham, E. Vasiliauskaite, A. Hinz, T. Strunskus, S. Wolf, M. Bonitz, H. Kersten, F. Faupel, Plasma based formation and deposition of metal and metal oxide nanoparticles using a gas aggregation source, Eur. Phys. J. D 72 (2018) 93.

[110] A. Vahl, J. Strobel, W. Reichstein, O. Polonskyi, T. Strunskus, L. Kienle, F. Faupel, Single target sputter deposition of alloy nanoparticles with adjustable composition via a gas aggregation cluster source, Nanotechnology 28 (2017) 175703.

[111] A.I. Ayesh, Size-selected fabrication of alloy nanoclusters by plasma-gas condensation, J. Alloy. Comp. 745 (2018) 299–305.

[112] Y.-H. Xu, J.-P. Wang, Direct gas-phase synthesis of heterostructured nanoparticles through phase separation and surface segregation, Adv. Mater. 20 (2008) 994–999.

[113] A. Mayoral, D. Llamosa, Y. Huttel, A novel Co@Au structure formed in bimetallic core@shell nanoparticles, Chem. Commun. 51 (2015) 8442–8445.

[114] D. Llamosa, M. Ruano, L. Martínez, A. Mayoral, E. Roman, M. García-Hernández, Y. Huttel, The ultimate step towards a tailored engineering of core-shell and core-shell-shell nanoparticles, Nanoscale 6 (2014) 3–6.

[115] L. Martínez, A. Mayoral, M. Espiñeira, E. Roman, F.J. Palomares, Y. Huttel, Core@shell, Au@TiO_x nanoparticles by gas phase synthesis, Nanoscale 9 (2017) 6463–6470.

[116] P. Grammatikopoulos, S. Steinhauer, J. Vernieres, V. Singh, M. Sowwan, Nanoparticle design by gas-phase synthesis, Adv. Phys. X. 0 (2016) 1–20.

[117] J. Hanuš, M. Vaidulych, O. Kylián, A. Choukourov, J. Kousal, I. Khalakhan, M. Cieslar, P. Solař, H. Biederman, Fabrication of Ni@Ti core–shell nanoparticles by modified gas aggregation source, J. Phys. D Appl. Phys. 50 (2017) 475307.

[118] F. Yin, Z.W. Wang, R.E. Palmer, Controlled formation of mass-selected Cu–Au core–shell cluster beams, J. Am. Chem. Soc. 133 (2011) 10325–10327.

[119] V. Singh, C. Cassidy, P. Grammatikopoulos, F. Djurabekova, K. Nordlund, M. Sowwan, Heterogeneous gas-phase synthesis and molecular dynamics modeling of Janus and core–satellite Si–Ag nanoparticles, J. Phys. Chem. C 118 (2014) 13869–13875.

[120] K. Wegner, P. Piseri, H.V. Tafreshi, P. Milani, Cluster beam deposition: a tool for nanoscale science and technology, J. Phys. D Appl. Phys. 39 (2006) R439–R459.

[121] S. Pratontep, S.J. Carroll, C. Xirouchaki, M. Streun, R.E. Palmer, Size-selected cluster beam source based on radio frequency magnetron plasma sputtering and gas condensation, Rev. Sci. Instrum. 76 (2005) 045103.

[122] V.N. Popok, I. Barke, E.E.B. Campbell, K.-H. Meiwes-Broer, Cluster-surface interaction: from soft landing to implantation, Surf. Sci. Rep. R 66 (2011) 347–377.

[123] M. Petr, O. Kylián, J. Hanuš, A. Kuzminova, M. Vaidulych, I. Khalakhan, A. Choukourov, D. Slavínská, H. Biederman, Surfaces with roughness gradient and invariant surface chemistry produced by means of gas aggregation source and magnetron sputtering, Plasma Process. Polym. 13 (2016) 663–671.

[124] J. Bansmann, S.H. Baker, C. Binns, J.A. Blackman, J.-P. Bucher, J. Dorantes-Dávila, V. Dupuis, L. Favre, D. Kechrakos, A. Kleibert, K.-H. Meiwes-Broer, G.M. Pastor, A. Perez, O. Toulemonde, K.N. Trohidou, J. Tuaillon, Y. Xie, Magnetic and structural properties of isolated and assembled clusters, Surf. Sci. Rep. 56 (2005) 189–275.

[125] J.A. Blackman (Ed.), Metallic Nanoparticles, Elsevier, Amsterdam, 2009.

[126] G.H. Takaoka, H. Ryuto, M. Takeuchi, Surface irradiation and materials processing using polyatomic cluster ion beams, J. Mater. Res. 27 (2012) 806–821.

[127] Y. Huttel (Ed.), Gas-phase Synthesis of Nanoparticles, Wiley-VCH, Weinheim, 2017.

[128] L. Ravagnan, G. Divitini, S. Rebasti, M. Marelli, P. Piseri, P. Milani, Poly(methyl methacrylate)–palladium clusters nanocomposite formation by supersonic cluster beam deposition: a method for microstructured metallization of polymer surfaces, J. Phys. D Appl. Phys. 42 (2009) 082002.

[129] P. Solar, O. Kylian, O. Polonskyi, A. Artemenko, D. Arzhakov, M. Drabik, D. Slavinska, M. Vandrovcova, L. Bacakova, H. Biederman, Nanocomposite coatings of Ti/C:H plasma polymer particles providing a surface with variable nanoroughness, Surf. Coat. Technol. 206 (2012) 4335–4342.

[130] T. Peter, S. Rehders, U. Schürmann, T. Strunskus, V. Zaporojtchenko, F. Faupel, High rate deposition system for metal-cluster/$SiO_xC_yH_z$–polymer nanocomposite thin films, J. Nanopart. Res. 15 (2013) 1710.

[131] V.N. Popok, E.E.B. Campbell, Beams of atomic clusters: effects on impact with solids, Rev. Adv. Mater. Res. 11 (2006) 19–45.

[132] R. Doering, Y. Nishi (Eds.), Handbook of Semiconductor Manufacturing Technology, CRC Press, Boca Raton, 2008.

[133] F. Aqra, A. Ayyad, Surface free energy of alkali and transition metal nanoparticles, Appl. Surf. Sci. 314 (2014) 308–313.

[134] B.J. Keene, Review of data for the surface tension of pure metals, Internat. Mater. Res. 38 (1993) 157–192.

[135] K.K. Nanda, A. Maisels, F.E. Kruis, Surface tension and sintering of free gold nanoparticles, J. Phys. Chem. C 112 (2008) 13488–14491.

[136] G.J. Kovacs, P.S. Vincett, Subsurface particle monolayer and film formation in softenable substrates: techniques and thermodynamic criteria, Thin Solid Films 111 (1984) 65–81.

[137] Z. Cao, M.J. Stevens, A.V. Dobrynin, Adhesion and wetting of nanoparticles on soft surfaces, Macromolecules 47 (2014) 3203–3209.

[138] R. Weber, I. Grotkopp, J. Stettner, M. Tolan, W. Press, Embedding of gold nanoclusters on polystyrene surfaces: influence of the surface modification on the glass transition, Macromolecules 36 (2003) 9100–9106.

[139] J.H. Teichroeb, J.A. Forrest, Direct imaging of nanoparticle embedding to probe viscoelasticity of polymer surface, Phys. Rev. Lett. 91 (2003) 016104.

[140] D. Qi, M. Ilton, J.A. Forrest, Measuring surface and bulk relaxation in glassy polymers, Eur. Phys. J. E 34 (2011) 56.

[141] Z. Fakhraai, J.A. Forrest, Measuring the surface dynamics of glassy polymers, Science 319 (2008) 600–604.

[142] K. Paeng, R. Richert, M.D. Ediger, Molecular mobility in supported thin films of polystyrene, poly(methylmethacrylate), and poly(2-vinyl pyridine) probed by dye reorientation, Soft Matter 8 (2012) 819–826.

[143] V.N. Popok, M. Hanif, F.A. Ceynowa, P. Foyan, Immersion of low-energy deposited metal clusters into poly(methyl methacrylate), Nucl. Instrum. Methods Phys. Res. B 409 (2017) 91–95.

[144] F.A. Ceynowa, M. Chirumamilla, V.A. Zenin, V.N. Popok, Arrays of size-selected metal nanoparticles formed by cluster ion beam technique, MRS Adv. 3 (2019) 2771–2776. https://doi.org/10.1557/adv.2018.427.

[145] R. Cardia, C. Melis, L. Colombo, neutral-cluster implantation in polymers by computer experiments, J. Appl. Phys. 113 (2013) 224307.

[146] C. Ghisleri, F. Borghi, L. Ravagnan, A. Podests, C. Melis, L. Colombo, P. Milani, Patterning of gold-polydimethylsiloxane (Au-PDMS) nanocomposites by supersonic cluster beam implantation, J. Phys. D Appl. Phys. 45 (2014) 015301.

[147] V.N. Popok, S.V. Prasalovich, E.E.B. Campbell, Nanohillock formation by impact of small low-energy clusters with-surfaces, Nucl. Instrum. Methods Phys. Res. B 207 (2003) 145–153.

[148] V.N. Popok, J. Samela, K. Nordlund, E.E.B. Campbell, Stopping of energetic argon cluster ions in graphite: role of cluster momentum and charge, Phys. Rev. B 82 (2010) 201403(R).

[149] V.N. Popok, Energetic cluster ion beams: modification of surfaces and shallow layers, Mater. Sci. Eng. R Rep. 72 (2011) 137–157.

[150] V.I. Shulga, P. Sigmund, Penetration of slow gold clusters through silicon, Nucl. Instrum. Methods Phys. Res. B 47 (1990) 236–242.

[151] Y. Yamamura, T. Muramoto, Time-evolution MC simulation of a few 100eV/atom cluster impacts, Radiat. Eff. Defects Solids 130–131 (1994) 225–233.

[152] T.P. Martin, Shells of atoms, Phys. Rep. 273 (1996) 199–241.

[153] H. Hartmann, V.N. Popok, I. Barke, V. von Oeynhausen, K.-H. Meiwes-Broer, Design and capabilities of an experimental setup based on magnetron sputtering for formation and deposition of size-selected metal clusters on ultra-clean surfaces, Rev. Sci. Instrum. 83 (2012) 073304.

[154] O. Kylian, J. Kratochvil, J. Hanuš, O. Polonskyi, P. Solař, H. Biederman, Fabrication of Cu nanoclusters and their use for production of Cu/plasma polymer nanocomposite thin films, Thin Solid Films 550 (2014) 46–52.

[155] S.M. Novikov, V.N. Popok, A.B. Evlyukhin, M. Hanif, P. Morgen, J. Fiutowski, J. Beermann, H.-G. Rubahn, S.I. Bozhevolnyi, Highly-stable monocrystalline silver clusters for plasmonic applications, Langmuir 33 (2017) 6062–6070.

[156] D. Tainoff, L. Bardotti, F. Tournus, G. Guiraud, O. Boisron, P. Melinon, Self-organization of size-selected bare platinum nanoclusters: toward ultra-dense catalytic systems, J. Phys. Chem. 112 (2008) 6842–6849.

[157] R. Alayan, L. Arnaud, M. Broyer, E. Cottancin, J. Lermé, S. Marhaba, J.L. Vialle, M. Pellarin, Organization of size-selected platinum and indium clusters soft-landed on surfaces, Phys. Rev. 76 (2007) 075424.

[158] I. Barke, H. Hartmann, D. Rupp, L. Fluckiger, M. Sauppe, M. Adolph, S. Schorb, C. Bostedt, R. Treusch, C. Peltz, S. Bartling, T. Fennel, K.-H. Meiwes-Broer, T. Moller, The 3D-architecture of individual free silver nanoparticles captured by X-ray scattering, Nat. Commun. 6 (2015) 6187.

[159] V.N. Popok, C.M. Jeppesen, P. Fojan, A. Kuzminova, J. Hanuš, O. Kylián, Comparative study of antibacterial properties of polystyrene films with TiOx and Cu nanoparticles fabricated using cluster beam technique, Beilstein J. Nanotechnol. 9 (2018) 861–869.

[160] J. Zhao, E. Baibuz, J. Vernieres, P. Grammatikopoulos, V. Jansson, M. Nagel, S. Steinhauer, M. Sowwan, A. Kuronen, K. Nordlund, F. Djurabekova, Formation mechanism of Fe nanocubes by magnetron sputtering inert gas condensation, ACS Nano 10 (2016) 4684–4694.

[161] J. Baj, J.-P. Wang, High-magnetic-moment core-shell-type FeCo-Au/Ag nanoparticles, Appl. Phys. Lett. 87 (2005) 152502.

[162] A. Bapat, C. Anderson, C.R. Perrey, C.B. Carter, S.A. Campbell, U. Kortshagen, Plasma synthesis of single-crystal silicon nanoparticles for novel electronic device applications, Plasma Phys. Control. Fusion 46 (2004) B97–B109.

[163] O. Kylián, J. Kratochvíl, M. Petr, A. Kuzminova, D. Slavínská, H. Biederman, J. Beranová, Ag/C:F Antibacterial and hydrophobic nanocomposite coatings, Funct. Mater. Lett. 10 (2017) 1750029.

[164] M. Maicu, R. Schmittgens, D. Hecker, D. Glöß, P. Frach, G. Gerlach, Synthesis and deposition of metal nanoparticles by gas condensation process, J. Vac. Sci. Technol. A 32 (2014) 02B113.

[165] O. Polonskyi, P. Solař, O. Kylián, M. Drábik, A. Artemenko, J. Kousal, J. Hanuš, J. Pešička, I. Matolínová, E. Kolíbalová, D. Slavínská, H. Biederman, Nanocomposite metal/plasma polymer films prepared by means of gas aggregation cluster source, Thin Solid Films 520 (2012) 4155–4162.

[166] S. Vasquez-Borucki, W. Jacob, C.A. Achete, Amorphous hydrogenated carbon films as barrier for gas permeation through polymer films, Diam. Relat. Mater. 9 (2000) 1971–1978.

[167] J. Hanuš, T. Steinhartová, O. Kylián, J. Kousal, P. Malinský, A. Choukourov, A. Macková, H. Biederman, Deposition of Cu/a-C:H nanocomposite films, Plasma Process. Polym. 13 (2016) 879–887.

[168] M. Vaidulych, J. Hanuš, T. Steinhartová, O. Kylián, A. Choukourov, J. Beranová, I. Khalakhan, H. Biederman, Deposition of Ag/a-C:H nanocomposite films with Ag surface enrichment, Plasma Process. Polym. 14 (2017) 1600256.

[169] P. Solař, O. Polonskyi, A. Olbricht, A. Hinz, A. Shelemin, O. Kylián, A. Choukourov, F. Faupel, H. Biederman, Single-step generation of metal-plasma polymer multicore@shell nanoparticles from the gas phase, Sci. Rep. 7 (2017) 8514.

[170] B. Balasubramanian, K.L. Kraemer, N.A. Reding, R. Skomski, S. Ducharme, D.J. Sellmyer, Synthesis of monodisperse TiO_2–Paraffin Core–Shell nanoparticles for improved dielectric properties, ACS Nano 4 (2010) 1893–1900.

[171] B. Balasubramanian, K.L. Kraemer, S.R. Valloppilly, S. Ducharme, D.J. Sellmyer, Cluster synthesis of monodisperse rutile-TiO_2 nanoparticles and dielectric TiO_2–vinylidene fluoride oligomer nanocomposites, Nanotechnology 22 (2011) 405605.

[172] O. Kylián, A. Kuzminova, M. Vaydulych, M. Cieslar, I. Khalakhan, J. Hanuš, A. Choukourov, D. Slavínská, H. Biederman, Core@shell Cu/hydrocarbon plasma polymer nanoparticles prepared by gas aggregation cluster source followed by in-flight plasma polymer coating, Plasma Process. Polym. 15 (2018) 1700109.

[173] A. Choukourov, P. Pleskunov, D. Nikitin, V. Titov, A. Shelemin, M. Vaidulych, A. Kuzminova, P. Solař, J. Hanuš, J. Kousal, O. Kylián, D. Slavínská, H. Biederman, Advances and challenges in the field of plasma polymer nanoparticles, Beilstein J. Nanotechnol. 8 (2017) 2002–2014.

[174] J. Schmelzer, S.A. Brown, A Wurl, M. Hyslop, R.J. Blaikie, Finite-size effects in the conductivity of cluster assembled nanostructures, Phys. Rev. Lett. 88 (2002) 226802.

[175] F.S. Teixeira, M.C. Salvadori, M. Cattani, I.G. Brown, Gold-implanted shallow conducting layers in polymethylmethacrylate, J. Appl. Phys. 105 (2009) 064313.

[176] M.G. Lukashevich, V.N. Popok, V.S. Volobuev, A.A. Melnikov, R.I. Khaibullin, V.F. Valeev, A. Wieck, V.B. Odzhaev, Magnetoresistive effect in PET films with iron nanoparticles synthesised by ion implantation, Open Appl. Phys. J. 3 (2010) 1–5.

[177] O. Kylián, J. Prokeš, O. Polonskyi, J. Čechvala, J. Kousal, J. Pešička, J. Hanuš, H. Biederman, Deposition and characterization of Pt nanocluster films by means of gas aggregation cluster source, Thin Solid Films 571 (2014) 13–17.

[178] A. Bellacicca, P. Milani, Supersonic cluster beam fabrication of a stretchable keyboard with multilayer electrical interconnects integrated with paper-based flexible devices, ArXive (1706) 08586.

[179] D. Tobjork, R. Osterbacka, Paper electronics, Adv. Mater. 21 (2011) 1935–1961.

[180] L.G. Bettini, A. Bellacicca, P. Piseri, P. Milani, Supersonic cluster beam printing of carbon microsupercapacitors on paper, Flex. Print. Electron. 2 (2017) 025002.

[181] M. Nogi, N. Komoda, K. Otsuka, K. Suganuma, Foldable nanopaper antennas for origami electronics, Nanoscale 5 (2013) 4395–4399.

[182] P.-K. Yang, Z.-H. Lin, K.C. Pradel, L. Lin, X. Li, X. Wen, J.-H. He, Z.L. Wang, Paper-based origami triboelectric nanogenerators and self-powered pressure sensors, ACS Nano 9 (2015) 901–907.

[183] M. Zheng, W. Li, M. Xu, N. Xu, P. Chen, M. Han, B. Xie, Strain sensors based on chromium nanoparticle arrays, Nanoscale 6 (2014) 3930–3933.

[184] Y. Tan, T. Santaniello, L.G. Bettini, C. Minnai, A. Bellacicca, R. Porotti, I. Denti, G. Fraone, M. Merlini, C. Lenardi, P. Milani, Electroactive ionic soft actuators with monolithically integrated gold nanocomposite electrodes, Adv. Mater. 29 (2017) 1606109.

[185] A. Lassesson, M. Schulze, J. van Lith, S.A. Brown, Tin oxide nanocluster hydrogen and ammonia sensors, Nanotechnology 19 (2008) 015502.

[186] T.F. Watson, D. Belic, P.Y. Convers, E.J. Boyd, S.A. Brown, The morphology of tin cluster assembled films and the effect of nitrogen, Eur. Phys. J. D 61 (2011) 81–85.

[187] J. van Lith, A. Lassesson, S.A. Brown, Hydrogen sensors based on percolation and tunneling in films of palladium clusters, Proc. SPIE 6800 (2008) 680009.

[188] B. Xie, L. Liu, X. Peng, Y. Zhang, Q. Xu, M. Zheng, T. Takiya, M. Han, Optimizing hydrogen sensing behavior by controlling the coverage in Pd nanoparticle films, J. Phys. Chem. 115 (2011) 16161–16166.

[189] E. Barborini, S. Vinati, M. Leccardi, P. Repetto, G. Bertolini, O. Rorato, L. Lorenzelli, M. Decarli, V. Guarnieri, C. Ducati, P. Milani, Batch fabrication of metal oxide sensors on micro-hotplates, J. Micromech. Microeng. 18 (2008) 055015.

[190] N. Pazos-Perez, W. Ni, A. Schweikart, R.A. Alvarez-Puebla, A. Fery, L.M. Liz-Marzan, Highly uniform SERS substrates formed by wrinkle-confined drying of gold colloids, Chem. Sci. 1 (2010) 174–178.

[191] T.Y. Jeon, D.J. Kim, S.-G. Park, S.-H. Kim, D.-H. Kim, Nanostructured plasmonic substrates for use as SERS sensors, Nano Converg. 3 (2016) 18.

[192] C. Ghisleri, M. Siano, L. Ravagnan, M.A.C. Potenza, P. Milani, Nanocomposite-based stretchable optics, Laser Photon. 7 (2013) 1020–1026.

[193] M.W. Majeski, I.L. Bolotin, L. Hanley, Cluster beam deposition of $Cu_{2-X}S$ nanoparticles into organic thin films, ACS Appl. Mater. Interfaces 6 (2014) 12901–12908.

[194] F. Family, T. Vicsek (Eds.), Dynamics of Fractal Surfaces, World Scientific, Singapore, 1991.

[195] A.L. Barabasi, H.E. Stanley, Fractal Concepts in Surface Growth, Cambridge University Press, 1995.

[196] F. Borghi, A. Podestà, C. Piazzoni, P. Milani, Growth mechanism of cluster-assembled surfaces: from submonolayer to thin-film regime, Phys. Rev. Appl. 9 (2018) 044016.

[197] D. Quéré, Non-sticking drops, Rep. Prog. Phys. 68 (2005) 2495–2532.

[198] B. Bhushan, Y.C. Jung, Natural and biomimetic artificial surfaces for superhydrophobicity, self-cleaning, low adhesion, and drag reduction, Prog. Mater. Sci. 56 (2011) 1–108.

[199] R.N. Wenzel, Resistance of solid surfaces to wetting by water, Ind. Eng. Chem. 28 (1936) 988–994.

[200] A.B.D. Cassie, S. Baxter, Wettability of porous surfaces, Trans. Faraday Soc. 40 (1944) 546–551.

[201] A. Podestà, G. Bongiorno, P.E. Scopelliti, S. Bovio, P. Milani, C. Semprebon, G. Mistura, Cluster-assembled nanostructured titanium oxide films with tailored wettability, J. Phys. Chem. C 113 (2009) 18264–18269.

[202] G.H. ten Brink, P.J. van het Hof, B. Chen, M. Sedighi, B.J. Kooi, G. Palasantzas, Control surface wettability with nanoparticles from phase-change materials, Appl. Phys. Lett. 109 (2016) 234102.

[203] G.H. ten Brink, N. Foley, D. Zwaan, B.J. Kooi, G. Palasantzas, Roughness controlled super-hydrophobicity on single nanometer length scale with metal nanoparticles, RSC Adv. 5 (2015) 28696–28702.

[204] P.E. Scopelliti, A. Borgonovo, M. Indrieri, L. Giorgetti, G. Bongiorno, R. Carbone, A. Podestà, P. Milani, The effect of surface nanometre-scale morphology on protein adsorption, PLoS One 5 (2010) e11862.

[205] C. Schulte, S. Rodighiero, M.A. Cappelluti, L. Puricelli, E. Maffioli, F. Borghi, A. Negri, E. Sogne, M. Galluzzi, C. Piazzoni, M. Tamplenizza, A. Podestà, G. Tedeschi, C. Lenardi, P. Milani, Conversion of nanoscale topographical information of cluster-assembled zirconia surfaces into mechano-transductive events promotes neuronal differentiation, J. Nanobiotechnol. 14 (2016) 18.

[206] M. Tamplenizza, C. Lenardi, E. Maffioli, S. Nonnis, A. Negri, S. Forti, E. Sogne, S. De Astis, M. Matteoli, C. Schulte, P. Milani, G. Tedeschi, Nitric oxide synthase mediates PC12 differentiation induced by the surface topography of nanostructured TiO_2, J. Nanobiotechnol. 11 (2013) 35.

[207] C. Schulte, M. Ripamonti, E. Maffioli, M.A. Cappelluti, S. Nonnis, L. Puricelli, J. Lamanna, C. Piazzoni, A. Podestà, C. Lenardi, G. Tedeschi, A. Malgaroli, P. Milani, Scale invariant disordered nanotopography promotes hippocampal neuron development and maturation with involvement of mechanotransductive pathways, Front. Cell. Neurosci. 10 (2016) 35.

[208] A. Galli, E. Maffioli, E. Sogne, S. Moretti, E.S. Di Cairano, A. Negri, S. Nonnis, G.D. Norata, F. Bonacina, F. Borghi, A. Podestà, F. Bertuzzi, P. Milani, C. Lenardi, G. Tedeschi, C. Perego, Cluster-assembled zirconia substrates promote long-term differentiation and functioning of human islets of Langerhans, Sci. Rep. 8 (2018) 9979.

[209] P. Solař, O. Kylián, A. Marek, M. Vandrovcová, L. Bačáková, J. Hanuš, J. Vyskočil, D. Slavínská, H. Biederman, Particles induced surface nanoroughness of titanium surface and its influence on adhesion of osteoblast-like MG-63 cells, Appl. Surf. Sci. 324 (2015) 99–105.

[210] O. Kylián, O. Polonskyi, J. Kratochvíl, A. Artemenko, A. Choukourov, M. Drábik, P. Solař, D. Slavínská, H. Biederman, Control of wettability of plasma polymers by application of Ti nano-clusters, Plasma Process. Polym. 9 (2012) 180–187.

[211] O. Kylián, M. Petr, A. Serov, P. Solař, O. Polonskyi, J. Hanuš, A. Choukourov, H. Biederman, Hydrophobic and super-hydrophobic coatings based on nanoparticles overcoated by fluorocarbon plasma polymer, Vacuum 100 (2014) 57–60.

[212] A. Kuzminova, A. Shelemin, O. Kylián, M. Petr, J. Kratochvíl, P. Solař, H. Biederman, From super-hydrophilic to super-hydrophobic surfaces using plasma polymerization combined with gas aggregation source of nanoparticles, Vacuum 110 (2014) 58–61.

[213] O. Kylián, A. Kuzminova, J. Hanuš, D. Slavínská, H. Biederman, Super-hydrophilic SiO_x coatings prepared by plasma enhanced chemical vapor deposition combined with gas aggregation source of nanoparticles, Mater. Lett. 227 (2018) 5–8.

[214] M. Petr, J. Hanuš, O. Kylián, J. Kratochvíl, P. Solař, D. Slavínská, H. Biederman, Superhydrophobic fluorine-free hierarchical coatings produced by vacuum based method, Mater. Lett. 167 (2016) 30–33.

[215] A. Shelemin, D. Nikitin, A. Choukourov, O. Kylián, J. Kousal, I. Khalakhan, I. Melnichuk, D. Slavínská, H. Biederman, Preparation of biomimetic nano-structured films with multi-scale roughness, J. Phys. D Appl. Phys. 49 (2016) 254001.

[216] M. Nosonovsky, B. Bhushan, Lotus versus rose: biomimetic surface effects, in: S. Zhang (Ed.), Material Letter, 2012, pp. 25–40.

[217] B. Bhushan, M. Nosonovsky, The rose petal effect and the modes of superhydrophobicity, Philos. Trans. R. Soc. A 368 (2010) 4713–4728.

[218] A. Choukourov, I. Melnichuk, A. Shelemin, P. Solař, J. Hanuš, D. Slavínská, H. Biederman, Plasma polymerization on mesoporous surfaces: n-hexane on titanium nanoparticles, J. Phys. Chem. C 119 (2015) 28906–28916.

[219] A. Choukourov, O. Kylián, M. Petr, M. Vaidulych, D. Nikitin, J. Hanuš, A. Artemenko, A. Shelemin, I. Gordeev, Z. Kolská, P. Solař, I. Khalakhan, A. Ryabov, J. Májek, D. Slavínská, H. Biederman, RMS roughness-independent tuning of surface wettability by tailoring silver nanoparticles with a fluorocarbon plasma polymer, Nanoscale 9 (2017) 2616–2625.

[220] P. Solař, O. Kylián, M. Petr, J. Hanuš, A. Choukourov, D. Slavínská, H. Biederman, Glancing angle deposition of silver promoted by pre-deposited nanoparticles, Plasma Process. Polym. 12 (2015) 486–492.

[221] O. Kylian, A. Shelemin, P. Solar, A. Choukourov, J. Hanus, M. Vaidulych, A. Kuzminova, H. Biederman, Plasma polymers: from thin films to nanocolumnar coatings, Thin Solid Films 630 (2017) 86–91.

[222] J. Kratochvíl, A. Kuzminova, O. Kylián, State-of-the-art, and perspectives of, silver/plasma polymer antibacterial nanocomposites, Antibiotics 7 (2018) 78.

[223] J. Kratochvíl, J. Štěrba, J. Lieskovská, H. Langhansová, A. Kuzminova, I. Khalakhan, O. Kylián, V. Straňák, Antibacterial effect of Cu/C:F nanocomposites deposited on PEEK substrates, Mater. Lett. 230 (2018) 96–99.

6

Cluster-assembled magnetic nanostructures and materials

Jeffrey E. Shield

DEPARTMENT OF MECHANICAL & MATERIALS ENGINEERING, NEBRASKA CENTER FOR MATERIALS AND NANOSCIENCE, UNIVERSITY OF NEBRASKA-LINCOLN, LINCOLN, NE, UNITED STATES

Chapter outline

6.1 Introduction

Magnetic materials have a wide range of applications, notably in consumer electronics and information storage, and are particularly critical in energy conversion technologies (e.g., motors and generators) [1,2]. Ferromagnetic materials can be broadly categorized as "soft," "hard," or "semi-hard" magnets, depending on coercivity (and, hence, applications). Soft magnetic materials are widely used in, for example, transformer cores and flux-directing applications, and are generally characterized by high permeability and magnetization, and very low coercivity; these combine for very low hysteretic losses. Hard magnetic materials are widely utilized in dc motors and generators, and are generally characterized by large coercivity. Due to the techniques necessary to generate a high coercivity (discussed below), the magnetization is usually lower than in soft magnetic materials. Semi-hard magnetic materials are materials whose polarity can be switched in a controllable manner; these materials are utilized in magnetic storage. In

Cluster Beam Deposition of Functional Nanomaterials and Devices. https://doi.org/10.1016/B978-0-08-102515-4.00006-4

each case, optimization of performance is much more complicated and relies on synergistic intrinsic and extrinsic properties. However, such detailed analysis is beyond the scope of this article.

In designing new magnetic materials, we are basically limited to materials whose magnetic moments spontaneously align in parallel (ferromagnetic or ferrimagnetic ordering), and, for earth-born applications, have a relatively high Curie temperature. Thus, our base elements from which to construct new materials are limited to Fe, Co, Ni, and sometimes Mn. Fe-based materials are favored because of the high magnetization and natural abundance. While pure Fe is relatively soft magnetically, much engineering is done to improve it for soft magnetic applications. Examples include the Fe-Si alloys commonly used for transformers, and nanocrystalline or amorphous materials useful for high-frequency applications, in particular. However, it is a significant challenge to create magnetic hardness necessary for permanent magnet applications. This paper will focus on developing hard (permanent) magnetic materials using cluster processing, creating new materials and novel nanostructures via inert gas condensation. There is other work in the literature related to fundamental magnetism of magnetic clusters [3].

6.2 Permanent magnetic materials

Throughout the past 100 years or so, advancements in permanent magnet materials have fueled many technological advancements, including information technology, miniaturization of machines, and electric vehicles. The most descriptive figure of merit for permanent magnets is the energy product, or density, and is essentially the work one can get out of a permanent magnet. The improvement in energy density has generally followed a trend of doubling every 14 years or so—the permanent magnet world's version of Moore's Law (Fig. 6.1). Currently, Nd-Fe-B-based permanent magnets possess the

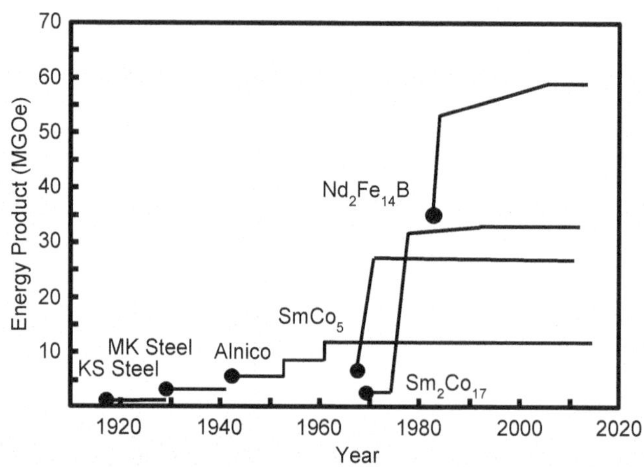

FIGURE 6.1 Chronological development of permanent magnets as measured by their energy density/product.

highest energy products of ~55 MGOe, approximately 86% of its theoretical maximum, but the trend of increasing the energy product has stalled. The development of a large energy product requires both high remanence (high magnetization) and high coercivity. High remanence/magnetization arises from a high percentage of large-magnetic-moment element in the alloy/compound (e.g., Fe), while high coercivity arises from a high anisotropy—essentially the energy required to rotate the magnetic moment from its "easy" or preferred direction. For permanent magnet materials, anisotropy is generally created in one of two ways. First, it can be generated through the shape of the magnetic material, or component. For example, a needle (made of Fe, for instance) magnetizes much more easily along the long axis than the short axis; this is driven by a reduction in magnetostatic energy. The alnico class of permanent magnets relies on shape anisotropy to generate coercivity, as the magnetic phase in the material forms as long needles.

Anisotropy is also generated from the crystal via magnetocrystalline anisotropy. Even in highly symmetric crystal structures such as cubic, the magnetization prefers to align in specific directions (although in cubic materials the differences between, for example, <001> or <111> are relatively small). However, asymmetric structures, notably hexagonal or tetragonal crystal structures, can have significant anisotropy. The best permanent magnet materials form in either hexagonal (e.g., $SmCo_5$) or tetragonal (e.g., $Nd_2Fe_{14}B$) crystal structures with the easy magnetization direction along the *c* axis; the high anisotropy in these compounds arise from strong spin-orbit coupling. One measure of anisotropy is the anisotropy constant "K." The maximum K due to shape anisotropy is about 10^4 J/m^3, and is proportional to the magnetization of the material and, of course, the shape. For comparison, K_1 (the first anisotropy constant) for magnetocrystalline anisotropy ranges from 10^3 J/m^3 to 10^7 J/m^3; for hcp Co, it is $5.3(10)^5$ J/m^3, while for $Nd_2Fe_{14}B$ it is $5(10)^6$ J/m^3 [4]. Generally, in uniaxial materials—materials with an easy magnetization direction along the *c* axis—the coercivity is proportional to K. A much more detailed treatise on magnetism can be found in other resources [4,5].

A few important length scales exist in magnetism, and interesting phenomena can occur when the scale of the structure approaches these length scales. For example, the *domain wall width* δ is related to intrinsic magnetic parameters A (the exchange stiffness) and the aforementioned anisotropy constant K_1 by

$$\delta = \pi(A/K_1)^{1/2} \tag{6.1}$$

This width, which separates regions of the material with different magnetization directions, ranges from 1 nm for highly anisotropic materials to 100 nm. The *exchange length* defines the transition from atomic exchange interactions to the regime where magnetostatic interactions dominate. This length is given by

$$l_{ex} = \left(A/\mu_0 M_s^2\right)^{1/2} \tag{6.2}$$

where $\mu_0 M_s$ is the saturation magnetization. This length scale is in the order of 1–2 nm for a broad range of materials [4]. The *single domain limit* is another important scale. This defines the grain/particle size below which it is not energetically favorable to form multiple domains. That is, the gain in reducing magnetostatic energy is not offset by the energy associated with creating a domain wall. The single domain limit is given by Ref. [4].

$$R_{SD} = 36(AK_1)^{1/2}/\mu_o M_s^2 \tag{6.3}$$

R_{SD} is strongly related to the intrinsic properties of the material, and ranges from the nanometer scale for softer materials to the micron scale for very hard materials.

Cluster-based processing allows these scales to be reached, providing a means to explore and design new materials. For example, exploring interactions across boundaries or between different magnetic phases when the scales of the phases approach or are below these fundamental length scales is possible, ultimately leading to the creation of new materials. One such example of this is the exchange-spring effect, which has created a new field of permanent magnetism—nanocomposite permanent magnets. Exchange-spring effects in ferromagnetic materials have the potential to improve properties of high-energy permanent magnets and revolutionize products in computer, automotive, oil, health services, and power generation (i.e., motors) industries [1]. This is true for both anisotropic magnets, where grain alignment produces the highest remanent magnetizations and thus energy densities, and isotropic magnets which offer ease of processing and energy products still superior to other classes of magnetic materials. Improving the magnetic properties of isotropic materials is also critical for further development of bonded magnets, the fastest growing segment of the permanent magnet market.

In exchange-spring magnets, the magnetic spins on either side of an interface are influenced by the magnetic spins across the interface (Fig. 6.2), and the exchange-coupled region is stabilized against demagnetization [6]. When one of the phases is a hard magnetic material, the spins in the adjoining phase across the interface can be "locked" in alignment with the first phase through these intergranular exchange interactions [7]. This is especially effective if the locked phase is a soft magnetic material, with a higher magnetization. This interaction enhances remanence and produces high energy products (Fig. 6.2C). Theoretical predictions suggest that isotropic nanocomposite permanent magnets may have an energy product comparable to the best current anisotropic materials [8], while the energy product of anisotropic permanent magnets may exceed 100 MGOe [7]. Early work considered the "ideal" scale of the structure to be twice the domain wall width of the hard magnetic phase, or approximately 2δ, to ensure complete exchange coupling [6]. Others have considered the anisotropy of the soft magnetic phase as well [9–11], leading to a slight relaxation in the required scale of the nanostructure. In either case, for exchange-spring nanocomposites based in the $RE_2Fe_{14}B/\alpha$-Fe system the critical soft-phase dimension for complete coupling is less than 10 nm. Larger grain sizes lead to a decrease in magnetic properties, as shown in both modeling and experimental studies [12–14]. The fraction of soft magnetic phase also strongly influences the magnetic behavior. While a higher fraction of soft magnetic phase increases the magnetization and, consequently, leads to a higher energy product, a concomitant decrease in coercivity occurs. The drastic loss of

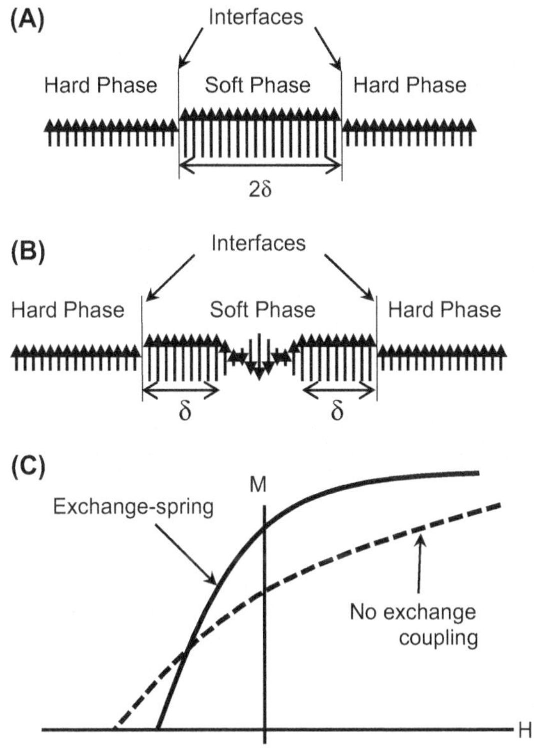

FIGURE 6.2 (A) Schematic representation of the magnetic moment distribution in a two-phase hard/soft system displaying exchange-spring behavior. The exchange-coupled region 2δ in the soft magnetic region resists reversal even in an applied field (pointing down in this case). (B) If the soft phase region is larger than 2δ, a portion is uncoupled and can reverse in a reverse applied field, ultimately leading to reversal of the entire system. (C) The effect of exchange-spring behavior on the M-H curve of an isotropic material. The remanence is enhanced, with a decrease in coercivity.

coercivity in experimental systems is the primary factor leading to the discrepancy between predicted and experimentally observed maximum energy products. The loss of coercivity limits the amount of soft magnetic phase that the nanocomposite structure can contain, thereby limiting the remanence. With a better understanding of factors that influence the magnetic reversal, and development of strongly exchange-coupled nanostructures, improved magnetic properties would be achieved.

Exchange-spring behavior was first observed in single-phase $Nd_2Fe_{14}B$ magnets with a nanoscale, isotropic grain structure produced by melt spinning [15]. Subsequently, two-phase systems that combine magnetically hard and soft phases were shown to have tremendous exchange-spring properties and outstanding potential as permanent magnet materials [16]. In two-phase magnets, the low anisotropy of the soft phase—commonly Fe—makes it much easier for its spins to be aligned with the hard magnetic phase. Combined with its higher magnetization values, the remanence values are greatly improved for two-phase magnets, leading to energy products that exceed the energy product of similar single-phase isotropic magnets with isolated grains by approximately 30% [17]. However, while promising, the experimentally observed energy products are much lower than theory predicts [7,12,13]. Challenges include creating a uniformly fine-scaled structure and maximizing soft magnetic phase fraction without creating uncoupled regions. Finally, in strongly coupled systems, magnetic reversal is easier. In strongly exchange-coupled material where the soft phase is completely coupled with the hard magnetic phase, reversal of the hard and soft phase occurs simultaneously. This has proven to be the case in strongly coupled hard/soft multilayers [18,19]. As the coupling strength decreases, reversal initiates in the soft magnetic layers [20–22]. In granular systems, experimental studies also indicate that reversal initiates in the soft magnetic phase [23]. Once reversal begins, the strong intergranular exchange coupling results in cooperative reversal (similar to domain expansion) throughout the material [24–26]. A model microstructure has been proposed that incorporates two length scales, one promoting exchange coupling between the soft and hard magnetic phases, and a longer one that eliminates cooperative reversal [27].

6.3 Cluster-assembled nanostructures and nanoparticles

Formation of metallic nanoparticles by inert gas condensation (IGC), in which metal atoms are either evaporated or sputtered into a cooled inert gas flow at relatively high pressure, has been well established in the last the two decades [28–30]. By repeated collisions with the carrier gas, the supersaturated metal vapor condenses and forms nanoparticles. This nanoparticle deposition technique can produce a very large range of mean nanoparticle sizes from 3 to 12 nm with deposition rates as high as 5 Å/s, and tremendous control over the nanoparticle size and size distribution. The nanoparticle size and size distribution are closely tied to the experimental conditions utilized, and a monodispersed size distribution is possible [28,31–33].

The University of Nebraska-Lincoln has constructed a cluster-deposition system with a sputtering-based gas-aggregation source, which is used for the preparation of magnetic nanoclusters and cluster-assembled nanocomposites [31,34,35]. Fig. 6.3 shows a 3D rendering of the cluster-deposition system. Briefly, the clusters are generated in the gas-aggregation tube; they go through two apertures and then are deposited onto the substrate. Because the atomic gas source in this system is DC magnetron sputtering, we are able to use any metallic alloy as a target, and thus can vary the composition of our clusters over a wide range. Targets can be either alloy targets or composite, the latter constructed by adding pellets of the other components along the "race track." Composite targets were used in work discussed here; they are more cost-effective and provide a broader range of compositions. Compositions are achieved by considering the sputtering yields of the elements, and then the correct area fraction of "plugs" are inserted around the race track. Fig. 6.4 shows an example of clusters produced in this system. The size distribution can be quantified by the ratio σ/d, where σ is the standard deviation and d the average cluster size. Ratios in the order of 0.1 are routinely obtained, and such low ratios are considered monodispersed. The system

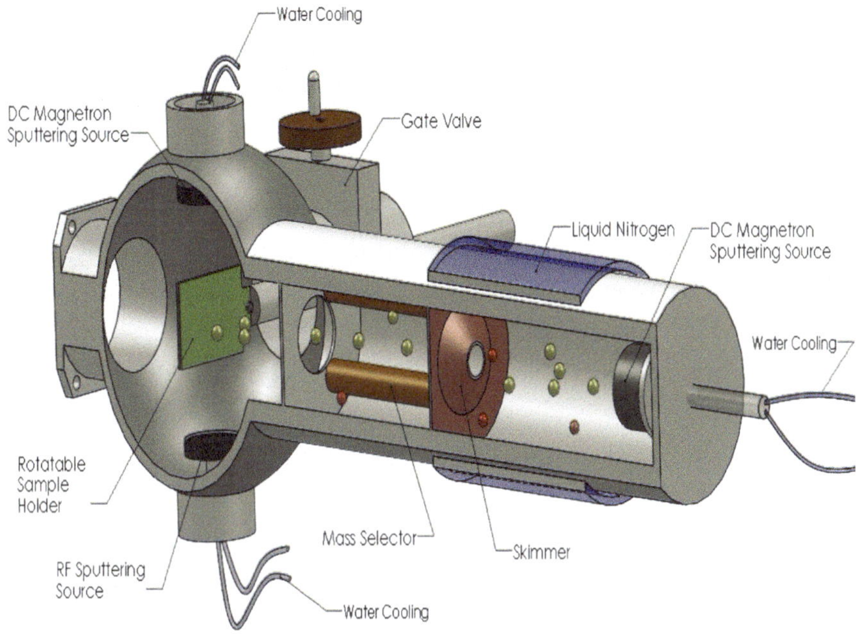

FIGURE 6.3 3D rendering of the cluster deposition chamber.

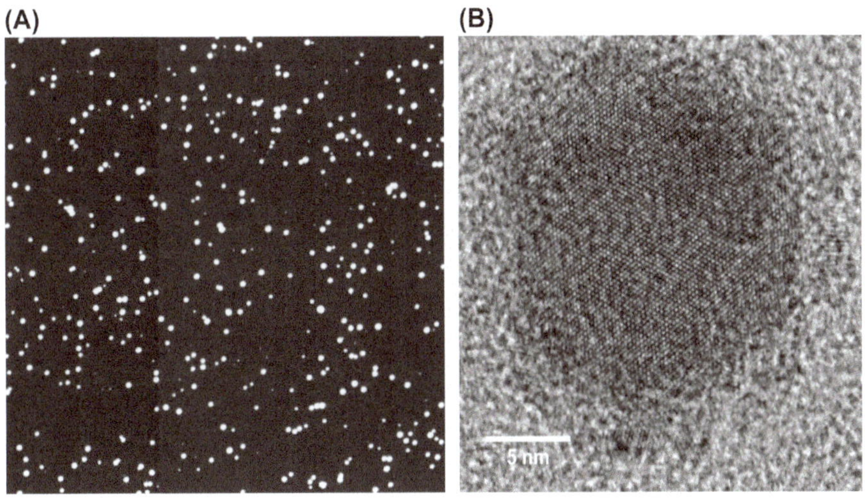

FIGURE 6.4 Representative clusters. (A) HAADF STEM image, and (B) high-resolution TEM image of a single cluster.

also contains a second conventional RF sputtering gun. By alternate deposition from the cluster source and any of the other sources, a nanocomposite consisting of clusters embedded in a matrix can be readily fabricated, or a protective covering can be applied to prevent oxidation for scanning/transmission electron microscopy investigation (Fig. 6.5).

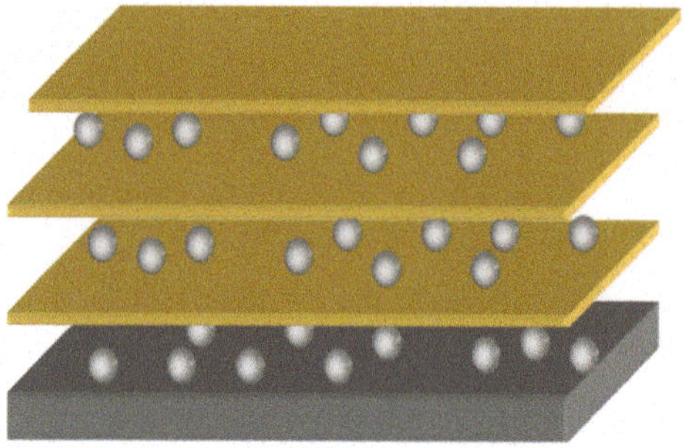

FIGURE 6.5 Alternate deposition of clusters (gray spheres) and thin films.

The combination of a cluster source and thin film source in the same chamber provides the opportunity to create novel nanostructures involving multiple magnetic phases and to investigate isolated clusters which may have unique phase distributions or atomic structures. Fig. 6.6 shows the variety of nanostructures and in-cluster structures possible. For the two-magnetic-phase structures, it is straightforward to control volume fraction and scale, especially with respect to ensuring full exchange coupling. Further, in-cluster structuring with multiple phases internal to a cluster allows exploration of phenomena at unprecedented length scales in 3D configurations (as opposed to thin film structures). Also, the far-from-equilibrium nature of the processing provides pathways to nonequilibrium structures such as dramatic extensions of solid solubility. Finally, isolated nanoclusters—insulated from one another by, for example, (mostly) inert SiO_2—allows thermal treatments where additional metastable states may be reached.

6.4 Nanocomposite permanent magnets: processing

The two-phase nanostructures required for bulk (as opposed to thin film) nanocomposite permanent magnets have been fabricated by several different methods. In the RE-Fe-B system (where RE is a rare earth), rapid solidification by melt spinning has effectively produced requisite nanostructures directly from the melt [16,17]. Melt spinning can also suppress crystallization completely, resulting in amorphous structures. A nanocomposite structure can then be produced by devitrification of the amorphous precursor [17]. However, in both instances control of the scale has proved difficult, although alloying additions have been used to reduce the scale [17]. A third technique commonly used to produce granular, nanostructured RE-Fe-B-based nanocomposites is mechanical milling [40]. However, in all three instances the resulting nanostructure is granular in nature, with a very limited soft magnetic phase fraction. More recently, cluster-assembly techniques have been used to produce nanocomposite Fe-Pt permanent magnets. Combining chemically synthesized FePt and Fe_3O_4, followed by reduction of the oxide and heat treatment resulted in a hard/

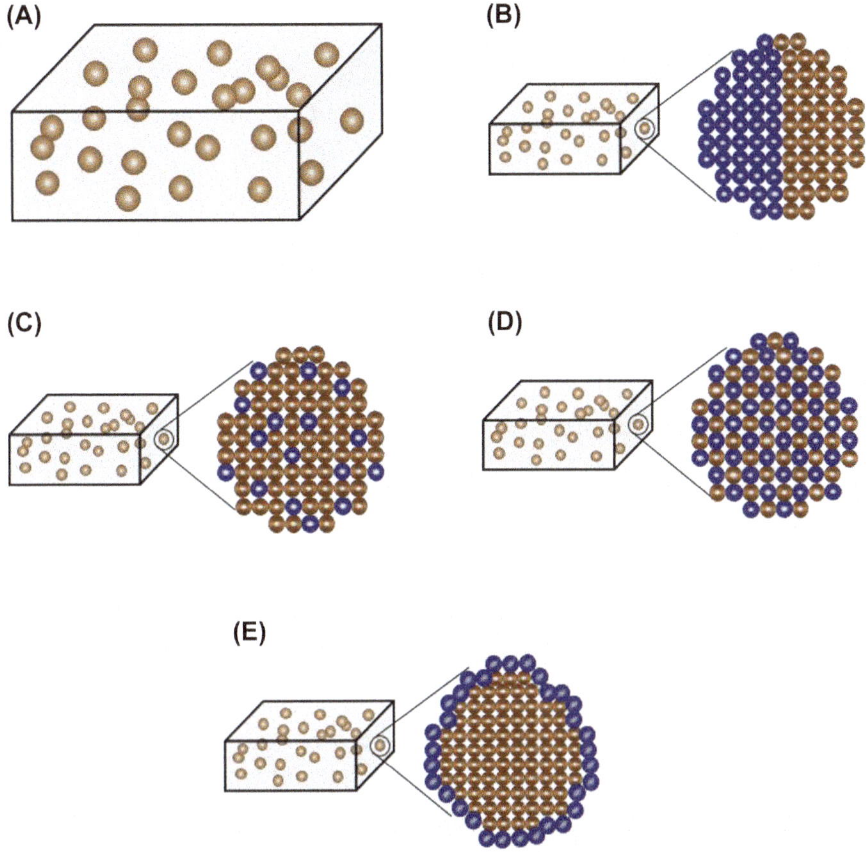

FIGURE 6.6 Various nanostructures and nanocluster structures possible with cluster-based processing. (A) Clusters embedded in a matrix to form a composite. The matrix would be functional—in this case, ferromagnetic. Insulating matrices can be used to isolate clusters to investigate structure and properties of uniquely structured clusters, including (B) two-phase Janus particles; (C) extended or nonequilibrium solid solutions; (D) unique atomic ordering; or (E) core/shell structures. All of these structures have been formed experimentally in the UNL group [33,36–39].

soft mixture of $L1_0$ FePt and Fe_3Pt [41]. Since the initial particle size was in the order of a few nanometers, the final structure was much finer than previously achieved. Cluster assembly provides a perfect pathway to create novel nanostructures and materials with scales in the order of important magnetic length scales. For example, composite structures (Fig. 6.6A) can be readily formed by combining the cluster source with a more conventional thin film deposition system. Here, either codeposition or alternatingly depositing from one source and then the other can create composite structures with the clusters embedded in another material. The matrix material can be either magnetically inert (e.g., SiO_2) or be another magnetic phase. By controlling the cluster deposition, the magnetic clusters can be well-separated from one another, minimizing magnetostatic interactions which allow the study of the magnetic behavior of essentially isolated clusters. Isolated nanoclusters can also be used to create novel multiphase materials, with two phases confined within a single nanocluster, with proper design of the cluster alloy composition and appropriate phase relationships (Fig. 6.6B). If the thin-film-deposited phase is a magnetic phase, a well-controlled, uniform nanostructure can be formed at the scale to explore interphase (and interfacial) magnetic interactions such as exchange-spring and exchange-bias behavior. Using this technique of co-assembly of clusters and thin films, nanocomposite $L1_0$ FePt and α-Fe were produced [36]. The soft magnetic phase fraction was limited in this case by Fe cluster-Fe cluster contact, the problem that plagues granular materials.

6.5 Exchange-spring nanocomposite structures

As discussed above, the exchange-spring phenomena offers the promise of achieving extraordinarily high energy products. However, predictions based on model systems have always far exceeded experimental results. This is likely due to a number of factors: (1) formation of the soft magnetic phase of the appropriate scale; (2) nonuniform nanostructures result in uncoupled regions, resulting in early reversal and the subsequent cascading reversal of the entire material; (3) the overall effect of Brown's paradox which lowers the coercivity significantly; (4) the strong intergranular (interphase) interactions lower the effective anisotropy constant, similar to the random anisotropy model used to describe nanostructured soft magnetic materials [42,43]. While at this time cluster-based processing cannot yet produce materials in the quantity necessary to make, for example, permanent magnet motors, it does offer the opportunity to explore the limits of what is achievable, and to gain fundamental understanding of the nanostructure/properties relationship.

One such study of cluster-assembled exchange-spring magnets was done by Rui et al. [36]. Here, they embedded Fe clusters in an FePt thin film by alternating deposition from the cluster source and a second sputtering system which deposited the FePt thin film. Fig. 6.7 shows an atomic force microscope (AFM) image of representative Fe clusters used to form the nanocomposites; X-ray diffraction revealed the structure to be bcc (i.e., the formation of α-Fe). Equiatomic Fe-Pt (FePt) can form the $L1_0$ tetragonal structure, which has excellent hard magnetic properties. Thus, combining the α-Fe clusters with $L1_0$ FePt results in hard/soft nanocomposite structures to exploit the exchange-spring phenomena. Further, by controlling the deposition times, the soft magnetic phase fraction can be controlled. However, the challenge comes from the fact that as-deposited FePt forms in the disordered A1 (fcc) structure and annealing is required to form the $L1_0$ structure. In this work, 600°C for 10 min produced optimum results.

Firstly, single-phase FePt films were produced. After annealing to form the $L1_0$ structure, the energy product was about 16 MGOe; since $L1_0$ FePt has a saturation magnetization of 14.1 kg, the theoretical maximum ($(BH)_{max(theory)} = 4\pi M_s^2/16$ for isotropic materials) would be about 12.4 MGOe for an isotropic material. Additionally,

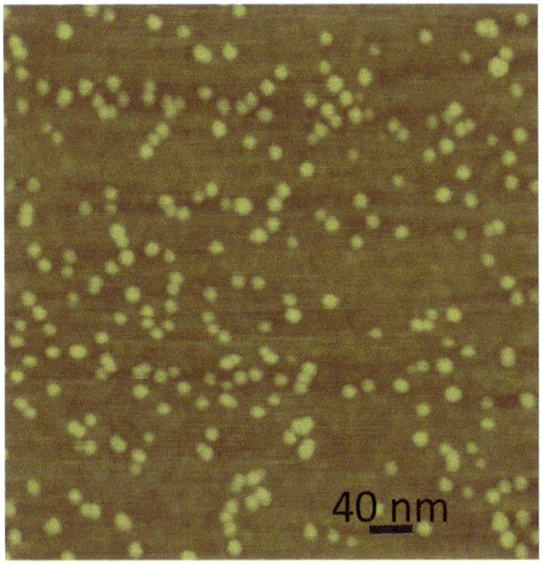

FIGURE 6.7 Atomic force microscopy image of Fe clusters. Mean size is about 8 nm.

the remanence was higher than predicted for noninteracting isotropic structures, which is $^{1}/_{2}$ M_S. The higher values here may indicate some texture developed during processing, or from the enhancement in remanence due to the nanocrystalline grain structure and FePt-FePt intergranular interactions (the original "Hi-Rem" single-phase permanent magnets [15]). Fe clusters were added to this structure, forming the α-Fe/FePt nanocomposite structures. While an increase in remanence was observed when Fe clusters were added to the microstructure, the coercivity dramatically decreased. Subsequently, there was an initial increase in energy product at 8 volume percent α-Fe, but then a large decrease occurred with increasing Fe cluster content (Fig. 6.8).

While the coercivity of the FePt film was relatively high, typically $L1_0$ FePt produces higher coercivities. Thus, a secondary annealing treatment was developed. A second heat treatment at 500°C for 10 and 20 min resulted in enhanced squareness in the second quadrant of the hysteresis loop (Fig. 6.9), dramatically improving the energy product of the nanocomposite structures to a maximum of 20.7 MGOe (Fig. 6.10). The improved energy product was driven by significant enhancement in coercivity. The improvement in magnetic performance was either due to improved $L1_0$ ordering, or the smoothing of the interface, enhancing coupling [44,45]. The heat treatment has the potential to destroy the two-phase nanostructure by dissolving the α-Fe clusters into the Fe-Pt solid solution. The nearness in composition, coupled with the nanoscale grain structure of the FePt film, provides challenges in the structural characterization. However, diffusion analysis revealed that total dissolution was unlikely [36].

While enhanced magnetic behavior was observed in these cluster + thin film systems, producing high energy products for isotropic systems, energy products were lower than

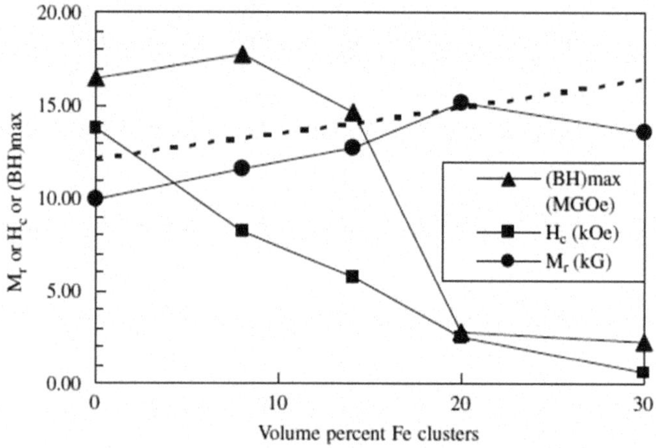

FIGURE 6.8 Magnetic properties of α-Fe/FePt nanocomposite as a function of cluster content. The dashed line indicates the theoretical energy product for a system considering the weighted average of the saturation magnetization of each phase [36].

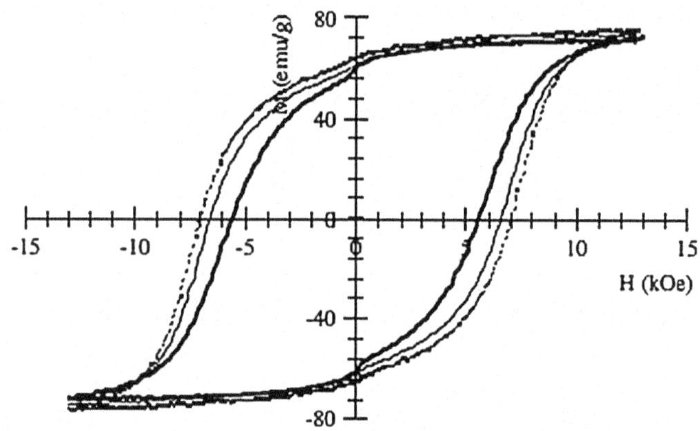

FIGURE 6.9 The effect of secondary heat treatment on the hysteretic behavior of the α-Fe/FePt nanocomposites with 8 volume percent Fe clusters. The solid line is after heat treatment at 600°C for 10 min. The long-dashed line (next highest coercivity) had an additional heat treatment at 500°C for 10 min, and the final dotted line had an additional heat treatment at 500°C for 20 min. Note that the kink in the second quadrant has been largely smoothed by the additional heat treatment [36].

predicted. The volume fraction of soft phase achieved was relatively low, the likely culprit for the lower-than-predicted performance. In particular, the coercivity decrease directly affected the energy product. The loss of coercivity is driven by a number of factors. However, first and foremost it is due to strongly interacting grains, which create "interaction domains." Consequently, the reversal, once initiated, can proceed by domain wall motion [27]. The addition of the soft magnetic phase both reduces the effective anisotropy of the system and provides sites for nucleation of reverse domains, which then can expand.

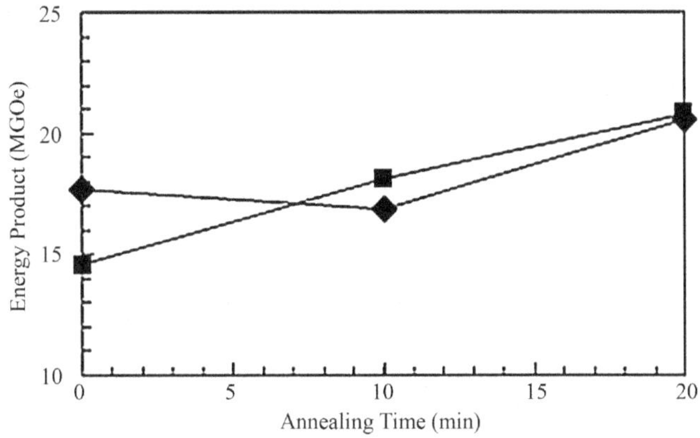

FIGURE 6.10 Energy product as a function of secondary heat treatment for 8 (◆) and 14 (■) volume percent Fe [36].

One possible remedy to eliminate domain wall motion is to create a micro/nanostructure with multiple length scales—one scale containing two strongly interacting phases, and a second to eliminate domain-controlled reversal (i.e., below the single-domain limit, R_{SD}). This structure can be constructed using clusters—two-phase clusters can be fabricated, and isolated from one another during codeposition (Fig. 6.6B). These Janus particles shown schematically in Fig. 6.6B are simplified; the actual two-phase morphology may be more complex, depending on the system. Here, we designed two-phase clusters in the Fe-Pt system by focusing on the Fe-rich side of the FePt $L1_0$ phase. This two-phase region consists of the hard magnetic $L1_0$ FePt and soft magnetic Fe_3Pt (or Fe-Pt solid solution if no $L1_2$ chemical ordering develops). A composite structure consisting of Fe-Pt clusters embedded in SiO_2 was fabricated for magnetic measurements, and a similar density of clusters were deposited directly onto a Cu grid with a graphite support film for TEM investigation. Single-phase $L1_0$ FePt was also fabricated to compare single-phase behavior with the two-phase nanoclusters. From the STEM results (Fig. 6.11), some cluster-cluster contact/overlap occurred, as after heat treatment cluster agglomeration was present. Here, the heat treatments were multistage at temperatures of 580°C, 550°C, and 600°C, which optimized magnetic properties [33,46]. Despite the agglomeration, regions were still largely isolated, creating a scale on the order of 20 nm which is well below the single-domain limit described by Eq. (6.3). Further, within these regions two distinct phases are visible with High-Angle Annular Dark Field (HAADF) imaging in STEM, which reveals atomic-number contrast (Fig. 6.11). Further, chemical analysis using energy dispersive X-ray spectroscopy revealed compositions relatively close to 1:1 and 3:1 Fe:Pt ratios for the bright and dark regions, respectively (points “1” and “2” in Fig. 6.11A). X-ray diffraction results confirmed the two phases [46] as well, and further was used to quantify the amount of hard and soft magnetic phases. Here, the soft phase fraction was about 55%—the highest soft magnetic phase fraction ever reported for hard/soft nanocomposite permanent magnets [46].

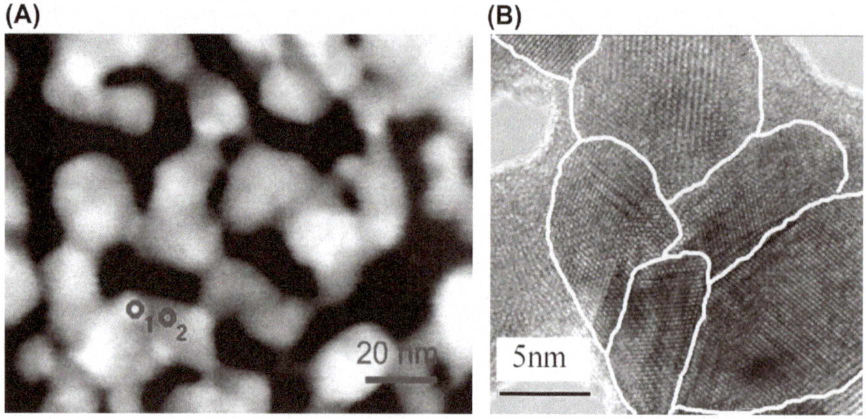

FIGURE 6.11 (A) HAADF STEM image of the heat-treated clusters. The black background is the carbon support film. The brightest regions are $L1_0$ FePt, while the medium gray regions are Fe_3Pt. The points marked 1 and 2 are where X-ray spectra were collected. (B) High-resolution TEM image of the clusters. The lines delineate different grains [46].

This high phase fraction of soft phase determined from X-ray diffraction is corroborated by the HAADF STEM image, where visually there is a large fraction of the medium gray phase—the Fe_3Pt. Image analysis proved difficult because of the low contrast between the FePt and Fe_3Pt phases. High-resolution TEM better revealed the different crystallographic regions (i.e., grains) (Fig. 6.11). Here, the scale of the grains is about 5 nm, small enough to result in full exchange coupling between the two phases.

The magnetic properties were also excellent. For comparison, the single-phase, similarly prepared $L1_0$ clusters displayed a coercivity of about 14 kOe with an energy product of 11.8 MGOe (Fig. 6.12). Note that the energy product was calculated based

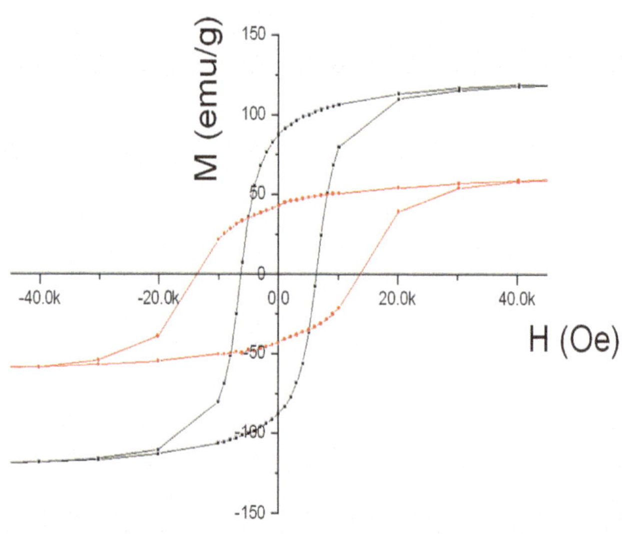

FIGURE 6.12 Hysteresis loops of single-phase $L1_0$ FePt clusters (*red circles*) and two-phase FePt and Fe_3Pt clusters (*black squares*).

only on the volume (mass) of the magnetic phase, and not the filler (SiO_2). Interestingly, as noted above, the theoretical energy product of isotropic $L1_0$ FePt is 12.4 MGOe; we are remarkably close to the theoretical maximum. For the two-phase structure, the hysteresis loop (Fig. 6.12) showed excellent coupling, indicated by the smooth demagnetization curve in the second quadrant. The coercivity was very good—6.5 kOe for the best composition in the two-phase region. Also, the energy product was 25.1 MGOe, the highest ever achieved for isotropic single-phase materials. Again, the volume considered was only that of the magnetic phase.

A simple micromagnetic model utilizing the intrinsic magnetic properties of $L1_0$ FePt and $L1_2$ Fe_3Pt was constructed, with magnetic properties calculated using object-oriented micromagnetic framework (OOMMF). There is remarkable agreement between the experimental and calculated results [46]. In the model system, the peak energy product was achieved at about 50 volume percent, and the value was comparable to our experimental results (Fig. 6.13). A model structure considering the Nd-Fe-B system was also evaluated. Nd-Fe-B magnets are the most commercially important permanent magnet material today, and possess intrinsic properties superior to Fe-Pt. Anisotropic magnets are available with energy products of about 50 MGOe, and the best isotropic materials have energy products of around 12–15 MGOe. Exchange-spring materials in this system have been extensively investigated, with the highest energy product around 23 MGOe [12]. Our model predicts an energy product of about 35 MGOe for isotropic structures (Fig. 6.13).

For a real material deployed in an application, some nonmagnetic phase along boundaries would lower the energy product. The purpose here is to show the potential of exchange-spring permanent magnets, and what could be achieved with ideal

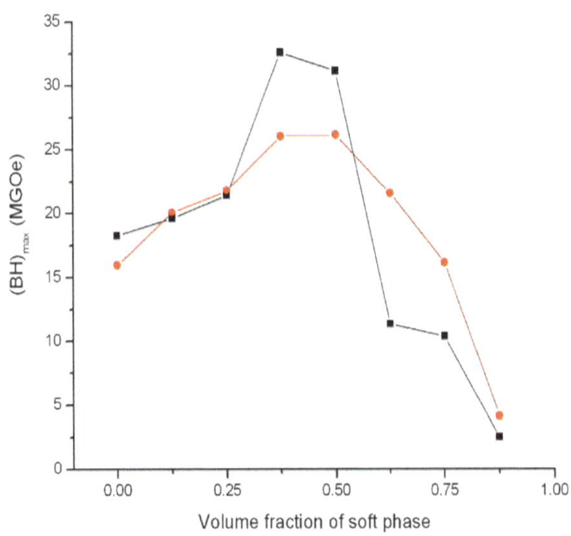

FIGURE 6.13 OOMMF models of Fe-Pt (*red circles*) and Nd-Fe-B (*black squares*) nanocomposite structures.

nanostructures—"perfect" exchange coupling, relatively uniform scale, eliminating domain wall nucleation and propagation, and so forth. The use of clusters to obtain these features was paramount to the study of the exchange-spring phenomena.

6.6 Solid solution extension

The far-from-equilibrium processing afforded by inert gas condensation can produce novel structures, notably extended solid solutions (Fig. 6.8C). Further, creating solid solutions in otherwise immiscible systems can be a pathway to new atomic structures—especially ordered structures (Fig. 6.8D). These structures in turn could lead to new magnetic materials. For instance, first-principles calculations suggest that the addition of heavy 5d transition metals such as W or Mo into Fe or Co could significantly enhance the magnetocrystalline anisotropy [47,48]. However, in equilibrium these systems have low or negligible solid solubility; for example, Fig. 6.14 shows the Co-rich section of the Co-W phase diagram, where there is negligible solubility of W in Co.

From our explorations of immiscible systems, we were able to create extended or complete solid solutions in a number of systems, although there is a strong size dependence on complete solid solution formation. In larger clusters, core/shell structures formed (Fig. 6.15) [39,49], while smaller clusters formed complete solid solutions. Interestingly, the transition is abrupt, and several systems displayed a transition at about

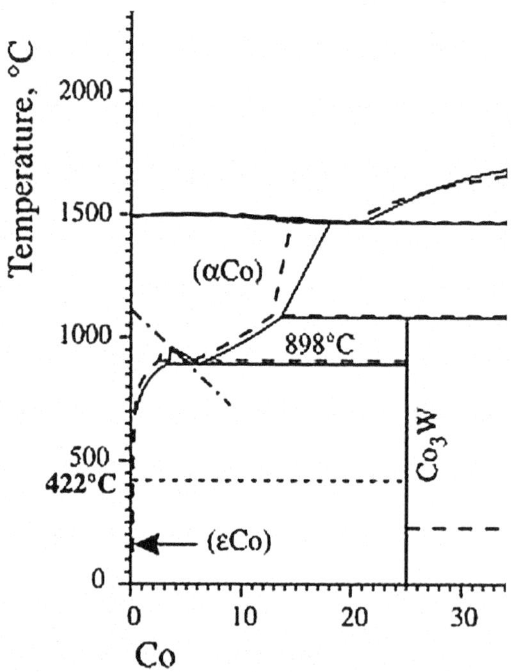

FIGURE 6.14 Co-rich section of the Co-W phase diagram (after [51]).

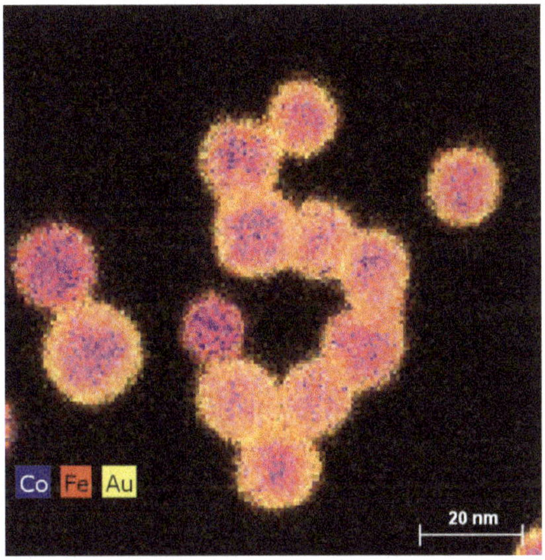

FIGURE 6.15 Core/shell clusters formed in the Fe(Co)-Au system. The lower surface energy Au forms on the surface.

8 nm (Fig. 6.16) [39]. The formation of a solid solution may be driven by surface energy considerations, where the energy required to form an interface is not offset by the decrease in free energy realized by forming two phases [50]. Additionally, the thermodynamics of nanosystems may lead to changes in the enthalpy (heat of mixing) in small systems, leading to conditions more favorable for mixing (solid solution formation).

The uptick in rare earth commodity prices and increasing demand for high-end Nd-Fe-B-based materials for wind turbine and electrical vehicle uses provided the impetus to search for rare earth-free permanent magnet materials. Co in the hcp form has a relatively high magnetocrystalline anisotropy constant, but generally not high enough for permanent magnetic applications. Thus, the means to inject higher 5d transition metal content into Co via inert gas condensation was explored, with the goal to increase the magnetocrystalline anisotropy [47,48]. The W content was controlled by the number of W plugs in the composite target. At high W content (above at least 12 atomic percent), two-phase hcp Co + Co_3W clusters were produced [52]. However, at 5 and 12 atomic percent W, single-phase clusters consisting of hcp Co with W dissolved into solid solution were produced (Fig. 6.17). The magnetic properties were measured at 10 K and 300 K (Fig. 6.18). The effective anisotropy constant, K_u, was determined from fitting the high-field data [53, 54]. The coercivity and K_u both increased between 5 and 12 atomic percent W, suggesting that the addition of 5d transition metal does indeed improve anisotropy (Fig. 6.19). The dramatic decrease at 23 atomic percent W is due to the presence of Co_3W, which has a low magnetocrystalline anisotropy. However, the high fraction of surface atoms in clusters tends to decrease intrinsic properties of a material due to broken symmetry. For example, clusters of hcp Co were observed to have lower

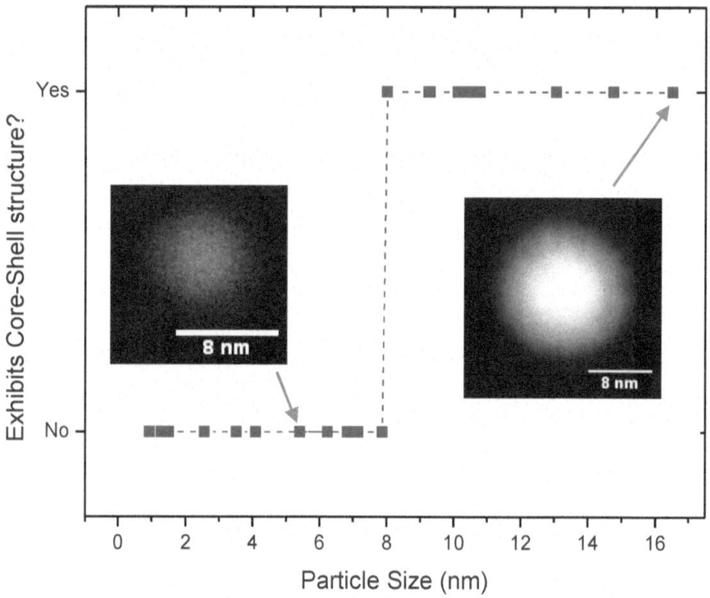

FIGURE 6.16 Size-dependence of core/shell formation in Fe-W clusters. The HAADF STEM images show core/shell behavior at sizes above 8 nm, and single-phase structures in smaller clusters.

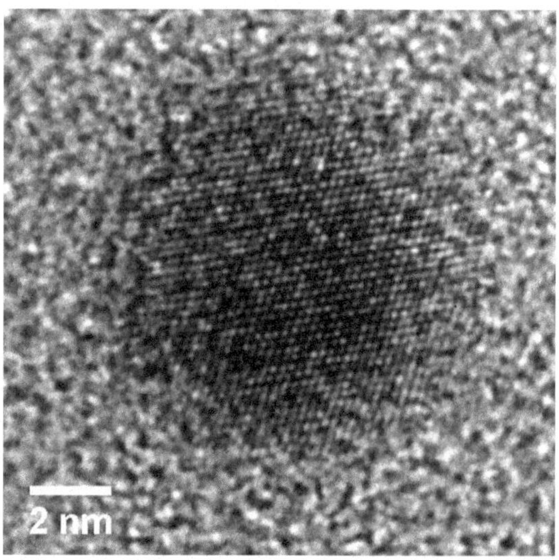

FIGURE 6.17 High-resolution TEM image of a Co-5 atomic percent cluster showing single-crystallinity. The fast Fourier transform indexed to the hcp structure.

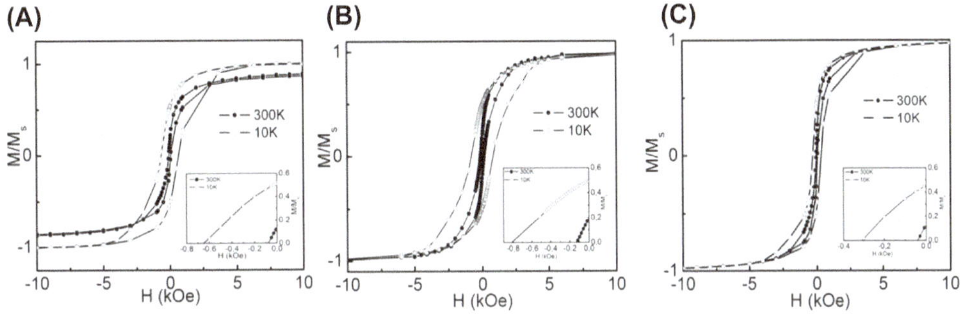

FIGURE 6.18 Hysteresis loops of the Co-W alloys at 10 and 300 K, with (A) 5, (B) 12, and (C) 23 atomic percent W.

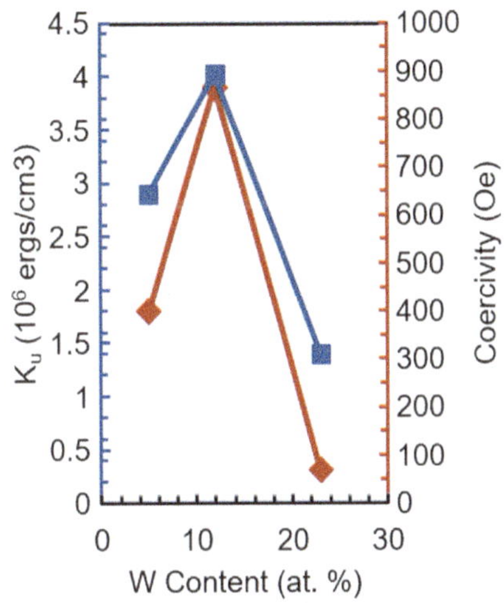

FIGURE 6.19 Coercivity (*red diamonds*) and effective anisotropy constant K_u (*blue squares*) of Co-W alloys.

magnetocrystalline anisotropy and saturation magnetization [55]. With these Co-W solid solution alloys, the gain realized by substituting additional W into the structure just offset the decrease arising due to the size of the system. Further, W, in a magnetic sublattice, carries a small magnetic moment that couples antiparallel with the Co, negatively influencing saturation magnetization. Nevertheless, increasing solid solubility of 5d transition metals in bulk systems may be a viable route for the development of future rare earth-free permanent magnets, given a suitable system is identified.

6.7 Summary

In this chapter, we have described the investigation of permanent magnet materials via inert gas condensation. Cluster-based processing has allowed the design and fabrication of well-controlled nanostructures, providing the opportunity to construct materials with unprecedented control over scale and uniformity. In particular, intracluster structuring—the formation of multiple phases within a single cluster—has led to structures in Fe-Pt with ~5 nm grain sizes. This has allowed us to explore the limits of exchange-spring permanent magnets in terms of maximum energy density, where we produced energy products more than double their single-phase counterparts. The nonequilibrium processing results in the formation of novel solid solutions—either creating solid solutions from immiscible components, or extending solid solubility—in order to make and explore new materials. One example was the increase of magneto-crystalline anisotropy in solid solution Co-W alloys.

Acknowledgments

The work presented here is due to the skill and technical abilities of a number of graduate students and collaborators. The graduate students include Dr. Xiangxin Rui, Mr. Jeremy Anderson, Dr. Pinaki Mukherjee, Dr. Farhad Golkar, and Dr. Mark (Mak) Koten. All have contributed to our understanding of cluster synthesis using inert gas condensation, and the structure and properties of novel materials. Key collaborators include Prof. David Sellmyer, Ralph Skomski, and Bala Balamurugan from the University of Nebraska–Lincoln and Matthew Kramer from Ames Laboratory. Financial support that enabled the research came from the National Science Foundation's Materials Research Science and Engineering Center program and the Department of Energy's EPSCoR and EERE programs.

References

[1] O. Gutfleisch, M.A. Willard, E. Bruck, C.H. Chen, S.G. Sankar, J.P. Liu, Magnetic materials and devices for the 21st century: stronger, lighter, and more energy efficient, Adv. Mater. 23 (2011) 821.

[2] M.J. Kramer, R.W. McCallum, I.E. Anderson, S. Constantinides, Prospects for non-rare earth permanent magnets for traction motors and generators, JOM 64 (2012) 752.

[3] V. Dupuis, A. Hillion, A. Robert, O. Loiselet, G. Khadra, P. Capiod, C. Albin, O. Boisron, D. Le Roy, L. Bardotti, F. Tournus, A. Tamion, Bottom-up strategies for the assembling of magnetic systems using nanoclusters, J. Nanopart. Res. 20 (2018) 128.

[4] R. Skomski, Nanomagnetics, J. Phys: Condens. Matter 15 (2003) R841.

[5] D.C. Jiles, Introduction to Magnetism and Magnetic Materials, third ed., CRC Press, Taylor & Francis, 2015.

[6] E.F. Kneller, R. Hawig, The exchange-spring magnet: a new material principle for permanent magnets, IEEE Trans. Magn. 27 (1991) 3588.

[7] R. Skomski, J.M.D. Coey, Giant energy product in nanostructured two-phase magnets, Phys. Rev. B 48 (15) (1993) 812.

[8] R. Fischer, T. Schrefl, H. Kronmuller, J. Fidler, Phase distribution and computed magnetic properties of high-remanent composite magnets, J. Magn. Magn. Mater. 150 (1995) 329.

[9] Z.S. Shan, J.P. Liu, V.M. Chakka, H. Zewng, J.S. Jiang, Energy barrier and magnetic properties of exchange-coupled hard-soft bilayer system, IEEE Trans. Magn. 38 (2002) 2907.

[10] X.G. Guo, J.S. Jiang, J.E. Pearson, S.D. Bader, J.P. Liu, Exchange-coupled Sm–Co/Nd–Co nanomagnets: correlation between soft phase anisotropy and exchange field, Appl. Phys. Lett. 81 (2002) 2029.

[11] J.P. Liu, N.M. Dempsey, P. de Rango (Eds.), Proceedings of the 18th Int'l. Workshop on High Performance Magnets and Their Applications pp. 679.

[12] J. Bernardi, T. Schrefl, J. Fidler, T. Rijks, K. deDort, V. Archambault, D. Pere, S. David, D. Givord, J.F. O'Sullivan, P.A.I. Smith, J.M.D. Coey, U. Czernik, M. Gronefeld, Preparation, magnetic properties and microstructure of lean rare-earth permanent magnetic materials, J. Magn. Magn. Mater. 219 (2000) 186.

[13] R. Fischer, T. Schrefl, H. Kronmuller, J. Fidler, Grain-size dependence of remanence and coercive field of isotropic nanocrystalline composite permanent magnets, J. Magn. Magn. Mater. 153 (1996) 35.

[14] H.A. Davies, Nanocrystalline exchange-enhanced hard magnetic alloys, J. Magn. Magn. Mater. 157/158 (1996) 11.

[15] R.W. McCallum, A.M. Kadim, G.B. Clemente, J.E. Keem, High performance isotropic permanent magnet based on Nd-Fe-B, J. Appl. Phys. 61 (1987) 3577.

[16] R. Coehoorn, D.B. Mooji, C. DeWaard, Meltspun permanent magnet materials containing Fe_3B as the main phase, J. Magn. Magn. Mater. 80 (1989) 101.

[17] G.C. Hadjipanayis, Nanophase hard magnets, J. Magn. Magn. Mater. 200 (1999) 373.

[18] I.A. Al-Omari, D.J. Sellmyer, Magnetic properties of nanostructured CoSm/FeCo films, Phys. Rev. B 52 (1995) 3441.

[19] M. Shindo, M. Ishizone, H. Kato, T. Miyazaki, A. Sakuma, Exchange-spring behavior in sputter-deposited α-Fe/$Nd_2Fe_{14}B$ multilayer magnets, J. Magn. Magn. Mater. 161 (1996) 1.

[20] E. Goto, N. Hayashi, T. Miyashita, K. Nakagawa, Magnetization and switching characteristics of composite thin magnetic films, J. Appl. Phys. 36 (1995) 2951.

[21] E.E. Fullerton, J.S. Jiang, S.D. Bader, Hard/soft magnetic heterostructures: model exchange-spring magnets, J. Magn. Magn. Mater. 200 (1999) 392.

[22] E.E. Fullerton, J.S. Jiang, M. Grimsditch, C.H. Sowers, S.D. Bader, Exchange-spring behavior in epitaxial hard/soft magnetic bilayers, Phys. Rev. B 58 (1998) 12193.

[23] Y. Gao, D. Shindo, A.K. Petford-Long, Nonuniform magnetic structure in $Nd_2Fe_{14}B/Fe_3B$ nanocomposite materials, J. Appl. Phys. 93 (2003) 8119.

[24] J. Zhou, A. Kashyap, Y. Liu, R. Skomski, D.J. Sellmyer, Domain-wall pinning at inhomogenities of arbitrary cross-sectional geometry, IEEE Trans. Magn. 40 (2004) 2940.

[25] D.C. Crew, E. Girt, D. Suess, T. Schrefl, K.M. Krishnan, G. Thomas, M. Guilot, Magnetic interactions and reversal behavior of $Nd_2Fe_{14}B$ particles diluted in a Nd matrix, Phys. Rev. B 66 (2002) 184418.

[26] J.E. Shield, B.B. Kappes, D.J. Branagan, J. Bentley, Chemical partitioning during crystallization and its effect on the microstructure and magnetic behavior of modified Nd-Fe-B permanent magnets, J. Magn. Magn. Mater. 246 (2002) 73.

[27] J.E. Shield, J. Zhou, S. Aich, V.K. Ravindran, R. Skomski, D.J. Sellmyer, Highly coercive rapidly solidified Sm-Co alloys, Appl. Phys. 99 (2006) 08B508.

[28] H. Haberland, M. Karrais, M. Mall, Y. Thurner, Thin films from energetic cluster impact: a feasibility study, J. Vac. Sci. Tech. A 10 (1992) 3266.

[29] S.H. Baker, S.C. Thornton, A.M. Keen, T.I. Preston, C. Norris, K.W. Edmonds, C. Binns, The construction of a gas aggregation source for the preparation of mass-selected ultrasmall metal particles, Rev. Sci. Instrum. 68 (1997) 1853.

[30] S.H. Baker, S.C. Thornton, K.W. Edmonds, M.J. Maher, C. Norris, C. Binns, The construction of a gas aggregation source for the preparation of size-selected nanoscale transition metal clusters, Rev. Sci. Instrum. 71 (2000) 3178.

[31] X. Rui, Z.G. Sun, Y. Xu, D.J. Sellmyer, J.E. Shield, Cluster-assembled exchange-spring nanocomposite permanent magnets, J. Appl. Phys. 97 (2005) 10K310.

[32] N.H. Hai, R. Lemoine, S. Remboldt, M. Strand, J.E. Shield, D. Schmitter, R.H. Kraus, M. Espy Jr., D.L. Leslie-Pelecky, Iron and cobalt-based magnetic fluids produced by inert gas condensation, J. Magn. Magn. Mater. 293 (2005) 75–79.

[33] X. Rui, Z. Sun, Y. Xu, D.J. Sellmyer, J.E. Shield, In-cluster-structured exchange-coupled magnets with high energy densities, Appl. Phys. Lett. 89 (2006) 122509.

[34] Y. Xu, Z.G. Sun, Y. Qiang, D.J. Sellmyer, Preparation and magnetic properties of CoPt and CoPt:Ag nanocluster films, J. Magn. Magn. Mater. 266 (2003) 164.

[35] Y. Xu, M.L. Yan, J. Zhou, Sellmyer, Magnetic properties of dilute FePt:C nanocluster films, J. Appl. Phys. 97 (2005) 10J320.

[36] X. Rui, Z. Sun, L. Yue, Y. Xu, D.J. Sellmyer, Z. Liu, D.J. Miller, J.E. Shield, High energy product exchange-spring FePt/Fe nanocomposite permanent magnets, J. Magn. Magn. Mater. 305 (2006) 76–82.

[37] P. Mukherjee, L. Zhou, M.J. Kramer, J.E. Shield, Formation of non-equilibrium solid solutions in nanoclusters, Appl. Phys. Lett. 102 (2013) 243103.

[38] P. Mukherjee, P. Manchanda, P. Kumar, L. Zhou, M.J. Kramer, A. Kashyap, R. Skomski, D.J. Sellmyer, J.E. Shield, Size-induced chemical and magnetic ordering in individual Fe-Au nanoparticles, ACS Nano 8 (2014) 8113–8120.

[39] M.A. Koten, P. Mukherjee, J.E. Shield, Core-shell nanoparticles driven by surface energy differences in the Co-Ag, W-Fe and Mo-Co systems, part. Part, Syst. Charact. 32 (2015) 848–853.

[40] P.G. McCormick, W.F. Miao, P.A.I. Smith, J. Ding, R. Street, Mechanically alloyed nanocomposite magnets, J. Appl. Phys. 83 (1998) 6256.

[41] H. Zeng, J. Li, J.P. Liu, Z.L. Wang, S. Sun, Exchange-coupled nanocomposite magnets by nano-particle self-assembly, Nature 420 (2002) 395.

[42] R. Alben, J.J. Becker, M.C. Chen, Random anisotropy in amorphous magnets, J. Appl. Phys. 49 (1978) 1653.

[43] G. Herzer, Grain structure and magnetism of nanocrystalline ferromagnets, IEEE Trans. Magn. 25 (1989) 3327.

[44] D.C. Crew, L.H. Lewis, J. Kim, K. Barmak, Magnetic signature of compositional gradient in exchange-spring films of CoPt/Co, J. Appl. Phys. 89 (2001) 7528.

[45] J.S. Jiang, J.E. Pearson, Z.Y. Liu, B. Kabius, S. Trasobares, D.J. Miller, S.D. Bader, D.R. Lee, D. Haskel, G. Srajer, J.P. Liu, A new approach for improving exchange-spring magnets, J. Appl. Phys. 97 (2005) 10K311.

[46] X. Rui, Z. Sun, R. Skomski, Y. Xu, D.J. Sellmyer, Y.Q. Wu, M.J. Kramer, J.E. Shield, Intra-cluster exchange-coupled high performance permanent magnets, J. Magn. Magn. Mater. 320 (2008) 2576.

[47] R. Skomski, V. Sharma, B. Balamurugan, J.E. Shield, A. Kashyap, D.J. Sellmyer, Anisotropy of doped transition-metal magnets, in: S. Kobe, P. McGuiness (Eds.), Proceedings of the 21st Workshop on Rare-Earth Permanent Magnets and their Applications, 2010, p. 55.

[48] A. Kashyap, P. Manchanda, P. Sahota, R. Skomski, J.E. Shield, D.J. Sellmyer, Anisotropy of W in Fe and Co, IEEE Trans. Magn. 47 (2011) 3336.

[49] P. Mukherjee, B. Balamurugan, J.E. Shield, D.J. Sellmyer, Direct gas-phase formation of complex core-shell and three-layer Mn-Bi nanoparticles, RSC Adv. 6 (2016) 92765–92770.

[50] P. Mukherjee, X. Jiang, Y. Wu, M.J. Kramer, J.E. Shield, Nucleation-suppressed phase stabilization in nanoclusters, J. Phys. Chem. C 117 (2013) 24071–24078.

[51] H. Baker (Ed.), Alloy Phase Diagrams, ASM International, Metals Park, OH, 1992.

[52] F. Golkar, M.J. Kramer, R. Skomski, D.J. Sellmyer, J.E. Shield, Solubility extension and phase formation in gas-condensed Co-W nanoclusters, J. Nanopart. Res. 15 (2013) 1638.

[53] E. Kneller, Ferromagnetism, Springer, Berlin, 1962.

[54] G.C. Hadjipanayis, D.J. Sellmyer, B. Brandt, Rare-earth-rich metallic glasses. I. Magnetic hysteresis, Phys. Rev. B 23 (1981) 3349.

[55] Y. Qiang, R.F. Sabriyanov, S.S. Jaswal, Y. Liu, H. Haberland, D.J. Sellmyer, Magnetism of Co nanocluster films, Phys. Rev. B 66 (2002) 064404.

7

Reawakening of plasmonic nanocomposites with the polarizonic reflective coloration: from metal to molecules

Mady Elbahri[1,*], Mehdi Keshavarz Hedayati[2], Shahin Homaeigohar[1], Moheb Abdelaziz[1]

[1]NANOCHEMISTRY AND NANOENGINEERING, DEPARTMENT OF CHEMISTRY AND MATERIALS SCIENCE, SCHOOL OF CHEMICAL ENGINEERING, AALTO UNIVERSITY, ESPOO, FINLAND; [2]DEPARTMENT OF ENGINEERING, DURHAM UNIVERSITY, DURHAM, UNITED KINGDOM

*CORRESPONDING AUTHOR

Chapter outline

*Corresponding author

Cluster Beam Deposition of Functional Nanomaterials and Devices. https://doi.org/10.1016/B978-0-08-102515-4.00007-6

7.1 Introduction

The optical nanocomposites have drawn the attention of the scientists worldwide within the last couple of decades owing to their immense potential for offering exotic optical properties [1–5]. However, this topic is not very new and archaeometric studies of the antiques of old nations, e.g., those of the first millennium BC, have shown that the ancient people despite not being aware of the scientific reasons of their observations had already made use of them for decoration and artistic purposes. The old colorful statues (e.g., the mask of Tutankhamen's mummy) (Fig. 7.1A), painted tiles (Fig. 7.1B), and colored (stained) glasses (e.g., the Roman Lycurgus cup), present a variety of colors (blue to red) stemming from the nanoparticles (NPs) and nanostructures embedded in a dielectric matrix. Recently, chemical and microscopic analyses of the 17th century Spanish and Italian lusters (Fig. 7.1C) and the 19th century Iranian painted tiles proved the existence of copper, gold, and silver NPs as the main coloring components [6–9]. Nevertheless, the physical explanation of the plasmonic phenomena was not clear before the first report of Maxwell Garnett in 1904 when he reviewed the work of his contemporary researchers on coloration of silver and gold particles and analyzed stained glasses (as a well-known nanocomposite) theoretically [10]. Afterward, Gustav Mie proposed an exact solution for Maxwell's equations and provided an easy way to interpret the optical properties of nanoparticles which has been in use till today.

Despite the wide applicability of such a theory, it cannot explain the origin and duality of the colors observed in the luster pottery from Susa, Tunis (Fig. 7.2A and B), for instance. Based on the reflection and interference effects, when the luster pottery is illuminated with white light, the color varies depending on the relative angle between the light source, the pottery surface, and the observer [11].

FIGURE 7.1 Plasmonic nanocomposites in a variety of old artifacts. (A) The mask of Tutankhamen's mummy. (B) Immortal soldiers at Darius' palace at Susa. *Wikipedia under a CC BY-SA 3.0 and CC BY 2.0 license, respectively.* (C) The Luster made by Mastro Giorgio (in 1528, Museo Civico della Ceramica di Gubbio, Italy). The transmission electron microscope (TEM) image, top right corner, shows nanoclusters of Cu distributed within the Au matrix. *Reproduced with permission P. Mazzoldi, C. Sada, A trip in the history and evolution of ion-exchange process, Materials Science and Engineering: B 149(2) (2008) 112–117.*

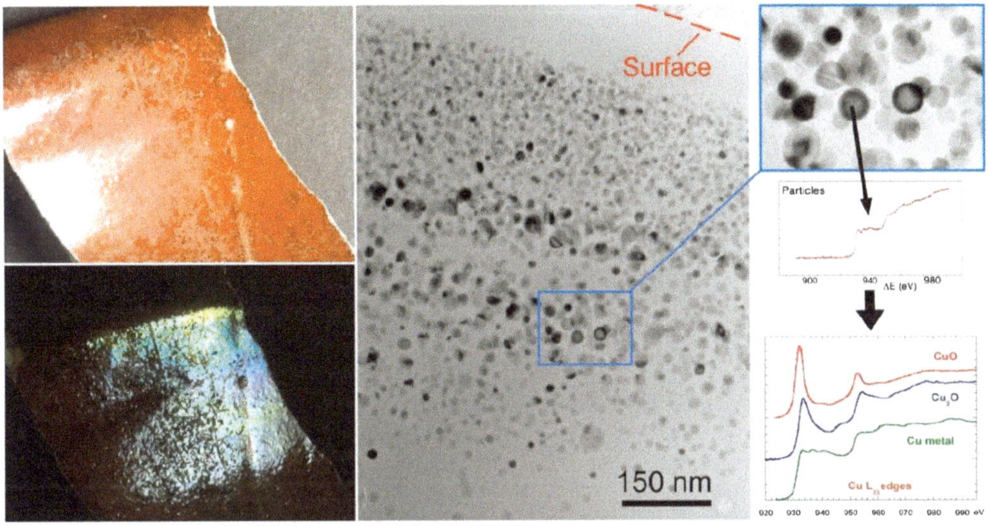

FIGURE 7.2 The lustered pottery found in Mesopotamia (Susa) (9th century CE). As seen here, the color shifts from red (top left) to blue/green (bottom left). Transmission electron microscopy (TEM) images (center and right) imply the nanoparticles' distribution mode. Electron energy loss spectroscopy (EELS) graphs prove that the nanoparticles are made of copper. *Reproduced with permission P. Sciau, Nanoparticles in ancient materials: the metallic lustre decorations of medieval ceramics, the delivery of nanoparticles, Intech (2012). Copyright 2012, IntechOpen.*

In our opinion, the abovementioned contradicting interpretations originate from mixing of several theories while ignoring the main difference between thermal emission/radiation and dipole radiation. For instance, the common wisdom is that qualitative color changes arise from energy dissipation and blackbody radiation (Planck's law). In the classic theory, Planck explains the blackbody radiation as the quantized thermal radiation emitting from the oscillating charged particles from a hot object's surface [12]. The thermal radiative color is based on Wien's displacement law, and accordingly the artificial plasmonic dipole absorber can be regarded as a diluted blackbody nanoheater. This would be most probably the case whenever the extinction solely is considered and/or when the electronic transition is involved.

Here, we challenge this general wisdom and provide a new insight by exploring the electromagnetic radiation that causes the specular reflection based colors from a plasmonic nanocomposite, Fig. 7.3.

This chapter is divided into five sections starting with an introduction. Next, the available models for explanation of the optical behavior of plasmonic nanocomposites will be presented. In this regard, the Mie, Maxwell Garnett, and Bruggeman theories are discussed. In the third section, the optical properties of the nanocomposites and the typical parameters affecting their plasmonic behavior are introduced. The reflective plasmonic nanocomposites as a new paradigm for design of optical devices are discussed in the fourth section. At last, a prospective overview of plasmonics in molecules that could inspire the future trend of the field is presented.

FIGURE 7.3 Schematic illustration of a plasmonic nanocomposite containing arbitrarily organized metal NPs that reflect unique coloration to the far field. *Reproduced with permission C. Etrich, S. Fahr, M.K. Hedayati, F. Faupel, M. Elbahri, C. Rockstuhl, Effective optical properties of plasmonic nanocomposites, Materials 7 (2) (2014) 727–741. Copyright 2014, MDPI.*

7.2 Theoretical considerations

7.2.1 The Mie theory

Since Faraday's work on gold colloids in the middle of 1800s, study on the optical properties of metal NPs has ever been of the main research interests of the scientists in the fields of physics and chemistry [14]. However, till onset of the 20th century when Gustav Mie proposed an exact solution for Maxwell's equation and analytically investigated the optical response of such a system, there have never been so many studies on the exploitation of the physical reason behind the coloration of metal colloids. Mie assumed that the particles and their surrounding media are both homogeneous and therefore he considered the optical constant of particles to be equal to that of the bulk [15]. Although this assumption works well for the particles bigger than 50 nm, there is a big doubt on utilizing the bulk value in calculation of the optical response of the metal/polymer nanocomposites possessing smaller nanoparticles [16]. For simplification, Mie developed his theory for spherical particles with a size smaller than the incident electromagnetic wave's wavelength. By taking into account these assumptions, the dipole moment can be calculated in electrostatics as follows:

$$P(\omega) = \alpha\varepsilon_0 E_0(\omega) = 4\pi\varepsilon_0 R^3 \frac{\varepsilon_m(\omega) - \varepsilon_h}{\varepsilon_m(\omega) + 2\varepsilon_h} E_0(\omega) \tag{7.1}$$

Where α is the polarizability of the sphere, R the nanoparticle's radius, E_0 the electric field strength of an incident electromagnetic wave, ε_0 the electric permittivity of vacuum, ε_m and ε_h are the relative complex electric permittivity of the metal and the dielectric host, respectively [14].

In case metal nanoparticles are embedded in a transparent medium (the imaginary part of the dielectric constant approaches zero), the absorption cross-section of the spherical nanoparticles can be derived as [14]:

$$\sigma(\omega) = \frac{\omega}{c\sqrt{\varepsilon_h}} Im[\alpha(\omega)] = 12\pi R^3 \frac{\omega\sqrt[3]{\varepsilon_h}}{c} \frac{\varepsilon_2(\omega)}{[\varepsilon_1(\omega) + 2\varepsilon_h]^2 + \varepsilon_2(\omega)^2} \quad (7.2)$$

Note that not only absorption but also scattering contributes to coloration of the nanocomposite. Therefore, extinction should be considered which is the summation of absorption and scattering cross-sections. For metallic NPs, extinction based on the Mie theory follows Eq. (7.3) [14]:

$$\sigma_{ext} = \frac{24\pi^2 R^3 \varepsilon_h^{\left(\frac{3}{2}\right)}}{\lambda} \frac{\varepsilon_2(\omega)}{[\varepsilon_1(\omega) + 2\varepsilon_h]^2 + \varepsilon_2(\omega)^2} \quad (7.3)$$

Obviously, extinction reaches its maximum if the denominator is at its minimum value which means:

$$[\varepsilon_1(\omega) + 2\varepsilon_h]^2 + \varepsilon_2(\omega)^2 \to 0 \quad (7.4)$$

If ε_2 is small in comparison with ε_1 or weakly dependent on ω, then Eq. (7.4) can be written as:

$$\varepsilon_1(\omega_{sp}) = -2\varepsilon_h \quad (7.5)$$

This means that the color of different metallic spheres is determined by the frequency (wavelength) at which the relative permittivity of the metal satisfies $\varepsilon_1 = -2\varepsilon_h$. On the other hand, we can see from this term that the resonance of nanoparticles strongly depends on their surrounding medium, i.e., any changes in the dielectric constant of the surrounding medium alter the metal's dielectric constant at which condition is fulfilled $\varepsilon_1 = -2\varepsilon_h$ [1].

Although one can extract some practical information out of Mie's method for spherical particles, it fails to describe the optical response of the particles with random shapes and geometry.

7.2.2 The T-matrix method

The T-matrix method [17] is widely used for determination of light scattering of nonspherical particles. In other words, it is a more general form of the Mie theory for calculation of optical properties of the clusters with complicated shapes. Because of the linearity of Maxwell's equation, in this method, the incident and scattered electromagnetic filed of axisymmetric particles with geometries such as cylinders or cuboids are related to each other by a transition matrix. Eq. (7.6) shows the relationship between the incident and scattered elements of the wave by a transition matrix where $a^{scat.}$ and $b^{scat.}$ are the scattered field and $a^{inc.}$ and $b^{inc.}$ represent the incident field coefficients, respectively. In fact, the size parameters of the particles besides the refractive index of the scattering medium and its surrounding determine the element of T-matrix. However, for nonspherical particles, it is also dependent upon the orientation of the particle with respect to the incident wave vector and polarization [18].

$$\begin{bmatrix} a_n^{sca.} \\ b_n^{sca.} \end{bmatrix} = \begin{pmatrix} T_{11} & T_{12} \\ T_{21} & T_{22} \end{pmatrix} \begin{bmatrix} a_n^{inc.} \\ b_n^{inc.} \end{bmatrix} \tag{7.6}$$

7.2.3 The discrete dipole approximation

The DDA also called as "coupled dipole" was proposed by Purcell and Pennypacker in the 70s. This approach is a general technique to calculate scattering and absorption of electromagnetic waves by particles of arbitrary geometry and composition. In this method, the scattering medium is replaced by a set of point dipoles [18]. Moreover, only the interaction of the dipoles with the incident wave and the neighboring dipoles is being considered. This reduces the solution of Maxwell's equations to an algebraic problem of many coupled dipoles [19,20]. A more detailed description of this method can be found in Ref. [21].

7.2.4 Effective medium theory

Optical properties of nanocomposites are routinely explained by several models such as the Maxwell Garnett (MG) theory and the Bruggeman effective medium approximation (EMA). The Maxwell Garnett approximation is a widely known method for calculating the bulk dielectric function of composite materials. It is practical when one of the constituents of a composite structure can be considered as a host matrix wherein the inclusions of the other component are incorporated. By this method, the field induced into the matrix by spherical or ellipsoidal particles is calculated precisely and its distortion because of the electrostatic interaction between the particles is also roughly estimated. The distortion is originated from charge dipoles and higher multipoles from the other surrounding particles. The induced dipole moments cause the longest range distortions and their average effect is incorporated in the MG estimation which results in a uniform field within all the particles [22]. In the linear case, the effective dielectric constant of the composite material is given by Ref. [23]:

$$\varepsilon_{eff} = \varepsilon_h \left(\frac{1 + 2\beta f}{1 - \beta f} \right) \tag{7.7}$$

$$\beta \equiv \frac{\varepsilon_i - \varepsilon_h}{\varepsilon_i + 2\varepsilon_h} \tag{7.8}$$

and where f denotes the filling factor of the embedded particles and ε_i and ε_h denote the linear dielectric constants of the inclusion and host materials, respectively. Fig. 7.3 shows the schematic of the typical structure which can be simulated by the MG approximation. Even though the collective response of nanoparticles is considered in the MG theory, their interaction is not precisely taken into account and that is why this

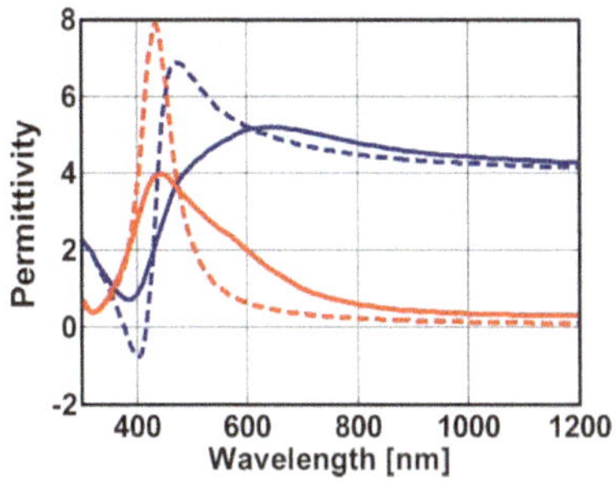

FIGURE 7.4 Graphs imply the effective permittivity of a plasmonic nanocomposite (20% metal nanoparticles [NPs]) (the real and imaginary parts are shown in blue and red, respectively), determined via full-wave finite-difference time-domain (FDTD) simulations (*solid lines*) and an Maxwell Garnett (MG) effective medium theory (*dashed line*). *Reproduced with permission C. Etrich, S. Fahr, M.K. Hedayati, F. Faupel, M. Elbahri, C. Rockstuhl, Effective optical properties of plasmonic nanocomposites, Materials 7 (2) (2014) 727–741. Copyright 2014, MDPI.*

theory can only be applied for simulation of the low filling factor composites [24]. In other words, when the volume fraction of the embedded particles exceeds 15%, the MG model fails to predict the optical response of the matter [25].

Recently, effective medium guided by full wave simulations has been applied for modeling of the nanocomposites, which can successfully predict the optical properties of the nanocomposites up to near percolation state (Fig. 7.4). In this approach, the rigorously calculated complex reflection and transmission coefficients from a slab of a nanocomposite are inverted to assign an effective permittivity [13].

7.2.5 The Bruggeman geometry

This geometry consists of two intermixed components possessing different linear and nonlinear optical properties. The linear optical properties are described by the following equation:

$$0 = f_a \frac{\varepsilon_a - \varepsilon_{eff}}{\varepsilon_a + 2\varepsilon_{eff}} + f_b \frac{\varepsilon_b - \varepsilon_{eff}}{\varepsilon_b + 2\varepsilon_{eff}} \tag{7.9}$$

where ε_a and ε_b stand for the dielectric function of component *a* and *b*, respectively, and *f* gives the filling factor of the component and ε_{eff} is the effective dielectric constant of the composite. Although the Bruggeman model better predicts the optical properties of the low filling factor composites, simulation of a 3D high filling factor nanocomposite with randomly arranged particles is not possible with this theory. In fact, both of these theories (MG and Bruggeman) are based on a critical assumption considering negligible particles' interactions due to their far distance and small size (much smaller than the incident wavelength). Therefore, they are unable to compute the optical response of the near percolating (high filling factor) composites. However, Granqvist et al. [26] showed that by including the dipole–dipole interaction in the Bruggeman theory (i.e., the extended model), the electrical properties of the near percolating composites can be predicted well. Despite such an enhancement, applicability of this model for prediction of optical properties of a percolating nanocomposite is doubtful.

7.2.6 Percolation threshold

In this section, we mainly focus on the percolation threshold in the optical nanocomposites, and we skip other aspects of the concept such as heat transfer, etc. Simply speaking, the percolation threshold corresponds to the minimum concentration at which an infinite cluster spans the space [27]. In other words, when the fraction of particles in a composite reaches a number where most of them are more or less interconnected, the percolation threshold is exceeded. Obviously, the optical behavior of the composite changes dramatically since the interaction of the nanoparticles is not any more negligible, and hence the effective medium theories (EMTs) are out of application. In fact, the exotic properties of the nanocomposite do not emerge as long as the filling factor of the nanocomposite is far below the PT. At the threshold, the electromagnetic confinement in the gap between the near percolating nanoparticles gives rise to some amazing effects such as a broadband resonance, a hot spot, a magnetic field in the visible frequencies among others, which are significantly different from those of their constituents [28].

Prediction of the optical properties of three-dimensional metal–dielectric nanocomposites is of important research topics in the optical society due to their diverse applications in optoelectronics [28]. In the next subsections, various methods for numerical analysis of optical responses of the plasmonic nanocomposites near the percolation threshold in different geometries and for a variety of applications will be presented and discussed.

As pointed out in the last sections, the EMTs fail to predict the optical behavior of a nanocomposite with random size and shape distribution (certainly near percolation). This is due to the strong interaction of the dipoles (electrical and magnetic field of the polarized particles) which are not considered in EMT. For periodic structures or well-defined geometries, numerical methods such as finite-element and finite-difference or even analytical methods have been applied. On the other hand, for the totally random structures, Finite-Difference Time-Domain (FDTD) technique has been implemented and promising results have been achieved [28]. It is worthy to note that very recently, FDTD was shown to be successful in recognition and guidance of the EMT calculation [13]. In this regard, the clusters' interaction was taken into consideration to simulate the existence of a high local electrical field between them, so-called "hot zones," that was not attainable by a simple EMT [13]. Fig. 7.5 shows amplitude distribution of the electric field in the cross-section of a silver plasmonic nanocomposite. The incident light wavelength of 430 nm matches the plasmon resonance wavelength of each nanoparticle, as represented by formation of intensive and steady fields within the nanoparticles. The excitation level of all nanoparticles is similar. At any locations, where several nanoparticles are in a short spacing, hot spots are formed. Accordingly, capacitive coupling occurs between the adjacent nanoparticles leading to an increase in the resonance strength and also redshifting of the resonance wavelength. Such a phenomenon has been investigated in detail for dimers or periodic assemblies of nanoparticles, and as discussed earlier its

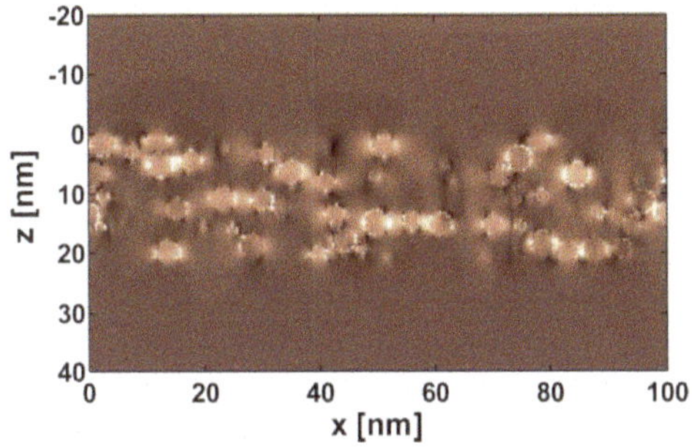

FIGURE 7.5 The electrical field amplitude distribution across the cross-section of a plasmonic nanocomposite (between z = 0 nm and z = 20 nm) (containing 20% of arbitrarily arranged separate silver nanoparticles [NPs] as small as 4.1 nm in a generic nondisperse dielectric matrix [$\varepsilon = 2.25$]). Note that an incident light with wavelength of 430 nm has been regarded in this study. *Reproduced with permission C. Etrich, S. Fahr, M.K. Hedayati, F. Faupel, M. Elbahri, C. Rockstuhl, Effective optical properties of plasmonic nanocomposites, Materials 7 (2) (2014) 727–741. Copyright 2014, MDPI.*

occurrence is also likely in amorphous systems [29–31]. As to be shown later, such extra resonances are crucial due to their large impact on the effective properties of the nanocomposite. Despite such importance, the coupling events are rarely accounted in the conventional homogenization theories.

The theory of optical phenomena in fractal clusters and percolation systems has been under development within the last couple of decades, in particular, by Berry [32], Stroud [33], Bergman [34], Fuchs and Claro [35], Devaty [36], Brouers [37], Niklasson [38], and by Markel, Stockman, Shalaev, and their coworkers [39–41]. In 1987, Shalaev et al. for the first time developed a theory for calculation of the optical properties of the fractal clusters (that can be considered as a random 3D composite) [42,43]. Experimental measurements later on proved the proposed theory of the nonlinear optical properties and broadband absorption of metallic clusters (in the case of a silver solution) [44]. In fact, Shalaev et al. attributed the high scattering intensity of the aggregated (percolated) fractal clusters to the existence of a high local field present between the clusters which they called it "hot zones." They mentioned that the enhancement is valid for subwavelength clusters as well [45]. In order to simulate the optical response of a real percolating composite with random particles size and shapes, Shalaev and coworkers developed a method to incorporate realistically the nongeometrical shape of the NPs in the calculation. They performed full-wave simulations using the 3D FDTD method, while extracting the shape geometry of the NPs from real scanning electron microscope (SEM) images [46]. They used a number of SEM images of real films as the inputs of the FDTD simulations. The images were converted to binary (black/white) ones with the help of a proper threshold to differentiate the areas on the film that were covered by metal from those by dielectric. To ensure sufficient statistical averaging, numerous SEM images were processed and binarized for each of the samples. Their results fit well with experiments even for the near percolation threshold. To predict the optical response of any composite (via theoretical simulation), they proposed a computer-generated image analysis (with FDTD) for a composite with random distribution of particles. The only difference of this latter method to the former one is that the nanocomposite structure with a certain filling factor and thickness should be generated virtually by a PC instead of taking images of a real sample. By this technique, one can also compute the field distribution along the NPs embedded into the matrix and hence fully analyze the optical response of a nanocomposite to an electromagnetic field [46].

7.3 Practical considerations

Plasmonic nanocomposites can be fabricated in different ways such as reduction of a metal salt [47], ball milling [48], plasma polymerization in conjunction with metal deposition [49,50], chemical vapor deposition (CVD) [51], coevaporation [52], and cosputtering [53,54]. Although the chemical-based methods have also been implemented for this purpose, difficult control of the composition as well as particle distribution [55] makes them unreliable for mass production. While physical vapor deposition (PVD) has proven to be very successful in fabrication of optical nanocomposites, wet chemical methods offer much better control of particle size and shape [55]. The PVD techniques are generally based on co- or tandem deposition of the metallic and organic components and formation of metallic nanoparticles by self-organization. The nanocomposites which have been produced by the PVD techniques are more applicable for modern utilities, thanks to their well integration into the microelectromechanical system (MEMS) and photovoltaic industries since 1980. To get more detailed information, the interested readers are referred to some useful review articles related to the production methods of nanocomposites (e.g., Refs. [48,55]).

Regardless of the production techniques, the optical properties of the nanocomposites can be generally understood and tuned through three main parameters of nanoparticles size and shape, volume fractions, and material type.

7.3.1 Plasmonic nanoparticle size

Among the different parameters which influence the interaction of a metal–dielectric nanocomposite and light, the size of the metallic particles plays an important role [56]. It is well known that the resonance peak of single nanoparticles is quite sharp (especially in the case of silver) and that attracts the attention of engineers for sensor applications. When the particles radius increases, the retardation effects become increasingly important and the radiative losses rise, as well. Additionally, the resonance peak intensity and bandwidths increase, and at the same time when the particles grow, the resonance peak position shifts toward lower frequencies (blueshift) [56]. It is worthy to note that in bigger particles, the contribution of scattering is higher than that of absorption. In other words, the absorption (loss) of the electromagnetic wave more likely happens in small (below 10 nm) particles [57]. The scattering and absorption cross-section can be calculated as follows:

$$\sigma_{sca} = \frac{2\lambda^2}{3\pi}\alpha^6\left|\frac{n^2-1}{n^2+2}\right|^2 \tag{7.10}$$

$$\sigma_{abs} = \frac{-\lambda^2}{3\pi}\alpha^3 Im\left\{\frac{n^2-1}{n^2+2}\right\}^2 \tag{7.11}$$

Where λ is the light wavelength, α is the size parameter, and n is the complex refractive index of the matter. Concluding from Eqs. (7.10) and (7.11), for a small particles size, the absorption is dominant (α^3), while for bigger particles, scattering is much stronger. This means that when the particles embedded into the matrix are small, the incident light more likely will be absorbed and dissipated as heat within the metals. It is worth mentioning that no matter the weight of contribution of scattering or absorption into extinction, when the particle's size increases, the corresponding resonance peak red-shifts and its bandwidth broaden.

7.3.2 Plasmonic nanoparticle filling factor

The volume ratio of metal particles to the whole nanocomposite (i.e., filling factor) affects the peak resonance of a plasmonic nanocomposite because of several reasons following. First, increment of the filling factor means that the number or size of the particles embedded in the matrix increases and results in broadening as well as shifting of the resonance to the lower frequencies [58]. The redshifting by large nanoparticles as previously discussed is attributed to the retardation effect. On the other hand, when the number of the particles raises, the optical response of the composite will be the collective response of different particles, therefore the resonance of the inclusions overlaps and makes the peak broader [25]. It is worthy to note that near the percolation threshold, not only the mentioned effects but also the multipolar and multiple particle interactions of the neighboring polarized particles further broaden and redshift the resonance peak [25].

7.3.3 Plasmonic nanocomposite's components

Intrinsic properties of the metallic constituent as well as the dielectric function of the surrounding environment mutate the resonance behavior of the plasmonic nanocomposites as well. As mentioned in Section 2.1, the particle plasmon resonance strongly depends on the type of the dielectric and the metal, and any changes in these parameters alter the peak position and its corresponding bandwidth [25]. For instance, the resonance peak of silver and gold with the average size of 5 nm in a polytetrafluoroethylene (PTFE) matrix appears at around 400 and 540 nm wavelength, respectively (provided that the filling factor of the composite does not exceed 20%) [54]. On the other hand, replacing the matrix of the composite with a matter possessing a higher refractive index leads to a shift of the plasmon resonance to a longer wavelength. This shift is attributed to the retardation effect of the matrix with a higher refractive index on the resonance of the metallic particles [59].

In addition to the abovementioned factors, the components of the composite are also determining on the resulting optical response of the system, as witnessed by several researches. For instance, Deki et al. [60] devised a composite thin film consisting of nanosized Au/Pd alloy particles dispersed in nylon 11. The films were prepared by a thermal relaxation technique combined with a coevaporation method. They showed that the absorbance band of the composite is originated from the gold's surface plasmon resonance (SPR), and therefore changing the ratio of Au/Pd alters the resonance intensity. In other words, the band structure of gold upon doping with more than 25% palladium changes so that the localized surface plasmon resonance (LSPR) of gold disappears. Mbhele et al. [61] reported fabrication of a silver–polyvinyl alcohol (PVA) nanocomposite by mixing a colloidal solution consisting of silver NPs with an aqueous solution of PVA in given ratios. Via this technique, various contents of silver NPs were incorporated into the PVA film after solvent evaporation. It was observed that the position of the absorbance band of silver in the solution and the polymer film differs. The shift to the lower frequencies and broadening of the surface plasmon absorption band upon inclusion of Ag nanoparticles into PVA was accredited to the probable agglomeration of the Ag nanoparticles and/or change of the dielectric properties of the surrounding environment. TEM analysis of the film also confirms the agglomeration of the particles in the solid polymer matrix. In a work reported by Zhang and Han [62], silver nanoparticle–polyacrylonitrile (PAN) composites with given particle sizes were synthesized by a simultaneous polymerization–reduction approach. They observed a direct relationship between the silver content of the nanocomposite and corresponding absorption intensity of the film. A clear redshift of the resonance peak was also observed from 413 to 439 nm when the size of the NPs increases. Based on their findings, the size distribution of the NPs embedded in the polymer matrix was very small that led to a narrow plasmon resonance band of the film. Employing a codeposition technique based on high-temperature evaporation, Faupel and coworkers [63] studied the optical behavior of several nanocomposites including Teflon AF/Ag, Teflon AF/Au, and Teflon AF/Cu. They implemented a mask during deposition to create a film composed of all the mentioned nanocomposites, i.e., a system with multiple plasmon resonance peaks. When exposed to the visible light, the system generates several colors whose origin is the resonance of each nanocomposite. Such a feature enables application of this system for multicolor filters. A polymer–metal nanocomposite with tunable refractive index was reported by Liu et al. [64]. In their work, silver nanoparticles were embedded in a polymerizable surfactant in order to realize gradient refractive index (GRIN) plastic rods. It was observed that the nanoparticles can raise the refractive index of the plastic rods effectively, thus changing the optical response of the system because of the correlation of plasmon resonance and refractive index. Schürmann et al. [54] cosputtered silver and PTFE for fabrication of a nanocomposite and subsequently investigated the optical and electrical properties of such a system. Cosputtering from two independent magnetron sources was implemented to develop the polymer–metal nanocomposite films. Electrical measurements showed that resistivity of the film significantly drops from 10^7 to $10^{-3}\ \Omega\ \mathrm{cm}$ upon exceeding the percolation threshold which was attributed to the interconnection of the particles at high filling factors resulting in a higher conductivity. The optical absorption in the visible region due to the SPRs was also strongly dependent on the metal content, represented by a redshift of the absorption peak from 405 nm to over 500 nm at higher silver contents. When the metal filling factor in the composite films raises up to near the percolation threshold, the gaps within single clusters become smaller. Moreover, due to the coalescence of the particles, the cluster's size and shape change, leading to a considerable redshift of the absorption maximum. Also, induced by the wider size distribution and more irregular shape of the clusters, a broadening effect of the resonance happens at higher filling factors. In another report by Faupel and co-workers [65], the optical properties of polymer–metal nanocomposites were tailored by changing their matrix. In this regard, the nanocomposite films containing Ag nanoparticles embedded in a polymer matrix of Teflon AF, poly(methyl methacrylate) (PMMA), and Nylon 6 were fabricated by vapor-phase codeposition in high vacuum. They showed that the plasmon resonance frequency varies in the visible region by increasing the Ag volume fraction from 4% to 80% (Figs. 7.6A–C). Based on their finding, not only the higher filling factors shift the plasmon resonance to the lower frequencies, but also the larger refractive index of the surrounding environment (e.g., Nylon 6 vs. PTFE) redshifts the resonance, in agreement with the Mie theory.

Pucci et al. [66] fabricated nanocomposites based on the vinyl alcohol–containing polymers and nanostructured gold by a UV photoreduction process. It was found that before irradiation, the film has a light yellow color originated from Au^{3+} ions, while it turns purple after 5 min of UV exposure, due to the presence of the surface plasmon absorption band at about 550 nm. The disappearance of the initial resonance peak upon

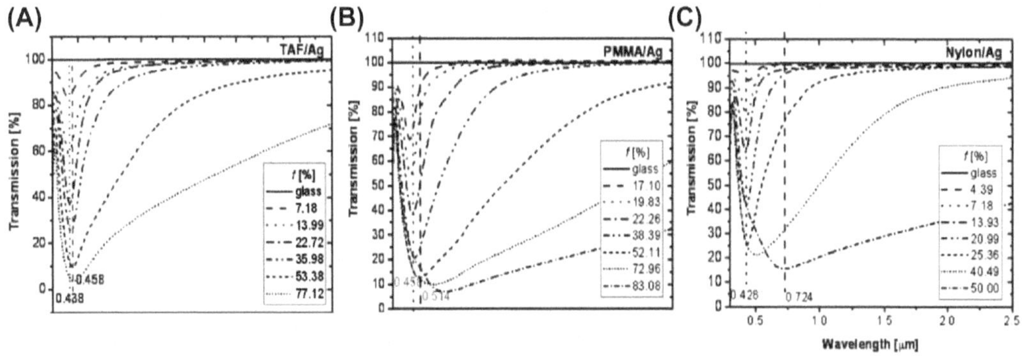

FIGURE 7.6 Transmission spectra of the plasmonic nanocomposites of (A) Teflon AF (TAF)/Ag, (B) Polymethyl methacrylate (PMMA)/Ag, and (C) Nylon/Ag with different metal filling factor. *Reproduced with permission H. Takele, H. Greve, C. Pochstein, V. Zaporojtchenko, F. Faupel, Plasmonic properties of Ag nanoclusters in various polymer matrices, Nanotechnology 17 (14) (2006) 3499. Copyright 2006, IOP Publishing Ltd.*

illumination indicates that the Au^{3+} ions are quickly converted to the nanostructured Au(0) even in a solid host matrix. However, the authors attributed the different absorption features (e.g., bandwidth, intensity and shift of the peak) at 5 and 10 min of irradiation to a probable increase in the number of the NPs with irradiation time. Chatterjee and Jewrajka [67] fabricated a nanocomposite made of metal alloys and polymer and studied its optical properties. For this sake, they synthesized [poly(2-(N,N-dimethylamino)ethylmethacrylate)]–b–poly(methylmethacrylate)–b–[poly(2-(N,N-dimethylamino)ethyl methacrylate)] (Mn = 45,000; 20–20 K; PDI = 1.2) block copolymer surfactant-stabilized amphiphilic gold–silver alloy nanoparticles (Au–Ag [PDMA–*b*–PMMA–*b*–PDMA]) in both water and an organic medium. They observed that the plasmon resonance peak of the alloy in the solid matrix redshifts compared to the particles suspended in a solution even though the particles size was equal in both matrices. They concluded that change of the refractive index could be the only reason for the shift. In another relevant study, Avasthi et al. [68] developed thin films of Ag–PET nanocomposite via atom beam cosputtering. The transmission measurement of the nanocomposite showed a dependency of the resonance intensity to the filling factor of the metallic particles. When the atomic concentration of the metal inside the film rises from 30% to 75%, the peak intensity declines from 28% to 16%. The authors attributed the narrow transmission at 320 nm to the absorption edge of the polymer and the high-energy absorption edge of the metal fractals in the nanocomposite. They also accredited the broad resonance of such a composite at high filling factors to the excitation of quadrupole resonance of the interconnected, coarse particles. Božanić et al. [69] fabricated a nanocomposite comprising silver (Ag) and silver sulfide (Ag_2S) nanoparticles in a sago starch matrix. The optical measurement of the aqueous solution of the sago starch–Ag nanocomposite revealed the existence of one and two distinctive SPR bands at low and high concentrations of the nanocomposite, respectively. As the authors state, in the case of the solutions with higher concentrations of the nanocomposite, the first peak is related to the silver NPs in water while the second one to the silver NPs in the partly dissolved starch matrix. Mishra et al. [70] studied a nanocomposite made of silver NPs and silica prepared by atom beam sputtering. Optical investigation of the nanocomposite revealed a blueshift in the SPR peak position along with an increase in full-width half-maximum (FWHM) when the filling factor of the film decreased. They attributed both the observed phenomena to the reduction in particle size upon a decrease in the metal fraction, leading to large electron scattering and hence broadening of the SPR absorption band. In addition, the large size distribution and interparticle distance of the embedded NPs contribute to broadening of the resonance peak. The gold–silica nanocomposite is another system fabricated and studied by Mishra et al. [71]. The plasmon resonance of the lower filling factor composite appears at a higher frequency, because of the small size of the silver particles. The Landau damping is one of the reasons for broadening of the resonance peak at smaller particle sizes. When the size of the NPs is extremely small (e.g., 1–4 nm), fast relaxation of surface plasmons occurs. The reason is direct relationship of the electrons' mean free path to the particle size. In addition, size dispersion of gold particles plays a role in the increment of the plasmon bandwidth of the nanocomposite. Jiang et al. [72] developed a silver–silica nanocomposite via a chemical route for optical applications. The silver nanoparticles were mounted onto the silica nanospheres via the ion-exchangeable poly(acrylic acid) (PAAc) ligands grafted on the silica surface. They showed that the LSPR of silvershifts to lower frequencies by increasing its concentration in the composite. The authors attributed the observed redshift (as small as 25 nm) to the size and concentration increment of the particles. Gold–zinc oxide nanocomposites fabricated by atom beam cosputtering and annealed at a high temperature were studied by Mishra et al. [73]. The annealing process was carried out in the temperature range of 200–600°C for 1 h. Upon annealing, the LSPR peak of the films becomes narrower and shifts to a longer wavelength which is attributed to the growth of the NPs at high temperatures. They confirmed this observation by X-ray diffraction (XRD) analysis where the change in interplanar spacing of gold was the sign of grain growth at elevated temperatures. Susha et al. [74] developed a highly luminescent metal–polymer nanocomposite via thermal decomposition of gold(I) dodecylthiolate. This precursor was previously synthesized by treating an ethanol solution of gold tetrachloroauric acid with an ethanol solution of 1-dodecanethiol. The authors found out that upon heat treatment, the films undergo a transition from the well-resolved spectrum of a gold(I) thiolate complex peaked at 390 nm to a broad plasmon peak of metallic gold at 600 nm. Additionally, the red emission of the sample appeared at 620–640 nm whether before or after the heat treatment. It was observed that after a prolonged heating, the emission is reduced due to the growth of the NPs. Mohapatra et al. [75] fabricated gold–silica core-shell nanocomposites by atom beam cosputtering. It was shown that increasing of the Si content of the film leads to an enhancement in intensity along with a redshift in the LSPR stemming from an increase in the refractive index of the matrix. In addition, annealing of the sample redshifts the LSPR peak and slightly shrinks the peak width. The authors accredited this observation to the growth of the nanoparticles upon annealing at high temperatures. Faupel and coworkers [76] made a nanocomposite with equal intensity double plasmon resonance (EIDPR) peak by coevaporation from three different sources. They realized EIDPR by using a sandwich geometry made of silver, TAF, and gold layers. They investigated the effect of different parameters such as thickness of each layer and the order of layers on the optical response of the system. They found out only at the optimum conditions, i.e., when silver is sandwiched between TAF films followed by an ultrathin gold cluster layer on top, EIDPR can be realized. On the other hand, when the order of silver and gold is changed, only one single peak appears which shows the domination of the top silver resonance compared to the gold middle layer. Based on their report, the dipole–dipole coupling is at the highest level for a TAF spacing of 3.3 nm, and increasing the spacer layer thickness reduces the coupling efficiency. In another report by Faupel and coworkers [77], the size of the embedded nanoparticles in a polymeric matrix was reduced by MeV ion irradiation. Irradiation of the films with 100 MeV Ag^{8+} ions at fluences from 3×10^{11} to 1×10^{12} ions/cm^2 resulted in the size reduction of the Ag NPs. Optical studies also confirmed that upon irradiation, the plasmon resonance bandwidth of the composite increases and at the same time the resonance shifts to a lower frequency. A further analysis by TEM proved that after ion irradiation, the nanoparticles shrink with a narrower size distribution.

Röder et al. [78] fabricated and studied nanocomposites comprising metal clusters (Ag, Au, Pd, and Cu) embedded in a polymer (polycarbonate, PC) matrix via the pulsed laser deposition technique. Their experimental works showed that the size and amount of the metal clusters are related to the number of laser pulses hitting the respective targets. For example, smaller particles and higher densities of clusters are achieved for Cu and Pd compared to Ag and Au due to their stronger reactivity with the polymer that lowers their diffusivity. In other words, in contrast to Cu and Pd, Ag and Au do not react with the polymer matrix, and therefore they can migrate and diffuse on the surface and make a bigger cluster by coalescence. Besides the detailed analysis of the fabrication technique, the nanocomposite's optical properties were also investigated. They observed that upon increasing the Ag NPs' size, the plasmon resonance peak redshifts up to 100 nm implying the notable change of the NPs' size with increasing the number of laser pulse. Xu et al. [79] developed a silver–TiO_2 nanocomposite and investigated its optical properties. In order to realize a multicolor optical device, they used magnetron sputtering for deposition along with a selective masking technique. Accordingly, they succeeded to create a gradient film made of Ag and dielectric, wherein the optical appearance of each part was different depending on its metal–dielectric ratio. They showed that when thickness of titanium dioxide raises from 0.5 to 7 nm (given the presence of 2 nm silver clusters), the sample's color switches from orange to dark blue corresponding to the wavelength of 460 and 570 nm, respectively. Additionally, in a multilayer system composed of several stacks of Ag/TiO_2/quartz, up to three resonance peaks can be realized due to the excitation of different plasmon resonances of the layers. Vodnik et al. [80] prepared a nanocomposite system by incorporation of the surface-modified Ag NPs in PMMA via in-situ radical polymerization of methyl methacrylate initiated by 2,2′-azobisisobutyronitrile. It was observed that along with a resonance band at around 420 nm (i.e., the plasmon resonance of Ag NPs), a shoulder peak appeared at higher frequencies. However, after dissolving the nanocomposite in chloroform, only the silver resonance peak remained. The authors concluded that the shoulder originated from the particles aggregate within the matrix and their calculations based on the Maxwell Garnet theory also supported this postulation. In other words, they showed that a combination of NPs with various sizes in a matrix can give rise to appearance of multiresonance plasmon peaks. Beyene et al. [81] developed a hybrid system composed of gold and SiO_2 by plasma-enhanced chemical vapor deposition (PECVD). In this work, PECVD is followed by the remote expanding thermal plasma (ETP) assisted rf magnetron sputtering method to deposit Au NPs on SiO_2 layers with given shapes, sizes, and size distributions. Thanks to an optimum controllability over the size and distribution of the gold particles, the plasmon resonance of the nanocomposite could also be tuned. Their experimental data showed that a resonance shift up to 100 nm can be achieved by raising the metal area fraction of the film from 24% to 85%. Tyurin et al. [82] devised plasmonic nanocomposites via a chemical technique. In their approach, silver–polyacrylonitrile (Ag–PAN) nanocomposites were in-situ synthesized by polymerization of acrylonitrile (AN) and simultaneous reduction of silver ions in the mixtures of silver nitrate ($AgNO_3$), AN, and UV photoinitiator (IN). Due to the excitation of plasmon resonance of silver at the visible frequency, the formation of Ag NPs during the polymerization could be visually tracked when the mixture's color shifts from colorless to yellow-brown by increase of the silver nitrate content. This color change was also revealed in the measured optical spectra by emergence of an absorption peak around 420 nm (the LSPR of Ag NPs). Liu et al. [83] also fabricated a metal–polymer nanocomposite by PVD. By cosputtering from gold and PTFE target sources, a homogeneous nanocomposite was made, as similarly done earlier by Faupel and coworkers [65]. Their experimental works showed that optical absorption of the nanocomposite is strongly dependent on the metal filling factor. Specifically, the absorption peak redshifts from 550 to over 700 nm at higher Au contents. In a recent work reported by Faupel and coworkers [84], a polymer–metal nanocomposite containing Ag1–xCux alloy particles was made by a vapor phase codeposition method. The method allows production of NPs with a controlled alloy composition, metal filling factor, and nanostructure in a protecting Teflon AF matrix. They showed that the plasmon resonances can be tuned in a wide range of frequencies upon changing the influential parameters of the nanocomposite such as alloy composition, metal filling factor, and nanostructure. Appearance of only one LSPR peak in a sample containing low filling factors and small particle sizes was attributed to the enhanced miscibility at the nanoscale. However, the double plasmon resonances revealed for bigger particles were in agreement with the phase separation expected from the phase diagram and were explained by formation of core-shell structures with Cu core and Ag shell. Annealing of the multicomponent film at 200°C showed that intensity of plasmon resonance of silver increases and shifts to a lower wavelength, whereas the Cu peak is damped in intensity and shifts to shorter frequencies. The authors attributed this observation to a further phase separation due to an enhanced diffusion and a more pronounced development of the core-shell structure at higher temperatures.

7.4 Reflective plasmonic nanocomposites

So far, research on plasmonic nanocomposites has been mainly focused on their absorption and scattering properties, and the reflection aspect has been rarely taken into account. In this regard, specular reflection effect could be of importance, particularly with respect to reflective coloration. Specular reflection has been generally described considering presence of a smooth and mirror-like surface as a vital prerequisite. In contrast, disordered plasmonic and dipolar media comprising weakly scattering building blocks which could be technologically very promising have not been considered or studied by now. Indeed, this effect has been shown and discussed as purely geometrical, neglecting the illuminated matter enclosed [85]. This assumption contradicts the electromagnetic dipole radiation concept that can explain the phenomenon, as recognized by several physicists [86–88]. We also believe that it is against the theory of dipole radiation elaborating various optical phenomena including the specular reflection. Such a principle was employed in the design of electromagnetically active materials for energy saving applications.

7.4.1 Scientific concept

When electromagnetic (EM) waves hit a resonating dipole (atom, molecule, and small nanoparticle) that is much smaller than their wavelength, the field is considered homogenous. This approximation, first introduced by Rayleigh [89], is called the quasistatic approximation. Accordingly, the electric field inside the dipole will drive the electron cloud relative to the positive core. The attraction between the negative charges and the core ions causes a restoring force that establishes oscillations as determined by Lorentz in the classical spring module (Fig. 7.7A). The dipole's oscillation creates an alternating field, leading to charge acceleration. In the simplest case, the accelerated charges result in the electrodynamically coupled electric and magnetic fields, thus radiation of photons away from the excited matter through its surface boundary. Accordingly, the dipole oscillation should be described by the dynamic electronic polarization with real and imaginary dielectric function, Fig. 7.7B, and not by the static one as usually thought.

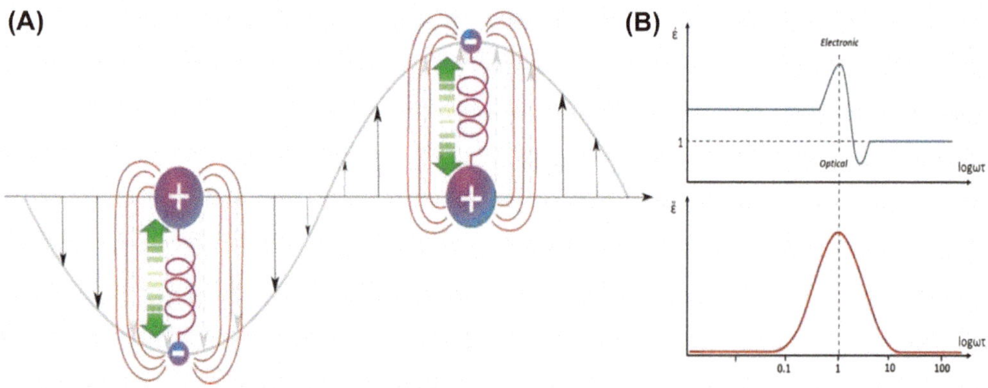

FIGURE 7.7 (A) Schematic illustrates an oscillating electric field moving the charges of a dipole relative to each other. (B) Variation of the dielectric function versus the frequency implying electronic polarization. *(A) Reproduced with permission M. Elbahri, A.U. Zillohu, B. Gothe, M.K. Hedayati, R. Abdelaziz, H.J. El-Khozondar, M. Bawa'aneh, M. Abdelaziz, A. Lavrinenko, S. Zhukovsky, S. Homaeigohar, Photoswitchable molecular dipole antennas with tailored coherent coupling in glassy composite, Light Sci. Appl. 4 (7) (2015) e316. Copyright 2015, Nature Publishing Group.*

Based on the spectroscopic conformity principle, the dynamic polarizability can be described as:

$$\alpha(\omega) = \frac{e^2}{m}\sum_n \frac{f_{n0}}{\omega_{n0}^2 - \omega^2 - i\omega\delta_{n0}} \tag{7.12}$$

where δ_{n0}, ω_{n0}, and f_{n0} are the damping constant, eigenfrequency and the oscillator strength. If the external field frequency equals one of the eigenfrequencies of the transition oscillators, so that the resonance condition, i.e., $|w - w_{n0}| \le w_{n0}$ is fulfilled and one resonant summand in the sum (Eq. 7.12) is maintained, the expression for resonant polarizability can be calculated from Eq. (7.12) as follows [91]:

$$\alpha_{res}(\omega) = \left(\frac{e^2}{2m\omega_{n0}}\right)\frac{f_{n0}}{\omega_{n0} - \omega - i\delta_{n0}/2} \tag{7.13}$$

Evidently, the resonant polarizability is a complex value whose real part can be either positive (for molecules and dielectric dipoles) or negative (for metals). Its imaginary part is associated with the damping constants of the transition oscillators and responsible for absorption of radiation. Whereas, the real part governs the reflection and refraction of an electromagnetic wave in a medium. Thus, at the resonance, two processes of absorption and specular reflection take place. Challenging the common belief, specular reflection is not dependent on the arrangements of dipoles, rather varies with the dynamic polarizability and thus can occur in the disordered glassy composite systems, as demonstrated by Elbahri et al. [90,91]. It is worthy to mention that specular reflection is associated with the dipolar resonance frequency and emerges by free (metal nanoparticles) and bound electrons of a dipole.

Indeed, specular reflection is a cooperative process. When an ensemble of atoms or dipoles are brought in close proximity, the electronic clouds of the oscillated dipoles act cooperatively. Here, the interference of all the induced waves, which are initiated from individual excited dipoles, leads to a coherent cooperative action. Fig. 7.8A and B visualize the unique properties of a single dipole and the coherent coupling of small dipoles in close proximity, respectively. When several dipoles are located in close proximity, their near field radiation patterns overlap and interfere.

Such mutual coupling gives rise to the radiation modes that are spatially and temporally coherent and differ strongly from the modes of an individual dipole. In the same way, an oscillating sheet of charges (dipolar composites) [90,92] produces electric and magnetic fields, undergoing a local change. Such a local change in the electric field induces, for instance, electric dipole and quadrupole moments, which eventually cause the specular reflection [93,94]. Based on this idea, we have demonstrated the first experimental proof of specular reflection at visible frequencies, which originates from and is controlled by photochromic molecular dipoles embedded in a polymeric matrix under dynamic light switching [90] as well as by metal nanoparticles in dielectric hosts [92]. With respect to the latter instance, reflective coloration of plasmonic nanocomposites composed of Ag nanoparticles embedded in a dielectric matrix was recently

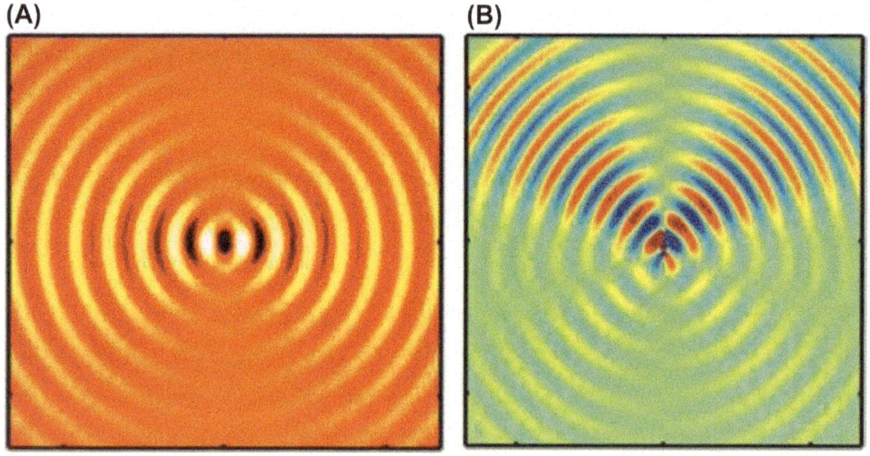

FIGURE 7.8 (A) The Hertzian electric dipole antennas concentrate the electrical current thereby emitting an electric field. (B) Six closely spaced Hertzian dipoles' interference and directed light radiation. *Reproduced with permission M. Elbahri, M. Abdelaziz, S. Homaeigohar, A. Elsharawy, M. Keshavarz Hedayati, C. Röder, M. El Haj Assad, R. Abdelaziz, Plasmonic metaparticles on a blackbody create vivid reflective colors for naked-eye environmental and clinical biodetection, Adv. Mater. 30 (4) (2018). Copyright 2018, John Wiley and Sons.*

shown by us, Fig. 7.9. The specular reflection, peaked at 442 nm (Fig. 7.9A), gives rise to polarization of the incident field thereby generating longitudinal and transverse waves. As an optical antenna, the electromagnetic field is dominated by electric and magnetic fields that are produced from charge separation and current. The total power radiation has two branches, namely the electric field and the transverse magnetic induction also known as transverse electric radiation. The electric dipole with vector lies in the plane (symmetric), while the magnetic dipole with vector lies out of the plane (antisymmetric). In fact, the only difference between them is the polarization. With respect to the s-polarized light reflection (Rs), the specular reflection peak emerges at 442 nm (Figs. 7.9A–C). Given that the oscillation orientation of electric dipoles is typically parallel to that of the reflected light, Rs raises dramatically with the incidence angle. A completely opposite behavior was recorded for p-polarized light (Fig. 7.9D and E), and the reflection peaked at 442 nm gradually vanished and instead a reflection dip appeared at the incident angle of 65 degrees. This performance is introduced as the Brewster wavelength phenomenon. The related experimental results are in a good agreement with those obtained by simulation (Fig. 7.9C and E), i.e., the thin film matrix method based on the effective optical properties of the nanocomposites. Based on the specular reflection behavior shown in Fig. 7.9A, the plasmonic nanocomposite should be bluish. Nevertheless, this color cannot be perceived by naked eyes. To address this issue, a black substrate is employed, thereby light transmission and diffuse scattering are hampered and the color stems merely from the specular reflection effect, as demonstrated in Fig. 7.9F.

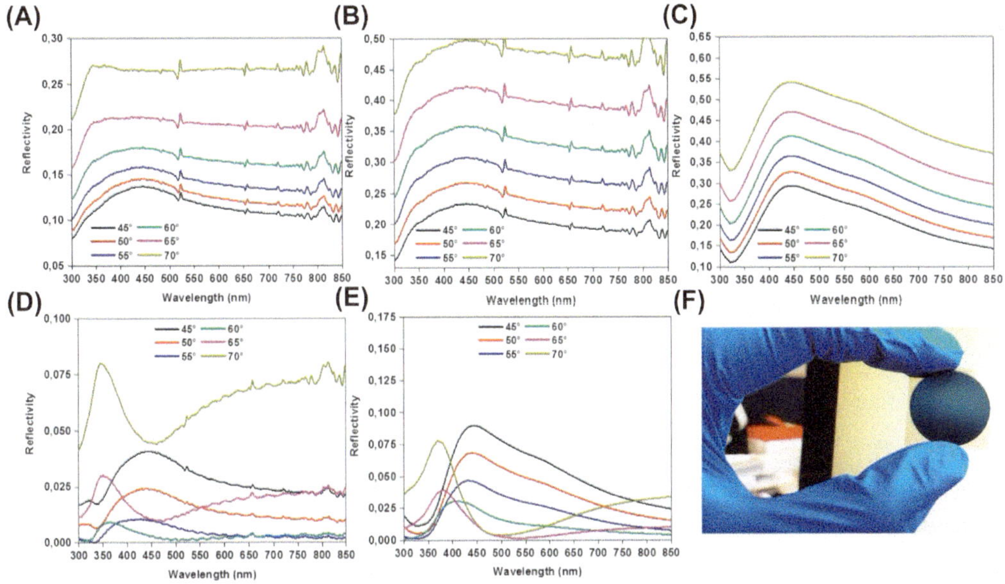

FIGURE 7.9 Specular reflection induced by a plasmonic nanocomposite. (A) The total specular reflection spectra of a plasmonic nanocomposite of Ag/SiO_2 illuminated under different oblique incidence angles. The (B) experimental- and (C) simulation-based s-polarized specular reflection spectra of the plasmonic nanocomposite illuminated under different oblique incidence angles. The (D) experimental- and (E) simulation-based p-polarized specular reflection spectra of the plasmonic nanocomposite illuminated under different oblique incidence angles. (F) The photograph shows a carbon tape overlaid by a thin layer of the nanocomposite. *Reproduced with permission M. Elbahri, M. Abdelaziz, S. Homaeigohar, A. Elsharawy, M. Keshavarz Hedayati, C. Röder, M. El Haj Assad, R. Abdelaziz, Plasmonic metaparticles on a blackbody create vivid reflective colors for naked-eye environmental and clinical biodetection, Adv. Mater. 30 (4) (2018). Copyright 2018, John Wiley and Sons.*

The reflective coloration discussed here emerges from a unique field based on cooperative radiations of plasmonic dipoles. To fully benefit from the fascinating properties of cooperatively and coherently interacting ultrafine particles and dipoles, they were mounted near/at a metallic mirror. This approach is efficient in validation and realization of the plasmonic interference effect for various sophisticated optical applications. For instance, the first plasmonic transparent conductor, Fig. 7.10A, was developed based on a hybrid, layered structure comprising a plasmonic nanocomposite film atop a metal film, Fig. 7.10B. As shown in Fig. 7.10C, utilizing the abovementioned plasmonic coupling and interference [95], an omnidirectional transparent metal can be created. Moreover, when the proximity is altered, e.g., by introducing a passive dielectric interlayer between the base metal and the composite overlay, the device can be evolved as a selective solar absorber [4,5,96,98], Fig. 7.10D, with complete absorption of visible light and black color, Fig. 7.10E. Replacing the passive interlayer with an active polymer composite (chromophore-doped polystyrene [PS]) results in the design of a smart molecular plasmonic cavity, characterized by weak and strong coupling in the off and on

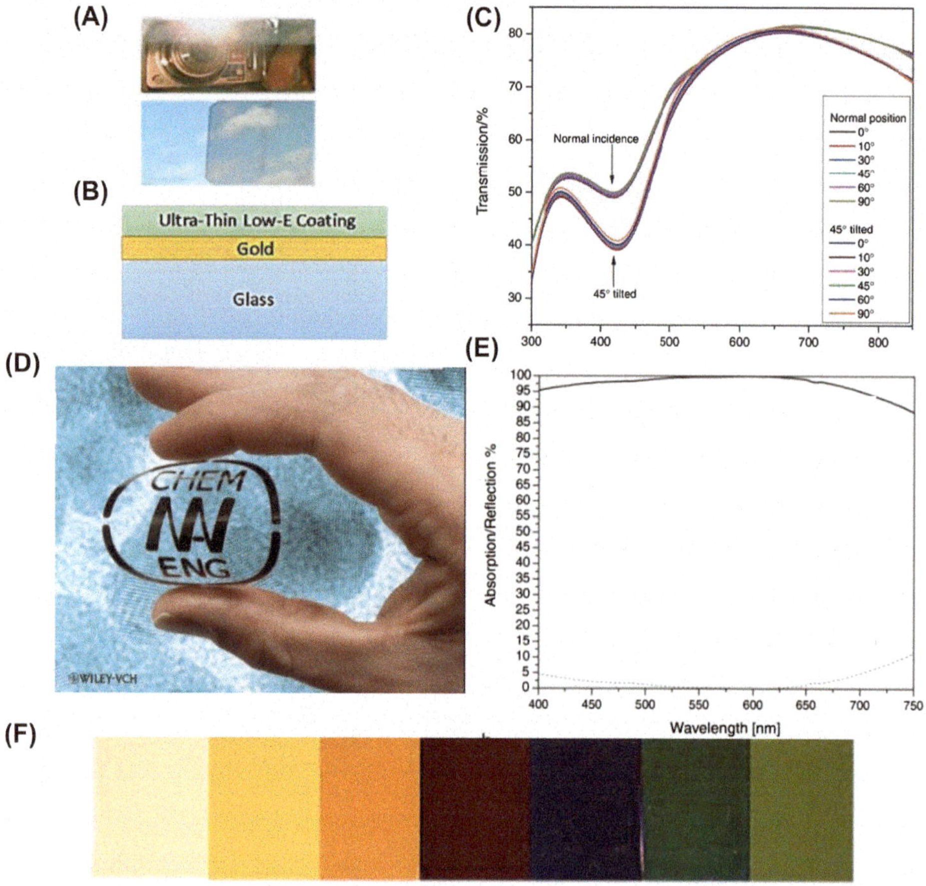

FIGURE 7.10 (A) Camera image of a 17 nm Au film as (top) uncoated and (bottom) coated with an Ag nanocomposite implying the transmission magnitude before and after coating. (B) Schematic illustration of the structure made for low-E coating and/or transparent conductor. (C) Angular sensitivity of the film's transmission property. (D) Camera image of a perfect black plasmonic absorber deposited on a flexible substrate. (E) The absorption (*solid line*) and reflection (*dashed line*) spectra of the black absorber at the visible frequencies. The absorber comprises the 20 nm nanocomposite atop a 20 nm silica spacer coated 100 nm gold film, (F) The strongly dispersive nature of above ultrafine nanocomposite is the basis of formation of a plasmonic rainbow. *(C) Reproduced with permission M. Elbahri, M.K. Hedayati, K. Chakravadhanula, V. Sai, M. Jamali, T. Strunkus, V. Zaporojtchenko, F. Faupel, An omnidirectional transparent conducting-metal-based plasmonic nanocomposite, Adv. Mater. 23 (17) (2011) 1993–1997.Copyright 2011, John Wiley & Sons. (D&E) Reproduced with permission M.K. Hedayati, M. Javaherirahim, B. Mozooni, R. Abdelaziz, A. Tavassolizadeh, V.S.K. Chakravadhanula, V. Zaporojtchenko, T. Strunkus, F. Faupel, M. Elbahri, Design of a perfect black absorber at visible frequencies using plasmonic metamaterials, Adv. Mater. 23 (45) (2011) 5410–5414. Copyright 2011, John Wiley & Sons. (F) Reproduced with permission M.K. Hedayati, S. Fahr, C. Etrich, F. Faupel, C. Rockstuhl, M. Elbahri, The hybrid concept for realization of an ultra-thin plasmonic metamaterial antireflection coating and plasmonic rainbow, Nanoscale 6 (11) (2014) 6037–6045. Copyright 2014, Royal Society of Chemistry.*

state, respectively [99]. As we showed, the strongly dispersive nature of such an ultrafine nanocomposite can be used for development of a plasmonic rainbow, Fig. 7.10F [97].

7.5 Outlook; plasmonics in molecule

As a new generation of plasmonic nanocomposites, incorporation of plasmonic molecules that undergo stimuli-induced reversible changes into polymeric hosts can boost the area of advanced, dynamic optics. According to the quasistatic approximation and similar to polarizability effect that generates plasmon in a metallic structure, electronic polarization of oscillating molecules allows collective electron displacement that can be further enhanced for the conjugated molecules containing delocalized electrons in the π–π^* orbitals. Here the molecules behave as a set of classical oscillators despite their electronic transitions. Owing to the change of the real part in the dynamic polarizability, specular radiation and the Brewster wavelength emerge. Based on this idea, we have demonstrated the first experimental proof of specular reflection at the visible frequencies, originating from the photochromic molecular dipoles embedded in a polymeric matrix under dynamic light switching [90]. In this context, the magnetic induction, which is a radiated field polarized within the plane of incidence, would be only observed under illumination with p-polarized light and hence is characterized by diminishing the radiation whenever the incident light becomes perpendicular to the axis of the oscillating dipole antenna (Fig. 7.11A). Hence, such conditions leads to a dipolar Brewster phenomenon featuring a reflection dip at a certain wavelength that depends on the

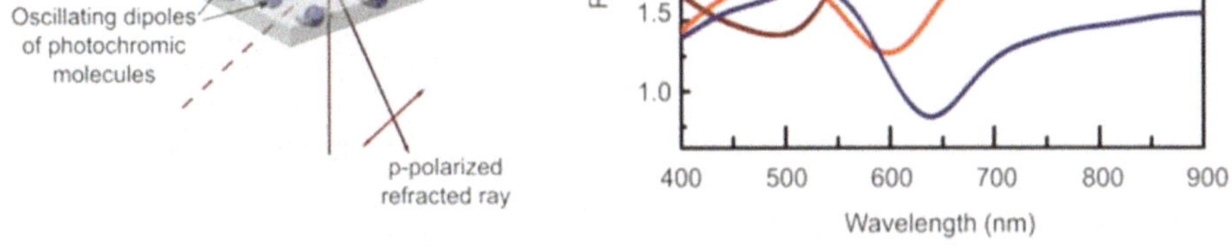

FIGURE 7.11 (A) Sketch illustrating that the oscillation dipoles (*blue spheres*) in the matrix material cause the Brewster phenomenon. (B) Ellipsometry results of UV light–exposed samples showing a significant, matrix-dependent change in the Brewster wavelength, observed at the Brewster angle of 65 degrees (p-polarized light). *Reproduced with permission M. Elbahri, A.U. Zillohu, B. Gothe, M.K. Hedayati, R. Abdelaziz, H.J. El-Khozondar, M. Bawa'aneh, M. Abdelaziz, A. Lavrinenko, S. Zhukovsky, S. Homaeigohar, Photoswitchable molecular dipole antennas with tailored coherent coupling in glassy composite, Light Sci. Appl. 4 (7) (2015) e316. Copyright 2015, Nature Publishing Group.*

mutual polarizability of the medium. Here, there is a mutual coupling between the dipole antenna generated by excitation and the permanent dipole of the surrounding matrix. Due to the dipolar coupling, a polarization is extrinsically and anisotropically induced that can govern the optical response. Based on this concept, the tailored molecular Brewster wavelength has been introduced. In other words, while the Brewster angle is determined by the refractive index at the interface, the Brewster wavelength is tailored by the dipolar coupling of the surrounding matrix and the matter enclosed. Thus, the surrounding media whose refractive index match (showing similar Brewster angles) can be easily distinguished based on the dipole moment (i.e., polarity) of their pristine matrix (the Brewster dip at different wavelengths). Fig. 7.11B shows the reflectance spectra of PS and PVDF (polyvinylidene fluoride) doped with SPO (spirooxazine) which were measured at 45 and 65 degrees. It is obvious that PVDF is strongly redshifted at the higher incident angle compared to the PS film. Although the refractive index of PS (1.58–1.61) is higher than PVDF's (1.42), the dipole moment of PS (0.2–0.3) is much weaker than PVDF's (2.1), indicating the stronger dipolar coupling with PVDF [90]. Thus, in general, it can be said that a new era of molecular plasmonic is blooming.

7.6 Summary

In this chapter, plasmonic nanocomposites were studied theoretically and practically to highlight their potential for advanced applications. Even though the topic is quite old, the recent progresses in nanotechnology and analytical techniques have given rise to emergence of a huge number of novel applications for this class of hybrid materials. The new concept of reflective plasmonics by such class of materials is at infancy but promising for scientists to realize some exotic properties not attainable by the plasmonic absorption of light. Besides the fascinating specular reflective properties of disordered plasmonic nanocomposites, the combination of these materials with thin metallic films opens up a new route for their implementation in energy area. As mentioned earlier, metallic nanoparticles in the vicinity of a film can trap energy in a small scale which holds promise for their use in thin film solar cells and sensors.

Since the properties of nanocomposites are dependent on the size, shape, type, and distribution of the embedded nanoparticles as well as the type of the surrounding environment, a large number of new nanocomposites can be designed and fabricated. Particularly, thanks to the plasmonic interference concept, new amazing functionalities for the plasmonic hybrid systems will be envisioned.

Last but not least, this chapter has also spotlighted a new generation of plasmonic nanocomposites based on plasmonic molecules that are believed to play a significant role in the future research in this area. The plasmonic molecules with the HOMO→LUMO-1 excitations bring about a dipolar polarizonic state along the longitudinal axis of the molecule hence induce an electron density oscillation between the ends of the molecule and create a plasmonic response in analogue with that in the metallic NPs. This oscillating dipole not only absorbs light but also specularly reflects the electromagnetic wave. This concept was validated for photoswitchable molecules and extended to metal plasmonic NPs by our group. In this regard, we are certain that a new family of active nano/molecular plasmonic composites will be devised that benefit from the precisely adjusted, engineered microscopic near field coupling for a customized macroscopic performance.

References

[1] W.A. Murray, W.L. Barnes, Plasmonic materials, Adv. Mater. 19 (22) (2007) 3771–3782.

[2] R.C. McPhedran, I.V. Shadrivov, B.T. Kuhlmey, Y.S. Kivshar, Metamaterials and metaoptics, NPG Asia Mater. 3 (11) (2011) 100.

[3] S. Homaeigohar, M. Elbahri, Switchable plasmonic nanocomposites, Adv. Optic. Mater. 7 (1) (2019) 1801101.

[4] M. Hedayati, F. Faupel, M. Elbahri, Review of plasmonic nanocomposite metamaterial absorber, Materials 7 (2) (2014) 1221.

[5] M.K. Hedayati, M. Elbahri, Review of metasurface plasmonic structural color, Plasmonics 12 (5) (2017) 1463–1479.

[6] J. Roqué, J. Molera, P. Sciau, E. Pantos, M. Vendrell-Saz, Copper and silver nanocrystals in lustre lead glazes: development and optical properties, J. Eur. Ceram. Soc. 26 (16) (2006) 3813–3824.

[7] J. Pérez-Arantegui, J. Molera, A. Larrea, T. Pradell, M. Vendrell-Saz, I. Borgia, B.G. Brunetti, F. Cariati, P. Fermo, M. Mellini, Luster pottery from the thirteenth century to the sixteenth century: a nanostructured thin metallic film, J. Am. Ceram. Soc. 84 (2) (2001) 442–446.

[8] I. Reiche, S. Röhrs, J. Salomon, B. Kanngießer, Y. Höhn, W. Malzer, F. Voigt, Development of a nondestructive method for underglaze painted tiles—demonstrated by the analysis of Persian objects from the nineteenth century, Anal. Bioanal. Chem. 393 (3) (2009) 1025–1041.

[9] P. Mazzoldi, C. Sada, A trip in the history and evolution of ion-exchange process, Materials Science and Engineering: B 149 (2) (2008) 112–117.

[10] J.M. Garnett, Colors in material glasses and metal films, Trans. Roy. Soc. 53 (1904) 385–420.

[11] P. Sciau, Nanoparticles in ancient materials: the metallic lustre decorations of medieval ceramics, the delivery of nanoparticles, Intech 115 (2012) 525–540.

[12] S. Blundell, K. Blundell, Concepts in Thermal Physics (Comprehensive Assessment of Water Management in Agriculture), Oxford University Press, USA, 2006.

[13] C. Etrich, S. Fahr, M.K. Hedayati, F. Faupel, M. Elbahri, C. Rockstuhl, Effective optical properties of plasmonic nanocomposites, Materials 7 (2) (2014) 727–741.

[14] K.L. Kelly, E. Coronado, L.L. Zhao, G.C. Schatz, The optical properties of metal nanoparticles: the influence of size, shape, and dielectric environment, J. Phys. Chem. B. 107 (3) (2003) 668–677.

[15] H. Takele, Optical and Electrical Properties of Metal-Polymer Nanocomposites Prepared by Vapor-Phase Co-evaporation, PhD thesis, CAU zu Kiel, Germany, 2008.

[16] S. Moiseev, Optical properties and some applications of plasmonic heterogeneous materials, metal, ceramic and polymeric composites for various uses, Intech (2011) 377–392.

[17] P.C. Waterman, Symmetry, unitarity, and geometry in electromagnetic scattering, Phys. Rev. D 3 (4) (1971) 825.

[18] J. Parsons, Nanoparticles and nanocomposites for display applications, PhD thesis, University of Exeter, UK, 2009.

[19] E.M. Purcell, C.R. Pennypacker, Scattering and absorption of light by nonspherical dielectric grains, Astrophys. J. 186 (1973) 705–714.

[20] B.T. Draine, P.J. Flatau, Discrete-dipole approximation for scattering calculations, J. Optic. Soc. Am. A 11 (4) (1994) 1491–1499.

[21] M.A. Yurkin, A.G. Hoekstra, The discrete dipole approximation: an overview and recent developments, J. Quant. Spectrosc. Radiat. Transf. 106 (1–3) (2007) 558–589.

[22] O. Levy, D. Stroud, Maxwell Garnett theory for mixtures of anisotropic inclusions: application to conducting polymers, Phys. Rev. B 56 (13) (1997) 8035.

[23] R.W. Boyd, R.J. Gehr, G.L. Fischer, J. Sipe, Nonlinear optical properties of nanocomposite materials, Pure Appl. Optic. J. Eur. Opt. Soc. A 5 (5) (1996) 505.

[24] R. Ruppin, Evaluation of extended Maxwell-Garnett theories, Optic Commun. 182 (4–6) (2000) 273–279.

[25] R. Del Coso, J. Requejo-Isidro, J. Solis, J. Gonzalo, C.N. Afonso, Third order nonlinear optical susceptibility of Cu: Al_2O_3 nanocomposites: from spherical nanoparticles to the percolation threshold, J. Appl. Phys. 95 (5) (2004) 2755–2762.

[26] C. Granqvist, O. Hunderi, Conductivity of inhomogeneous materials: effective-medium theory with dipole- dipole interaction, Phys. Rev. B 18 (4) (1978) 1554.

[27] M.B. Isichenko, Percolation, statistical topography, and transport in random media, Rev. Mod. Phys. 64 (4) (1992) 961.

[28] M.D. Thoreson, J. Fang, A.V. Kildishev, L.J. Prokopeva, P. Nyga, U.K. Chettiar, V.M. Shalaev, V.P. Drachev, Fabrication and realistic modeling of three-dimensional metal-dielectric composites, J. Nanophotonics 5 (1) (2011) 051513.

[29] E. Hao, G.C. Schatz, Electromagnetic fields around silver nanoparticles and dimers, J. Chem. Phys. 120 (1) (2004) 357–366.

[30] C. Rockstuhl, T. Scharf, A metamaterial based on coupled metallic nanoparticles and its band-gap property, J. Microsc. 229 (2) (2008) 281–286.

[31] M.K. Hedayati, A. Zillohu, T. Strunskus, F. Faupel, M. Elbahri, Plasmonic tunable metamaterial absorber as ultraviolet protection film, Appl. Phys. Lett. 104 (4) (2014) 041103.

[32] M.V. Berry, I.C. Percival, Optics of fractal clusters such as smoke, Opt. Acta Int. J. Optics 33 (5) (1986) 577–591.

[33] I. Zabel, D. Stroud, Metal clusters and model rocks: electromagnetic properties of conducting fractal aggregates, Phys. Rev. B 46 (13) (1992) 8132.

[34] D.J. Bergman, D. Stroud, Physical Properties of Macroscopically Inhomogeneous Media, Solid State Physics, Elsevier, 1992, pp. 147–269.

[35] F. Claro, R. Fuchs, Collective surface modes in a fractal cluster of spheres, Phys. Rev. B 44 (9) (1991) 4109.

[36] R. Devaty, Scaling theory applied to far-infrared absorption by metal-insulator composite materials, Phys. Rev. B 44 (2) (1991) 593.

[37] F. Brouers, D. Rauw, J. Clerc, G. Giraud, Self-similar properties of the far-infrared and optical absorption of fractal metallic clusters, Phys. Rev. B 49 (20) (1994) 14582.

[38] G. Niklasson, Fractal aspects of the dielectric response of charge carriers in disordered materials, J. Appl. Phys. 62 (7) (1987) R1–R14.

[39] V. Markel, L. Muratov, M.I. Stockman, T.F. George, Theory and numerical simulation of optical properties of fractal clusters, Phys. Rev. B 43 (10) (1991) 8183.

[40] V.M. Shalaev, R. Botet, Optical free-induction decay in fractal clusters, Phys. Rev. B 50 (17) (1994) 12987.

[41] M.I. Stockman, L.N. Pandey, L.S. Muratov, T.F. George, Optical absorption and localization of eigenmodes in disordered clusters, Phys. Rev. B 51 (1) (1995) 185.

[42] V. Shalaev, M. Stockman, Optical properties of fractal clusters (susceptibility, surface enhanced Raman scattering by impurities), Sov. Phys. JETP 65 (1987) 509–522.

[43] A. Butenko, V. Shalaev, M. Shtokman, Giant impurity nonlinearities in the optics of fractal clusters, Zh. Eksp. Teor. Fiz. 94 (1988) 107–124.

[44] S.G. Rautian, V.P. Safonov, P.A. Chubakov, V.M. Shalaev, M.I. Stockman, Surface enhanced parametric scattering of light by silver clusters, JETP Lett 47 (1988) 243–246.

[45] V.M. Shalaev, R. Botet, R. Jullien, Resonant light scattering by fractal clusters, Phys. Rev. B 44 (22) (1991) 12216.

[46] U.K. Chettiar, P. Nyga, M.D. Thoreson, A.V. Kildishev, V. Drachev, V. Shalaev, FDTD modeling of realistic semicontinuous metal films, Appl. Phys. B 100 (1) (2010) 159–168.

[47] L. Tröger, H. Hünnefeld, S. Nunes, M. Oehring, D. Fritsch, Structural characterization of catalytically active metal nanoclusters in poly (amide imide) films with high metal loading, J. Phys. Chem. B 101 (8) (1997) 1279–1291.

[48] K. Günther-Schade, H. Castricum, H. Ziegler, H. Bakker, F. Faupel, Free volume changes in mechanically milled PS and PC studied by positron annihilation lifetime spectroscopy (PALS), Polym. Eng. Sci. 44 (7) (2004) 1351–1359.

[49] J. Hong, E. Kay, S.X. Wang, Granular magnetic cobalt metal/polymer thin film system, IEEE Trans. Magn. 32 (5) (1996) 4475–4477.

[50] L. Martinu, H. Biederman, J. Zemek, Metal doped polymer films prepared by simultaneous plasma polymerization of tetrafluoromethane and evaporation of gold, Vacuum 35 (4–5) (1985) 171–176.

[51] N. Bahlawane, K. Kohse-Höinghaus, T. Weimann, P. Hinze, S. Röhe, M. Bäumer, Rational design of functional oxide thin films with embedded magnetic or plasmonic metallic nanoparticles, Angew. Chem. Int. Ed. 50 (42) (2011) 9957–9960.

[52] A. Biswas, Z. Marton, J. Kanzow, J. Kruse, V. Zaporojtchenko, F. Faupel, T. Strunskus, Controlled generation of Ni nanoparticles in the capping layers of teflon AF by vapor-phase tandem evaporation, Nano Lett. 3 (1) (2003) 69–73.

[53] H. Biederman, RF sputtering of polymers and its potential application, Vacuum 59 (2–3) (2000) 594–599.

[54] U. Schürmann, W. Hartung, H. Takele, V. Zaporojtchenko, F. Faupel, Controlled syntheses of Ag–polytetrafluoroethylene nanocomposite thin films by co-sputtering from two magnetron sources, Nanotechnology 16 (8) (2005) 1078.

[55] F. Faupel, V. Zaporojtchenko, T. Strunskus, M. Elbahri, Metal-polymer nanocomposites for functional applications, Adv. Eng. Mater. 12 (12) (2010) 1177–1190.

[56] U. Kreibig, B. Schmitz, H. Breuer, Separation of plasmon-polariton modes of small metal particles, Phys. Rev. B 36 (9) (1987) 5027.

[57] D.W. Hahn, Light scattering theory, University of Florida, Gainesville, 2009.

[58] H. Takele, U. Schürmann, H. Greve, D. Paretkar, V. Zaporojtchenko, F. Faupel, Controlled growth of Au nanoparticles in co-evaporated metal/polymer composite films and their optical and electrical properties, Eur. Phys. J. Appl. Phys. 33 (2) (2006) 83–89.

[59] J. Trice, C. Favazza, R. Sureshkumar, H. Garcia, R. Kalyanaraman, Design and optimization of plasmonic- based metal-dielectric nanocomposite materials for energy applications, Nanophotonic Mater. V, Int. Soc. Optics Photonics (2008) 70300M.

[60] S. Deki, K. Akamatsu, Y. Hatakenaka, M. Mizuhata, A. Kajinami, Synthesis and characterization of nano- sized gold-palladium bimetallic particles dispersed in polymer thin film matrix, Nanostruct. Mater. 11 (1) (1999) 59–65.

[61] Z. Mbhele, M. Salemane, C. Van Sittert, J. Nedeljković, V. Djoković, A. Luyt, Fabrication and characterization of silver– polyvinyl alcohol nanocomposites, Chem. Mater. 15 (26) (2003) 5019–5024.

[62] Z. Zhang, M. Han, One-step preparation of size-selected and well-dispersed silver nanocrystals in polyacrylonitrile by simultaneous reduction and polymerization, J. Mater. Chem. 13 (4) (2003) 641–643.

[63] A. Biswas, O. Aktas, U. Schürmann, U. Saeed, V. Zaporojtchenko, F. Faupel, T. Strunskus, Tunable multiple plasmon resonance wavelengths response from multicomponent polymer-metal nanocomposite systems, Appl. Phys. Lett. 84 (14) (2004) 2655–2657.

[64] J.H. Liu, D.S. Wu, K.Y. Tseng, Fabrication and characterization of GRIN plastic rods containing silver nanoparticles with novel surfmers, Macromol. Chem. Phys. 205 (16) (2004) 2205–2213.

[65] H. Takele, H. Greve, C. Pochstein, V. Zaporojtchenko, F. Faupel, Plasmonic properties of Ag nanoclusters in various polymer matrices, Nanotechnology 17 (14) (2006) 3499.

[66] A. Pucci, M. Bernabò, P. Elvati, L.I. Meza, F. Galembeck, C.A. de Paula Leite, N. Tirelli, G. Ruggeri, Photoinduced formation of gold nanoparticles into vinyl alcohol based polymers, J. Mater. Chem. 16 (11) (2006) 1058–1066.

[67] U. Chatterjee, S.K. Jewrajka, Synthesis of block copolymer-stabilized Au–Ag alloy nanoparticles and fabrication of poly (methyl methacrylate)/Au–Ag nanocomposite film, J. Colloid Interface Sci. 313 (2) (2007) 717–723.

[68] D. Avasthi, Y. Mishra, D. Kabiraj, N. Lalla, J. Pivin, Synthesis of metal–polymer nanocomposite for optical applications, Nanotechnology 18 (12) (2007) 125604.

[69] D. Božanić, V. Djoković, J. Blanuša, P. Nair, M. Georges, T. Radhakrishnan, Preparation and properties of nano-sized Ag and Ag_2S particles in biopolymer matrix, Eur. Phys. J. E 22 (1) (2007) 51–59.

[70] Y. Mishra, S. Mohapatra, D. Kabiraj, B. Mohanta, N. Lalla, J. Pivin, D. Avasthi, Synthesis and characterization of Ag nanoparticles in silica matrix by atom beam sputtering, Scripta Mater. 56 (7) (2007) 629–632.

[71] Y. Mishra, S. Mohapatra, D. Avasthi, D. Kabiraj, N. Lalla, J. Pivin, H. Sharma, R. Kar, N. Singh, Gold–silica nanocomposites for the detection of human ovarian cancer cells: a preliminary study, Nanotechnology 18 (34) (2007) 345606.

[72] L. Jiang, Z. Wu, D. Wu, W. Yang, R. Jin, Controllable embedding of silver nanoparticles on silica nanospheres using poly (acrylic acid) as a soft template, Nanotechnology 18 (18) (2007) 185603.

[73] Y. Mishra, S. Mohapatra, R. Singhal, D. Avasthi, D. Agarwal, S. Ogale, Au–ZnO: a tunable localized surface plasmonic nanocomposite, Appl. Phys. Lett. 92 (4) (2008) 043107.

[74] A.S. Susha, M. Ringler, A. Ohlinger, M. Paderi, N. LiPira, G. Carotenuto, A.L. Rogach, J. Feldmann, Strongly luminescent films fabricated by thermolysis of Gold– thiolate complexes in a polymer matrix, Chem. Mater. 20 (19) (2008) 6169–6175.

[75] S. Mohapatra, Y. Mishra, D. Avasthi, D. Kabiraj, J. Ghatak, S. Varma, Synthesis of gold-silicon core-shell nanoparticles with tunable localized surface plasmon resonance, Appl. Phys. Lett. 92 (10) (2008) 103105.

[76] V. Chakravadhanula, M. Elbahri, U. Schürmann, H. Takele, H. Greve, V. Zaporojtchenko, F. Faupel, Equal intensity double plasmon resonance of bimetallic quasi-nanocomposites based on sandwich geometry, Nanotechnology 19 (22) (2008) 225302.

[77] Y.K. Mishra, V.S.K. Chakravadhanula, U. Schürmann, H. Kumar, D. Kabiraj, S. Ghosh, V. Zaporojtchenko, D.K. Avasthi, F. Faupel, Controlled reduction of size of Ag nanoparticles embedded in teflon matrix by MeV ion irradiation, Nucl. Instrum. Methods Phys. Res. Sect. B Beam Interact. Mater. Atoms 266 (8) (2008) 1804–1809.

[78] J. Röder, J. Faupel, H.-U. Krebs, Growth of polymer–metal nanocomposites by pulsed laser deposition, Appl. Phys. A 93 (4) (2008) 863–867.

[79] G. Xu, C.-M. Huang, Q. Liang, M. Tazawa, P. Jin, Wavelength multiplexing and tuning in nano- Ag/ dielectric multilayers, Appl. Phys. A 94 (3) (2009) 525–530.

[80] V. Vodnik, D. Božanić, E. Džunuzović, J. Vuković, J. Nedeljković, Thermal and optical properties of silver–poly (methylmethacrylate) nanocomposites prepared by in-situ radical polymerization, Eur. Polym. J. 46 (2) (2010) 137–144.

[81] H.T. Beyene, F.D. Tichelaar, P. Peeters, I. Kolev, M.C. van de Sanden, M. Creatore, Hybrid sputtering- remote PECVD deposition of Au nanoparticles on SiO_2 layers for surface plasmon resonance-based colored coatings, Plasma Process. Polym. 7 (8) (2010) 657–664.

[82] A. Tyurin, G. De Filpo, D. Cupelli, F. Nicoletta, A. Mashin, G. Chidichimo, Particle size tuning in silver- polyacrylonitrile nanocomposites, Express Polym. Lett. 4 (2) (2010) 71–78.

[83] F. Liu, S. Shang, Y. Duan, L. Li, Electrical and optical properties of polymer-Au nanocomposite films synthesized by magnetron cosputtering, J. Appl. Polym. Sci. 123 (5) (2012) 2800–2804.

[84] H. Beyene, V. Chakravadhanula, C. Hanisch, T. Strunskus, V. Zaporojtchenko, M. Elbahri, F. Faupel, Vapor phase deposition, structure, and plasmonic properties of polymer-based composites containing Ag–Cu bimetallic nanoparticles, Plasmonics 7 (1) (2012) 107–114.

[85] M.A. Biot, Some new aspects of the reflection of electromagnetic waves on a rough surface, J. Appl. Phys. 28 (12) (1957) 1455–1463.

[86] M. Kerker, The Scattering of Light and Other Electromagnetic Radiation: Physical Chemistry: A Series of Monographs, Academic Press, 2013.

[87] A. Movchan, C.G. Poulton, L.C. Botten, N. Nicorovici, R.C. McPhedran, Noncommuting limits in electromagnetic scattering: asymptotic analysis for an array of highly conducting inclusions, SIAM J. Appl. Math. 61 (5) (2001) 1706–1730.

[88] M.A. Biot, Generalized boundary condition for multiple scatter in acoustic reflection, J. Acoust. Soc. Am. 44 (6) (1968) 1616–1622.

[89] L. Rayleigh, of Cauchy's, Theory of reflection by opaque bodies, Phil. Mag. 42 (1871) 441.

[90] M. Elbahri, A.U. Zillohu, B. Gothe, M.K. Hedayati, R. Abdelaziz, H.J. El-Khozondar, M. Bawa'aneh, M. Abdelaziz, A. Lavrinenko, S. Zhukovsky, S. Homaeigohar, Photoswitchable molecular dipole antennas with tailored coherent coupling in glassy composite, Light Sci. Appl. 4 (7) (2015) e316.

[91] V. Astapenko, Polarization Bremsstrahlung on Atoms, Plasmas, Nanostructures and Solids, Springer Science & Business Media, 2013.

[92] M. Elbahri, M. Abdelaziz, S. Homaeigohar, A. Elsharawy, M. Keshavarz Hedayati, C. Röder, M. El Haj Assad, R. Abdelaziz, Plasmonic metaparticles on a blackbody create vivid reflective colors for naked-eye environmental and clinical biodetection, Adv. Mater. 30 (4) (2018).

[93] S. Martin, An Introduction to Ocean Remote Sensing, Cambridge University Press, New York, USA, 2004.

[94] H. Fujiwara, Spectroscopic Ellipsometry: Principles and Applications, John Wiley & Sons, Tokyo, Japan, 2007.

[95] M. Elbahri, M.K. Hedayati, K. Chakravadhanula, V. Sai, M. Jamali, T. Strunkus, V. Zaporojtchenko, F. Faupel, An omnidirectional transparent conducting-metal-based plasmonic nanocomposite, Adv. Mater. 23 (17) (2011) 1993–1997.

[96] M.K. Hedayati, M. Javaherirahim, B. Mozooni, R. Abdelaziz, A. Tavassolizadeh, V.S.K. Chakravadhanula, V. Zaporojtchenko, T. Strunkus, F. Faupel, M. Elbahri, Design of a perfect black absorber at visible frequencies using plasmonic metamaterials, Adv. Mater. 23 (45) (2011) 5410–5414.

[97] M.K. Hedayati, S. Fahr, C. Etrich, F. Faupel, C. Rockstuhl, M. Elbahri, The hybrid concept for realization of an ultra-thin plasmonic metamaterial antireflection coating and plasmonic rainbow, Nanoscale 6 (11) (2014) 6037–6045.

[98] M.K. Hedayati, F. Faupel, M. Elbahri, Tunable broadband plasmonic perfect absorber at visible frequency, Appl. Phys. A 109 (4) (2012) 769–773.

[99] M.K. Hedayati, M. Javaheri, A.U. Zillohu, H.J. El-Khozondar, M.S. Bawa'aneh, A. Lavrinenko, F. Faupel, M. Elbahri, Photo-driven super absorber as an active metamaterial with a tunable molecular-plasmonic coupling, Adv. Optic. Mater. 2 (8) (2014) 705–710.

8

Cluster decoration of semiconductor nanostructures toward gas sensors and biosensors

Alexander James Porkovich, Eric Danielson

NANOPARTICLES BY DESIGN UNIT, OKINAWA INSTITUTE OF SCIENCE AND TECHNOLOGY, OKINAWA, JAPAN

Chapter outline

8.1 Introduction

8.1.1 Semiconductors

Semiconductors are a well-known and important class of materials, featuring unique electrical conduction properties. A semiconductor typically has a conductivity between a conductor and an insulator. This is because the electrons in a semiconductor

Cluster Beam Deposition of Functional Nanomaterials and Devices. https://doi.org/10.1016/B978-0-08-102515-4.00008-8

crystal are ionically bound to the atoms within the lattice, as opposed to metallic conductors in which electrons exist in a "cloud." This is to say that electrons are not locally bound to any single atom in the crystal, and thus able to move freely throughout the conductor. Unlike insulators, however, a semiconductor can allow the propagation of electric current through their crystal structure via "charge carriers." These carriers take the form of either unbound electrons, or positive "holes," created by electron vacancies in the ionic bonds within the crystal lattice. A semiconductor in which electrons act as charge carriers is said to be n-type (negative charge), while a semiconductor in which charge is carried by holes is p-type (positive charge). Semiconductors can be fabricated from both types, which will result in a p-n junction. This junction will only allow propagation of current in only one direction, as propagation in the other direction results in the recombination of electrons and holes.

There are multiple ways by which charge carriers can be injected into semiconductors. Semiconductors can be doped with materials which creates a direct injection of either holes or electrons, depending on the dopant and the semiconductor itself. Additionally, unlike an insulator, charge carriers in a semiconductor can be excited into a conductive state with an external energy source.

This is best understood using band theory, a concept which describes the conductivity in terms of the difference in energy between the valence and conduction bands of a material. The valence band represents the energy of a material when all the electrons are bound in ionic bonds, while the conduction band represents the energy of a material in which the electrons are mobile. The energy difference between these bands, known as the band gap, represents the amount of energy required to excite an electron from the valence to the conductance band. Metallic conductors have a small or nonexistent band gap, insulators have a large band gap of at least several electronvolts, and semiconductors feature band gaps in between.

Traditionally semiconductors have come from the group 14 elements of the periodic table, specifically silicon and geranium. The main advantage of these elements is that they have a valence of 4, meaning that these elements can be doped with a group 13 element to create a p-type semiconductor, or with a group 15 element to create an n-type semiconductor. Compounds formed from elements of group 13 and 15 (such as gallium arsenide) have also been used to fabricate semiconductors. Other group 4 elements such as carbon are semiconducting in certain structural conformations, while acting as an insulator in others.

Generally, oxidation of metals was thought to turn a metal into an insulating ceramic. However, many low band gap metal oxides have been discovered, which has led to an increase in research interest, as metal oxide–based semiconductors offer the desired properties at a potentially cheaper price due to their relative abundance and easier processing required for metal oxides as opposed to elemental or compound semiconductors.

Semiconductor metal oxides and compound materials usually are naturally either p- or n-type. However, under some conditions, these materials can be made to switch conduction mechanisms.

These properties have led to semiconductors becoming indispensable in a diverse range of fields including computing and electronics, optics and photonics, energy production, and sensors. As seen above, semiconductor conductivity is sensitive to surface charge and energy fluctuations, once a certain energy threshold is reached.

8.1.2 Semiconductive nanostructures as sensors

The effect of these properties is that the conductivity of semiconductive materials is dependent on the surrounding energy and charge flux. Changes in the composition of light, chemicals, biomolecules, and/or gases surrounding a semiconductor can be measured directly by monitoring current.

Semiconductive photodetectors and other semiconductive photonic devices are beyond the scope of this chapter. The discussion will instead be focused on how charges surrounding the semiconductor (from gases, chemicals, or biomolecules) cause resistance fluctuations via the creation of depletion regions and charge injection.

The effect on resistance of a semiconductive sensor depends on the nature of the charge surrounding the sensor, in addition to whether the sensor itself is conducting via a p- or n-type mechanism. Fig. 8.1 outlines the effect on the resistance of each type of material by changes in charge. When an excess of negative charge is brought close to the surface, positive charges are induced in the conduction band. The excess electrons in an

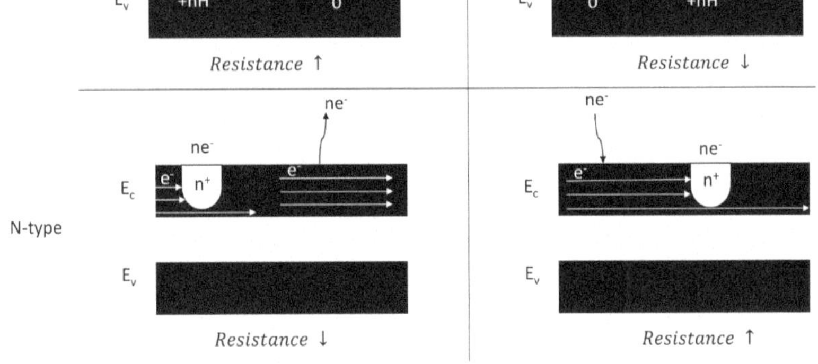

FIGURE 8.1 A summarization of the effect of surface charges on the conduction (E_c) and valence (E_v) bands of either a p- or n-type semiconductor. Positive holes are represented as +nH (where n is simply an integer multiplier to indicate the number of charges), electrons are represented by e^- or ne^- (again n means multiple charges), and the recombination of holes and electrons is represented by 0.

n-type semiconductor recombine with or are driven away from the surface by the induced positive charges, creating a subsequent increase in resistance. This area affected by the induced positive charge is known as a *depletion layer*. Alternatively, in p-type semiconductors, this positive charge attracts electrons from the valence band. The resulting positive holes are then able to carry a current, decreasing the resistance of the semiconductor. When negative charge is removed from the surface, the depletion layer in the conduction band decreases causing the conduction band to allow more charge carriers through. At the same time, a decrease in this layer means more electrons will be in the valence band, reducing the number of positive holes and increasing resistance of a p-type mechanism.

Typically charge-sensitive semiconductor-based sensors have a simple geometry. The semiconductive sensing elements are connected to conductive electrodes forming a simple circuit. The operation of the sensor features either a constant bias being applied across the semiconductor, or a constant current being held by the circuit. From this the resistance can be determined by measuring the reciprocal phenomena and applying Ohm's law.

One of the main advantages of nanostructures is the high surface-to-volume ratio. By increasing the amount of material which is located on the surface of a sensing element, the sensor becomes more sensitive in terms of response and kinetics as a higher percentage of the material is at the surface, able to interact with charges and form depletion regions. The current industry standard in semiconductor sensors are thin films; however, 2-D sheets, 1-D nanowires and tubes, and 0-D nanoparticles and clusters are all active and promising areas of research into future sensing devices [1,2].

Semiconductor nanowire devices are fabricated using a variety of techniques, usually divided into bottom-up or top-down methods. Using top-down methods, nanowires are fabricated using subtractive methods, such as wet or reactive ion etching to sculpt a nanowire out of the bulk semiconductor material. Ion-implantation doping is often used to increase the conductivity of the sculpted nanowire. The physical properties, orientation to the substrate plane, and substrate location of a nanowire can all be precisely controlled using standard CMOS industry techniques. It is relatively easy to integrate these nanowires into electronic devices for sensing. However, top-down fabrication requires many complicated instruments and the resolution of lithography steps limits the size of the nanowires.

Bottom-up methods start with the growth of the nanowire via additive self-assembly processes, usually via chemical vapor deposition (CVD). The growth of the nanowire is catalyzed using metal nanoclusters or nanoparticles via the vapor-liquid-solid (VLS) mechanism. The vapor-phase precursor of the nanowire material is passed over the liquid alloy of the metal nanocluster and one of the constituent elements of the nanowire, while they are heated above the eutectic point. The liquid alloy becomes supersaturated with the nanowire precursor and undergoes nucleation to form a liquid/solid interface directing the growth of the semiconductor nanowire. Nanowire properties such as length and composition are determined by the CVD parameters and the nanowire diameter is defined by using different sized metal clusters, down to the nanometer scale. More complicated structures such as kinked, coaxial modulated, and branched nanowires can also be generated using this technique. This is a possible application of cluster beam deposition, to produce nanoparticles that will generate nanowires of a particular size and structure via the VLS mechanism.

As nanowires formed using bottom-up methods typically have random positions and substrate orientations, large-scale assembly techniques are needed to arrange nanowires on a substrate for device fabrication. Possible methods use microfluidic flow techniques, electric or magnetic field-assisted alignment, contact printing, nanocombing, and even bubble-blowing. Once placed in an ordered manner, the nanowires can be electrically contacted using standard lithography and metal evaporation techniques to fabricate useful electronic sensors.

In this chapter we will explore the decoration of these types of sensors with cluster beam, including the purpose of decorations, the advantages of decoration via cluster beam methods, and the interaction of nanostructured supports with cluster beam deposited clusters.

8.2 Cluster decoration of nanostructures for biosensing

Biosensing refers to the detection of the type and concentration of a specific analyte that is relevant in biological properties. Possible analytes of interest can vary widely, from simple compounds such as glucose, sequences of nucleic acid (DNA or RNA), specific proteins, or antibodies, to entire virus particles or bacterial cells. Therefore, to construct a useful biosensor, we need to utilize a specific bioreceptor that will bind or interact only with the desired analyte to generate a signal. Like their target analytes, possible bioreceptors also vary widely and include antibodies, enzymes, nucleic acids, etc. Decoration of nanostructures with metal clusters can provide a useful platform for many types of bioreceptors and enhance sensitivity of nanostructure-based biosensors.

Fabricating electronics using nanostructures provides many advantages in device performance. Moving to the nanoscale allows for faster response time, greater power efficiency, and room for many devices on the same platform. Nanowire biosensor devices have attracted significant attention due to their high surface-to-volume ratio and similar size to many biologically relevant analytes [3–5]. Many semiconducting materials such as Si or metal oxides are biocompatible, nontoxic, and stable in a biological environment. The sensitivity of nanowire biosensors is extremely high, enabling detection in the range from micromoles to picomoles, even up to the limit of single molecules or particles. This improved sensitivity has direct biomedical applications, as the detection of certain proteins or nucleic acids at extremely low concentrations is very important in the early diagnostics of cancer and other disorders.

8.2.1 Field effect transistor biosensors

Field effect transistors (FETs) are 3-terminal semiconductors in which current flowing between a source and drain electrode can be modulated by the applied voltage on a third electrode, the gate. The gate voltage electrostatically controls the number charge carriers in the channel between the source and drain. In nanowire FETs, a nanowire composed of a semiconducting material acts as the channel between the source and drain. The presence of an electric field from a charged particle close to the nanowire can also modulate the number of charge carriers in the nanowire, just like the applied gate voltage. As the nanowire's diameter is comparable to or smaller than the Debye length in solution, the presence of a single charged biomolecule can induce a significant charge accumulation or depletion region within the nanowire that substantially changes the current flow [4]. Nanowire FET biosensors have been previously studied for real-time, label-free detection of a variety of biomolecules such as proteins, nucleic acids, and viruses [3,6–10]. A schematic diagram for a typical nanowire biosensor and its sensing mechanism is shown in Fig. 8.2.

In order to selectively recognize only the desired biomolecule, the nanowire surface is functionalized with receptors such as antibodies or single-strand DNA probes. Large arrays of nanowire FETs functionalized with different receptors can sense multiple biomolecules simultaneously from the same sample and have been incorporated into microfluidic devices as an important milestone to clinical testing. However, these biological receptors do not bind directly the NW surface, alkoxysilane derivatives are attached to the hydroxyl groups on the native semiconductor oxide and present functional amino, thiol, or aldehyde groups to the solution [4]. This additional layer increases the insulation between the solution and NW channel, degrading the performance and detection limit of the sensor. Protonation and deprotonation of the amino and silanol

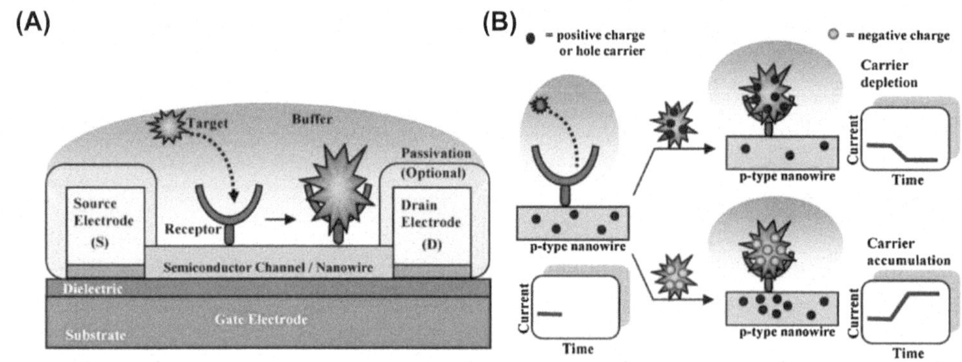

FIGURE 8.2 (A) Illustration of a nanoscale FET biosensor with a cross-sectional view. Target molecules can be recognized by the receptor on the channel surface. (B) Demonstration of the effect on the current in a p-type nanowire after a positively or negatively charged molecule binds to the receptor. *Reproduced from K.-I. Chen, B.-R. Li, and Y.-T. Chen, Silicon nanowire field-effect transistor-based biosensors for biomedical diagnosis and cellular recording investigation, Nano Today 6 (2) (2011) 131–154.*

groups on the oxide layer can also modulate the surface charge and therefore the NW conductance, limiting NW biosensor use in high ionic strength solutions like blood [11]. Functionalization with metal nanoparticles, especially gold, offers a way to bypass these restrictions.

Gold nanoparticles are also biocompatible and have been widely used as biomarkers and versatile molecular delivery platforms [12]. The interaction between AuNPs and receptors such as thiolated DNA or antibodies is energetically very favorable, forming a strong Au-S bond and straightening the molecule away from the nanowire surface for better sensing ability. Decoration of the NW with AuNPs also increases its surface area-to-volume ratio and therefore the biosensor sensitivity. The conductivity of the gold can also enhance electron transfer between the charged biomolecule and the NW surface.

Ryu et al. present the use of an AuNP decorated SiNW biosensor for label-free DNA detection [13]. SiNWs with 80 nm width and 500 nm length were fabricated using a top-down etching method on a p-type Si wafer. AuNP decoration was by depositing a 4 nm Au wetting layer using RF sputtering and post annealing at 500°C, although cluster beam deposition could achieve similar results. After fabrication, the devices were functionalized with a 10 μM solution of single stranded breast cancer DNA. The sensing ability and specificity of these devices were tested by measuring the change in the device current upon exposure to complementary and noncomplementary DNA strands (Fig. 8.3).

As the Si is p-type, the binding of the negatively charged DNA to the AuNP induces a higher charge density in the NW and the current increases. Upon introduction of the complementary strand, the DNA hybridizes into the double-stranded form and the current increases further. The device is also reusable since the strands can be separated by immersion in hot water, returning the current to its original level. The device is highly

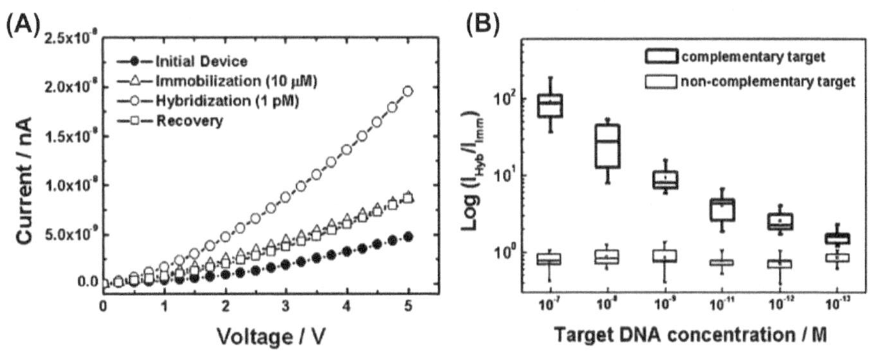

FIGURE 8.3 (A) Id-Vd characteristics of the AuNP decorated SiNW devices with the probe DNA strand at 10 μM, the complementary strand at 1 pM, and after recovery treatment. (B) Current ratio of the functionalized and hybridized device currents upon exposure to complementary and noncomplementary DNA strands at different concentrations. *Reproduced from S.-W. Ryu, C.-H. Kim, J.-W. Han, C.-J. Kim, C. Jung, H.G. Park, and Y.-K. Choi, Gold nanoparticle embedded silicon nanowire biosensor for applications of label-free DNA detection, Biosens. Bioelectron. 25 (9) (2010) 2182–2185.*

sensitive, showing a measurable current increase down to a 0.1 pM solution of the complementary strand [13]. It is also selective, as the noncomplementary strand does not bind to the strand on the AuNPs to change the device current.

AuNP decorated SiNWs have also been used for protein sensing. Presnova et al. compared different SiNW functionalization methods, including AuNPs, for sensing prostate specific antigen (PSA); an important indicator for prostate cancer in human serum [11]. The SiNW devices were fabricated using a top-down etching method and then functionalized with various alkoxysilane derivatives. AuNPs were attached to some of these SiNWs using a liquid deposition method. Monoclonal antibodies that specifically bind to the PSA were then introduced and bound to the surface of the AuNPs, or to the active thiol and amine groups of the alkoxysilane layer. Compared with the alkoxysilane functionalized SiNW devices, the AuNP decorated devices had a much larger current response upon exposure to PSA (Fig. 8.4).

The AuNP decorated NW biosensors could detect PSA in buffer over a range of seven orders of magnitude in concentration, down to a minimum of 0.7 fM [11]. This is two orders of magnitude greater than the established enzyme-linked immunoabsorbent assay (ELISA) method often used in medical diagnostics. Samples measured using both ELISA and the NW biosensor showed a good correlation of results. Devices functionalized using alkoxysilanes had a much lower sensitivity.

When the PSA binds to the antibodies attached to the NW, its negative charge induces a higher charge density in the p-type SiNW, increasing the device current. The strength of this electric charge is inversely proportional to the square of the distance to the NW surface; therefore, even a small separation (3–4 nm) caused by the alkoxysilane

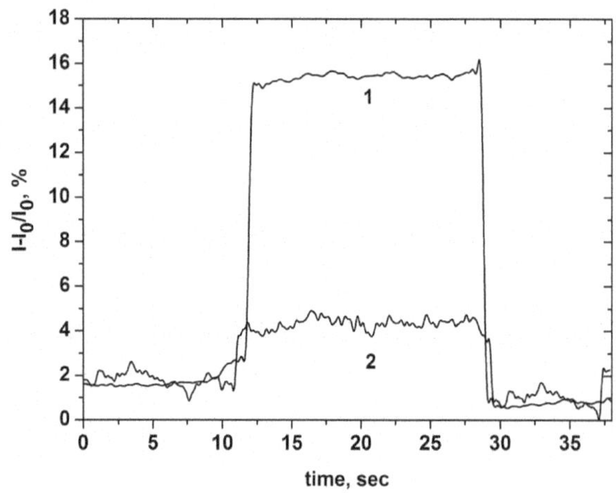

FIGURE 8.4 Plots of normalized current change versus time for NW FETs modified with GOPS-SH/5 nm AuNPs (a) and APTMS/PDITC (2), exposed to PSA in human serum diluted 1/100 in 0.01 x PBS (pH 8.0). *Reproduced from G. Presnova, D. Presnov, V. Krupenin, V. Grigorenko, A. Trifonov, I. Andreeva, O. Ignatenko, A. Egorov, and M. Rubtsova, Biosensor based on a silicon nanowire field-effect transistor functionalized by gold nanoparticles for the highly sensitive determination of prostate specific antigen, Biosens. Bioelectron. 88 (2017) 283–289.*

derivative can lead to a large difference in current response [11]. In addition, when the antibodies are immobilized on the AuNPs, their active sites are oriented toward the nanoparticles. The PSA binds close to the nanoparticle surface, and the conductive gold enhances charge transport to the NW. The rapid change in current also indicates that binding takes place close to the surface. These benefits of AuNP decoration could be enhanced by depositing the nanoparticles directly on the NW using cluster beam deposition, rather than using an intervening layer.

8.2.2 Electrochemical biosensors

Redox reactions are fundamental in many biochemical processes and often produce products that are significant markers of disease or health. Amperometric biosensors can exploit these processes to induce an electrochemical reaction sequence at the surface of an electrode, producing a measurable current signal. The electrode surface is functionalized with enzymes which catalyze the redox reaction, enabling direct electron from the solution. Nanomaterials have been extensively employed in this area of research, as their structure increases the available surface area for the adsorption of enzymes and enhances the electron transfer rate [14–16]. Noble metal nanoparticles are particularly useful due to their unique electronic and catalytic properties. However, there are drawbacks to using enzymes, especially their cost, chemical instability, and temperature sensitivity [17]. They may also interact with other oxidizable species within the solution. Researchers have recently moved toward using nonenzymatic biosensors, where direct oxidations occur at the surface of the electrode. For these types of biosensors, metal nanoparticles are often needed to introduce catalytic sites onto the less reactive bulk electrode material.

Kwon et al. describe the use of AuNPs on SiNW electrodes to construct an electrochemical biosensor for bovine serum albumin (BSA) [18]. The SiNWs were fabricated using a chemical etching process and then decorated with AuNPs using a self-assembled monolayer of aminopropyltriethoxysilane (APTES). However, the SiNW electrode needed to be heat treated at 800°C in vacuum to transfer the AuNPs to the SiNW surface. This liquid deposition and heating procedure was repeated to decorate the SiNW with a sufficient number of AuNPs [18]. This decoration could be more easily achieved using cluster beam deposition of nanoparticles. A three-electrode electrochemical biosensor was constructed using the decorated SiNWs as the working electrode, with a platinum wire mesh as the counter electrode and a saturated Ag/AgCl electrode as the reference electrode. The device's biosensing capability was tested in different concentrations of BSA in a phosphate buffered solution (PBS) using cyclic voltammetry (Fig. 8.5).

In contrast to undecorated sensors, the AuNP decorated SiNW have a clear response to the BSA, with a maximum sensitivity of 1 μM BSA. The oxidation peak at 1.1 V is related to the cleavage of the sulfur double bonds on the BSA molecule [18]. The AuNP/SiNW sensor produces a well-behaved linear response to BSA concentration, clearly showing that this type of decorated nanostructure can be used to make effective nonenzymatic biosensors.

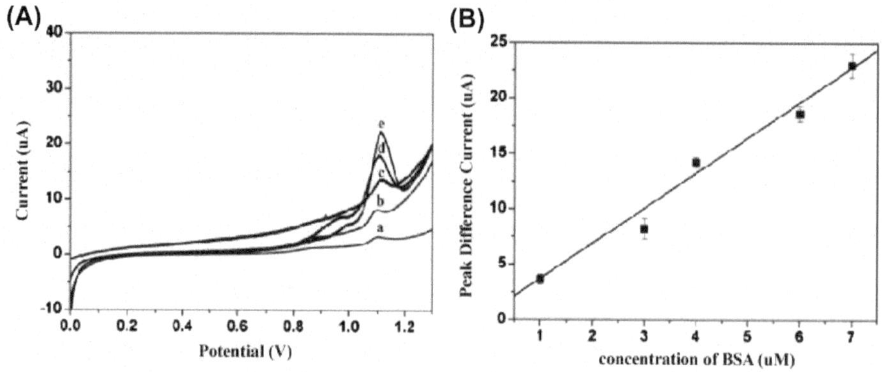

FIGURE 8.5 (A) Linear sweep voltammograms of different concentrations of BSA (1–7 μM) in PBS (pH 7.4) using the twice AuNP decorated SiNW sensor. (B) Peak current versus concentration of BSA in PBS (pH 7.4) at 1.1 V. *Reproduced from D.H. Kwon, H.H. An, H.-S. Kim, J.H. Lee, S.H. Suh, Y.H. Kim, and C.S. Yoon, Electrochemical albumin sensing based on silicon nanowires modified by gold nanoparticles, Appl. Surf. Sci. 257 (10) (2011) 4650–4654.*

8.2.2.1 Glucose sensing

Diabetes is one of the major diseases worldwide, especially in developed countries. Fast and accurate measurement of glucose levels is essential for managing this disease, as well as early diagnosis of other metabolic disorders. Glucose sensors are among the most successful biosensors and are based on the biological process:

$$\text{Glucose} + O_2 \xrightarrow{\text{GOx}} \text{Gluconolactone} + H_2O_2$$

Since the amount of glucose is proportional to that of produced H_2O_2, the glucose concentration can be determined from measuring the current derived from the electrochemical reaction of H_2O_2. This reaction is catalyzed by the enzyme glucose oxidase (GOx), and biosensors using nanostructured electrodes functionalized with GOx have been demonstrated [16,19]. However, other groups have produced nonenzymatic biosensors using metal nanoparticle decorated nanostructures.

Liao et al. present the use of Ag nanoparticles on glass carbon electrodes (GCEs) to use in a hydrogen peroxide sensor [20]. The AgNPs were deposited onto the GCEs using cluster beam deposition from a magnetron sputtering system. The GCE was used in a three-electrode electrochemical biosensor with a platinum wire as the counter electrode and a saturated Ag/AgCl electrode as the reference electrode. In a 0.05M phosphate buffer solution (PBS) with a certain concentration of H_2O_2, the reaction mechanism on the AgNP decorated electrode is:

$$H_2O_2 + e^- \xrightarrow{\text{Ag}} OH_{(ads)} + OH^-$$

$$H_2O_2 + H^+ + OH_{(ads)} + e^- \xrightarrow{\text{Ag}} 2OH_{(ads)} + H_2O$$

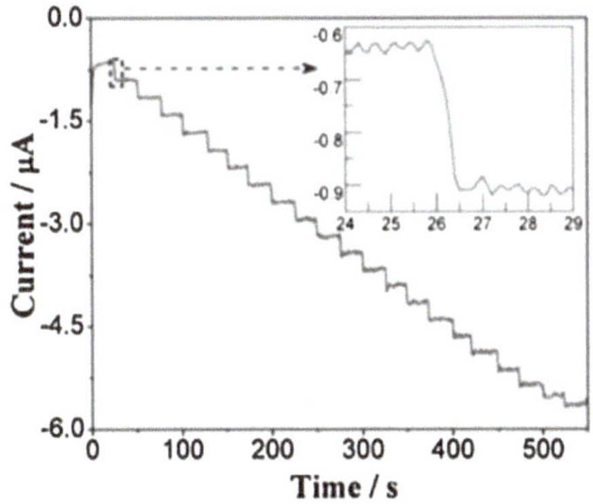

FIGURE 8.6 Amperometric response of the AgNP/GCE sensor upon successive addition of H_2O_2 into gently stirred 0.05M PBS (pH 7.4) at −0.4 V. *Reproduced from K. Liao, P. Mao, Y. Li, Y. Nan, F. Song, G. Wang, and M. Han, A promising method for fabricating Ag nanoparticle modified nonenzyme hydrogen peroxide sensors, Sens. Actuators B Chem. 181 (2013) 125–129.*

where the reduction of the first H_2O_2 molecule provides an initial source of $OH_{(ads)}$ to allow the reduction of the second H_2O_2 molecule at a faster rate. This process can only take place efficiently with the Ag catalysis, and the AgNPs provide a high number of possible catalysis sites per unit area. Fig. 8.6 shows the amperometric response of the sensor to increasing H_2O_2 concentration in PBS.

The sensor shows quick response to the concentration change and a linear relationship between the catalytic current and the H_2O_2 concentration is easily derived. The sensitivity of these devices was 63 μA/mM with a detection limit of 1.3×10^{-6} M at a signal-to-noise ratio of 3 [20]. This is higher than many other metal, oxide, and enzyme-based electrode devices. Liao et al. also showed the dependence of nanoparticle morphology on the catalytic properties of these devices, as seen below in Fig. 8.7.

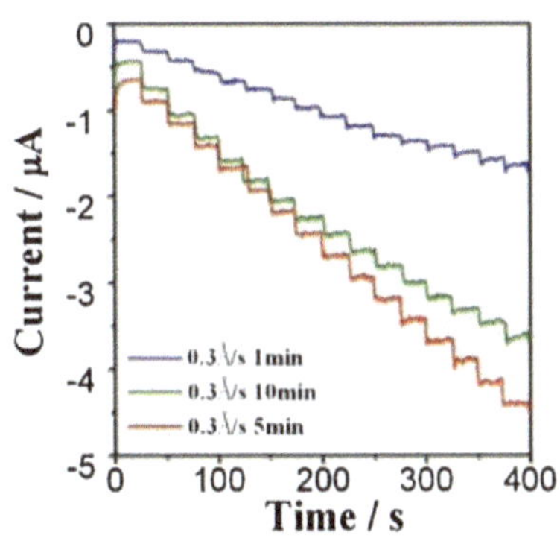

FIGURE 8.7 Amperometric responses of AgNP/GCE sensors with different deposition times upon successive addition of H_2O_2 into gently stirred 0.05M PBS (pH 7.4) at −0.4 V. *Reproduced from K. Liao, P. Mao, Y. Li, Y. Nan, F. Song, G. Wang, and M. Han, A promising method for fabricating Ag nanoparticle modified nonenzyme hydrogen peroxide sensors, Sens. Actuators B Chem. 181 (2013) 125–129.*

The diameter of the sputtered AgNPs was 6 nm with depositions of 1 and 5 min; however, this increased to 13 nm at 10 min deposition due to aggregation and coalescence. Nanoparticle coverage for the different deposition times was estimated to be 15%, 60%, and 75%, respectively. The reduction peak current increases markedly after increasing deposition time to 5 min, but there is a slight decrease at 10 min due to the decreased catalytic properties of the larger AgNPs [20]. Cluster beam deposition is necessary to decorate electrodes with nanoparticles that are closely spaced but sufficiently isolated from each other for optimal electrochemical biosensors.

Lin et al. present the use of zinc oxide (ZnO) nanowires modified with Au nanoparticles on Au electrodes in a nonenzymatic glucose sensor that measures the effect of direct glucose oxidation [17]. A ZnO seed layer was deposited on the electrodes using RF magnetron sputtering; then nanowires were grown using the VLS growth mechanism from Zn metal powder and Ar/O_2 gas flow at 600°C. Nanowires with average lengths and diameters of 4.5 μm and 300 nm, respectively, grew at a high density on the electrode. The nanowires were then decorated with gold nanoparticles using a liquid deposition method; however, cluster beam deposition could also be used with this fabrication procedure. A conventional three-electrode sensor was used, with the AuNP/ZnO NW/Au as the working electrode, a gold thin film electrode on glass as the counter electrode, and a standard Ag/AgCl electrode as the reference. These sensors were then exposed to a 0.1M NaOH solution containing 20 mM of glucose and compared with similar devices without AuNP decoration. Examining the CVs of these two types of devices in Fig. 8.8, we see that as the potential becomes more positive, the oxidation of intermediates is catalyzed and direct oxidation of glucose first occurs at zinc vacancies on the ZnO NW surface [17]. The AuNPs further oxidize the glucose leading to a more positive potential scan and a peak at 0.85 V (Fig. 8.8).

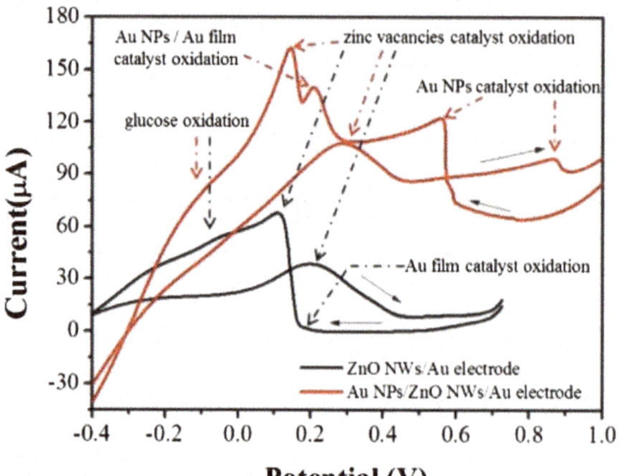

FIGURE 8.8 CVs of ZnO NWs/Au electrode (*black line*) and AuNPs/ZnO NWs/Au electrode (*red line*) in 0.1M NaOH containing 20 mM glucose at 10 mV/s. *Reproduced from S.-Y. Lin, S.-J. Chang, and T.-J. Hsueh, ZnO nanowires modified with Au nanoparticles for nonenzymatic amperometric sensing of glucose, Appl. Phys. Lett. 104 (19) (2014) 193704.*

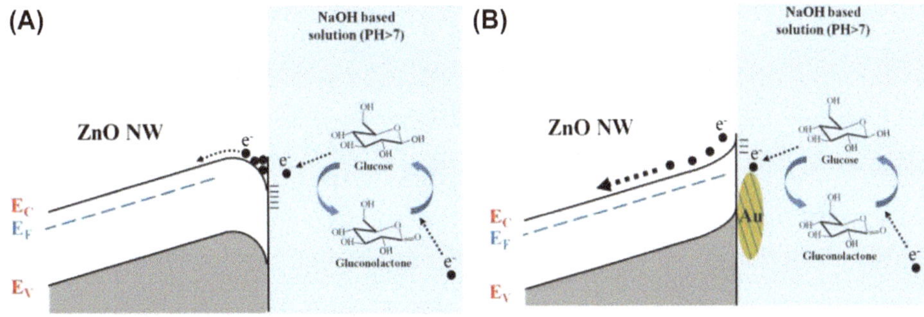

FIGURE 8.9 Schematic energy band diagrams of ZnO NWs and AuNP/ZnO NWs electrodes during electrocatalysis at an operated potential. *Reproduced from S.-Y. Lin, S.-J. Chang, and T.-J. Hsueh, ZnO nanowires modified with Au nanoparticles for nonenzymatic amperometric sensing of glucose, Appl. Phys. Lett. 104 (19) (2014) 193704.*

The reduction of gold oxide on the AuNPs in the negative scan direction also cause the obvious current peak at 0.56 V. Although glucose sensitivity is seen in both decorated and undecorated ZnO NW electrodes, a significant enhancement of the sensitivity is obtained with AuNP decoration. When CVs were repeated at different concentrations of glucose in the nM range, peak currents of devices with AuNPs were consistently larger.

Amperometric measurements were also used to distinguish sensing ability of the two types of electrode. The AuNP decorated sensor increased in current by 60 μA upon introducing 1.5 mM of glucose into the 0.1M NaOH solution, while the undecorated device showed no change [17]. Over 30 mM of glucose was needed to produce a similar current response in the undecorated device. The mechanism behind this increased sensitivity is shown in Fig. 8.9.

When the n-type ZnO contacts the NaOH solution, large numbers of electrons collect at the surface of the ZnO, causing downward band bending and reducing the current response during the electrocatalysis of glucose. With the addition of the AuNP, the slightly higher work function of the Au induces upward band bending and prevents electron collection at the surface. This is a clear demonstration of how the decoration of semiconductor nanostructures with metal nanoparticles can have significant effects on the biosensing capability of these materials.

8.3 Cluster decoration of nanostructures for gas sensing

8.3.1 How do resistive gas sensors work

Resistive gas sensors generally consist of two (or in low and high resistance sensors, four) electrical contacts, bridges by a variable resistance element, as shown in Fig. 8.10. The resistive elements can include thin films, nanoclusters, nanowires, or nanosheets.

Resistive gas sensors utilize the effect of a chemical interaction to deviate an electrical signal. There are multiple ways this can occur. One mechanism is for the analyte gas to reversibly diffuse into a metallic sensitive element. One example is hydrogen gas sensors,

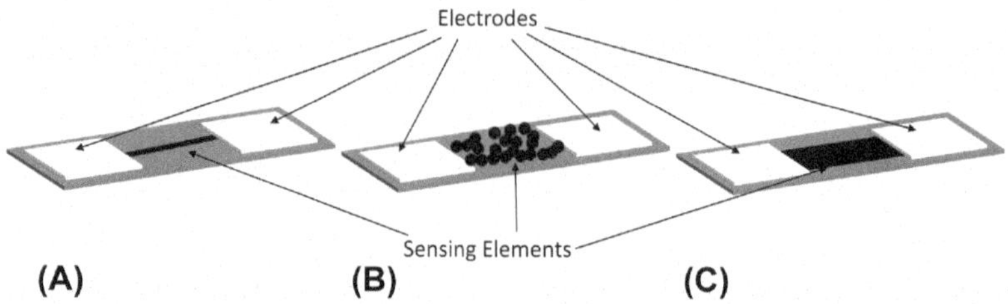

FIGURE 8.10 A series of generic geometries of semiconductive gas sensors. The substrates are typically an insulating layer, while electrode materials are typically gold and platinum (metals which have good conductivity and tend not to corrode). The black elements are the MOS sensing structure as (A) nanowires/nanotubes, (B) layers of percolating particles, and (C) thin films.

which consist of palladium nanowires as a sensing element bridging two electrodes. Palladium has a strong affinity for hydrogen, dissociating the hydrogen molecules before forming palladium hydride [21]. This results in a difference in resistance as the palladium becomes more, or less, hydrated.

Another type of resistive gas sensor is based on resistance changes related to tunneling currents between a gas-phase cluster beam assembled film. Again, an example of this is a hydrogen gas sensor based on palladium clusters [22]. The clusters are deposited between two electrodes until a point just before the clusters reach percolation. This allows low tunneling current to pass through the sensitive element. As hydrogen gas interacts with the palladium nanoclusters in the manner described above, the clusters' size increases, which results in a decrease in resistance as the tunneling distance decreases [22]. Interestingly, higher coverage nanoparticle films are more sensitive to low pressures of hydrogen, while lower coverage films provide more sensitivity in high pressure hydrogen environments [23]. This is because lower coverage films are more reactive (the clusters have a smaller mean diameter, presenting a higher surface area-to-volume ratio) as demonstrated by the fast reaction kinetics [24]. However, if there is a lower pressure, there is a smaller probability of interaction between the clusters and gas molecules. Therefore, having more clusters increases the chance of interaction. Similar hydrogen sensors have been made from palladium-nickel alloys. This changes the pressure of the alpha-to-beta phase transition in Pd-H, which effects operation of the sensor. The clusters are formed using two separate targets and magnetrons cosputtered into the condensation gas, eventually coalescing into alloy clusters, the concentration of which can be controlled. All the alloys are less sensitive than pure Pd clusters; however, an atomic concentration of 16% Ni allowed the clusters good sensing properties while appropriately decreasing the pressure when the alpha-to-beta phase transition occurred in Pd-H [25].

Finally, there are semiconductor-based gas sensors, especially metal oxide–based semiconductors. These sensors work through the physisorption of gas molecules onto the semiconductor surface. Once in close interaction, the gas molecules can dissociate and chemisorb onto the surface. Any charge on these immobilized species will then affect the semiconductor sensor by the same mechanism shown in Fig. 8.1 and detailed in Section 8.1.2.

While some sensors based on more traditional semiconductors such as carbon nanotubes and silicon nanowires will be discussed briefly in terms of gas sensors, most sensors featured below will be MOS [1]. In this group of gas sensors, the interaction of oxygen with the surface is of great importance. Chemisorbed oxygen species are negatively charged, leading to negative surface charges. On p-type MOS, this results in a decrease of resistance as oxygen species increase, while n-type's resistance increases as oxygen surface concentration increases (Fig. 8.1). Zinc oxide, tin oxide, and tungsten oxide are common n-type semiconductors, while cupric oxide (CuO) is an example of p-type semiconductor.

The basic concept of the MOS gas sensor is that analyte gases interrupt this equilibrium of absorbed surface oxygen species. The analyte gas is physiosorbed on the surface of the semiconductor, before being chemisorbed onto the surface. The captured gas molecules are then able to interact with the chemisorbed surface oxygen. Depending on the type of gas, the surface oxygen is reduced or further oxidized, as the gas reacts with the surface molecules. For instance, carbon monoxide is known to reduce the surface of metal oxides (such as CuO), as carbon dioxide is a much more stable molecule, summarized by the equation:

$$CO_{(g)} + O^-_{(ads)} \xrightarrow{CuO_{(s)}} CO_{2(g)} + e^- \tag{8.1}$$

On the other hand, nitrogen dioxide can oxidize the metal oxide surface further, as nitric oxide is released from the surface. Eq. 8.2 summarizes this reaction as:

$$NO_{2\,(g)} + e^- \xrightarrow{CuO_{(s)}} NO_{(g)} + O^-_{(ads)} \tag{8.2}$$

Therefore, the concentration of analyte gas is related directly to the subsequent changes in resistance that these reactions cause to the semiconductor.

The typical MOS gas sensor is operated at an elevated temperature. This energy allows more stable gas molecules to interact with the surface of MOS gas sensors. In the case of most semiconductors, the resistance decreases as the temperature surrounding the semiconductor increases. The reason for this phenomenon is that the semiconductor has more energy to close the band gap, which creates more charge carriers. This is the case for p-type MOS sensors; however, n-type sensors can demonstrate an increase in resistance with increasing resistance, as the oxygen absorption on the surface increases. For n-type sensors, it is a balancing act between increasing surface absorption of gases, while maintaining the necessary conductivity.

Current state of the art with regards to MOS gas sensors is thin-film based. For decades now however, the aim of research has been to transition to nanostructured semiconductors (particularly single or crystalline nanowires, nanotubes, and nanostructured thin films illustrated in Fig. 8.10). This improves reaction kinetics by increasing the surface area and decreasing the volume. Subsequently, the reaction on the surface effects the conductivity in the sensor much faster. It can also reduce the amount of energy needed in heated sensors as the sensing element has lower thermal mass. Finally, the increased area increases the probability of reactions with the sensor itself.

However, there are still drawbacks in integrating these elements into working sensors, including their size, the fact that many still require some significant external energy source, and the fact that these sensors, and MOS sensors especially, are highly sensitive to other gases. Any gas which removes oxygen in the case of a MOS sensor or introduces a charge to the surface will affect the conductivity of the sensor, which means that the sensor will respond to any gas, not simply the target analyte. Nanoclusters are a major avenue of research to counter these issues.

8.3.2 What clusters add to gas sensors

A great advantage in adding clusters to semiconductive sensors is surface sensitization. Metallic clusters with a specific affinity toward an analyte gas can adsorb and dissociate the target gas. This is the spillover effect, whereby the gas is dissociated on the cluster and subsequently able to interact with semiconductor surface [26,27]. The result of this is an increase in signal, with respect to the analyte gas versus other interfering species, known as the specificity or selectivity of the sensor [28].

An example of this is the work by Kong et al., where semiconductive p-type carbon nanotubes were chosen as a sensing element for hydrogen gas [29]. This nanotube was decorated with Pd nanoclusters, created by thermally evaporating a thin 5 Å film. Depositions carried out in this way do not result in the formation of a full uniform film but instead in isolated Pd islands of cluster growth. As described earlier, Pd has a high affinity to hydrogen gas and as the gas molecules absorb onto the Pd surface, the work function of the Pd nanoparticles changes. The work function is the energy required for an electron to be removed from the atomic surface of a solid. Interactions with molecules on the solid surface elevates or lowers this energy. Thus, the hydrogen adsorbing on the surface lowers this energy. This leads to electrons injecting into the nanotubes and recombining with the charge carrying holes [29]. Therefore, as hydrogen concentration increases, the resistance of the sensor increases proportionally.

This result also highlights that in addition to catalytic effects, the clusters themselves have an electronic effect on the surface of the semiconductor. The creation of nano-Schottky barriers between the cluster and semiconductors have been demonstrated by Kolmakov et al. for tin oxide nanowires and nanobelts decorated with Pd [26]. The Pd was decorated using a similar vapor growth method as Kong et al., where the nanoparticles are formed by nucleation on the surface as more Pd is deposited [26]. The current through the nanowires and nanobelts was monitored in situ during the Pd deposition. As the amount of Pd increased (and the subsequent size of clusters increased), the current decreased, demonstrating the formation of these depletion regions on the tin oxide nanostructure [26].

The theory of catalysis is that a catalyst reduces the activation energy of a reaction. In MOS-based gas sensors, the sensing element is heated, usually to temperatures between 200 and 300°C, to improve the reaction of chemisorbing analyte gases onto the surface of the MOS. It should therefore be possible to lower the required energy of reaction with catalytic clusters. Joshi et al. demonstrated this using a solution-based deposition method to decorate ZnO nanowire-based CO sensor with Au colloidal nanoclusters [30]. The sensor was shown to work at a temperature of 200°C, while being inactive at room temperature. This is due to ZnO not being able to chemisorb O^- or O^{2-} without external energy. When the Au clusters were deposited on the ZnO sensor element, however, the sensor was operational at room temperature as the Au can produce chemisorbed O^- or O^{2-} species without this energy. It is also important to consider the size of the clusters being deposited. A similar scheme facilitated by Hwang et al. using Ag clusters (grown through thermal evaporation) to sensitize tin oxide nanowires tested the effect of different sizes of clusters (increasing the time of the Ag deposition). It was found that smaller Ag clusters had a strong increase in response, while larger clusters decreased the sensitivity of the sensor, as the larger Ag clusters decreased the area of the nanowire which could chemisorb gases [31].

Of course, if the operating temperature of the Au decorated nanowires is increased, then the sensors' sensitivity can be increased [32]. Chang et al. have demonstrated that the response of the sensors as a percentage of their conductivity in air increases with temperature, until 250°C, before decreasing as temperature is further increased to 350°C. This decrease is ascribed to the higher temperatures interfering with the exothermic absorption of CO onto the ZnO surface. The reason that a high sensitivity is desirable is that the sensor can have a higher resolution, allowing sample differences in gas concentration to be measured [32].

These different features that clusters add can also work in one final way, notably the electronic nose. This is a concept which is designed to mimic the human sense of smell. Humans do not possess a scent receptor for each aromatic compound we can detect. Instead, we have arrays of different receptors which produce a wide range of responses. These responses form patterns from which the olfactory area of the brain can determine different aromatic compounds or, put simply, smells. The goal of the electronic nose (E-nose) is to move away from finding sensing elements specific to one certain gas and instead use a select array of different sensing elements to create a multitude of signals, which when processed, will be able to identify the target gases and concentrations.

An early example of this was created by Baik et al., using arrays of tin oxide nanowire sensors [33]. Some arrays consisted of undecorated tin oxide nanowires, while others featured tin oxide nanowires decorated with thermally evaporated Pd clusters. One final array of tin oxide nanowires was decorated with similar thermally evaporated Ag clusters. Different concentrations of hydrogen, carbon monoxide, and ethylene were tested separately on the array, while each array response was measured. Once the responses were recorded, a matrix-based classification technique known as *linear discriminant analysis (LDA)* was applied to the data. This technique classifies results into similar classes based on multiple variables. The basic idea is that each variable is treated as a dimension (temperature, concentration, gas type, decoration, etc.), and as the analysis is applied, it compresses the data, determining the dimensions which show the best separation between classes. The work by Baik was able to show good separation between the three gases in just two dimensions [33].

Cluster beam deposition has also been used to build the sensitive elements of gas sensors themselves. Bismuth nanoclusters are deposited using gas-phase aggregation until the percolation limit is reached. This creates a sensing element, which has a much higher surface area-to-volume ratio than a traditional thin film [34].

Bismuth and antimony clusters have also been used to build nanostructured nanowires. Partridge et al. etched "V-grooves" using lithography processes, into silicon wafers before creating a silicon dioxide passivation layer. Contacts were then deposited before the nanoclusters were deposited from a gas-aggregated cluster beam. The clusters with higher kinetic energy were able to bounce around and between the edges of the "V-groves," eventually assembling a nanowire electrical connection [35]. A similar technique has also been used to fabricate tin oxide nanowires [36]. In this case, the width of the nanowire can be controlled, which means the effect of the depletion layer can be increased or decreased by changing the thickness.

Alternatively, cluster beam deposition has been used to make nanostructured thin films from tin oxide nanoparticles. Kennedy et al. created a sensor for ethanol [37], hydrogen, and methane [38] using evaporation and gas-aggregation to produce tin nanoclusters, which can be size-selected and oxidized in-flight before the cluster is deposited on the substrate. This allows a measure of control over the structure of the film, leading to control over sensitivity [37]. As these clusters were made from a MOS sensor, the tin oxide was operated at temperatures between 200 and 300°C. It was found that tin oxide in $SnO_{1.8}$ stoichiometry was achieved in more cases when the clusters were annealed in-flight as opposed to SnO_2 if a thin film of tin particles was oxidized [38]. It was also found that the deposition parameters had a significant effect on the film, more so than the structure of the tin oxide nanoparticles. ZnO films, which have been doped with cobalt, have also been assembled via a similar method using magnetron sputtering gas-phase aggregation [39].

Similar humidity sensors have been developed based on nanostructured carbon films with high porosity created using supersonic cluster beam deposition (SCBD). This form of sputtering-based gas-phase aggregation growth uses multiple apertures (skimmer) to form an aerodynamic lens which filters cluster size, as opposed to the more traditional mass quadrupole filter system. This allows control of high flux beams of clusters, enabling faster deposition rates, more suited to nanostructured thin films. Bruzzi et al. developed a resistive and capacitive setup to achieve this result [40]. In the resistive configuration, the nanostructured carbon film was deposited on top of the separated electrodes, while the capacitive sensor had one gold electrode under the active carbon layer (rear) and one on top of the layer (front). The resistive sensor was not active until relative humidity reached 60%, while the capacitive sensor had a range of 10% to 80%

relative humidity. The main reason for this is that the active layer absorbing humidity in the resistive sensor is the top of the film, which has less effect on the current flowing through the nanostructured carbon film. The capacitive sensor, however, has the front electrode also in contact with the active layer, producing a higher sensitivity to humidity [40]. This technique is well known in the assembly of gas sensors, as it can create films with tunable resistivity. Barborini et al. showed that increasing the thickness of a thin film created in this way increases the resistance, as there is less contact forming between clusters in the film [41].

This technique has also been used by Zhao et al. to produce hydrogen gas sensors based on thin, porous tungsten oxide films decorated with Pd nanoclusters [42,43]. Pd decoration was achieved by depositing a 2.5 nm thin film using magnetron sputtering. Clusters were formed as islands of Pd atoms began nucleating together on the surface. If deposition is stopped before this nucleation begins to percolate into a film, then the growth of nanoclusters is achieved. The underlying thin film consists of cluster beam deposited metallic tungsten. This layer was oxidized producing the MOS sensitive element. It was observed that functionalization with Pd not only improved sensitivity to hydrogen, but also aided in the recovery of the sensor post response. The presence of the Pd assisted the dissociation and diffusion of oxygen into the film at an operating temperature of 80°C. It was also found that a thin film of 11.2 nm was preferable to a thicker 150 nm film. This is explained as the hydrogen can diffuse easily through the entire thicker film; however, oxygen is able to only diffuse to about 11.2 nm deep into the film. Hence a thicker film will not be able to recover after the first-time hydrogen is detected [42,43].

Like Kennedy's tin oxide films, it was found that different oxidization conditions produced a different film. At temperatures of 250°C the oxidized film was less porous than at 700°C. Due to this the 700°C films were more responsive and faster to respond as more pores allowed the hydrogen (and palladium deposited) more surface area to react, than the 250°C films, which had most of the Pd assisted layer on the film surface [44]. These sensors have been demonstrated to work at room temperature at concentrations below 60 ppm. In addition, because of the room temperature operation, and the Pd decoration, the sensor has excellent selectivity against volatile organic compounds (VOC), specifically methanol, ethanol, isopropanol, formaldehyde, and acetone [45,46]. Comparing films prepared by SCBD to those which were created by reacting sputter of tungsten metal in a 20%Ar:80% O_2 environment showed that while the reactively sputtered WO_3 could be more accurate and respond faster, they could not operate in a continuous process due to their poor recovery time in general, unlike the SCBD assembled films, which recovered orders of magnitude faster [47].

Zhao and Wong et al. have continued these studies with similar sensors made from SCBD zinc oxide [48,49]. The process deposited Zn metal which was then oxidized at 500°C, producing a percolating network of approximately 5 nm thick ZnO. The temperature 500°C was chosen as the oxidizing temperature, as it produced a sensor which produced the highest photocurrent, with the most stability [48]. ZnO is typically used as its band gap of 3.37 eV makes it sensitive to UV light. Exposing ZnO to UV light provides enough energy for the charge carriers to begin to populate the conduction band, and the resultant current is known as *photocurrent*. This film is then coated with a hydrophobic polymer-coating and used as a sensor for O_2 in a variety of environments, including dry atmosphere, humid atmosphere, and when immersed in water. As more oxygen is adsorbed on the surface, the photocurrent decreases, due to the charge layer depletion caused by increasing oxygen concentration [48].

These ZnO films have also been decorated with a 5 nm Pd thin film, resulting in Pd cluster decorated ion, for hydrogen gas sensing [49]. In this case, sensing was performed at temperatures of 20°C and 80°C, with and without UV illumination. The 80°C was much more sensitive to hydrogen, and responded much faster, like in previous studies. However, the UV illumination had a big effect on the ability of the sensor to recover at room temperature. The effect was present but highly reduced at 80°C. In addition, at 80°C with UV illumination, the sensor response was much more reproducible and consistent. This is not fully studied but attributed to alleviating the constant "cracking" of Pd due to repeated hydrogenation/dehydrogenation that sensing requires [49].

Finally, films made by depositing clusters of iron cubes made from magnetron sputtering and gas phase aggregation have been used as sensors for NO_2 [50]. These sensors operate at temperatures of 200°C allowing them a measure of selectivity against other gases such as CO, as the cubes lack the required energy to chemisorb the CO at this temperature. The nanocubes themselves have demonstrated the Kirkendall effect, which is the diffusion of oxygen to the core being slower than the outward diffusion of the iron to the surface during oxidation, creating a void in the center of the cubes. The effect of this has not yet been studied on gas sensing; however, it is expected that the space charge region would extend to the entire iron oxide film, as there is no iron-iron oxide interface in individual clusters [50].

8.3.3 Cluster beam decoration in gas sensing

Cluster beam deposition already has great utility in the gas sensing space, as can be seen in the previous section. Methods such as SCBD are more tailored to the creation of nanostructured films, while other gas-aggregation schemes, built in a way where a mass filter is present to select the size of clusters, are more useful in the creation of monodisperse clusters. Clusters produced in this way are better served for the decoration of preformed semiconductive gas sensing structures. Most decoration techniques involve either growth of the nanocluster on the surface, or a physical attachment with a colloidal nanocluster. Nanoclusters grown on the surface are not necessarily well controlled, while chemically grown nanoclusters have a variety of surfactants and ligands, which limits interactions between the clusters and the substrate (resulting in physical weaker binding), as well as clusters and the gases.

For example, SnO_2 nanoclusters were deposited on TiO_2 nanoflowers using the gas-aggregation cluster beam deposition method [51]. The clusters were compared against hydrothermally grown SnO_2 similar TiO_2 nanoflowers. The clusters deposited through gas-aggregation showed less agglomeration, and more control over the size of individual clusters [51].

In addition to the above example, some sensors have already been reported which use gas-phase aggregation to decorate a semiconductive element. Ayesh et al. have described a hydrogen sensor, which features a SnO_2 semiconductive nanostructured percolating film deposited using gas-phase aggregation [52]. The clusters are deposited as Sn metal and oxidized over a period of 24 h. Following this, Pd nanoclusters were also grown using gas-aggregation and deposited on top of the SnO_2 layer. This decorating layer of Pd nanoclusters is also a percolating film. In this case, the proposed mechanism is that Pd draws oxygen from the SnO_2 layer, resulting in a decrease in resistance as Pd becomes PdO. As hydrogen interacts with the sensor, the Pd is reduced back to a metallic state and the oxygen surrounds the SnO_2 clusters, resulting in a corresponding increase in resistance [52].

In this way, oxygen is used as charge transfer; however, Ayesh et al. have demonstrated a similar sensor using charge transfer of electrons [53,54]. Two sensors have been demonstrated for Hg^{2+} ions in liquid, one based on a graphene sensing element decorated with Au gas-aggregated nanoclusters [53], and another based on graphene oxide and similarly grown Ag nanoclusters [54]. In both cases a percolating layer of size-controlled nanoclusters are deposited. As Hg^{2+} ions flow past the sensor, Au and Ag bind with the ions (due to their generally high binding energy and absorption enthalpy of Au and Ag), which results in an increase in resistance due to the decrease in mobility of electrons across the graphene sheets. The sensors themselves are more selective to Hg^{2+} than other heavy metal ions such as Cr^{2+}, Cu^{2+}, Zn^{2+}, Cd^{2+}, Fe^{3+}, CO^{2+}, and K^{+}, and work at concentrations as low as 0.05 ppb.

Another recent sensor based on gas-aggregated nanocluster decoration of nanostructured ZnO has been used for the detection of 4-aminophenol (pAP), also in liquid [55]. In this case the clusters were grown through laser ablation and gas-aggregation, rather than magnetron sputtering and gas-aggregation. This resulted in clusters of sizes about 2–3 nm. Au, Pt, and bimetallic AuPt nanoclusters were used to decorate liquid grown ZnO nanowires. Impedance spectroscopy, cyclic voltammetry, and chronoamperometry were used to determine the response of the sensor. It was found that Au decoration was not able to determine pAP from phosphate buffer solution (PBS) using impedance spectroscopy or cyclic voltammetry. However, Pt and AuPt decorated nanowires showed strong responses, while AuPt showed better responses to pAP through amperometry. This was explained as a synergistic effect, with Pt offering better catalysis and Au providing stronger electron transport [55].

So far, most of the materials that have been discussed in this chapter are n-type MOS sensors, particularly, ZnO, WO_3, and SnO_2. However, there has been some recent work with p-type MOS decorated with gas-aggregated nanoclusters, specifically copper oxide. Copper oxide is a p-type semiconductor, which has interesting properties in its cuprous and cupric forms. Cuprous oxide (Cu_2O) is less electrically conductive but has the smallest known photon emission band, which is of great interest in quantum photonics. Cupric oxide (CuO) on the other hand is more stable, with greater electrical conductivity, and has been used as the basis of MOS sensors [56]. One major advantage of copper oxide is that it can be synthesized in nanowire configurations through low temperature thermal oxidation. While tin and zinc oxide nanowires are better studied, due to their high electrical conductivity and well-defined crystal structures, and tungsten oxide is gaining interest due to its resistance to humidity, they are not as simple to synthesize as copper oxide.

Heating copper in an oxygen atmosphere will result in the growth of micron-long nanowires after a few hours. These nanowires can form at temperature as low as 335°C, which is a temperature which would be compatible with silicon technology used in complementary metal-oxide-semiconductor (CMOS) sensors, which currently form the state of the art. At present, sensors developed using other materials require growth of the nanowires off-chip (due to higher growth temperatures), before they are dispersed onto the silicon substrate. This results in less control of where the nanowire structures are located, which is problematic when trying to design an integrated circuit (Fig. 8.11).

Steinhauer et al. have successfully developed a method of growing copper oxide nanowires on top of a microhotplate, which is a platform that could lead to the integration of nanowires onto a CMOS chip [56]. The basic concept of this sensor is shown in Fig. 8.12. The sensor is based on an integrated interdigitated electrode-microhotplate design, which releases a multilayer membrane from the Si substrate using etching to improve the heat transfer from the heating element to the electrode. The multilayers are SiN-Pt microheater-SiO_2-Pt electrode, and thin films of copper are deposited via e-beam evaporation on top of separated electrodes. An approximate 3 μm gap remains between the copper, a gap which is bridged by the growth of copper oxide nanowires as the copper thin films are thermally oxidized in synthetic air (80% N_2 and 20% O_2), using the thermal energy applied by the Pt heater. This resulted in discrete increases in conductance related to the formation of nanowire bridges. This sensor showed responses to concentration of CO as low as 1 ppm when it was not exposed to humid air; however, on exposure to approximately 30% relative humidity, the sensitivity decreased dramatically [56].

Pd nanoclusters grown through gas-aggregation and deposited via cluster beam have been used to alleviate this issue with CuO nanowires [57]. CuO nanowires were grown thermally on Cu foil, which was then ultrasonicated in isopropanol, before being spin coated on SiO_2. Four electrode contacts were deposited on a single CuO nanowire located and e-beam patterned using a scanning electron microscope. The Pd nanoclusters were then deposited on top of this substrate, and subsequently the CuO nanowire. The nanoclusters measured by the quadrupole mass filter had a mean size of 5 nm, which showed high correlation measurements of the clusters deposited on Si substrates for analysis using atomic force microscopy. Fig. 8.12 shows the response of the nanowire to CO gas in a 50% relative humidity environment, operating at a local temperature of 350°C. It also shows the difference in response between decorated and

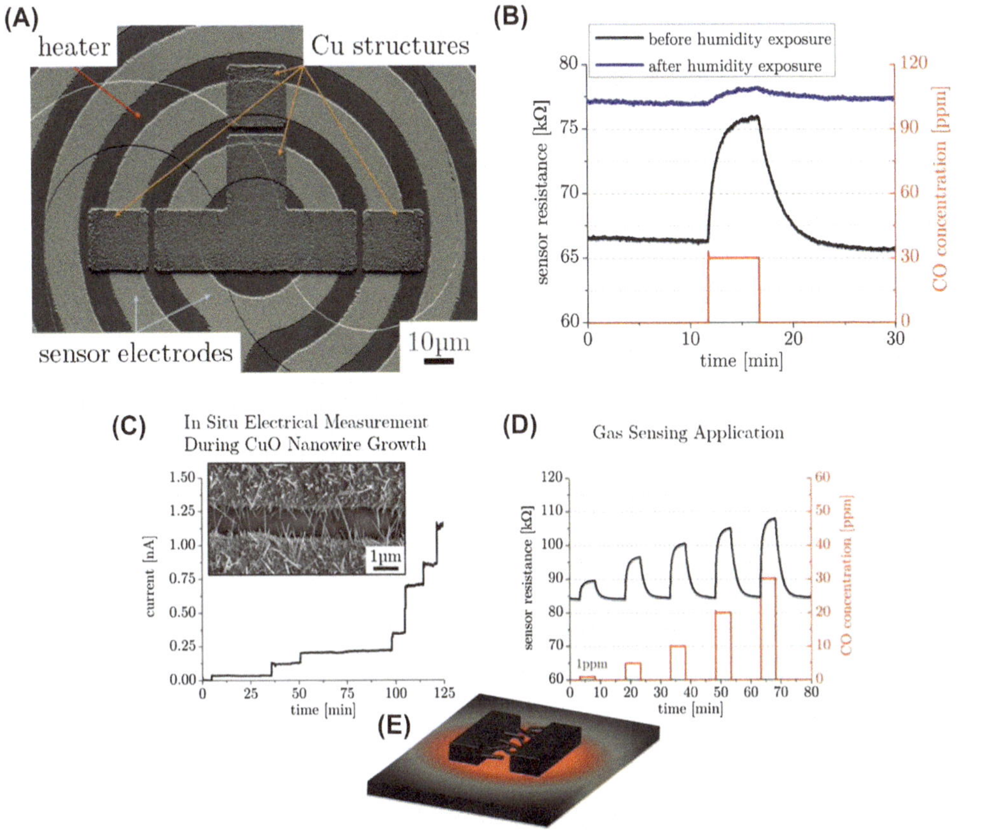

FIGURE 8.11 (A) SEM image of the nanowire integrated chip. Pt heater is separated from electrodes by a SiN layer, (B) shows the difference in response to 30 ppm of CO (in dry synthetic air with an operating temperature of 325°C) between the nanowires before and after exposure to 30% relative humidity, (C) shows nanowires grown between electrodes, in SEM image and with the increase in current between electrodes during oxidation, while (D) shows the response of the sensor to various concentrations of CO (operating in dry air at 325°C), and finally, (E) is a schematic showing the localized heating of the nanowires due to the microhotplate. *Adapted with permission from S. Steinhauer, A. Chapelle, P. Menini, and M. Sowwan, Local CuO nanowire growth on micro-hotplates: In situ electrical measurements and gas sensing application, ACS Sens. 1 (5) (2016) 503–507. Copyright 2016 American Chemical Society.*

undecorated CuO nanowires. It can be seen from these images that decoration of the nanowires using the Pd nanoclusters has a significant improvement on the signal of the sensor in a humid environment [57].

The reasons for this improvement still need further research, with many possible causes, including surface sensitization to CO by Pd nanoclusters. However, given this is a more complex environment than many of the previous examples in this chapter (added hydroxyl groups due to greater humidity), it requires further study to understand how the clusters affect competition between H_2O and CO molecules and the chemisorbed oxygen on the nanowire surface [57].

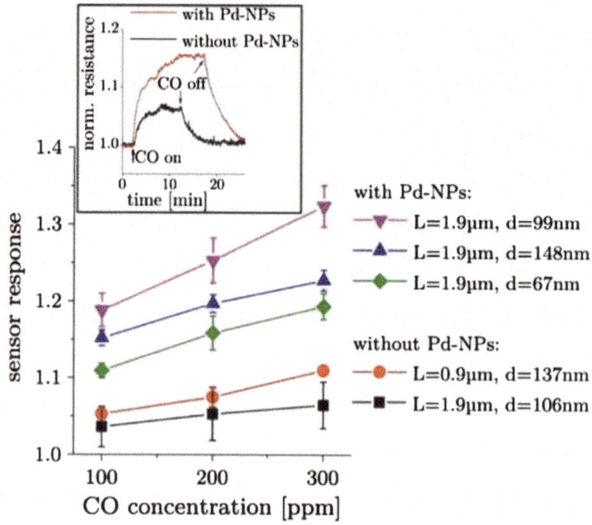

FIGURE 8.12 The difference in response between Pd decorated and undecorated CuO nanowires in 50% relative humidity at an operating temperature of 350°C. Inset shows comparison of a single response to CO under said conditions. *Reproduced from S. Steinhauer, V. Singh, C. Cassidy, C. Gspan, W. Grogger, M. Sowwan, and A. Köck, Single CuO nanowires decorated with size-selected Pd nanoparticles for CO sensing in humid atmosphere, Nanotechnology 26 (17) (2015) 175502.*

Still this is an example of the proven utility of gas-phase aggregation cluster beam deposition, which could be applied directly to the CuO nanowires on the chip shown in Fig. 8.11. This technique also compliments the design of the device as a way to functionalize devices based on silicon technology, without the need for solvents or excessive temperatures, conditions that chip manufacturers look to avoid.

8.3.4 Semiconductor-cluster interaction

Since gas aggregated nanoclusters have no stabilizing ligands or coatings, there will likely be direct interaction with the semiconductive substrate on which they are deposited. In addition to interactions between the nanoclusters and substrate on deposition, there are interactions between the nanocluster and the support in the sensing atmosphere at elevated temperatures.

High resolution transmission electron microscopy (HRTEM), especially when coupled to in situ transmission electron microscopy (TEM) and environmental transmission electron microscopy (E-TEM) techniques, has allowed the probing of materials on the atomic scale. What in situ TEM and E-TEM have added is the capacity to investigate the way material structure and composition change at increased temperatures and gas pressure values. In addition, some TEM holders support biasing of samples being imaged during TEM. These advances have allowed the imaging of MOS-based gas sensors in low pressure atmosphere, as they are operating within the column of a TEM.

A demonstration of this technique was published by Steinhauer et al. in which CO gas sensing was performed in situ on SnO_2 nanowires, both pristine and decorated with Fe-Pd cubes, in an O_2 environment [58]. The corresponding changes to morphology and composition could then be directly observed. The SnO_2 nanowires were synthesized through spray pyrolysis and oxidation on a SnO_2 seeded Si wafer with a photolithographic template [59]. Once grown, the nanowires were dispersed in isopropanol and then drop coated on a membrane-based TEM heating chip. These chips featured patterned tungsten contacts, and e-beam deposition of Pt was performed in a focused ion beam scanning electron microscope (FIB-SEM), binding SnO_2 nanowires which bridged the tungsten electrodes. The chip was then imaged in an E-TEM, with the electrical measurements taken of the biased nanowire, at different temperatures and pressures of O_2. Additionally, the electrical response of the pristine nanowire to 100 ppm CO diluted in O_2, at an operating pressure of 20 mbar and temperature of 250°C was studied [58].

It is important to note that imaging was not performed while the sensor was in operation. This is because biased MOS-based sensors operating in a TEM column can interact with the electron beam and gas molecules in the chamber, which increases the uncertainty in any gas measurements conducted on the current version of these membrane heaters [60]. Despite this issue, the technique is still superior performing the experiments ex situ and then reinserting the sample into the microscope. The changes which have occurred during the time the beam has been turned off and then on can mostly be ascribed to the experiment instead of possible outside contaminations possible in ex situ conditions.

Once the in situ SnO_2 was characterized in the TEM, it was decorated using gas-aggregated cluster beam deposition. Three separate sputtering targets (two Fe and one Pd) were used to grow and deposit bimetallic cubic nanoclusters [58]. The sensor was once again measured in situ with CO at the same conditions as the pristine nanowire, and it was found that functionalization of the nanowire with the cubic clusters improved the response of the sensor to CO gas. The images collected from the TEM showed that the SnO_2 nanowire had not shown any great change; however, the Fe-Pd clusters showed a dramatic change [58]. As Fig. 8.13 demonstrates, after the decorated nanowire was heated in oxygen, the cube showed a void partly filled with another structure (found to be Pd using electron energy loss spectroscopy (EELS)). The void formation has been demonstrated before in cubic Fe nanoclusters, as seen in the previous section, which is explained by the Kirkendall effect [50]. What is of interest is that the Pd has apparently also diffused to the void and created Pd clusters. Fig. 8.13A shows an image from STEM mode of an as deposited cube. It is clear from here that when the cluster is deposited, it shows a uniformity in the contrast and structure, suggesting a uniform alloy. The full significance of the effect of this change is thus far not understood, however, it is now possible to track structural, composition and morphological changes to certain events, allowing the creation of more detailed explanatory models.

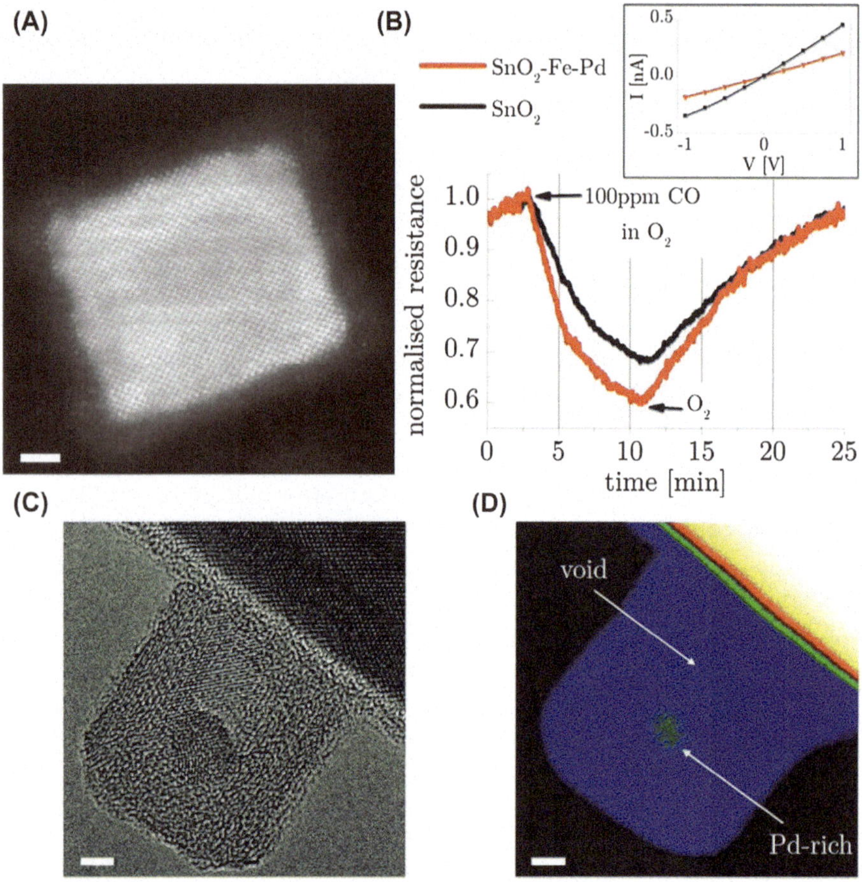

FIGURE 8.13 (A) Shows an STEM image of the Fe-Pd cubes as deposited. (B) Is a comparison of the in situ response of a bare SnO_2 nanowire and a SnO_2 nanowire decorated with Fe-Pd cubes, operated in a 20 mbar O_2 environment at a temperature of 250°C. Inset shows the difference in I-V characteristics of a bare and decorated nanowire. (C) Shows an HRTEM image of an Fe-Pd cube on the SnO_2 nanowires after a measurement. The void and small Pd cluster can be seen. (D) The false color of the HRTEM image shows the difference between Fe, Pd, void, and the SnO_2 nanowire. All scale bars are 2 nm. *Reproduced from S. Steinhauer, J. Vernieres, J. Krainer, A. Köck, P. Grammatikopoulos, and M. Sowwan, In situ chemoresistive sensing in the environmental TEM: probing functional devices and their nanoscale morphology, Nanoscale 9 (22) (2017) 7380–7384.*

Steinhauer et al. recently used the information gained from compositional and structural changes in nanocluster-nanowire systems in oxygen atmosphere to build a model system, which was then used to elucidate the driving forces behind the observed changes [61]. In this case Pd nanoclusters were grown through gas-phase aggregation cluster beam deposition and deposited onto CuO nanowires. Using an E-TEM, the Pd decorated nanowires were oxidized in situ at temperatures of or above 300°C and in a 20 mbar O_2 atmosphere. It was found that Pd nanoclusters generally appear to oxidize from the CuO-Pd interface [61]. Typically, the nanoclusters maintained their general

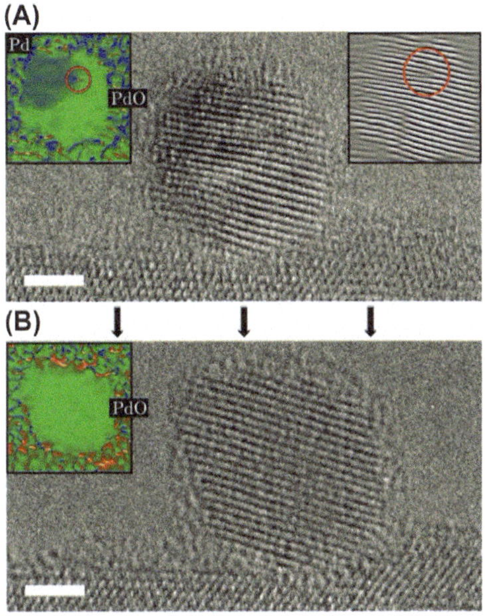

FIGURE 8.14 (A) Pd cluster on a CuO nanowire surface following in situ thermal oxidation at a pressure of 20 mbar O_2 and temperature of 300°C for 1 h. Left inset shows GPA strain analysis of the Pd 111 spatial frequency in relation with PdO 101, while the right inset shows the Bragg-filtered image (*red circle* highlights a dislocation at the Pd-PdO interface). (B) The same nanocluster following an additional thermal oxidation treatment at the same pressure as (A) and a temperature of 350°C for 1 h, after which point it is fully oxidized to PdO, seen from the left inset showing a homogenous phase in the GPA analysis. Scale bar represents 2 nm in both images. *Reprinted with permission from S. Steinhauer, J. Zhao, V. Singh, T. Pavloudis, J. Kioseoglou, K. Nordlund, F. Djurabekova, P. Grammatikopoulos, and M. Sowwan, Thermal oxidation of size-selected Pd nanoparticles supported on CuO nanowires: the role of the CuO–Pd interface, Chem. Mater. 29 (14) (2017) 6153–6160. Copyright 2017 American Chemical Society.*

morphology after oxidation to single crystal palladium oxide (PdO); however, some instances showed onionlike structures, indicating subsurface defects in the nanoparticle. Analysis of the fast-Fourier transform (FFT) of the TEM images showed that the oxidized nanoclusters appeared to align along a preferential direction on the CuO lattice, that is PdO [100] was parallel to CuO [110]. Fig. 8.14 shows the growth of PdO, with Pd metal still lattice still visible at the top of the cluster, and PdO lattice making up most of the rest of the cluster from the lattice up (Fig. 8.14A). Fig. 8.14B shows the completely oxidized cluster, while the insets show geometric phase analysis (GPA) of the corresponding cluster. GPA shows the presence of dislocation between Pd metal and PdO.

With this information, it was possible to build models of Pd clusters on CuO supports and use ab initio density probability theory (DFT) to calculate whether there were specific atomic planes of metallic Pd where oxygen would most likely enter the lattice, and whether this would change with Pd on a CuO surface [61]. In fact, it was found that the presence of the CuO substrate decreases the energy barrier for oxygen into the Pd lattice by approximately one third. It was also observed that the presence of an extra oxygen atom at the interface significantly reduces the energy barrier between PdO [100] and adjacent Pd [100].

This research clearly indicates the importance of the interaction between gas-aggregated cluster beam deposited nanoclusters and the semiconductive nanostructure interface. In this case, the clusters are well attached to the surface; however, it is also necessary to understand the effect this interaction has on the cluster itself [61].

8.4 Outlook

As this chapter shows, the work currently conducted utilizing gas-aggregated cluster beam deposition for the purposes of decoration is still an emerging area. The attraction of a solvent less deposition method, which allows some degree of control of the clusters, is of great interest to those in the semiconductor and integrated circuit fabrication industry. The main advantages are strong interactions between the nanoclusters and the substrate anchoring the clusters and increasing electronic interaction between cluster and semiconductor. This is just the tip of the proverbial iceberg. Gas-phase aggregation of clusters has been demonstrated to grow much more interesting clusters than most of those discussed here, including multicore-shell structures [62–64], alloys [65–68], decorated [69,70] and metastable "ukidama"[71] structured clusters. These cluster materials may demonstrate interesting synergistic effects with the semiconductor surface, as shown by the transformation of Fe-Pd cubes on nanowire substrates. Additionally, as more insight is gained on controlling the morphology of clusters (i.e., cubes), a whole new avenue of decoration will be opened with controlling which crystal planes are exposed for gas sensing or functionalization. Finally, with clusters being able to assemble nanodimensional semiconductor structures, bimetallic deposition may open whole new ways of decorating nanostructures via postdeposition treatment. For example, a percolating film of oxidized Fe-Pd nanoclusters would have an entirely different electronic structure than oxidized Fe nanoclusters, due to the Pd nanocluster in the void, which could drastically affect the gas sensing properties.

References

[1] N.S. Ramgir, Y. Yang, M. Zacharias, Nanowire-based sensors, Small 6 (16) (2010) 1705–1722.

[2] A.I. Ayesh, Metal/metal-oxide nanoclusters for gas sensor applications, J. Nanomater. 2016 (2016) 17. Article ID 2359019.

[3] F. Patolsky, C.M. Lieber, Nanowire nanosensors, Mater. Today 8 (4) (2005) 20–28.

[4] Z. Wei, D. Xiaochuan, M.L. Charles, Advances in nanowire bioelectronics, Rep. Prog. Phys. 80 (1) (2017) 016701.

[5] K.-I. Chen, B.-R. Li, Y.-T. Chen, Silicon nanowire field-effect transistor-based biosensors for biomedical diagnosis and cellular recording investigation, Nano Today 6 (2) (2011) 131–154.

[6] Y. Cui, Q. Wei, H. Park, C.M. Lieber, Nanowire nanosensors for highly sensitive and selective detection of biological and chemical species, Science 293 (5533) (2001) 1289.

[7] F. Patolsky, G. Zheng, O. Hayden, M. Lakadamyali, X. Zhuang, C.M. Lieber, Electrical detection of single viruses, Proc. Natl. Acad. Sci. U.S.A 101 (39) (2004) 14017–14022.

[8] J.-i. Hahm, C.M. Lieber, Direct ultrasensitive electrical detection of DNA and DNA sequence variations using nanowire nanosensors, Nano Lett. 4 (1) (2004) 51–54.

[9] G. Zheng, F. Patolsky, Y. Cui, W.U. Wang, C.M. Lieber, Multiplexed electrical detection of cancer markers with nanowire sensor arrays, Nat. Biotechnol. 23 (10) (2005) 1294–1301.

[10] A. Zhang, C.M. Lieber, Nano-Bioelectronics, Chem. Rev. 116 (1) (2016) 215–257.

[11] G. Presnova, D. Presnov, V. Krupenin, V. Grigorenko, A. Trifonov, I. Andreeva, O. Ignatenko, A. Egorov, M. Rubtsova, Biosensor based on a silicon nanowire field-effect transistor functionalized by gold nanoparticles for the highly sensitive determination of prostate specific antigen, Biosens. Bioelectron. 88 (2017) 283–289.

[12] M.S. Khan, G.D. Vishakante, H. Siddaramaiah, Gold nanoparticles: a paradigm shift in biomedical applications, Adv. Colloid Interface Sci. 199 (2013) 44–58.

[13] S.-W. Ryu, C.-H. Kim, J.-W. Han, C.-J. Kim, C. Jung, H.G. Park, Y.-K. Choi, Gold nanoparticle embedded silicon nanowire biosensor for applications of label-free DNA detection, Biosens. Bioelectron. 25 (9) (2010) 2182–2185.

[14] J. Wang, Nanomaterial-based electrochemical biosensors, Analyst 130 (4) (2005) 421–426.

[15] J. Wang, Electrochemical biosensors: towards point-of-care cancer diagnostics, Biosens. Bioelectron. 21 (10) (2006) 1887–1892.

[16] H. Wang, X. Wang, X. Zhang, X. Qin, Z. Zhao, Z. Miao, N. Huang, Q. Chen, A novel glucose biosensor based on the immobilization of glucose oxidase onto gold nanoparticles-modified Pb nanowires, Biosens. Bioelectron. 25 (1) (2009) 142–146.

[17] S.-Y. Lin, S.-J. Chang, T.-J. Hsueh, ZnO nanowires modified with Au nanoparticles for nonenzymatic amperometric sensing of glucose, Appl. Phys. Lett. 104 (19) (2014) 193704.

[18] D.H. Kwon, H.H. An, H.-S. Kim, J.H. Lee, S.H. Suh, Y.H. Kim, C.S. Yoon, Electrochemical albumin sensing based on silicon nanowires modified by gold nanoparticles, Appl. Surf. Sci. 257 (10) (2011) 4650–4654.

[19] T. Kong, Y. Chen, Y. Ye, K. Zhang, Z. Wang, X. Wang, An amperometric glucose biosensor based on the immobilization of glucose oxidase on the ZnO nanotubes, Sens. Actuators B Chem. 138 (1) (2009) 344–350.

[20] K. Liao, P. Mao, Y. Li, Y. Nan, F. Song, G. Wang, M. Han, A promising method for fabricating Ag nanoparticle modified nonenzyme hydrogen peroxide sensors, Sens. Actuators B Chem. 181 (2013) 125–129.

[21] F. Yang, D.K. Taggart, R.M. Penner, Joule heating a palladium nanowire sensor for accelerated response and recovery to hydrogen gas, Small 6 (13) (2010) 1422–1429.

[22] J.v. Lith, A. Lassesson, S.A. Brown, M. Schulze, J.G. Partridge, A. Ayesh, A hydrogen sensor based on tunneling between palladium clusters, Appl. Phys. Lett. 91 (18) (2007) 181910.

[23] B. Xie, L. Liu, X. Peng, Y. Zhang, Q. Xu, M. Zheng, T. Takiya, M. Han, Optimizing hydrogen sensing behavior by controlling the coverage in Pd nanoparticle films, J. Phys. Chem. C 115 (32) (2011) 16161–16166.

[24] B. Xie, M. Zheng, F. Liu, X. Peng, G. Wang, M. Han, Fast response characteristics of hydrogen sensors based on Pd nanoparticle films with controlled coverage, J. Nanoparticle Res. 15 (1746) (2013).

[25] L. Sun, M. Chen, X. Peng, B. Xie, M. Han, The effects of Ni contents on hydrogen sensing response of closely spaced Pd–Ni alloy nanoparticle films, Int. J. Hydrogen Energy 41 (2) (2016) 1341–1347.

[26] A. Kolmakov, D.O. Klenov, Y. Lilach, S. Stemmer, M. Moskovits, Enhanced gas sensing by individual SnO_2 nanowires and nanobelts functionalized with Pd catalyst particles, Nano Lett. 5 (4) (2005) 667–673.

[27] J.M. Baik, M.H. Kim, C. Larson, C.T. Yavuz, G.D. Stucky, A.M. Wodtke, M. Moskovits, Pd-sensitized single vanadium oxide nanowires: highly responsive hydrogen sensing based on the metal-insulator transition, Nano Lett. 9 (12) (2009) 3980–3984.

[28] S. Rane, S. Arbuj, S. Rane, S. Gosavi, Hydrogen sensing characteristics of Pt–SnO_2 nano-structured composite thin films, J. Mater. Sci. Mater. Electron. 26 (6) (2015) 3707–3716.

[29] J. Kong, M.G. Chapline, H. Dai, Functionalized carbon nanotubes for molecular hydrogen sensors, Adv. Mater. 13 (18) (2001) 1384–1386.

[30] R.K. Joshi, Q. Hu, F. Alvi, N. Joshi, A. Kumar, Au decorated zinc oxide nanowires for CO sensing, J. Chem. Phys. C 113 (36) (2009) 16199–16202.

[31] I.-S. Hwang, J.-K. Choi, H.-S. Woo, S.-J. Kim, S.-Y. Jung, T.-Y. Seong, I.-D. Kim, J.-H. Lee, Facile control of C_2H_5OH sensing characteristics by decorating discrete Ag nanoclusters on SnO_2 nanowire networks, ACS Appl. Mater. Interfaces 3 (8) (2011) 3140–3145.

[32] S.-J. Chang, T.-J. Hsueh, I.-C. Chen, B.-R. Huang, Highly sensitive ZnO nanowire CO sensors with the adsorption of Au nanoparticles, Nanotechnology 19 (17) (2008) 175502.

[33] J.M. Baik, M. Zielke, M.H. Kim, K.L. Turner, A.M. Wodtke, M. Moskovits, Tin-oxide-nanowire-based electronic nose using heterogeneous catalysis as a functionalization strategy, ACS Nano 4 (6) (2010) 3117–3122.

[34] J. Schmelzer Jr., S.A. Brown, A. Wurl, M. Hyslop, R.J. Blaikie, Finite-size effects in the conductivity of cluster assembled nanostructures, Phys. Rev. Lett. 88 (22) (2002).

[35] J.G. Partridge, S.A. Brown, A.D.F. Dunbar, R. Reichel, M. Kaufmann, C. Siegert, S. Scott, R.J. Blaikie, Templated-assembly of conducting antimony cluster wires, Nanotechnology 15 (9) (2004) 1382–1387.

[36] E.J. Boyd, S.A. Brown, The size dependence of tin oxide atomic cluster nanowire field effect transistors, Nanotechnology 20 (42) (2009) 425201.

[37] M.K. Kennedy, F.E. Kruis, H. Fissan, B.R. Mehta, S. Stappert, G. Dumpich, Tailored nanoparticle films from monosized tin oxide nanocrystals: particle synthesis, film formation, and size-dependent gas-sensing properties, J. Appl. Phys. 93 (1) (2003) 551.

[38] M.K. Kennedy, F.E. Kruis, H. Fissan, H. Nienhaus, A. Lorke, T.H. Metzger, Effect of in-flight annealing and deposition method on gas-sensitive SnOx films made from size-selected nanoparticles, Sens. Actuators B 108 (1–2) (2005) 62–69.

[39] S.-F. Zhao, C.-H. Yao, Q. Lu, F.-Q. Song, J.-G. Wan, G.-H. Wang, Cluster-assembled cobalt doped ZnO nanostructured film prepared by low energy cluster beam deposition, Trans. Nonferrous Metals Soc. China 19 (6) (2009) 1450–1453.

[40] M. Bruzzi, S. Miglio, M. Scaringella, G. Bongiorno, P. Piseri, A. Podesta, P. Milani, First study of humidity sensors based on nanostructured carbon films produced by supersonic cluster beam deposition, Sens. Actuators B 100 (1–2) (2004) 173–176.

[41] E. Barborini, G. Corbelli, G. Bertolini, P. Repetto, M. Leccardi, S. Vinati, P. Milani, The influence of nanoscale morphology on the resistivity of cluster-assembled nanostructured metallic thin films, New J. Phys. 12 (2010) 073001.

[42] M. Zhao, J. Huang, C.-W. Ong, Feasibility of H_2 sensors composed of tungsten oxide nanocluster films, Int. J. Hydrogen Energy 38 (35) (2013) 15559–15566.

[43] M. Zhao, J.X. Huang, C.W. Ong, Diffusion-controlled H_2 sensors composed of Pd-coated highly porous WO_3 nanocluster films, Sens. Actuators B 191 (2014) 711–718.

[44] M. Zhao, M.H. Wong, J.X. Huang, C.W. Ong, Ultrathin percolated WO_3 cluster film and its resistive response to H_2, J. Alloy. Comp. 612 (2014) 163–169.

[45] M. Zhao, J.X. Huang, C.W. Ong, Room-temperature resistive H_2 sensing response of Pd/WO_3 nanocluster-based highly porous film, Nanotechnology 23 (31) (2012) 315503.

[46] M. Zhao, C.W. Ong, Improved H_2-sensing performance of nanocluster-based highly porous tungsten oxide films operating at moderate temperature, Sens. Actuators B 174 (2012) 65–73.

[47] M. Zhao, J. Huang, C.-W. Ong, Preparation and structure dependence of H_2 sensing properties of palladium-coated tungsten oxide films, Sens. Actuators B 177 (2013) 1062–1070.

[48] M.H. Wong, N.H. Ng, M. Zhao, K.S. Lau, H.C. Man, J.T. Cheung, C.W. Ong, Room-temperature polymer-coated supersonic cluster beam deposited ZnO film for O_2 gas and dissolved O_2 sensing, Sens. Actuators B 251 (2017) 1068–1075.

[49] M. Zhao, M.H. Wong, H.C. Man, C.W. Ong, Resistive hydrogen sensing response of Pd-decorated ZnO "nanosponge" film, Sens. Actuators B 249 (2017) 624–631.

[50] J. Vernieres, S. Steinhauer, J. Zhao, A. Chapelle, P. Menini, N. Dufour, R.E. Diaz, K. Nordlund, F. Djurabekova, P. Grammatikopoulos, M. Sowwan, Gas phase synthesis of multifunctional Fe-based nanocubes, Adv. Funct. Mater. 27 (11) (2017) 1605328.

[51] A. Kusior, K. Kollbek, K. Kowalski, M. Borysiewicz, T. Wojciechowski, A. Adamczyk, A. Trenczek-Zajac, M. Radecka, K. Zakrzewska, Sn and Cu oxide nanoparticles deposited on TiO_2 nanoflower 3D substrates by inert gas condensation technique, Appl. Surf. Sci. 380 (2016) 193–202.

[52] A.I. Ayesh, S.T. Mahmoud, S.J. Ahmad, Y. Haik, Novel hydrogen gas sensor based on Pd and SnO_2 nanoclusters, Mater. Lett. 128 (2014) 354–357.

[53] A.I. Ayesh, Z. Karam, F. Awwad, M.A. Meetani, Conductometric graphene sensors decorated with nanoclusters for selective detection of Hg^{2+} traces in water, Sens. Actuators B 221 (2015) 201–206.

[54] A.R. Said, K. Said, F. Awwad, N.N. Qamhieh, S.T. Mahmoud, M.A. Meetani, S. Tariq, A.I. Ayesh, Design, fabrication, and characterization of Hg^{2+} sensor based on graphite oxide and metallic nanoclusters, Sens. Actuators 271 (2018) 270–277.

[55] G. Fiaschi, S. Cosentino, R. Pandey, S. Mirabella, V. Strano, L. Maiolo, D. Grandjean, P. Lievens, Y. Shacham-Diamand, A novel gas-phase mono and bimetallic clusters decorated Zno nanorods electrochemical sensor for 4-aminophenol detection, J. Electroanal. Chem. 811 (2018) 89–95.

[56] S. Steinhauer, A. Chapelle, P. Menini, M. Sowwan, Local CuO nanowire growth on microhotplates: In situ electrical measurements and gas sensing application, ACS Sens. 1 (5) (2016) 503–507.

[57] S. Steinhauer, V. Singh, C. Cassidy, C. Gspan, W. Grogger, M. Sowwan, A. Köck, Single CuO nanowires decorated with size-selected Pd nanoparticles for CO sensing in humid atmosphere, Nanotechnology 26 (17) (2015) 175502.

[58] S. Steinhauer, J. Vernieres, J. Krainer, A. Köck, P. Grammatikopoulos, M. Sowwan, In situ chemoresistive sensing in the environmental TEM: probing functional devices and their nanoscale morphology, Nanoscale 9 (22) (2017) 7380–7384.

[59] E. Brunet, T. Maier, G.C. Mutinati, S. Steinhauer, A. Köck, C. Gspan, W. Grogger, Comparison of the gas sensing performance of SnO_2 thin film and SnO_2 nanowire sensors, Sens. Actuators B 165 (1) (2012) 110–118.

[60] S. Steinhauer, Z. Wang, Z. Zhou, J. Krainer, A. Köck, K. Nordlund, F. Djurabekova, P. Grammatikopoulos, M. Sowwan, Probing electron beam effects with chemoresistive nanosensors during in situ environmental transmission electron microscopy, Appl. Phys. Lett. 110 (9) (2017) 094103.

[61] S. Steinhauer, J. Zhao, V. Singh, T. Pavloudis, J. Kioseoglou, K. Nordlund, F. Djurabekova, P. Grammatikopoulos, M. Sowwan, Thermal oxidation of size-selected Pd nanoparticles supported on CuO nanowires: the role of the CuO–Pd interface, Chem. Mater. 29 (14) (2017) 6153–6160.

[62] V. Singh, C. Cassidy, F. Abild-Pedersen, J.-H. Kim, K. Aranishi, S. Kumar, C. Lal, C. Gspan, W. Grogger, M. Sowwan, Engineering high-performance Pd core–MgO porous shell nanocatalysts via heterogeneous gas-phase synthesis, Nanoscale 7 (32) (2015) 13387–13392W.

[63] M. Benelmekki, M. Bohra, J.-H. Kim, R.E. Diaz, J. Vernieres, P. Grammatikopoulos, M. Sowwan, A facile single-step synthesis of ternary multicore magneto-plasmonic nanoparticles, Nanoscale 6 (7) (2014) 3532–3535.

[64] M. Benelmekki, J. Vernieres, J.-H. Kim, R.-E. Diaz, P. Grammatikopoulos, M. Sowwan, On the formation of ternary metallic-dielectric multicore-shell nanoparticles by inert-gas condensation method, Mater. Chem. Phys. 151 (2015) 275–281.

[65] J. Vernieres, M. Benelmekki, J.-H. Kim, P. Grammatikopoulos, J.-F. Bobo, R.E. Diaz, M. Sowwan, Single-step gas phase synthesis of stable iron aluminide nanoparticles with soft magnetic properties, Appl. Phys. Lett. Mater. 2 (11) (2014) 116105.

[66] M. Bohra, P. Grammatikopoulos, R.E. Diaz, V. Singh, J. Zhao, J.-F. Bobo, A. Kuronen, F. Djurabekova, K. Nordlund, M. Sowwan, Surface segregation in chromium-doped NiCr alloy nanoparticles and its effect on their magnetic behavior, Chem. Mater. 27 (9) (2015) 3216–3225.

[67] M. Bohra, V. Singh, P. Grammatikopoulos, E. Toulkeridou, R.E. Diaz, J.-F. Bobo, M. Sowwan, Control of surface segregation in bimetallic NiCr nanoalloys immersed in Ag matrix, Sci. Rep. 6 (2016) 19153.

[68] M. Bohra, P. Grammatikopoulos, V. Singh, J. Zhao, E. Toulkeridou, S. Steinhauer, J. Kioseoglou, J.-F. Bobo, K. Nordlund, F. Djurabekova, M. Sowwan, Tuning the onset of ferromagnetism in heterogeneous bimetallic nanoparticles by gas phase doping, Phys. Rev. Mater. 1 (6) (2017) 066001.

[69] V. Singh, C. Cassidy, P. Grammatikopoulos, F. Djurabekova, K. Nordlund, M. Sowwan, Heterogeneous gas-phase synthesis and molecular dynamics modeling of janus and core–satellite Si–Ag nanoparticles, J. Phys. Chem. C 118 (25) (2014) 13869–13875.

[70] C. Cassidy, V. Singh, P. Grammatikopoulos, F. Djurabekova, K. Nordlund, M. Sowwan, Inoculation of silicon nanoparticles with silver atoms, Sci. Rep. 3 (2013) 3083.

[71] P. Grammatikopoulos, J. Kioseoglou, A. Galea, J. Vernieres, M. Benelmekki, R.E. Diaz, M. Sowwan, Kinetic trapping through coalescence and the formation of patterned Ag–Cu nanoparticles, Nanoscale 8 (18) (2016) 9780–9790.

9

Nanoparticle-based biomedical sensors

Antonio Tricoli, Renheng Bo

RESEARCH SCHOOL OF ENGINEERING, THE AUSTRALIAN NATIONAL UNIVERSITY, CANBERRA, ACT, AUSTRALIA

Chapter outline

9.1 Introduction

In the past decades, the rapidly rising medical expenditures in many developed countries have motivated the development of novel medical technologies that may help reduce medical costs while providing better health outcomes. Novel biomedical sensing technologies that can be miniaturized and integrated in point-of-care and wearable devices have the potential to revolutionize the understanding, early-stage detection, and management of diseases leading to prevention and better management of numerous diseases and of some of their most severe side effects, and more effective personalized medicine. The potential impact to the medical system is significant, with a reduction of the costs required for the treatment of the rapidly growing population affected by numerous chronic diseases, and an increase of life quality.

Cluster Beam Deposition of Functional Nanomaterials and Devices. https://doi.org/10.1016/B978-0-08-102515-4.00009-X

Nanoparticles are playing a major role as the functional building blocks of a rapidly growing family of medical sensing devices. The very high surface-to-volume atom ratio of these zero-dimensional nanomaterials is particularly beneficial to detect very small concentration of medically important analytes and transduce their detection in electrical or optical signals. Despite numerous scientific achievements, the commercial development of nanoparticle-based medical sensing devices is still in its early stage, requiring further improvements in nanomaterial and device design to achieve sufficient long-term stability, selectivity, and, often, sensitivity.

In addition to the careful engineering of the nanoparticle surface and bulk composition at the atomic scale, the fabrication of highly performing sensing devices requires carefully controlling the grain to grain morphology and nanoparticle film architecture to a size ranging between several micrometers and few millimeters. Optimizing the mass, heat, and electron transfers as well as the light transport through such complex hierarchical nano-macro structures is a multiscale engineering problem, which requires novel nanofabrication tools. In this respect, the gas-phase self-assembly of nanoparticle and cluster films, such as in cluster-beam deposition and thermophoretic-driven processes, is particularly promising. The former approach allows the rapid fabrication of nanoparticle films with well-controlled and reproducible morphological properties of very large surfaces, providing a tool for their advanced and rational design. In this chapter, we will review some of the recent achievements in the engineering of nanoparticle-based medical sensors made by the gas-phase self-assembly of nanoparticles and clusters. We will present state-of-the-art fabrication approaches and their implementation for the production of three types of the most common and promising biomedical sensor technologies, namely chemoresistive, electrochemical, and optical sensors. We will conclude by providing a summary of current opportunities offered by state-of-the-art technologies, and future research directions required for the development and commercialization of nanoparticle-based medical sensing technologies.

9.2 Design of nanoparticle networks for biomedical sensors

Three-dimensional nanoparticle networks are of significant interest for biomedical sensing and have potential for integration in wearable technologies, thanks to their unique properties such as a large and accessible surface area and very high porosity of up to 98% [1–5]. The design and engineering of such nanoparticle networks is being increasingly studied for both research and industrial purposes. Among promising approaches, deposition of nanoparticle aerosols from the gas-phase and the resulting self-assembly of nanoparticle films (e.g., by cluster beam deposition [6] or thermophoresis [1,5,7–12]) are being increasingly applied for the fabrication of 3D nanostructured architectures [3,13]. However, to further ameliorate the design of nanoparticle-based biomedical sensors, the fundamental understanding of the nanoparticle self-assembly and restructuring process as well as prediction of the resulting structure is essential. In the upcoming sections we will provide a review of the fundamentals and state-of-the-art processes for the nanofabrication of nanoparticle networks with specific focus on approaches that have demonstrated potential for biomedical sensing applications.

9.2.1 Nanoparticle self-assembly dynamics and scaling laws

The underlying nanoparticle self-assembly dynamics is among the key factors determining the morphology resulting from the selected gas-phase synthetic process such as thermophoretic deposition process, often based on the flame synthesis of nanoparticle aerosols [4], and ballistic-driven cluster beam deposition [6]. The different nanoparticle self-assembly regimes are usually described in terms of the scaled ratio between the constant displacement induced by an advective velocity and the stochastically distributed one caused by Brownian motion, namely Péclet (Pe) [13]. Generally, three regimes (Fig. 9.1A) have been identified: the highly porous (ca. 98%) diffusion regime, where the nanoparticle self-assembly is dominated by Brownian motion; the more dense (ca. 86% porous) ballistic regime, where the initial trajectory of the nanoparticles is mostly preserved and determines the resulting film structure; the transition regime where both mechanisms lead to a range of film morphologies and porosities ranging from ca. 86% −98% [13]. More complex morphologies are possible by depositing nonspherical porous particles (Fig. 9.1B–E) such as fractals that can be composed of a number (N_p) of initial particles with a specific size (d_p) and fractal dimensions (D_f) [7]. However, recent findings suggest that such regimes may not be able to describe the self-assembly dynamics of ultrafine nanoparticles (e.g., with size below 100 nm at standard room temperature and pressure conditions) [3,7,13–17]. For instance, the structural properties resulting from the self-assembly of 50 and 5 nm nanoparticle aerosols have been recently modeled by Langevin dynamics simulations [3]. These results (Fig. 9.1F and G) revealed that in the ballistic regime, the nanoparticle film structure converges toward a predetermined architecture, which scales linearly with the particle size while preserving its key features such as the film porosity. The morphology in the ballistic regime can be described as a set of parallel agglomerates growing in the vertical direction, which resemble an inverted multicone structure. However, in the diffusion regime the resulting morphology has a complex nonlinear relationship with the particle size. Notably, the density of the resulting film increases near the surface by more than two times with the nanoparticle size decreasing from 100 to 1 nm.

A particle size–dependent mechanism has been recently suggested for the self-assembly of such ultrafine nanoparticle films during deposition from the aerosol phase. For the self-assembly of 50 nm nanoparticles, two distinct behaviors were observed in the diffusion and ballistic regimes (Fig. 9.2A and F). In the diffusion regime, it was observed that within a square area with side of 20 times the particle diameter, a prevailing nanoparticle agglomerate arises within a thickness of 10 times the particle

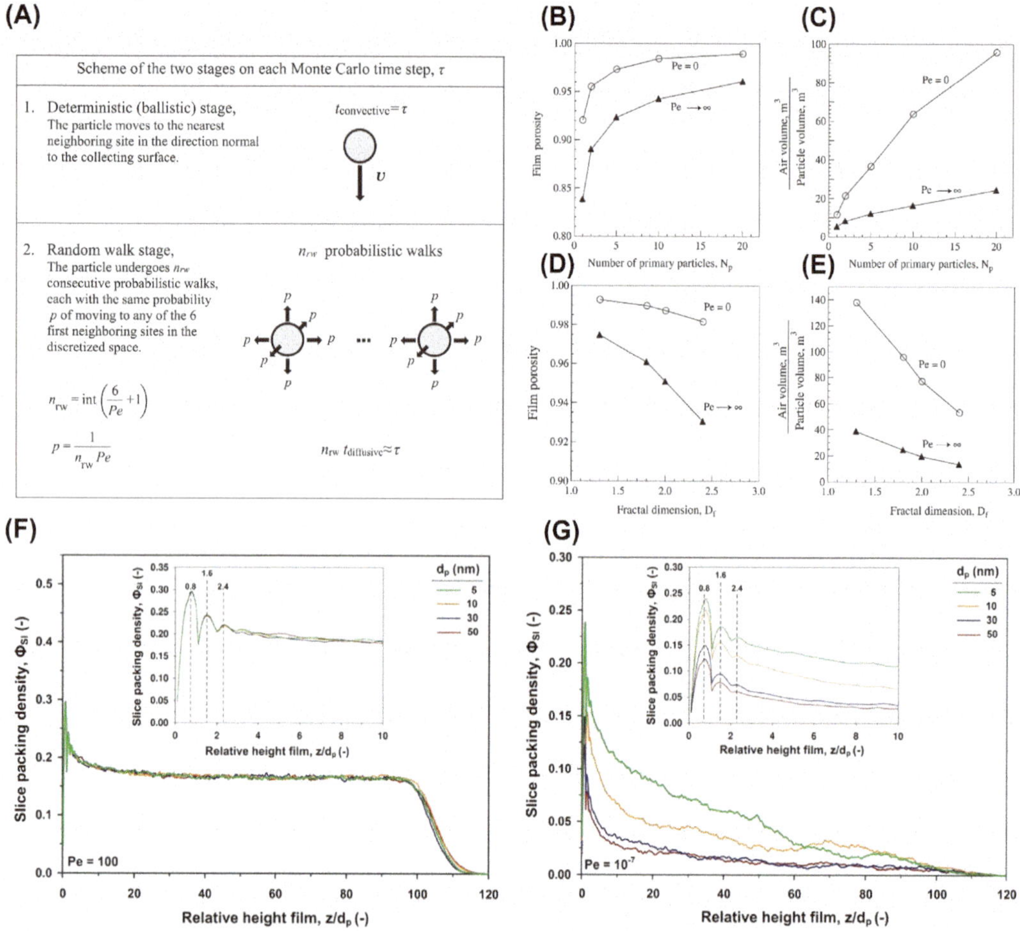

FIGURE 9.1 (A) Two-stage Monte Carlo simulation. Deterministic stage (1) and random-walk stage (2) (with n_{rw} consecutive probabilistic walks) [13]. (B and C) Limiting porosities as a function of the number of primary particles in an agglomerate for Pe = 0 and Pe → ∞ ($D_f = 1.8$, $d_p = 10$ nm). The film porosities increased with agglomerate size (larger N_p). Deposition in the diffusion limit resulted in higher film porosities compared to the ballistic limit. The air to particle volume ratio steadily increased with the number of primary particles in an agglomerate [7]. (D and E) Limiting porosities as a function of agglomerate fractal dimension for the two limiting Péclet numbers (Pe = 0 and Pe → ∞) for agglomerates with $N_p = 20$ and $d_p = 10$ nm. In both cases the films were more porous for smaller fractal dimensions. The air to particle volume ratio steadily decreased with the agglomerate fractal dimension [7]. (F and G) Representative slice packing density profiles (F) in the ballistic and (G) diffusion regimes as a function of the particle diameter from 50 to 5 nm. Three main density maxima (F, G, insets) are determined at 0.8, 1.6, and 2.4 relative film thicknesses (z/d_p). The magnitude of these maxima is (F) independent of the particle diameter in the ballistic regime while (G) increases by more than 100% with decreasing particle size in the diffusion regime [3]. *(A) Reproduced with permission J.L. Castillo, S. Martin, D. Rodriguez-Perez, A. Perea, P.L. Garcia-Ybarra, Morphology and nanostructure of granular materials built from nanoparticles, KONA Powder Part. J. 31 (2014) 214–233. Copyright 2014, Hosokawa Powder Technology Foundation. (B and C) Reproduced with permission L. Mädler, A.A. Lall, S.K. Friedlander, One-step aerosol synthesis of nanoparticle agglomerate films: simulation of film porosity and thickness, Nanotechnology 17 (2006) 4783. Copyright 2006, IOP Publishing Ltd. (D and E) Reproduced with permission L. Mädler, A.A. Lall, S.K.*

diameter. This prevailing agglomerate would start shielding the minor agglomerates effectively stopping their growth. Eventually, this leads to the establishment of a single agglomerate at a thickness of 20 times the particle diameter and a branchlike nanotexture morphology (Fig. 9.2A) [3,10,11]. In contrast, in the ballistic regime, no prevailing agglomeration behavior was observed leading to a constant density and morphologies (Fig. 9.2F) [18].

In stark contrast, modeling of the self-assembly dynamics of 5 nm particles led both in the diffusion and ballistic regimes to ballisticlike morphologies (Fig. 9.2B, C and G). Notably, further investigation revealed that the established Péclet-scaling law cannot be applied to predict the morphology resulting from the self-assembly of ultrafine nanoparticles in the diffusion regime [3]. Computational results show that in the diffusion regime the surface density of the first particle layer increases by more than 100% with decreasing particle size from 50 to 5 nm. Thereafter, the film density decreases continuously from an initial maximum within the first layers near the substrate surface and reaches to a minimum on the top layers of the networks. These theoretical findings have been confirmed by experimental results obtained by the deposition of flame-made TiO_2 nanoparticle aerosols and formation of mono- and multilayer morphologies [3]. As a result, a randomly oriented ballistic regime was proposed for the self-assembly of ultrafine nanoparticles in the diffusion regime. A simplified analytical model was proposed to predict the initial surface density (SD) of such films for a particle size ranging from 1 to 1000 nm [3].

$$SD = \frac{\pi}{4\left[\pi + K_{RPF}(2\overline{X} - \left[\overline{X}\sqrt{1-\overline{X}^2} + sin^{-1}(\overline{X})\right]\right]} \tag{9.1}$$

where SD is the surface density, K_{RPF} is a constant, proportional to the additional path swept by a particle due to the stochastic nature of diffusion, and *X* is the average horizontal movement of a particle due to Brownian displacement.

The departure from the diffusion regime into the randomly oriented ballistic regime has been observed for particle size below 100 nm. Different from previous works [7,13–17], in this model, an S-shaped densification of the first particle layer has been observed with the particle size decreasing from 1000 to 1 nm. The most rapid variation was observed within a particle size range of 5–100 nm. Recent nanoparticle films synthesized by thermophoretic-driven deposition of nanoparticle-aerosols [1,11,12] seem to be in agreement with this model (Fig. 9.2D and E), providing a tool for the accurate design of nanoparticle-based films and devices.

Friedlander, One-step aerosol synthesis of nanoparticle agglomerate films: simulation of film porosity and thickness, Nanotechnology 17 (2006) 4783. Copyright 2006 IOP Publishing Ltd. (F and G) Reproduced with permission N. Nasiri, T.D. Elmøe, Y. Liu, Q.H. Qin, A. Tricoli, Self-assembly dynamics and accumulation mechanisms of ultra-fine nanoparticles, Nanoscale 7 (2015) 9859–9867. Copyright The Royal Society of Chemistry 2015.

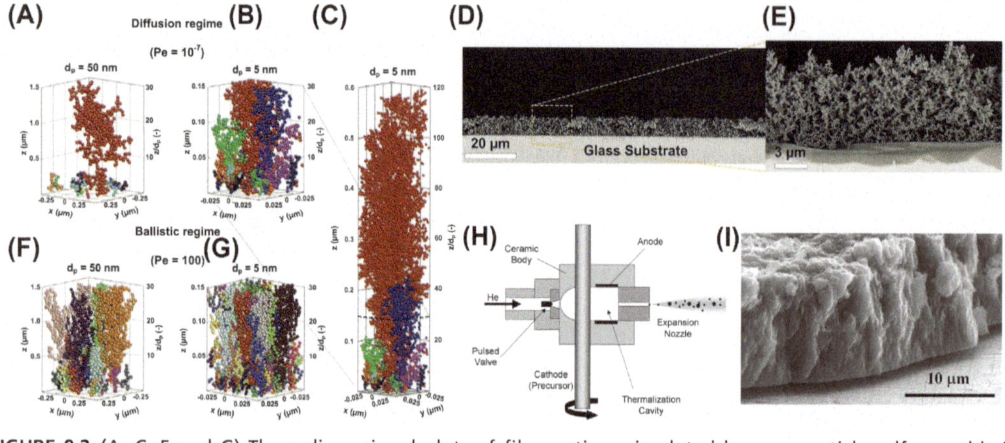

FIGURE 9.2 (A, C, F and G) Three-dimensional plots of film sections simulated by nanoparticle self-assembly in the diffusion (A, B, C) and ballistic (F, G) regimes for particles of 50 nm (A, F) and 5 nm (B, C, G) in diameter. Each contiguous set of particles (surface-bond agglomerate) was labeled with a color [3]. (D and E) Cross-sectional SEM images of exemplary films deposited within diffusion regime. (H) Schematic representation of a PMCS [6]. (I) Scanning electron micrograph of a thick film formed by carbon nanoparticle deposition using CBD with PMCS [6]. *(A, C, F and G) Reproduced with permission N. Nasiri, T.D. Elmøe, Y. Liu, Q.H. Qin, A. Tricoli, Self-assembly dynamics and accumulation mechanisms of ultra-fine nanoparticles, Nanoscale 7 (2015) 9859–9867. Copyright The Royal Society of Chemistry 2015. (D and E) Reproduced with permission Nasiri, N., Bo, R., Wang, F., Fu, L. & Tricoli, A. Ultraporous Electron-Depleted ZnO Nanoparticle Networks for Highly Sensitive Portable Visible-Blind UV Photodetectors. Advanced Materials 27, 4336–4343 (2015). Copyright 2015, WILEY-VCH Verlag GmbH & Co. KGaA, Weinheim. (H and I) Reproduced with permission K. Wegner, P. Piseri, H.V. Tafreshi, P. Milani, Cluster beam deposition: a tool for nanoscale science and technology, J. Phys. D Appl. Phys. 39 (2006) R439. Copyright 2006, IOP Publishing Ltd.*

9.2.2 Emerging nanofabrication technologies for nanoparticle-based sensors

Nanoparticle-based sensors have been synthesized by various methods including sol-gel [19], CVD [20], hydrothermal [21], PLD [22], sputtering [23], and thermophoretic deposition of flame-made nanoparticle aerosols [24,25] and cluster beam deposition [26]. Operating at the opposite ends of the nanoparticle/cluster aerosol deposition regimes, the latter two approaches offer numerous advantages [2]. For instance, cluster beam deposition, working in the ballistic regime (Fig. 9.2H and I), is a rapid and precise nanofabrication technology for the patterning of nanoparticle networks providing a highly accessible surface area, which is of particular benefit for the design of solid-state gas sensors, as well as superior control over the microscale topography of the devices. At the start of 21st century, supersonic cluster beam-deposited (SCBD) nanostructured carbon films have been employed for humidity sensors [27]. Following these promising results, SCBD nanostructured TiO_2 thin films have been demonstrated as chemical sensors [28] and Pd/WO_3 has been reported for room-temperature H_2 sensing [29]. In 2015, SCBD Ag nanoparticles have been utilized as biomedical sensors for biogenic amines [30], and in 2017 ZnO thin films have been used for O_2 gas and dissolved O_2 sensing [31]. Recently, cluster-assembled nanostructures have been proposed as promising materials for biomedical applications [32] indicating that such nanostructured materials can be utilized for the fabrication of biocompatible devices in the near future.

Thermophoretic deposition of flame-made nanoparticle aerosols, instead, works in the diffusion regime (Fig. 9.2D and E) [1,11], providing a promising approach for the omnidirectional coating of surfaces [33]. Furthermore, as discussed in the previous chapter, the diffusion-driven self-assembly results of the nanoparticles result in higher film porosity (98%), which can be advantageous for the rapid penetration of the gas molecules in and out of the nanostructured sensing material. The first demonstration of the application of this approach for gas sensing was presented in 2006 showing that Pt-doped SnO_2 nanoparticle networks could be utilized to detect particle per million concentrations of CO [34]. Despite the advantageous open and bimodal meso-macro pore structure of such films, a major disadvantage was their very poor mechanical [35] and thermal stability [10]. These limitations were overcome in 2008 by inducing a rapid in-situ flame-annealing step upon deposition [35], and by introducing very small amount of a secondary thermochemical stable phase such as SiO_2 in the sensing material. This led to the design of numerous highly performing devices based on a range of Si-modified metal oxide semiconductors including the deposition of multilayer and multifunctional sensing films [24] and of highly selective metastable phases such as the ε-phase of WO_3 for breath acetone detection. Recently, these 3D nanoparticle networks have demonstrated potential for sensing of UV-light [1,5,8,9,36] and fabrication of nanoscale heterojunctions [8]. Therefore, as a promising candidate for fabrication of well-controlled 3D nanotextures, cluster beam deposition and thermophoretic deposition of flame-made nanoparticle aerosols are attracting significant research and commercial attention for the design of miniaturized and wearable nanoparticle-based sensor technologies.

9.3 Emerging miniaturized and wearable sensor technologies

The radical technological developments of the past decades have brought dramatic ameliorations to our lifestyle and life quality. For instance, life expectancy at birth has climbed to more than 80 years in many developed countries [37]. This is attributed to several causations including matured mass production of essential pharmaceuticals, development of noninvasive medical diagnosis technologies such as magnetic resonance and computed tomography [38], novel surgical procedures, and improved sterilization, as well as access to clean water and nutrition. However, despite these achievements, many developing countries are still suffering from substandard health outcomes with life expectancies of 50–60 years. This is also due to the high cost of many modern healthcare approaches [37]. Recent OECD data shows that, even in developed countries, healthcare is a significant financial expenditure, peaking up to 17% of the gross domestic product (GDP) in the United States of America [37]. Therefore, novel low-cost solutions for early-stage medical diagnosis and treatment of diseases are highly required to curb the growth of medical expenditures.

Most recently, the emergence of miniaturized and/or wearable sensor technologies for continuous or frequent monitoring of important human biomarkers is regarded as a promising approach to improve our understanding of diseases, achieve their very early detection, possibly before end organ damage, and provide a better, more informed, treatment, which is personalized to the individual response. This would possibly result in lower incidence of chronic diseases and related complications, and at the same time increase the efficiency of expensive medical treatments such as precision medicine, with the potential of reducing health expenditures while improving outcome [37].

The need for miniaturized and wearable technologies for preventive and personalized medicine is motivating significant multidisciplinary efforts in sensor and analytical chemistry research [39,40]. Based on the applications, these emerging technologies can be classified into several groups including the body fluids sensors and volatile biomarker analyzers (Fig. 9.3D and E) [4,37,41–44]. Body fluids, such as blood, have been utilized for disease detection for decades; therefore, body fluid sensors are one of the most established technologies progressing toward miniaturized diagnostic. For examples, treatment of diabetes includes invasive blood sampling, which is required for measurement of blood sugar levels. A lot of effort is focused on translating from such invasive sensing technologies to noninvasive ones that rely on easily accessibly body fluids like sweat [45–47]. This brings upon numerous research requirements. For instance, volatile biomarker sensing is impelled by the limited current understanding of the breath composition and its correlation to different physical health conditions. Furthermore, developing miniaturized devices for the detection of ultralow biomarker concentrations down to the particle per billion (ppb) level in complex gas mixtures as the breath is also highly required [37]. With respect to the latter, efforts are focusing on the nanofabrication of the sensing component through bionanotechnology approaches and on their integration in miniaturized electronic devices. As such, advanced fabrication approaches to integrate micrometer-sized detectors in on-chip electronic system with ultralow energy consumption are being rapidly developed. Gas-phase approaches, such as the discussed thermophoretic nanoparticle aerosol and cluster beam deposition techniques, are promising candidates, which satisfy numerous requirements including scalability, high sensitivity, and flexibility of the synthesis approach. In the next sections, we will introduce and compare three promising biomedical sensor technologies including chemoresistive, electrochemical, and optical devices. We will focus on the use of nanoparticle films for detectors emphasizing recent achievements of nanoparticle aerosol- and cluster-based nanofabrication methods.

9.3.1 Chemoresistive sensors

Chemoresistive gas sensors based on metal oxide semiconductor nanoparticles have been widely studied and utilized for numerous applications. They offer some advantages over other approaches including simplicity of the device design, low cost, low power consumption, as well as a high potential for miniaturization [48] making them attractive

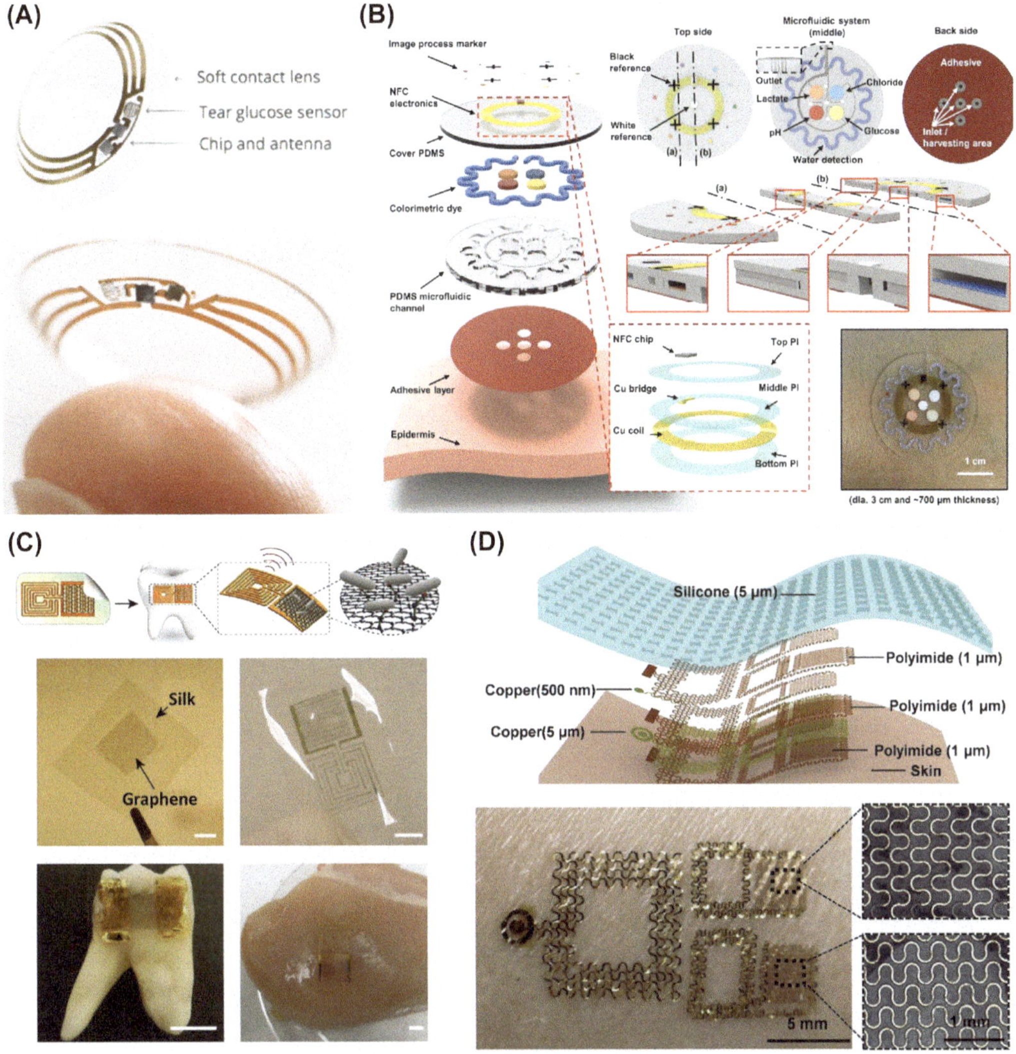

FIGURE 9.3 (A) Prototype of a contact lens sensor. (B) Schematic illustration of an epidermal microfluidic sweat monitoring device and an enlarged image of the integrated near-field communication (NFC) system. Illustration of the top, middle, and back sides of the device. The reference color (white and black) markers are on the top side, along with the NFC electronics. The microfluidic channels with colorimetric assay reagents (water, lactate, chloride, glucose, and pH) are in the middle. The bottom side consists of a uniform layer of adhesive bonded to the bottom surfaces of the PDMS-enclosed microchannels, with openings that define sweat access and inlets that connect to these channels. Cross-sectional diagrams of the cuts are defined by the dashed lines and shown in the top side. Optical image of a fabricated device mounted on the forearm [44]. (C) A schematic and real image of a wireless tattoo-based resistive sensor for *Staphylococcus aureus* [41]. (D) A multimodal wireless epidermal sensor. Exploded view schematic diagram of the sensor on the skin. Images of a sensor on a water-soluble tape and the image of a sensor integrated directly on the skin, after removal of the tape [42]. *(A) Reproduced with permission N.M. Farandos, A.K. Yetisen, M.J. Monteiro, C.R. Lowe, S.H. Yun, Contact lens sensors in ocular diagnostics, Adv. Healthc. Mater. 4 (2015) 792–810. Copyright 2014, WILEY-VCH Verlag GmbH & Co. KGaA, Weinheim. (B) Reproduced with permission A. Koh, et al., A soft, wearable microfluidic device for the capture, storage, and*

for commercial purposes. The mechanisms of these chemoresistive sensors can be briefly explained and classified based on n-type and p-type semiconductors. More details can be found in numerous review articles on metal oxide chemoresistive gas sensors [2,49]. For n-type semiconductors, where electrons are the majority carriers, the conductivity of the semiconductor is increased upon contact with a reducing analyte and decreased upon contact with an oxidizing analyte. While for p-type semiconductors an opposite behavior is observed. The sensing behaviors of n-type SnO_2 toward ethanol and CO can be used as an exemplary well-investigated model for chemoresistive sensors. There are two general descriptions of the sensing mechanisms, namely oxygen-vacancy model (reduction–reoxidation mechanism) [50,51] and ionosorption [52–55]; both of these models can be used to describe most of the experimental observations.

Based on the ionosorption model, the oxygen ionosorption on the SnO_2 surface traps electrons reducing both the free carrier concentrations and their mobility (Fig. 9.4) [2]. Upon receiving a reducing analyte, such as ethanol, CO, or H_2, on its surface, the concentration of absorbed oxygen species on the SnO_2 (Fig. 9.4A) decreases due to the complete or partial oxidation of the analyte gas molecules (Fig. 9.4B). This leads to reduced concentrations of both trapped electrons and scattered centers, thereby increasing the overall conductivity of SnO_2. The sensing reactions for ethanol (Eqs. 9.2, 9.3) and CO (Eq. 9.4) on the surface of SnO_2 can be simplified as follows:

$$C_2H_5OH_{(g)} + O^-_{(ad)} \rightleftharpoons CH_3CHO_{(g)} + H_2O_{(g/ad)} + e^- \tag{9.2}$$

$$H_{2(g)} + O^-_{(ad)} \rightarrow H_2O_{(g)} + e^- \tag{9.3}$$

$$CO_{(g)} + O^-_{(ad)} \rightarrow CO_{2(g)} + e^- \tag{9.4}$$

where analytes are in gas phase (g) and react with the absorbed oxygen species (ad) on surfaces, while the oxidizing analytes perform oppositely.

There is an increasing focus on the development of chemoresistive sensors for the detection and analysis of volatile organic compounds (VOCs) for the monitoring of the human metabolism. In particular, detection of biomarkers in the human breath and skin perspirations dated back to mid-1800 [37] and is among the most promising noninvasive approaches for disease diagnostics and monitoring. However, the concentration of biomarkers contained in human breath and perspiration are, often, in the ppb-ppt range, challenging the development of sufficiently sensitive and miniaturized detection technologies. Breath sensing has the advantage of having to detect several orders of magnitude higher VOC concentrations than for skin perspiration sensing [37].

colorimetric sensing of sweat, Sci. Transl. Med. 8 (2016) 366ra165. Copyright 2016 The Authors, some rights reserved; exclusive licensee American Association for the Advancement of Science. (C) Reproduced with permission A.J. Bandodkar, J. Wang, Non-invasive wearable electrochemical sensors: a review, Trends Biotechnol. 32 (2014) 363–371. Copyright Elsevier Ltd. (D) Reproduced with permission X. Huang, et al., Materials and designs for wireless epidermal sensors of hydration and strain, Adv. Funct. Mater. 24 (2014) 3846–3854. Copyright 2014, WILEY-VCH Verlag GmbH & Co. KGaA, Weinheim.

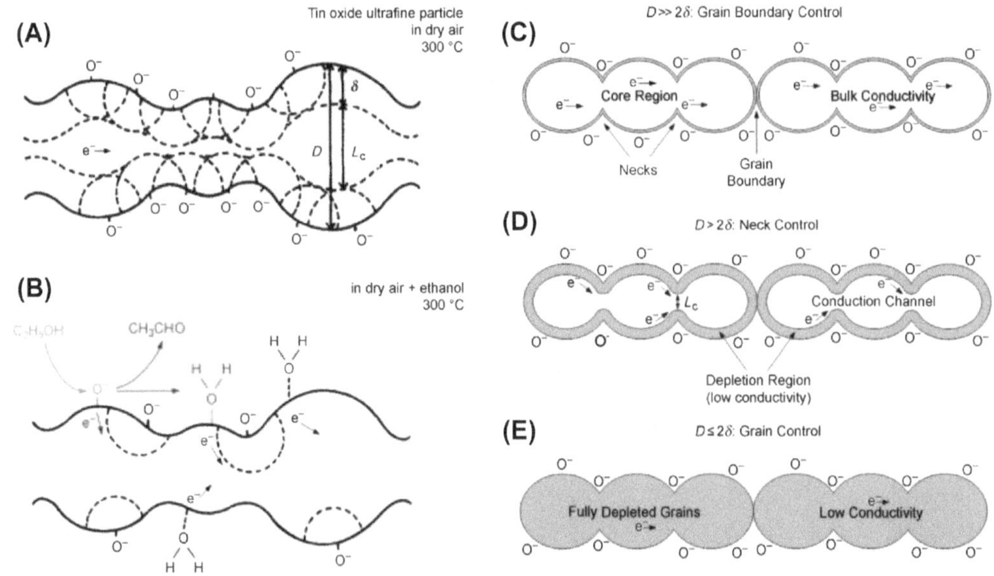

FIGURE 9.4 Schematic model of oxygen ionosorption [2] on the SnO_2 surface in (A) pure dry air and (B) with ethanol. The adsorbed oxygen species (A) trap electrons from SnO_2 forming ions that scatter electrons within the Debye length δ of the oxide (3 nm for SnO_2 at 250°C). This decreases the electron concentration and mobility resulting in a reduction of the SnO_2 conductivity. The reaction between ethanol and the ionosorbed oxygen (B) decreases the concentration of these scattering centers (O^-) and releases the trapped electrons thereby increasing the SnO_2 conductivity. The mechanism controlling the conductivity change and its magnitude depends mainly on the ratio between grain size (D) and Debye length. If $D \gg \delta$ (C), the depletion of the surface between the grain boundaries controls the conductivity variation. In this case low sensitivity to the analyte is expected as only a small part of the semiconductor is affected by interaction with the analyte. If $D > 2\delta$ (D), a conduction channel with high conductivity exists but its wideness (L_c) is controlled by the surface concentration of oxygen ions leading to moderate sensitivities. If $D \leq 2\delta$ (E), the whole grain is depleted and changes in the surface oxygen concentration affects the whole semiconductor resulting in high sensitivity. *Reproduced with permission A. Tricoli, M. Righettoni, A. Teleki, Semiconductor gas sensors: dry synthesis and application, Angew. Chem. Int. Ed. 49 (2010) 7632–7659. Copyright 2010, Wiley-VCH Verlag GmbH & Co. KGaA, Weinheim.*

Considering the very low limit of detection (ppb) of chemoresistive nanoparticle-based gas sensors, such devices are of significant potential for the design of miniaturized breath diagnostic units for healthcare.

Diverse approaches have been employed for the fabrication of such sensors [2]. Among many methods pioneered in the past decades, gas-phase-based synthesis and self-assembly approaches such as thermophoretic aerosol deposition [1–5,11,56] and cluster beam deposition [26–28] allow the one-step and scalable fabrication of nanoparticle films with well-controlled structural properties such as porosity and thickness. In 2008, Tricoli et al. [10] reported a generally applicable strategy for the design of highly sensitive and stable nanoparticle-based chemoresistive sensors (Fig. 9.5A and B). A Si-doped SnO_2 nanoparticle network was synthesized by flame synthesis and aerosol deposition, and used to measure ethanol with a sensitivity down to 100 ppb. It was found

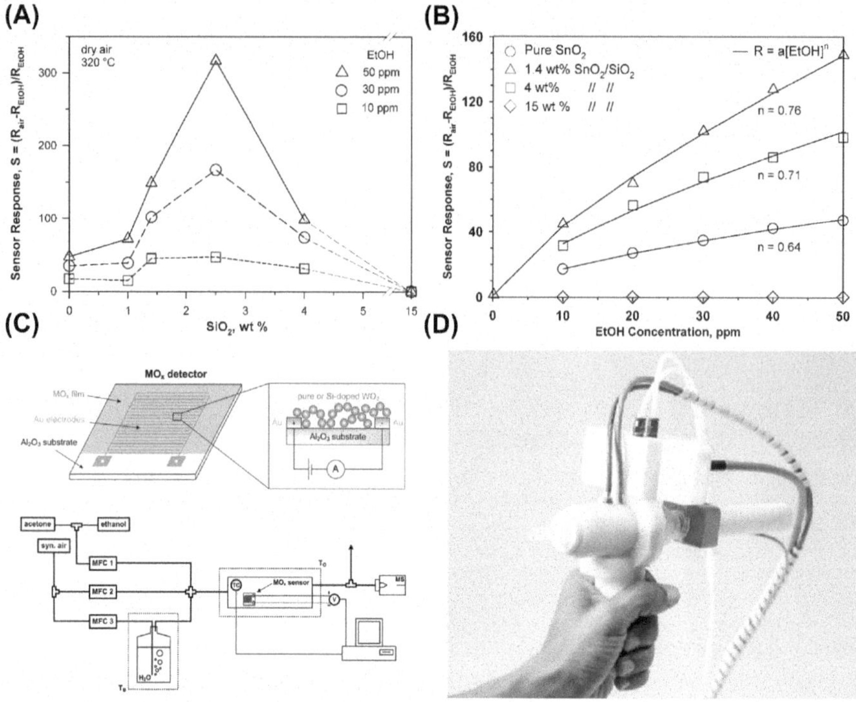

FIGURE 9.5 (A) Sensor response to 10 (*squares*), 30 (*circles*), and 50 (*triangles*) ppm EtOH as a function of the SiO_2 content. A maximum response is found at 2.5 wt% SiO_2 content [10]. (B) Sensor response as a function of EtOH concentration for various SiO_2-doped sensors. The 1.4 wt% SiO_2-doping (*triangles*) shows the highest response indicating the existence of an optimal SiO_2-doping content. These responses can be fitted (*lines*) by a power law. At 15 wt% SiO_2 content resulted in an insulator [10]. (C) Detector schematic: a metal oxide (MO_x) film is deposited onto a sensor substrate consisting of an Al_2O_3 support with interdigitated Au electrodes. Sensor characterization setup: the acetone and ethanol flows are dosed by mass flow controllers. A humidified air flow from a water bubbler ($T_B = 20°C$) is mixed with dry air to obtain the desired relative humidity. The detector is kept in an oven ($T_O = 325°C–500°C$) and connected to a voltmeter to measure the film resistance. The exhaust flow is then analyzed by a mass spectrometer (MS) [57]. (D) Picture of the operating handheld sampler and sensing unit [59]. *(A and B) Reproduced with permission A. Tricoli, M. Graf, S.E. Pratsinis, Optimal doping for enhanced SnO2 sensitivity and thermal stability, Adv. Funct. Mater. 18 (2008) 1969–1976. Copyright 2008, WILEY-VCH Verlag GmbH & Co. KGaA, Weinheim. (C) Reproduced with permission M. Righettoni, A. Tricoli, S.E. Pratsinis, Si: WO3 sensors for highly selective detection of acetone for easy diagnosis of diabetes by breath analysis, Anal. Chem. 82 (2010) 3581–3587. Copyright 2010, American Chemical Society. (D) Reproduced with permission M. Righettoni, et al., Monitoring breath markers under controlled conditions, J. Breath Res. 9 (2015) 047101. Copyright 2015, IOP Publishing Ltd.*

that an optimal SiO_2 doping content of 2.5 wt% resulted in 2–10 times higher sensitivity than pure SnO_2 sensors, as well an enhanced thermal stability of up to 900°C versus the 300°C of the pure metal oxide. In 2010, following this approach, Righettoni et al. [57] reported a Si-doped WO_3 acetone sensor with a metastable ε-phase, and achieving the highly selective detection of acetone down to 20 ppb at the 90% relative humidity. These findings were further applied to real human breath analysis and confirmed by proton transfer reaction (PTR-MS) demonstrating the first portable acetone breath analyzer [58]. Recently, a handheld version of this technology has been presented (Fig. 9.5C–E) [59].

In 2011, Barborini et al. [60] reported a series of cluster-assembled nanostructured films including Mo, W, and Nb oxides demonstrating chemoresistive gas sensing at 300°C–400°C with performance comparable to the best performing state-of-the-art chemoresistive devices and with superior miniaturization potential. In 2018, Borghi et al. [61] presented a thorough investigation and review of such cluster-assembled nanostructures, critically presenting the potential of SCBD for the fabrication of breath analyzers.

9.3.2 Electrochemical sensors

Electrochemical sensors are among the most common technologies for measurement of biomarkers in body fluids including readily available sweat, saliva, and tears. Furthermore, their miniaturization, thanks to planar cell technology and the nano-texturing of the electrodes, is enabling their integration in wearable and flexible devices that can be used for continuous monitoring of biomarkers from sweat and saliva [37]. Fig. 9.6 shows a simplified schematic of the working principle of such electrochemical sensors [62]. The most common design consists of working (WE), counter (CE), and reference (RE) electrodes. Usually the targeted biomarker reacts on the working electrode, resulting in a current generated between the working and counter electrodes, where the opposite redox reaction takes place [63,64]. The generated electrochemical current is proportional to the biomarkers' concentration and available reaction sites on the WE [65]. Various nanostructures are textured on the working electrodes' surface to increase the accessible surface and sites for the redox reactions. As a result, an increased signal-to-noise ratio as well as an enhanced detection limit can be expected. Furthermore, functionalization of these nanotextured surfaces with selective groups (e.g., enzymes) enable to very selectively target specific analytes and achieve low limit of detections. Tuning the potential between the WE and CE enables to tune the activation energy further, thus improving selectivity.

Numerous technologies have been employed to nanostructure the working electrode surfaces such as hydrothermal deposition [21], chemical vapor deposition [20], and cluster beam deposition [26]. A review of electrochemical sensors for miniaturized biomarker analysis has been published recently [37]. In 2009, Decarli et al. [66] reported a novel synthesis method, based on pulsed microplasma clusters, demonstrating the deposition of various metal oxides, which opened a new flexible approach for fabrication of such electrochemical active materials. Later, Fiaschi et al. [67] reported an electrochemical 4-aminophenol sensor based on Au/Pt decorated ZnO nanorods using the laser ablation cluster beam deposition (LACBD) achieving a sensitivity of 14.31 μA/mM cm^2 and a limit of detection of 3.6 μM. Based on these metallic clusters, Rahmani et al. [68] reported an enzymeless electrochemical methyl parathion (MP) sensor based on bovine serum albumin (BSA) templated Au-Ag bimetallic clusters revealing the linear dynamic

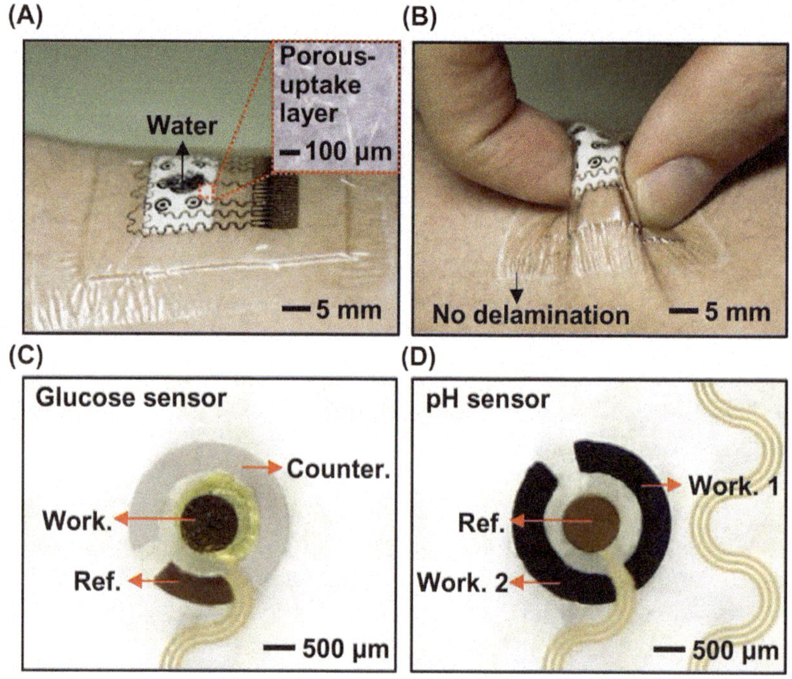

FIGURE 9.6 Exemplary schematic representations of electrochemical sensors. (A) Optical image of a wearable sweat analysis patch with a sweat-uptake layer and a waterproof band. The inset shows the magnified view of the porous sweat-uptake layer. (B) Optical image of the sweat analysis patch under deformation. (C) Optical image (*top*) and calibration curve (*bottom*) of a glucose sensor. (D) Optical image (*top*) and calibration curve (*bottom*) of the pH sensor [62]. *(A–D) Reproduced with permission H. Lee, et al., Wearable/disposable sweat-based glucose monitoring device with multistage transdermal drug delivery module, Science advances 3 (2017) e1601314. Copyright The Authors, some rights reserved; exclusive licensee American Association for the Advancement of Science. Distributed under a Creative Commons Attribution NonCommercial License 4.0 (CC BY-NC).*

ranges of MP between 0.02 to 8.0 and 8.0 to 200 mmol L^{-1} with a detection limit of 8.2 nmol L^{-1}. Recently, Wang et al. [69] reported a nonenzymatic electrochemical glucose sensor employing the three-dimensional kenafstem-derived carbon (3D-KSC)-loaded C_3O_4 nanoclusters showing a wide linear range (0.088–7.0 mM) and low detection limit of 26 μM. Most recently, Raileanu, Alexandra et al. [26] reported an amperometric thiocholine sensor based on cluster-assembled zirconia modified electrodes achieving a detection limit between 0.25 and 1.3 μM under a working potential of ca. 0.8 V.

Despite several promising results, the application of electrochemical sensing technology to ultraminiaturized wearable devices requires overcoming several challenges. These include the development of flexible and robust electrodes that can be utilized over long periods (weeks–months); further reducing the relatively high operation voltages of above 0.4 V [37], which currently requires the integration of large power supplies in millimeter scale wearable devices; the biocompatibility of the various active components that need to be placed in relatively close contact with the skin or other organs. As a result, a significant research focus is directed toward self-powered and contactless devices.

9.3.3 Optical sensors

Among numerous emerging biosensor technologies, optical sensors are particularly promising, thanks to the possibility to achieve selective room-temperature and real-time detection of small concentrations of analytes with highly miniaturized and often contactless devices. The potential of nanotechnology to enable more sensitive, smaller footprint sensors and integrate them in portable equipment has promoted a drastic development in the field of optical biosensors with applications ranging from noninvasive diagnostic and healthcare to preventive and personalized medicine [70,71]. The basic concept implemented in a variety of ways in optical biosensors is to produce a detectable light signal proportional to the concentration of the target analyte. An efficient way is provided by localized surface plasmon resonance (LSPR). The mechanism of LSPR devices relies on the interaction of light with subwavelength metallic nanoparticles, which triggers a collective resonant behavior of the metal-free electron. The latter drastically increases the light absorption and scattering in the nanoparticles [72]. As a consequence of this coupling, nanostructured metallic clusters tightly confine and enhance the electromagnetic field in the very proximity of the scattering center (~30 nm) [73,74]. This mechanism is highly suitable for sensing applications as molecules present within the plasmonic evanescent volume change the local refractive index (RI), and thus can be detected through shifts in the LSPR spectral peak [75]. The ability to transduce RI in measurable spectral shift has enabled plasmonic sensors to be used for biomolecule detection in liquid environment because of the higher RI contrast. The urgent need of detecting biomolecules interaction for disease detection and early diagnoses has paved the way for different liquid-based LSPR sensors.

Since the pioneering work of Mirkin et al. [76], functionalization of plasmonic nanoparticles with binding probes that can selectively bind or capture the target analyte is often employed in LSPR sensors. This work demonstrated the detection of femtomole concentrations of oligonucleotide through specific interaction with single stranded DNA-decorated gold nanoparticles. When gold nanoparticles are functionalized with DNA complementary sequences, base pairing induces aggregation of the particle. Thus, the organic biomolecule induces a highly selective linking with complementary ssDNA resulting in a visible color change due to plasmonic interaction and clustering in solution. Following a similar functionalization concept (Fig. 9.7A), a vast diversity of biochemical molecules, ranging from cancer biomarkers to viruses and bacteria, has been detected through LSPR sensors based on metallic nanocluster [77]. Through a variety of surface chemistries, including chemisorption, covalent attachment, or affinity-based

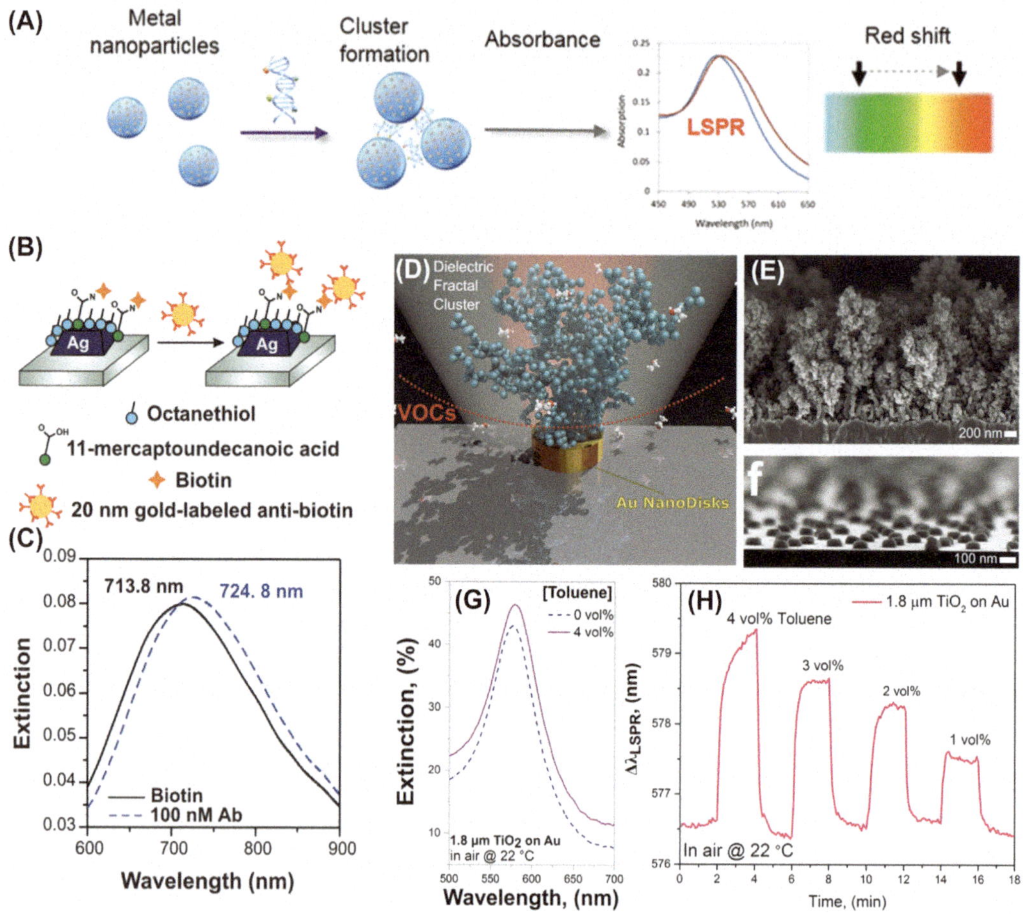

FIGURE 9.7 (A) Schematic image of the formation of metallic clusters and their functionalization with biomolecules, for optical linear sensing applications. Upon the interaction with the analyte, the typical plasmonic absorption spectrum shifts toward longer wavelengths. (B) Schematic of the gold nanoparticles functionalization with antibody. In this particular case, biotin is covalently linked to the NP using a coupling agent (1-ethyl-3-(3-dimethylaminopropyl) carbodiimide HCl linker). (C) Extinction spectra of the before (*solid black line*) and after (*dashed blue line*) the binding event of antibiotin shows a net redshift of the plasmonic peak. (D) schematic rendering of a TiO_2 dielectric cluster integrated onto an Au LSPR metasurface employed for room temperature VOCs sensing. (E–F) Cross-section SEM of the engineered fractal cluster with thickness of 1.8 μm and the resonant metasurfaces made by Au nanodisks with in-plane diameter of 100 nm. (G) Static gas sensing measurement upon purging 4 vol% of toluene at room temperature. As expected, the LSPR extinction spectrum shows a redshift due to the interaction with the volatile molecules. (H) Dynamic response of the LSPR peak position over time to different cycles of air/toluene with decreasing concentrations at room temperature. *(A) Copyright 2017 Journal of Biological Engineering. Reproduced with permission B. Chowdhury, I.-H. Cho, J. Irudayaraj, Technical advances in global DNA methylation analysis in human cancers, J. Biol. Eng. 11 (2017) 10. (B and C) Copyright 2010, Journal of Physical Chemistry C. Reproduced with permission from W.P. Hall, S.N. Ngatia, R.P. Van Duyne, LSPR biosensor signal enhancement using nanoparticle-antibody conjugates, J. Phys. Chem. C 115 (2011) 1410–1414. https://doi.org/10.1021/jp106912p. (D–H) Copyright 2018, WILEY-VCH Verlag GmbH & Co. KGaA, Weinheim. Reproduced with permission Z. Fusco, et al., Nanostructured dielectric fractals on resonant plasmonic metasurfaces for selective and sensitive optical sensing of volatile compounds. Adv. Mater. (2018) 1800931.*

interactions, many receptors such as antibodies [78], enzymes [79], peptides [80], or aptamers [81] have been immobilized on plasmonic nanostructures. Real-time colorimetric assays were then performed for the spectral detection of different analytes. Although the functionalization strategies are now well established, those sensors are limited to work in physiological environment to avoid the degradation of the receptors' biological activity. In addition, often the limits of detection are too low to detect very early-stage disease, making those optical sensors suitable for medical therapeutic tests.

As discussed in the previous sections, a promising approach for the very early diagnoses of illnesses is to develop noninvasive strategies based on the detection of biomarkers produced by the human body, which are often also present in the vapor phase. For instance, exhaled gases such as acetone (C_3H_6O) or ammonia (NH_3) from the human breath are correlated to diabetes type-I and renal/kidney diseases [82–84], respectively. However, application of LSPR sensors to measure volatile molecules in the gas-phase has proven to be very challenging, due to the large gas volume and the low concentration of molecules near the plasmonic surface. This lack of sensitivity of LSPR-based sensors toward small refractive index changes, which is characteristic of gas sensing, has proven a major limiting factor for their application in breath analysis and other noninvasive approaches based on volatile biomarkers. Several researches have achieved improvements in the gas sensitivity by shaping plasmonic structure in sophisticated geometries to fabricate optical nanoantennas with the purpose of enhancing the near-electromagnetic-field and reducing the linewidth of the plasmonic resonance [85,86].

An alternative approach is to nanotexture the plasmonic materials with dielectric thin films. Sputtering, electron-beam deposition, and sol-gel methods are the most common ways to embed metallic particles inside dielectric films, achieving high-metal doped nanocomposite samples [87–92]. Contrary to the direct LSPR sensing, here the metal nanoparticles act as catalytic centers for molecule-interactions and the optical property changes are often due to charge-transfer processes between the dielectric and the metal nanostructures, which often happens at high temperature, limiting their applicability [87,93,94]. Very recently, the application of 3D dielectric nanocluster architecture has shown potential to overcome some of these challenges, thanks to their controllable and ideal properties, such as specific surface area (SSA), porosity, crystallinity, and optical properties [9,35,57]. More specifically, integration of fractal clusters of dielectric nanoparticles onto plasmonic Au metasurfaces has been demonstrated to achieve room-temperature selective and sensitive detection of gaseous volatile organic compounds (VOCs) [4] (Fig. 9.7B–D). The clusters, comprised of crystalline TiO_2 nanoparticles with diameter of ca. 10 nm, were self-assembled in the diffusion regime by thermophoretic deposition of flame-made nanoparticle aerosols, resulting in a 98% porous three-dimensional fractal architecture. The thickness of the dielectric fractal didn't affect the plasmonic properties (i.e., resonant wavelength) up to at least thickness of $\sim 3\ \mu m$ (Fig. 9.7E). In addition, the high porosity and SSA of these nanoparticle-based devices are highly beneficial for gas molecule sensing, contributing both to increase in the number of active adsorption sites and the rapid reversible adsorption of gas molecules.

These nanoparticle-based optical LSPR sensors were tested at room temperature toward different VOCs, namely ethanol, acetone, and toluene, using standard UV-Vis read-out spectroscopy (Fig. 9.7F and G). Notably, the integration of fractal cluster permitted the measurement of VOCs concentration as low as 0.1 vol%, equivalent to refractive index changes of less than 8×10^{-6}. Furthermore, under standard humidity and temperature condition, the response of these LSPR devices to several VOCs was in line with the Hamaker theory indicating the formation of a thin liquid layer of ~0.4 nm on the fractal-metal system. This is high enough to raise the refractive index by at least an order of magnitude, permitting their optical detection. This condensation mechanism was also utilized for the selective detection of three tested VOCs permitting the identification of the different analytes as a function of the response rate change with increasing device surface temperature (Fig. 9.7H). These findings pave the way for the development of a new class of optical devices for noninvasive biomedical diagnostics, extending the use of already established optical sensor technologies for the sensing of liquid biomolecules to the gas-phase.

9.4 Summary

In this chapter, we have presented a review of the design, fabrication, and use of nanoparticle- and cluster-based materials for biomedical sensing applications. The self-assembly of materials from the gas-phase has been proposed as a scalable and rapid approach for the nanofabrication of a range of sensing materials ranging from nanostructured semiconductors to plasmonic metal-based metasurfaces. A key advantage of gas-phase nanofabrication approaches has been determined in their ability to result in highly reproducible 3D architectures, which can be predetermined by selecting an appropriate self-assembly regime. The fundamental mechanisms of nanoparticle and clusters self-assembly from the gas-phase have been summarized defining conditions for the fabrication of highly porous (ca. 98%) or denser (ca. 86%) nanoparticle films by decreasing the relative contribution of Brownian motion from the diffusion to the ballistic regime.

Exemplary nanoparticle-film morphologies obtained in these regimes have been presented along emerging nanofabrication tools such as thermophoretic deposition of flame-made nanoparticle aerosols in the diffusion regime, and cluster beam deposition in the ballistic regime. Focusing on these flexible approaches, the progress in the application of nanoparticle-based films for biomedical sensing have been presented for three major sensing technologies, namely chemoresistive, electrochemical, and optical sensors. The benefits provided by nanoparticle films with well-defined 3D morphologies have been defined in terms of the available surface sites for interaction with the target analyte, beneficial pore structure to achieve a fast reaction and diffusion kinetics throughout the film, and more complex interactions between nanocomposites materials such as 3D distributed dielectric and plasmonic nanoparticles. These advantages have translated into significant achievements in the corresponding sensing technologies resulting in lower limit of detection, smaller sensing materials, and novel approaches for the selective identification of multiple analytes.

9.5 Outlook

Three-dimensional nanoparticle/cluster films are rapidly emerging as a promising material for a broad range of sensing technologies ranging from the more established chemoresistive devices to innovative optical nanosensor architectures. The progress in the development of widely applicable approaches for the rapid and scalable nanofabrication of well-controlled 3D morphologies of nanoparticle- and cluster-based materials is supporting the investigation of the nanoscale interactions of biomedically relevant molecules with such nanostructured materials. The findings are driving the design of nanostructured optoelectronic and electrochemical detectors for miniaturized and wearable biomedical sensors. Translating these promising technologies to commercial products is ongoing, and is in turn supporting the further development of advanced nanoparticle and cluster fabrication tools such as supersonic cluster beam deposition and flame synthesis.

References

[1] N. Nasiri, R. Bo, F. Wang, L. Fu, A. Tricoli, Ultraporous electron-depleted ZnO nanoparticle networks for highly sensitive portable visible-blind UV photodetectors, Adv. Mater. 27 (2015) 4336–4343.

[2] A. Tricoli, M. Righettoni, A. Teleki, Semiconductor gas sensors: dry synthesis and application, Angew. Chem. Int. Ed. 49 (2010) 7632–7659.

[3] N. Nasiri, T.D. Elmøe, Y. Liu, Q.H. Qin, A. Tricoli, Self-assembly dynamics and accumulation mechanisms of ultra-fine nanoparticles, Nanoscale 7 (2015) 9859–9867.

[4] Z. Fusco, et al., Nanostructured dielectric fractals on resonant plasmonic metasurfaces for selective and sensitive optical sensing of volatile compounds, Adv. Mater. (2018) 1800931.

[5] R. Bo, et al., Low-voltage high-performance UV photodetectors: an interplay between grain boundaries and debye length, ACS Appl. Mater. Interfaces 9 (2017) 2606–2615.

[6] K. Wegner, P. Piseri, H.V. Tafreshi, P. Milani, Cluster beam deposition: a tool for nanoscale science and technology, J. Phys. D Appl. Phys. 39 (2006) R439.

[7] L. Mädler, A.A. Lall, S.K. Friedlander, One-step aerosol synthesis of nanoparticle agglomerate films: simulation of film porosity and thickness, Nanotechnology 17 (2006) 4783.

[8] N. Nasiri, R. Bo, L. Fu, A. Tricoli, Three-dimensional nano-heterojunction networks: a highly performing structure for fast visible-blind UV photodetectors, Nanoscale 9 (2017) 2059–2067.

[9] N. Nasiri, et al., Tunable band-selective UV-photodetectors by 3D self-assembly of heterogeneous nanoparticle networks, Adv. Funct. Mater. 26 (2016) 7359–7366. https://doi.org/10.1002/adfm.201602195.

[10] A. Tricoli, M. Graf, S.E. Pratsinis, Optimal doping for enhanced SnO_2 sensitivity and thermal stability, Adv. Funct. Mater. 18 (2008) 1969–1976.

[11] A. Tricoli, A.S. Wallerand, M. Righettoni, Highly porous TiO_2 films for dye sensitized solar cells, J. Mater. Chem. 22 (2012) 14254–14261.

[12] A. Tricoli, N. Nasiri, H. Chen, A.S. Wallerand, M. Righettoni, Ultra-rapid synthesis of highly porous and robust hierarchical ZnO films for dye sensitized solar cells, Sol. Energy 136 (2016) 553–559.

[13] J.L. Castillo, S. Martin, D. Rodriguez-Perez, A. Perea, P.L. Garcia-Ybarra, Morphology and nanostructure of granular materials built from nanoparticles, KONA Powder Part. J. 31 (2014) 214–233.

[14] D. Rodríguez-Pérez, J.L. Castillo, J.C. Antoranz, Relationship between particle deposit characteristics and the mechanism of particle arrival, Phys. Rev. E 72 (2005) 021403.

[15] P. Kulkarni, P. Biswas, A Brownian dynamics simulation to predict morphology of nanoparticle deposits in the presence of interparticle interactions, Aerosol Sci. Technol. 38 (2004) 541–554.

[16] T.D. Elmøe, A. Tricoli, J.-D. Grunwaldt, S.E. Pratsinis, Filtration of nanoparticles: evolution of cake structure and pressure-drop, J. Aerosol Sci. 40 (2009) 965–981.

[17] J.A.H. Dreyer, et al., Simulation of gas diffusion in highly porous nanostructures by direct simulation Monte Carlo, Chem. Eng. Sci. 105 (2014) 69–76. https://doi.org/10.1016/j.ces.2013.10.038.

[18] H. Keskinen, A. Tricoli, M. Marjamäki, J.M. Mäkelä, S.E. Pratsinis, Size-selected agglomerates of SnO_2 nanoparticles as gas sensors, J. Appl. Phys. 106 (2009) 084316.

[19] A.M. Whelan, in: Sol-Gel Materials for Energy, Environment and Electronic Applications, Springer, 2017, pp. 121–153.

[20] L. Durrer, et al., SWNT growth by CVD on Ferritin-based iron catalyst nanoparticles towards CNT sensors, Sens. Actuators B Chem. 132 (2008) 485–490.

[21] H.-C. Chiu, C.-S. Yeh, Hydrothermal synthesis of SnO_2 nanoparticles and their gas-sensing of alcohol, J. Phys. Chem. C 111 (2007) 7256–7259.

[22] A.P. Caricato, A. Luches, R. Rella, Nanoparticle thin films for gas sensors prepared by matrix assisted pulsed laser evaporation, Sensors 9 (2009) 2682–2696.

[23] R.K. Joshi, S. Krishnan, M. Yoshimura, A. Kumar, Pd nanoparticles and thin films for room temperature hydrogen sensor, Nanoscale Res. Lett. 4 (2009) 1191.

[24] A. Tricoli, S.E. Pratsinis, Dispersed nanoelectrode devices, Nat. Nanotechnol. 5 (2010) 54.

[25] L. Mädler, H. Kammler, R. Mueller, S.E. Pratsinis, Controlled synthesis of nanostructured particles by flame spray pyrolysis, J. Aerosol Sci. 33 (2002) 369–389.

[26] A. Raileanu, et al., An amperometric sensor for thiocholine based on cluster-assembled zirconia modified electrodes, J. Nanosci. Nanotechnol. 18 (2018) 6905–6912.

[27] M. Bruzzi, et al., First study of humidity sensors based on nanostructured carbon films produced by supersonic cluster beam deposition, Sens. Actuators B Chem. 100 (2004) 173–176.

[28] T. Mazza, et al., Libraries of cluster-assembled titania films for chemical sensing, Appl. Phys. Lett. 87 (2005) 103108.

[29] M. Zhao, J.X. Huang, C.W. Ong, Room-temperature resistive H_2 sensing response of Pd/WO_3 nanocluster-based highly porous film, Nanotechnology 23 (2012) 315503.

[30] C. Marega, et al., Polyvinyl alcohol electrospun nanofibers containing Ag nanoparticles used as sensors for the detection of biogenic amines, Nanotechnology 26 (2015) 075501.

[31] M.H. Wong, et al., Room-temperature polymer-coated supersonic cluster beam deposited ZnO film for O_2 gas and dissolved O_2 sensing, Sens. Actuators B Chem. 251 (2017) 1068–1075.

[32] A. Galli, et al., Cluster-assembled zirconia substrates promote long-term differentiation and functioning of human islets of Langerhans, Sci. Rep. 8 (2018) 9979.

[33] W.S. Wong, et al., Omnidirectional self-assembly of transparent superoleophobic nanotextures, ACS Nano 11 (2017) 587–596.

[34] L. Mädler, et al., Direct formation of highly porous gas-sensing films by in situ thermophoretic deposition of flame-made Pt/SnO_2 nanoparticles, Sens. Actuators B Chem. 114 (2006) 283–295.

[35] A. Tricoli, et al., Micropatterning layers by flame aerosol deposition-annealing, Adv. Mater. 20 (2008) 3005–3010. https://doi.org/10.1002/adma.200701844.

[36] N. Nasiri, et al., Structural engineering of nano-grain boundaries for low-voltage UV-photodetectors with gigantic photo-to dark-current ratios, Adv. Opt. Mater. 4 (2016) 1787–1795.

[37] A. Tricoli, N. Nasiri, S. De, Wearable and miniaturized sensor technologies for personalized and preventive medicine, Adv. Funct. Mater. 27 (2017) 1605271.

[38] C.D. Mathers, G.A. Stevens, T. Boerma, R.A. White, M.I. Tobias, Causes of international increases in older age life expectancy, Lancet 385 (2015) 540–548.

[39] A.P. Turner, Biosensors: sense and sensibility, Chem. Soc. Rev. 42 (2013) 3184–3196.

[40] J.R. Windmiller, J. Wang, Wearable electrochemical sensors and biosensors: a review, Electroanalysis 25 (2013) 29–46.

[41] A.J. Bandodkar, J. Wang, Non-invasive wearable electrochemical sensors: a review, Trends Biotechnol. 32 (2014) 363–371.

[42] X. Huang, et al., Materials and designs for wireless epidermal sensors of hydration and strain, Adv. Funct. Mater. 24 (2014) 3846–3854.

[43] N.M. Farandos, A.K. Yetisen, M.J. Monteiro, C.R. Lowe, S.H. Yun, Contact lens sensors in ocular diagnostics, Adv. Healthc. Mater. 4 (2015) 792–810.

[44] A. Koh, et al., A soft, wearable microfluidic device for the capture, storage, and colorimetric sensing of sweat, Sci. Transl. Med. 8 (2016), 366ra165.

[45] B. Schazmann, et al., A wearable electrochemical sensor for the real-time measurement of sweat sodium concentration, Anal. Methods 2 (2010) 342–348.

[46] A.J. Bandodkar, et al., Epidermal tattoo potentiometric sodium sensors with wireless signal transduction for continuous non-invasive sweat monitoring, Biosens. Bioelectron. 54 (2014) 603–609.

[47] Z. Sonner, et al., The microfluidics of the eccrine sweat gland, including biomarker partitioning, transport, and biosensing implications, Biomicrofluidics 9 (2015) 031301.

[48] M. Righettoni, A. Amann, S.E. Pratsinis, Breath analysis by nanostructured metal oxides as chemoresistive gas sensors, Mater. Today 18 (2015) 163–171.

[49] G. Korotcenkov, Metal oxides for solid-state gas sensors: what determines our choice? Mater. Sci. Eng., B 139 (2007) 1–23.

[50] J.N. Zemel, Theoretical description of gas-film interaction on SnO_x, Thin Solid Films 163 (1988) 189–202.

[51] X. Wang, L. Andrews, S. Riedel, M. Kaupp, Mercury is a transition metal: the first experimental evidence for HgF_4, Angew. Chem. 119 (2007) 8523–8527.

[52] S.R. Morrison, The Chemical Physics of Surfaces, Springer Science & Business Media, 2013.

[53] H. Ogawa, M. Nishikawa, A. Abe, Hall measurement studies and an electrical conduction model of tin oxide ultrafine particle films, J. Appl. Phys. 53 (1982) 4448–4455.

[54] N. Barsan, U. Weimar, Conduction model of metal oxide gas sensors, J. Electroceram. 7 (2001) 143–167.

[55] U. Weimar, Understanding the fundamental principles of metal oxide based gas sensor. The example of CO sensing with SnO_2 sensors in the presence of humidity, J. Condens. Matter Phys. 15 (2003) 813–839.

[56] H. Chen, R. Bo, T. Tran-Phu, G. Liu, A. Tricoli, One-step rapid and scalable flame synthesis of efficient WO_3 photoanodes for water splitting, ChemPlusChem 83 (2018) 569–576.

[57] M. Righettoni, A. Tricoli, S.E. Pratsinis, Si: WO_3 sensors for highly selective detection of acetone for easy diagnosis of diabetes by breath analysis, Anal. Chem. 82 (2010) 3581–3587.

[58] M. Righettoni, et al., Breath acetone monitoring by portable Si: WO_3 gas sensors, Anal. Chim. Acta 738 (2012) 69–75.

[59] M. Righettoni, et al., Monitoring breath markers under controlled conditions, J. Breath Res. 9 (2015) 047101.

[60] E. Barborini, et al., Nanostructured refractory metal oxide films produced by a pulsed microplasma cluster source as active layers in microfabricated gas sensors, Jpn. J. Appl. Phys. 50 (2011) 01AK01.

[61] F. Borghi, A. Podestà, C. Piazzoni, P. Milani, Growth mechanism of cluster-assembled surfaces: from submonolayer to thin-film regime, Phys. Rev. Appl. 9 (2018) 044016.

[62] H. Lee, et al., Wearable/disposable sweat-based glucose monitoring device with multistage transdermal drug delivery module, Sci. Adv. 3 (2017) e1601314.

[63] W.-S. Wang, W.-T. Kuo, H.-Y. Huang, C.-H. Luo, Wide dynamic range CMOS potentiostat for amperometric chemical sensor, Sensors 10 (2010) 1782–1797.

[64] S. Azzouzi, et al., A novel amperometric biosensor based on gold nanoparticles anchored on reduced graphene oxide for sensitive detection of L-lactate tumor biomarker, Biosens. Bioelectron. 69 (2015) 280–286.

[65] P. Bollella, et al., Beyond graphene: electrochemical sensors and biosensors for biomarkers detection, Biosens. Bioelectron. 89 (2017) 152–166.

[66] M. Decarli, et al., Integration of a technique for the deposition of nanostructured films with MEMS-based microfabrication technologies: application to micro gas sensors, Microelectron. Eng. 86 (2009) 1247–1249.

[67] G. Fiaschi, et al., A novel gas-phase mono and bimetallic clusters decorated Zno nanorods electrochemical sensor for 4-aminophenol detection, J. Electroanal. Chem. 811 (2018) 89–95.

[68] T. Rahmani, A. Hajian, A. Afkhami, H. Bagheri, A novel and high performance enzyme-less sensing layer for electrochemical detection of methyl parathion based on BSA templated Au–Ag bimetallic nanoclusters, New J. Chem. 42 (2018) 7213–7222.

[69] L. Wang, et al., Three-dimensional macroporous carbon/hierarchical Co_3O_4 nanoclusters for nonenzymatic electrochemical glucose sensor, Appl. Surf. Sci. 402 (2017) 47–52.

[70] R.T. Hill, Plasmonic biosensors, Wiley Interdiscip. Rev. Nanomed. Nanobiotech. 7 (2014) 152–168.

[71] P. Damborský, J. Švitel, K. Jaroslav, Optical biosensors, Essays Biochem. 60 (2016) 91–100.

[72] J.N. Anker, et al., Biosensing with plasmonic nanosensors, Nat. Mater. 7 (2008) 442–453.

[73] A.J. Haes, S. Zou, G.C. Schatz, R.P. Van Duyne, Nanoscale optical biosensor: short range distance dependence of the localized surface plasmon resonance of noble metal nanoparticles, J. Phys. Chem. B 108 (2004) 6961–6968.

[74] A.J. Haes, S. Zou, G.C. Schatz, R.P. Van Duyne, A nanoscale optical biosensor: the long range distance dependence of the localized surface plasmon resonance of noble metal nanoparticles, J. Phys. Chem. B 108 (2004) 109–116.

[75] A. Tittl, H. Giessen, N. Liu, Plasmonic gas and chemical sensing, Nanophotonics 3 (2014).

[76] R. Elghanian, J.J. Storhoff, R.C. Mucic, R.L. Letsinger, C.A. Mirkin, Selective colorimetric detection of polynucleotides based on the distance-dependent optical properties of gold nanoparticles, Science 277 (1997) 1078–1081. https://doi.org/10.1126/science.277.5329.1078.

[77] B. Chowdhury, I.-H. Cho, J. Irudayaraj, Technical advances in global DNA methylation analysis in human cancers, J. Biol. Eng. 11 (2017) 10.

[78] W.P. Hall, S.N. Ngatia, R.P. Van Duyne, LSPR biosensor signal enhancement using nanoparticle-antibody conjugates, J. Phys. Chem. C 115 (2011) 1410–1414. https://doi.org/10.1021/jp106912p.

[79] D. Aili, M. Mager, D. Roche, M.M. Stevens, Hybrid nanoparticle–liposome detection of phospholipase activity, Nano Lett. 11 (2011) 1401–1405. https://doi.org/10.1021/nl1024062.

[80] C. Guarise, L. Pasquato, V. De Filippis, P. Scrimin, Gold nanoparticles-based protease assay, Proc. Natl. Acad. Sci. USA 103 (2006) 3978–3982. https://doi.org/10.1073/pnas.0509372103.

[81] S.Y. Oh, et al., Development of gold nanoparticle-aptamer-based LSPR sensing chips for the rapid detection of *Salmonella typhimurium* in pork meat, Sci. Rep. 7 (2017) 10130. https://doi.org/10.1038/s41598-017-10188-2.

[82] S.J. Davies, P. Španěl, D. Smith, Breath analysis of ammonia, volatile organic compounds and deuterated water vapor in chronic kidney disease and during dialysis, Bioanalysis 6 (2014) 843–857. https://doi.org/10.4155/bio.14.26.

[83] G. Neri, et al., Real-time monitoring of breath ammonia during haemodialysis: use of ion mobility spectrometry (IMS) and cavity ring-down spectroscopy (CRDS) techniques, Nephrol. Dial. Transplant. 27 (2012) 2945–2952. https://doi.org/10.1093/ndt/gfr738.

[84] W. Chen, et al., Biochemical pathways of breath ammonia (NH_3) generation in patients with end-stage renal disease undergoing hemodialysis, J. Breath Res. 10 (2016) 036011.

[85] N. Liu, M.L. Tang, M. Hentschel, H. Giessen, A.P. Alivisatos, Nanoantenna-enhanced gas sensing in a single tailored nanofocus, Nat. Mater. 10 (2011) 631–636.

[86] J.M. Bingham, J.N. Anker, L.E. Kreno, R.P. Van Duyne, Gas sensing with high-resolution localized surface plasmon resonance spectroscopy, J. Am. Chem. Soc. 132 (2010) 17358–17359. https://doi.org/10.1021/ja1074272.

[87] D. Buso, et al., Au and NiO nanocrystals doped into porous sol-gel SiO(2) films and the effect on optical CO detection, Nanotechnology 17 (2006) 2429–2433.

[88] D. Buso, et al., Self-assembled gold nanoparticle monolayers in sol–gel matrices: synthesis and gas sensing applications, J. Mater. Chem. 19 (2009) 2051.

[89] D. Buso, M. Post, C. Cantalini, P. Mulvaney, A. Martucci, Gold nanoparticle-doped TiO_2 semiconductor thin films: gas sensing properties, Adv. Funct. Mater. 18 (2008) 3843–3849.

[90] E. Della Gaspera, et al., Au nanoparticles in nanocrystalline TiO_2–NiO films for SPR-based, selective H_2S gas sensing, Chem. Mater. 22 (2010) 3407–3417.

[91] E. Della Gaspera, et al., CO optical sensing properties of nanocrystalline ZnO–Au films: effect of doping with transition metal ions, Sens. Actuators B Chem. 161 (2012) 675–683.

[92] E. Della Gaspera, A. Martucci, Sol-gel thin films for plasmonic gas sensors, Sensors 15 (2015) 16910–16928.

[93] M. Ando, Recent advances in optochemical sensors for the detection of H_2, O_2, O_3, CO, CO_2 and H_2O in air, TrAC Trends Anal. Chem. 25 (2006) 937–948.

[94] M. Ando, T. Kobayashi, S. Iijima, M. Haruta, Optical CO sensitivity of Au–CuO composite film by use of the plasmon absorption change, Sens. Actuators B Chem. 96 (2003) 589–595.

10

Cluster-assembled nanostructured materials for cell biology

Carsten Schulte

INTERDISCIPLINARY CENTRE FOR NANOSTRUCTURED MATERIALS AND INTERFACES (CIMaINa) and Department of Physics, Università degli Studi di Milano, Milan, Italy

Chapter outline

10.1 Introduction

For decades and until today conventional in vitro cell culture and also most of the experimentation have been routinely performed utilizing plastic or glass surfaces as substrates which are basically flat at the nanoscale and stiff. Although it has been long-known that modulations of the surface chemistry and roughness, e.g., by reactive gas plasma treatment, improve the adhesive properties of commercially available polystyrene-based tissue culture dishes or flasks toward cells [1,2], they still do not sufficiently represent the natural cellular milieu.

These canonical cell culture substrates neglect important physical characteristics and dimensionalities (i.e., stiffness/elasticity and nanotopography) of the physiological microenvironment cells encounter in vivo, i.e., the extracellular matrices (ECM). Although the composition and features of ECM differ enormously from tissue to tissue (just think of the rigid mineralized bone matrix, the soft brain tissue, or the transparent cornea as some very diverse examples), it is always built up by a combination of

Cluster Beam Deposition of Functional Nanomaterials and Devices. https://doi.org/10.1016/B978-0-08-102515-4.00010-6

nanoscale components secreted by the cells, i.e., proteins (e.g., collagens, laminins, fibronectin, vitronectin) and polysaccharides (usually attached to ECM glycoproteins or proteoglycans) [3–6]. These building blocks self-assemble into complex meshworks characterized by fibrils, fibers, pores, and asperities that determine an ECM structure and surface nanotopography (Fig. 10.1A), which is not flat from the cell's perspective [5,6].

The roughness parameter R_q of tissue culture plastics, even after the mentioned plasma treating, rarely exceeds a few nanometers root mean square (RMS) (usually below 5 nm RMS) [1]. Although systematic and comparative characterizations of ECM nanotopographies along different tissues are yet missing, the roughness values of common tissue culture plastic or glass surfaces appear to be rather flat compared to known topographies of natural ECM (Fig. 10.1A) [5,6]. In addition, significant variations between different manufacturers have been found which can have an impact on the reproducibility of experiments [1].

In recent years, cell biologists started to understand the essential relevance of biophysical ECM cues at the nanoscale and their strong influence on cellular adhesion, morphology, and decision making by an intricate signaling pathway called *mechanotransduction* (Fig. 10.1A). Therefore numerous and versatile biomaterial approaches attempt to draw the in vitro conditions nearer to the in vivo situation by better mimicking the structural properties of the ECM. In this context, biomaterials with nanotopographical features become increasingly interesting for cell biological research and tissue engineering–based biomedical applications due to their potential in providing a reproducible control over cellular functions and behavior [6–11]. However, despite this potential the fabrication of nanostructured biomaterials that are compatible with the needs of modern cell biology and regenerative biomedicine certainly remains a challenging task, largely due to the demanding interdisciplinary requirements [6,8,10]. The nanofabrication techniques implemented for these kinds of approaches are usually tools that have been originally developed for microelectromechanical systems manufacturing, such as photo- or electron beam lithographical technologies. They often show technical limitations hampering the adaption to the necessities of cell biological applications, in particular, concerning scale-up capacities, cost-efficiency, and/or controllability of the feature size and dimensionality [10].

10.2 Fabrication of cluster-assembled substrates by supersonic cluster beam deposition to mimic nanotopographical features of the extracellular matrix

In this framework, supersonic cluster beam deposition (SCBD) represents a technological approach with the potential to fabricate, in a controllable and reproducible manner at a large scale, nanostructured surfaces with characteristics mimicking the nanotopographical features of extracellular matrices (Fig. 10.1B) [16].

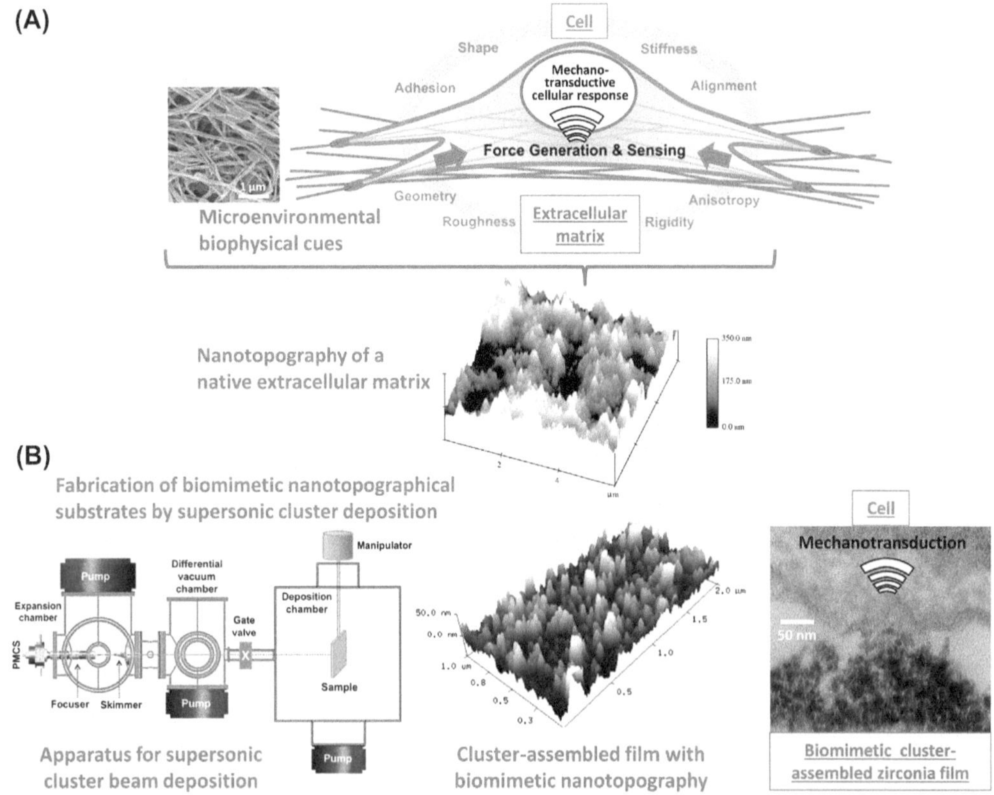

FIGURE 10.1 Resemblance of the surface nanotopography between native extracellular matrix and cluster-assembled nanostructured films produced by SCBD. (A) The cell is influenced by various structural and mechanical parameters of the extracellular matrix (ECM) it interacts with (on the left an example of the structural features of a hydrogel made from brain-derived decellularized ECM is shown). Image reproduced with permission from Ref. [12]. Copyright 2018, Springer Nature. The sensing of the biophysical cues from the microenvironment provokes a mechanotransductive response of the cell that can impact on its adhesive structures, morphology, and mechanical properties (image adapted with permission from Ref. [7]. Copyright 2011, Royal Society of Chemistry). The biophysical features of the ECM determine its nanotopography (below an example of a native ECM (in this case the basement membrane of a rhesus macaque corneal epithelium) is demonstrated. Image reproduced with permission from Ref. [13]. Copyright 1999, Springer Nature). (B) The cartoon (*left*, adapted from Ref. [14]. Copyright 2015, AIP) illustrates the working principle of the apparatus for supersonic cluster beam deposition (further information can be found in Section 10.2), which enables the fabrication of cluster-assembled nanostructured films (*middle*, image from Ref. [15]) that mimic the nanotopography of extracellular matrices (compare with the image above). Cells can interact with these nanotopographical features which modulate cellular mechanotransductive processes and behavior (image adapted from Ref. [15]).

The SCBD technique utilized for this purpose consists of igniting an argon plasma jet in the stagnation cavity of a pulsed microplasma cluster source, which ablates atoms and ions from a target material (in this context made of biocompatible materials such as zirconium or titania) that thermalize with the gas and condense to nanoclusters. The formed nanoclusters/gas mix then exits through an aerodynamic focuser and a nozzle into an expansion chamber collimating into a seeded supersonic cluster beam. The central portion of this cluster beam passes through a skimmer into the deposition chamber, where the nanoparticles impinge on a substrate (e.g., glass or plastic) placed into the beam trajectory (Fig. 10.1B). Depending on the deposition time in this ballistic regime a nanostructured film grows obeying simple scaling laws. This allows an accurate control over the nanoscale surface morphology with respect to its roughness parameter. More specific details about SCBD, e.g., regarding further technical and operating aspects or nanoparticle tuning modes, can be found in Refs. [14,17,18].

These cluster-assembled nanostructured films produced by SCBD hold a great promise as biomimetic substrates for exigent primary cell cultures (shown, e.g., already for neuronal cells [19] and β-cells [20]), stem cell differentiation, tissue engineering, as well as organ-on-chip devices. To explain why the nanoscale properties of the films are relevant for cell biology, the next paragraph will introduce the molecular mechanisms by which cells perceive and interpret biophysical nanometric cues in their microenvironment and how they respond to them. Research dissecting how cellular interaction with cluster-assembled nanostructured materials impacts on significant cell biological processes, such as mechanotransduction and cell differentiation, will be presented in the following paragraphs. Furthermore, it will be discussed how these effects of the cluster-assembled nanotopographical substrates can be exploited in future to create intelligently engineered biomaterials based on cluster-assembling capable to guide cellular behavior and fate.

10.3 The cell world at the nanoscale: how cells sense, interpret, and respond to nanotopographical cues in their microenvironment by mechanotransductive processes

In recent years, it became more and more evident that the ECM serves not only as a simple passive anchorage point for the cells, but instead provides important biophysical cues to the cells that enforce specific and directive effects on cell behavior (Fig. 10.1A) [3–6]. A crucial objective of significant biomaterials has thus to be the mimicry of essential ECM structural features on the nanoscale [6,8–11,16] (Fig. 10.1A and B).

To understand the potential of biomimetic nanostructured surfaces in controlling cellular processes and decision making, it is necessary to have a closer look at what is known about the actual molecular mechanisms by which cells sense, interpret, and respond to their microenvironments. The cell's perception of biophysical information from the microenvironment and its conversion into appropriate cellular responses, defined as mechanosensing and mechanotransduction, are in fact remarkably intricate and dynamic (Fig. 10.1A) [21–26].

The principal adhesive structures and mechanosensors of the cell are integrin adhesion complexes (IAC). These multiprotein complexes (Fig. 10.2) represent the major cellular interface with the microenvironment. These structures integrate (therefore the name integrin was chosen for these proteins) the cellular outside with its inside, by interpreting the nature and configuration of the ECM and regulating concerted and complex cellular actions and signaling cascades accordingly [22–26].

The hierarchical formation of these complexes starts with nascent adhesions when the eponymous transmembrane integrin receptors adhere extracellularly to appropriate cues present in the microenvironment and begin to cluster [25–28]. Intracellularly, in these clusters a so-called molecular clutch engages, i.e., the key protein for this process, talin, links forces generated through actin polymerization and actomyosin contraction (resulting in a retrograde actin flow) to the integrins. The level of force loading within the engaged molecular clutch and the composition and dimension of the IAC depend directly on several structural microenvironmental factors, in particular, the substrate rigidity and/or spatial distribution of integrin ligands [25–29]. The latter aspect is highly relevant with respect to nanostructured surfaces assembled by clusters, as the nanotopography influences the spatial organization of potential nanometric adhesion sites for the integrins, i.e., their dimensions and distances between each other (Figs. 10.1B and 10.2) [9,10,15].

The force loading in the molecular clutch dictates the extent of talin stretching/activation and consequentially whether cryptic binding sites for other proteins are exposed that are hidden in the inactive closed talin conformation [25,26]. The stretched talin thus reinforces the molecular clutch by enabling the recruitment of further IAC proteins [31], starting, e.g., with vinculin [27] and subsequently other numerous adaptor and/or signaling proteins (e.g., LIM domain proteins (e.g., paxillin and zyxin), α-actinin, focal adhesion kinase, src, integrin-linked kinase), which transform the nascent adhesions into IAC and signaling hubs [22–24,26,32,33]. This maturation of IAC, first to focal complexes and then to focal adhesions [25–28], leads to the formation of multiprotein molecular machines with a horizontally stratified nanoarchitecture, composed by an extracellular receptor/matrix binding layer, an integrin signaling layer, a force transduction layer, and an actin regulatory layer, the latter attached to actin stress fibers [27,34] (Fig. 10.2). The IAC composition is highly dynamic and very complex, potentially comprising >200 different protein constituents [22–24,33]. Alterations in the IAC composition modulate cellular signaling and can induce a remodeling of the cytoskeleton and the chromatin in the nucleus which are connected by the Linker of Nucleoskeleton and Cytoskeleton complex (LINC complex) [21,26]. These actions eventually mediate the mechanotransductive impact on cellular behavior and fate, by modulating the activity of mechanosensitive transcription factors (e.g., YAP [35]) and mechanical hotspots for genome regulation [36] (Fig. 10.3). Although many details remain still elusive, these processes were found to be strongly involved in virtually all cell biological processes, in particular, in cell differentiation and migration.

For this complexity of mechanosensing and transduction, although only simplified and insinuated here, knowledge is still insufficient about how the IAC composition is

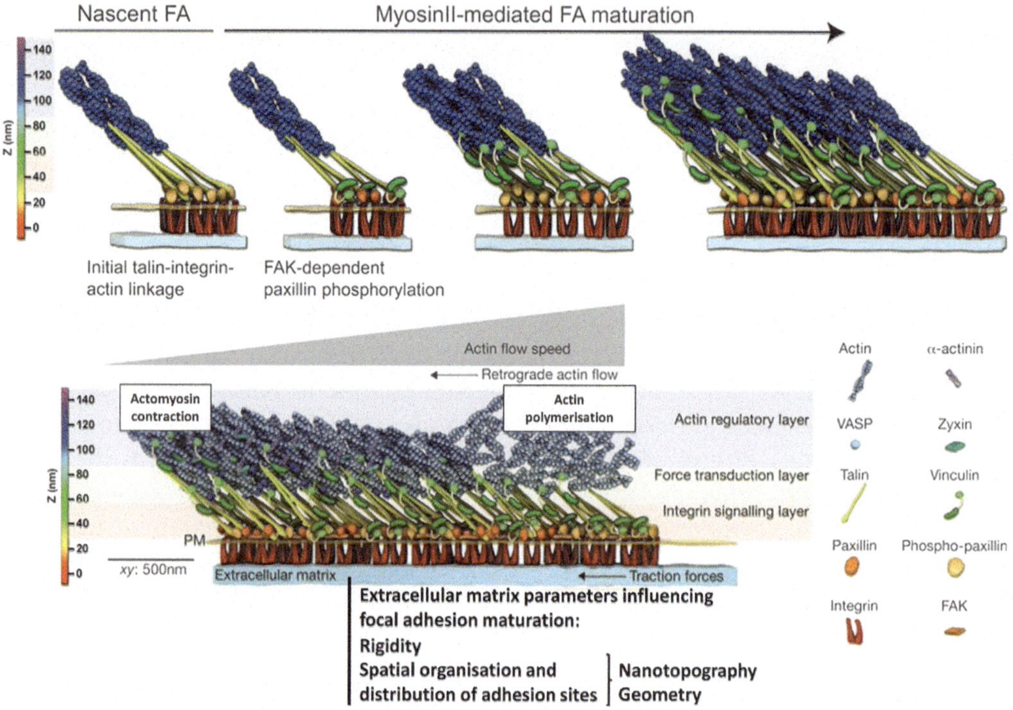

FIGURE 10.2 Formation and nanoarchitecture of integrin adhesion complexes and influencing microenvironmental factors. The cartoon (adapted with permission with elements from Refs. [27,30]. Copyright 2015, Springer Nature) illustrates the hierarchical formation of integrin adhesion complexes (IACs) starting from nascent adhesions in which the integrins are linked through talin to the filamentous actin. The forces generated by the actomyosin contraction and actin polymerization stretch talin and enable a reinforcement and maturation of the complexes to focal adhesion by the recruitment of further adaptor (e.g., vinculin and paxillin) and signaling proteins (e.g., FAK). The dimension and composition of the IAC depend on structural and mechanical parameters of the extracellular matrix (further details on focal adhesion maturation can be found in Section 10.3).

determined by the microenvironment and how this controls (patho) physiological cellular processes [22,23,25,26,35]. However, an accurate understanding of the cell's perception of its nanoscale environment and how the information is converted into cellular responses is inevitable to achieve a reliable control over these mechanotransductive processes through engineered biomaterial cues [5,6,8–11,16].

Cluster-assembled nanostructured surfaces can be useful in this context in two ways. On the one hand, they can contribute to unravel these processes by providing cellular microenvironments with controllable and reproducible nanotopographies to study nanotopography-dependent mechanotransductive responses and processes. On the other hand, if designed in a smart way, the effect of the cluster-assembled nanostructured surfaces on mechanotransductive processes can be exploited to guide and control cellular behavior (Fig. 10.1B). In the next paragraphs research taking steps toward both approaches will be outlined.

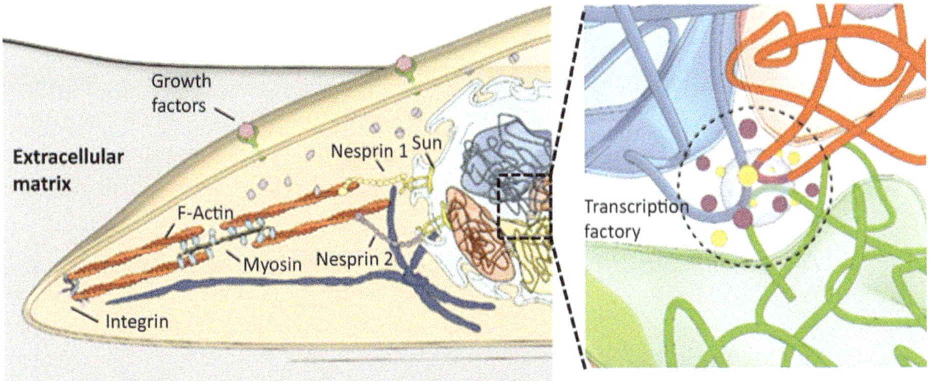

FIGURE 10.3 Mechanotransductive sequence from cellular microenvironment to the nucleus. The cartoon shows the intracellular connection from the integrins (that are attached to the ECM) to the nucleus via the cytoskeleton and components of the LINC complex. The force transmission and mechanotransductive sequence impact on the spatial chromatin organization and transcription factors modulating eventually the cellular program. *Adapted with permission from C. Uhler, G.V. Shivashankar, Chromosome intermingling: mechanical hotspots for genome regulation, Trends Cell Biol. 27 (11) (2017) 810–819. Copyright 2017 Elsevier.*

10.4 Cluster-assembled nanostructured surfaces as biomimetic substrates and their impact on mechanotransductive processes and cellular decision making

The initial inspiration to implement cluster-assembled nanostructured films produced by SCBD as substrates in a cell biological framework was based on the observation that their nanotopographical characteristics of granularity, porosity, and asperities resembled those found in extracellular matrixes [37] (Fig. 10.1). The deposition of clusters made from transition metal oxides, such as titania or zirconia, was chosen to guarantee biocompatibility, as these materials are already widely used in implantology (in particular, in dentistry and orthopedics) precisely because of this favorable feature [38,39].

In fact, in initial studies with different types of cells from versatile biological backgrounds, cell lines (MG-63, U2OS, immortalized fibroblasts) and primary cells (human melanocytes, human myogenic CD133 + progenitors, rat microglia) were found to be able to adhere and grow on these nanostructured surfaces [37,40–42], which gave a first indication that these materials have indeed a potential as substrates in cell-based applications.

A transmission electron microscopy–based analysis of the interface between neuron-like PC12 cells and nanostructured zirconia films (examples in Figs. 10.1B and 10.4) demonstrated that these cells interact mostly with the apical part of the asperities of the nanotopographical surfaces (with a roughness of 15 nm RMS), restricting the width of these adhesion sites to 53.2 ± 48.0 nm in average, compared to 90.2 ± 93.2 nm on flat (<1 nM RMS) zirconia surfaces [15]. This is particularly interesting in the context of integrin-mediated cell adhesion because this dimension is in the range of the integrin clusters (50–100 nm, containing 20–50 integrins) that assemble as loosely aggregated modular building blocks, the focal adhesions, and define their nanoscale organization [28]. Also the distance between these nanoscopic adhesion sites on the nanostructured substrate (average: 99.1 ± 101.4 nm, median: 60.4 ± 33.4 nm) oscillates around the threshold values shown to be critical for the formation of focal adhesions (≅ 60–70 nm) [9,26,29,43]. In fact, on these nanotopographical surfaces smaller adhesive structures in the dimension of focal complexes or growth cone point contacts assembled and actin stress fiber formation was low compared to flat surfaces where focal adhesions matured and stress fibers bundled [15] (Fig. 10.4). Congruently, a similar effect on the focal adhesion and stress fiber formation was found in human pancreatic β-cells [20]. The cytoskeletal organization and configuration dictate the biomechanical properties of cells [25,28]. In line with the observation of reduced stress fiber formation, measurements of the cytoskeleton mechanics of living PC12 cells showed that they were softer when they were interacting with cluster-assembled zirconia surfaces compared to flat zirconia [15] (Fig. 10.4).

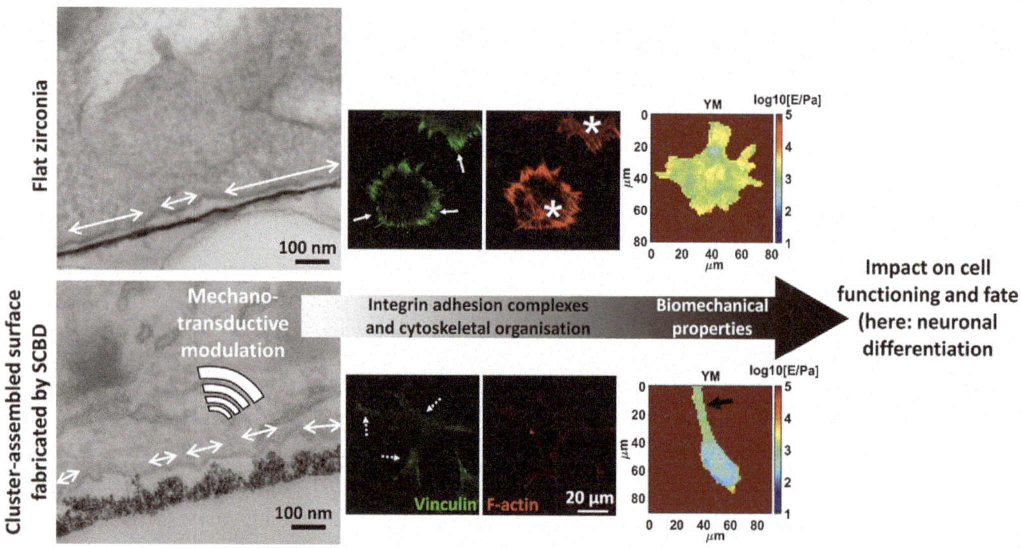

FIGURE 10.4 Mechanotransductive response to cluster-assembled nanostructured materials. On the left, transmission electron images of cellular interaction sites between PC12 cells and flat or cluster-assembled nanostructured zirconia surfaces (with a roughness of 15 nm RMS) are shown. The *white arrows* illustrate the dimension of the adhesion sites, indicating smaller dimensions of the nanometric adhesion sites on the nanotopographical surface. The cell/nanotopography interaction leads to further modulations within the mechanotransductive sequence, such as smaller integrin adhesion complexes (vinculin staining in green appears punctate, indicating focal complexes, examples marked by the *dashed line arrows*, images obtained by total internal reflection microscopy) and the absence of actin stress fibers (f-actin staining in red in epifluorescence), compared to the flat zirconia surface where the cells have mature focal adhesions (examples indicated by *arrows with continuous line*) and actin stress fibers (*asterisks*). The cells on the cluster-assembled substrate are furthermore softer (the color code of the images at the right side indicates the Young's modulus measured by atomic force microscopy) and show the outgrowth of a neurite (*black arrow*). *Figure adapted with permission from C. Schulte, A. Podestà, C. Lenardi, G. Tedeschi, P. Milani, Quantitative control of protein and cell interaction with nanostructured surfaces by cluster Assembling, Acc. Chem. Res. 50 (2) (2017) 231–239. Copyright 2016, American Chemical Society.*

These effects of cell interaction with the nanostructured zirconia substrates along the mechanotransductive sequence have a strong influence on the cellular program, functioning and differentiative behavior in versatile cell biological frameworks. In neuronal cells (PC12 cells and hippocampal neurons) cluster-assembled zirconia surfaces with an appropriate roughness foster essential differentiation events, such as the outgrowth of neurites (Figs. 10.4 and 10.5A), synaptogenesis, and the maturation of neuronal network activity and communication (Fig. 10.5B) [15,19].

Also ex vivo human pancreatic β-cells' survival was enhanced on nanostructured zirconia substrates and their functioning and differentiation (insulin packaging and glucose-stimulated insulin secretion), Fig. 10.6) was promoted [20].

The capacity of SCBD to produce these nanostructured surfaces on a large macroscopic scale allowed furthermore an assessment of modulations in the cellular program and signaling pathways by (phospho)proteomic approaches, which require the yield of big amounts of cellular material from the different experimental conditions [16]. Proteomics represents a powerful technique that enables a large-scale comparison of protein expression levels.

For all the different tested cell types (PC12 cells, rat hippocampal neurons, and human pancreatic β-cells) the obtained proteomic data confirmed consistently the observed effects with respect to cell differentiation also on the protein level [15,19,20,44]. In the hippocampal neurons, e.g., many proteins involved in vesicle transport and important in synaptic transmission are upregulated in the cells interacting with the nanostructured surface, such as Rab family proteins, synapsins, AP180, and syntaxin-1b, as well as markers for neuronal differentiation, e.g., GAP43, NCAM, or N-Cadherin [19] (Fig. 10.7). In pancreatic β-cells grown on cluster-assembled biomaterial a reactome analysis showed an enrichment of proteins related to carbohydrate and glucose catabolic processes [20].

Moreover and in line with the altered mechanotransductive processes, the proteomes of the cells growing on nanostructured cluster-assembled materials congruently showed strong changes in regard to proteins related to integrin adhesion complexes and/or cytoskeletal organization compared to the respective canonical standard substrate [15,19,20,44]. In hippocampal neurons a gene annotations analysis indicated the participation of axon guidance, cell adhesion, and integrin signaling pathways in response to the interaction with the cluster-assembled nanotopographies that promote neuronal maturation. Known components of the IAC (such as, e.g., LIM domain-containing proteins, among them also zyxin) or proteins partaking in IAC-related processes are differentially expressed in the neurons grown on the nanostructured substrates that foster neuron maturation [19]. Comparable abundance of modulations related to IAC proteins were found in PC12 and β-cells [15,20,44]. Interestingly, in an ingenuity pathway analysis comparing PC12 cells on nanostructured surfaces with different roughness parameters (15 vs. 25 nm RMS), integrin-linked kinase (ILK) signaling appeared as affected by the interaction with the topographical features (Fig. 10.8) [44]. The ILK pathway, a key regulatory pathway for cell adhesion and cytoskeletal dynamics, has been shown to be susceptible to different integrin ligand densities

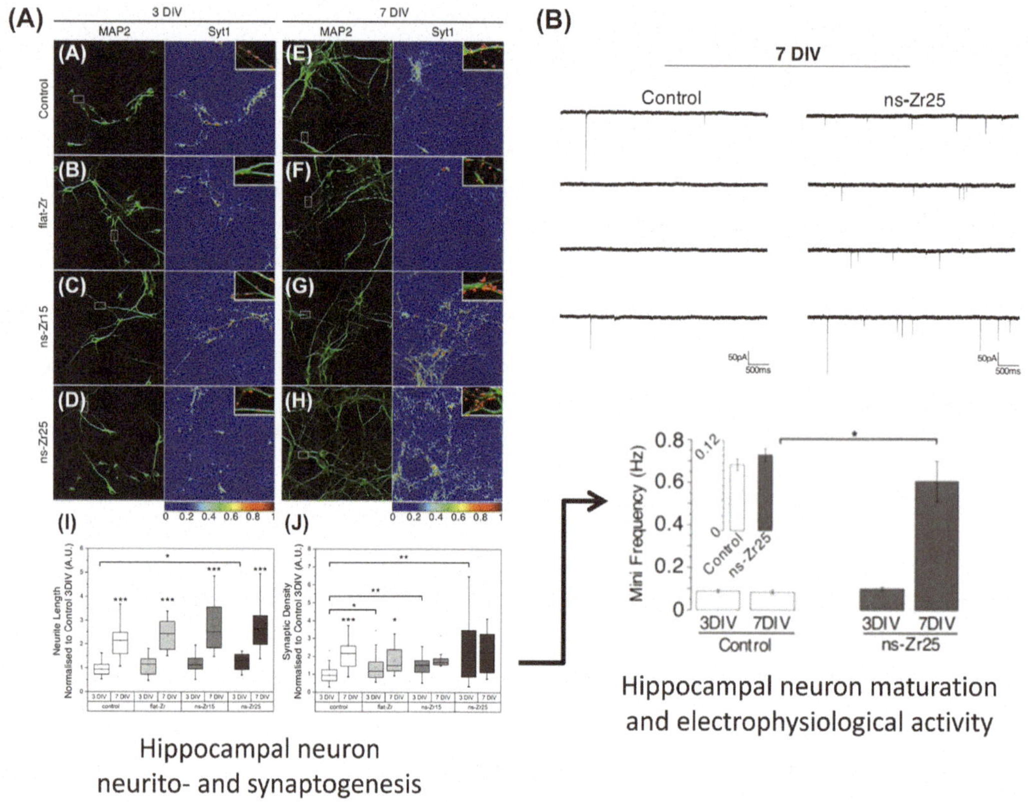

FIGURE 10.5 Impact of cellular interaction with cluster-assembled nanostructured materials on neuron functioning and behavior. (A) The panel shows confocal microscopic images of neurite outgrowth visualized by microtubule-associated protein 2 (MAP2) staining in green; left images in (A–D and E–H), and synaptogenesis visualized by synaptotagmin 1(Syt1) staining presented in color code corresponding to the fluorescence intensity; right images in (A–D and E–H), of neonatal rat hippocampal neurons cultured on different substrate conditions for 3 or 7 days (glass coated with poly-ornithine and diluted matrigel (control), flat zirconia (flat-Zr), or nanostructured cluster-assembled zirconia with two different roughness parameter 15 nm RMS and 25 nm RMS (ns-Zr15 and ns-Zr25, respectively)) and the according quantifications (I, J). The graphs demonstrate an acceleration of neuritogenesis (I) and synaptogenesis (J) in particular on ns-Zr25 after 3 days in culture (DIV). (B) These effects lead to an earlier neuron maturation and electrophysiological network activity (higher frequency of spontaneous miniature postsynaptic currents (minis) recorded by whole-cell patch clamp) present after 7 days and not visible in the control condition. *The images are from C. Schulte, M. Ripamonti, E. Maffioli, M.A. Cappelluti, S. Nonnis, L. Puricelli, et al., Scale invariant disordered nanotopography promotes hippocampal neuron development and maturation with involvement of mechanotransductive pathways, Front. Cell. Neurosci. 10 (2016) 267.*

of the substrate [45]. Also alterations for proteins and pathways (in particular, related to Rho GTPase signaling) involved in the regulation of cytoskeletal and nuclear organization were abundant and evident in PC12 cells, hippocampal neurons, and β-cells upon interaction with the nanostructured surfaces [15,20,44].

Altogether, these multifarious impacts along the whole mechanotransductive sequence and on cellular differentiative processes demonstrate the potential of cluster-

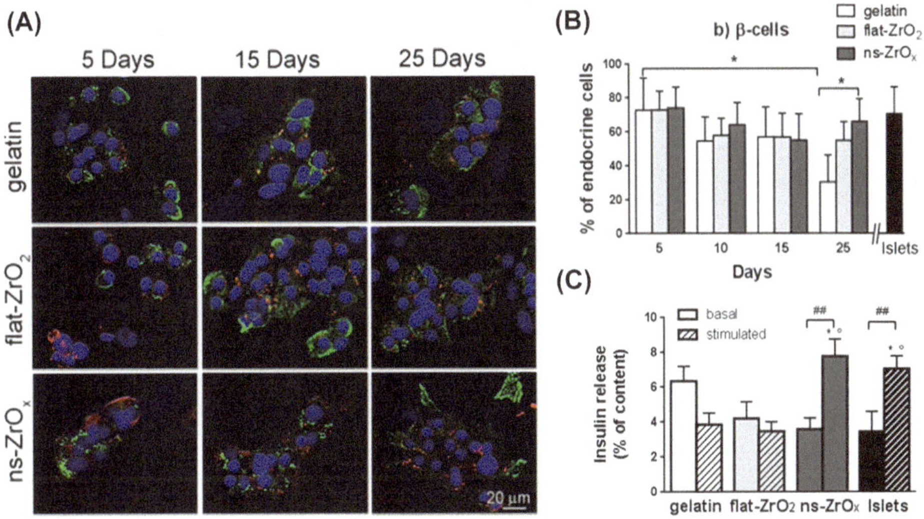

FIGURE 10.6 Impact of cellular interaction with cluster-assembled nanostructured materials on β-cell functioning and behavior. (A) The panel (confocal images of human Langerhans islets stained for insulin in red and chromogranin A in green, the nucleus is marked by DAPI in blue) and (B) the according quantification of the insulin-positive β-cells demonstrate the better long-term preservation of β-cell differentiation (after 25 days still at the level of freshly isolated islets) if they are grown on the nanostructured zirconia surfaces (ns-ZrO_x), compared to gelatin (canonical cell culture condition) or flat zirconia (flat-ZrO_2) substrates. (C) The high glucose-stimulated insulin release (measured by ELISA immunoassays) of the β-cells in the ns-ZrO_x condition (after 20 days) shows a similar extent as the one of the young islets, confirming again the stronger preservation of the β-cell differentiation and functioning. *The images are all taken from A. Galli, E. Maffioli, E. Sogne, S. Moretti, E.S. Di Cairano, A. Negri, et al., Cluster-assembled zirconia substrates promote long-term differentiation and functioning of human islets of Langerhans, Sci. Rep. 8 (1) (2018) 9979.*

assembled nanotopographical films as materials for cell biological applications. Furthering our understanding of the details regarding the molecular mechanisms underlying the instructive influence of the microenvironmental nanotopography on cellular behavior could help to direct cell fate and decision making in a predictable manner by precisely engineered biomimetic cluster-assembled substrates. Due to the demonstrated capacities, conceivable applications of these nanostructured materials are manifold and versatile, e.g., as reliable tools for profound mechanobiological studies as already outlined here, or in future as platforms with differentiation-promotive properties for stem cell culture, tissue engineering, and organ-on-a-chip approaches, or for drug testing.

10.5 Micropatterning of cluster-assembled nanostructured surfaces to control geometric partitioning of cell adhesion and interactions

The geometrical arrangements of cellular environments are of fundamental importance for cell interactions, behavior, and functioning in many cell biological in vivo contexts.

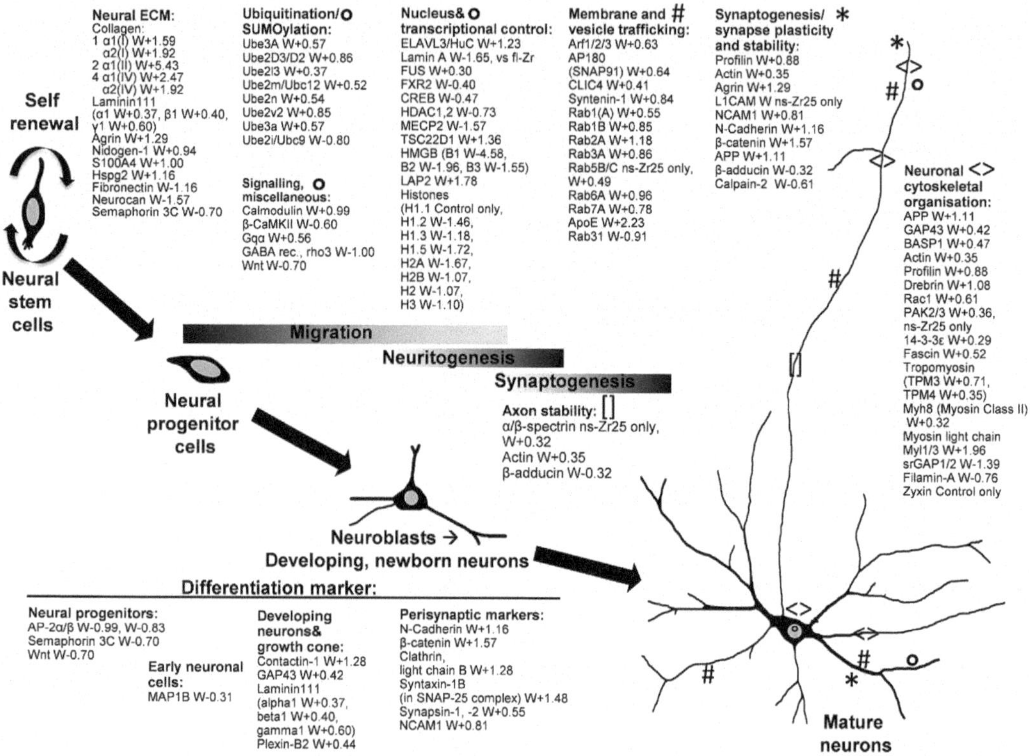

FIGURE 10.7 Modulations of the hippocampal neuron proteome after neuron/nanotopography interaction. The graphic lists selected examples of proteins that are differentially expressed in rat hippocampal neurons interacting for 3 days with the cluster-assembled nanostructured surfaced (compared to the control condition, i.e., glass coated with polyornithine and diluted matrigel). Many proteins with important roles in neuronal differentiation and/or functioning are modulated in their expression level confirming the promotion of neurito- and synaptogenesis and network maturation by the neuron/nanotopography interaction also at the proteome level. In addition, modulations of proteins known to be involved in the regulation of mechanotransduction and/or cytoskeletal organization are numerous. The number behind the protein names provides the corresponding Welch t-test difference. *The image is adapted from C. Schulte, M. Ripamonti, E. Maffioli, M.A. Cappelluti, S. Nonnis, L. Puricelli, et al., Scale invariant disordered nanotopography promotes hippocampal neuron development and maturation with involvement of mechanotransductive pathways, Front. Cell. Neurosci. 10 (2016) 267.*

An extremely fascinating example is the remarkable computational capacities of the brain that emerge from the highly organized, modular architecture of the neuronal cell localization and connectivity, i.e., the brain connectome [46,47].

A randomly forming network between ex vivo primary brain cells grown in canonical cell culture does not reflect the structured physiological network topology which hampers the study of neuronal cell communication and information processing. The accurate recapitulation of the in vivo neuronal circuitry in in vitro cell culture, i.e., brain-on-chip approaches, is therefore one of the major challenges in cellular neuroscience. It requires a precise micrometric control over the spatial organization of neuronal cell adhesion and connectivity by engineering micropatterned cell-adhesive substrates. To date, most

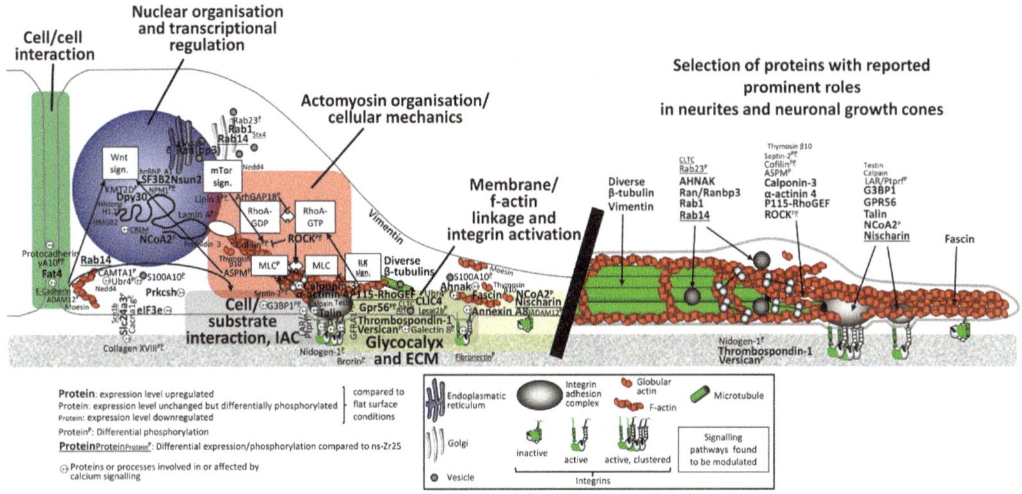

FIGURE 10.8 Substrate nanotopography-sensitive cellular signaling hubs. The cartoon summarizes changes of protein expression and phosphorylation in the neuron-like PC12 cells upon interaction with the cluster-assembled nanostructured zirconia surfaces. In particular, proteins are highlighted and grouped that have a potential connection to the mechanotransductive signaling sequence, categorizing them into specific cell biological functions, such as cell/cell interaction (*green box*), glycocalyx and ECM (*box with gray patterned filling*), cell substrate interaction and IAC (*gray box, gray oval represents the IAC*), membrane/f-actin linkage and integrin activation (*yellow box*), actomyosin organization/cellular mechanics (*red box*), as well as nuclear organization and transcriptional regulation (*blue circle symbolizing the nucleus*). On the right, proteins with specific known function in neuronal outgrowth are reported. *The image is from E. Maffioli, C. Schulte, S. Nonnis, F. Grassi Scalvini, C. Piazzoni, C. Lenardi, et al., Proteomic dissection of nanotopography-sensitive mechanotransductive signalling hubs that foster neuronal differentiation in PC12 cells, Front. Cell. Neurosci. 11 (2017). https://doi.org/10.3389/fncel.2017.00417.*

approaches to achieve predetermined neuronal network configurations (with areas for neuronal cell body adhesion and paths for neurite outgrowth) are based on microcontact printing of adhesive molecules (e.g., poly-lysines or ECM proteins) on surfaces passivated for cell adhesion, or engineering of microstructured platforms [48–52].

Considering the abovementioned mechanotransductive effects of the microenvironmental nanotopography fostering neuronal network formation and maturation, integrating the cluster-assembled ECM nanotopography-mimicking films into a micropatterned topology has a strong potential in the neuroscience field [51,53,54].

This type of micropatterning of cell adhesion-permissive areas with nanotopographies produced by SCBD on elsewhere cell-repellent surfaces has been achieved recently by the following procedure (Fig. 10.9) [55]. First, the glass surfaces were made repulsive for (neuronal) cell adhesion by a functionalization with copolymer specifically developed for neural patterns (i.e., PAcrAm-g-PMOXA, NH_2, Si [56]) that covalently binds to hydroxylated glass forming a very thin (~ 1 nm) monomolecular layer. The deposition of the zirconia nanoclusters on these functionalized antifouling glass surfaces is then implemented through stencil metal masks with predetermined apertures (obtained by laser cutting). As a distinctive feature, e.g., compared to the microcontact printing

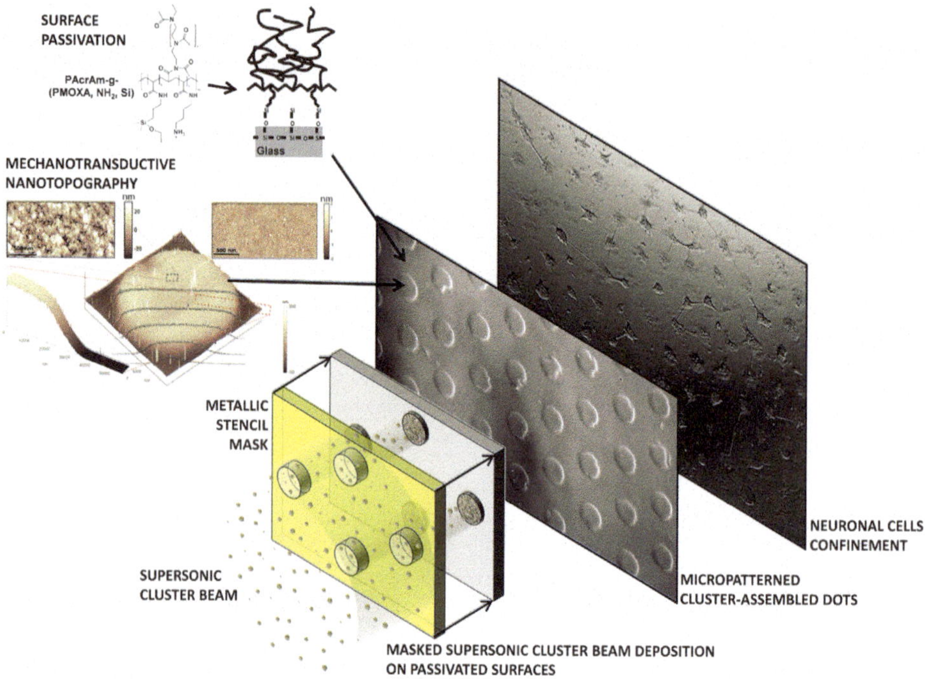

FIGURE 10.9 Micropatterning of cluster-assembled nanostructured surfaces to control neuronal cell adhesion. The graphic outlines the different steps to achieve micropatterned cluster-assembled nanostructured surfaces that enable cell confinement in a predetermined topology. The glass surfaces are passivated with the copolymer PAcrAm-g-(PMOXA, NH_2, Si) covalently binding to the glass. This functionalization makes the surface cell-repellent. On this passivated glass surface the nanostructured zirconia films are assembled in a micropatterned manner by supersonic cluster beam deposition through metallic stencil masks with predefined apertures (obtained by laser cutting). The zones with the cell-adhesive cluster-assembled films possess nanotopographical features that can influence mechanotransductive processes (as outlined before). Further details can be found in Section 10.5. *The figure reproduces with permission elements from C. Schulte, J. Lamanna, A.S. Moro, C. Piazzoni, F. Borghi, M. Chighizola, et al., Neuronal cells confinement by micropatterned cluster-assembled dots with mechano-transductive nanotopography, ACS Biomater. Sci. Eng. (2018). https://doi.org/10.1021/acsbiomaterials.8b00916. Copyright 2018, American Chemical Society.*

approaches, this technique permits to independently configure the topology of the micropattern and the biomimetic nanotopographical surface features (in terms of the roughness parameter).

The proof-of-principle of this approach has been demonstrated with micropatterns that consist of circular dots with a diameter of 150 μm and two different dot-to-dot interdistances (300 and 500 μm). The effective confinement of cells on these micropatterns has been shown for PC12 cells and primary hippocampal neurons [55]. The primary neurons build up also functional networks within the dots that communicated with neighboring neuron populations if they were connected by spontaneous neurite/axon bridging that occurred between adjacent dots (in case of 300 μm interdistance) (Fig. 10.10) [55]. These kinds of micropatterns based on the SCBD technique have a

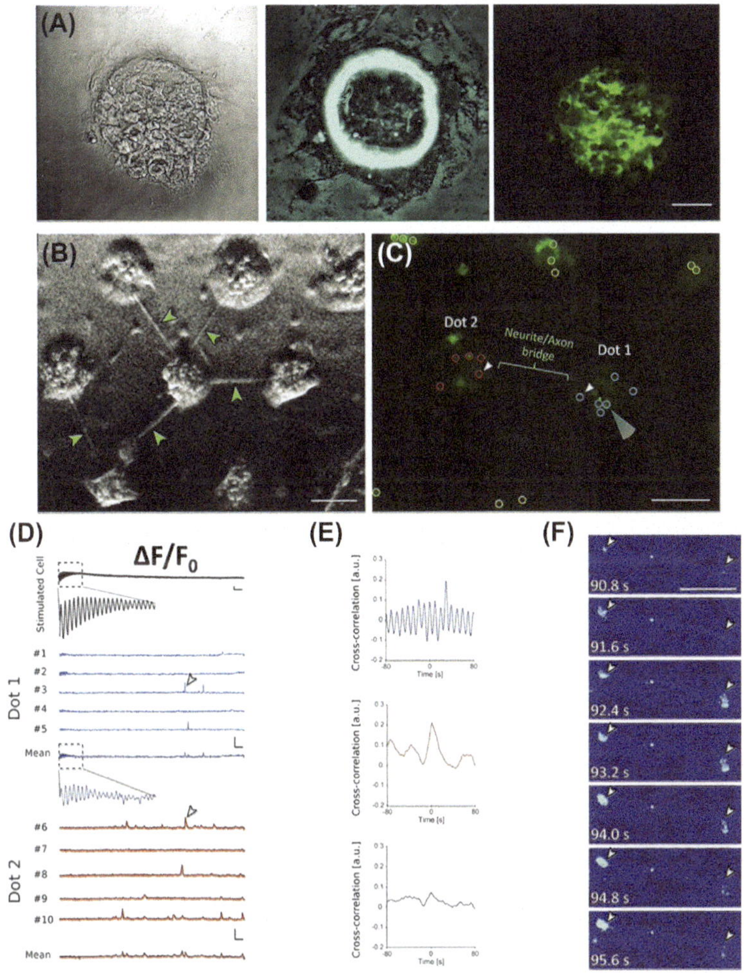

FIGURE 10.10 Functional neuronal network formation on micropatterns with mechanotrasductive nanotopographies. (A) On these micropatterned cluster-assembled zirconia substrates primary rat hippocampal cells grow in a confined manner (example shows cells after 7 days in culture in phase contrast (*left*), confocal reflection (*middle*), and NeuN staining in green (*right*)). Scale bar = 50 μm. (B) The images show confined neuronal populations on cluster-assembled dots with a diameter of 150 μm and dot-to-dot interdistances of 300 μm after 10 days in culture. Scale bar = 150 μm. Neurite/axon bridging between adjacent neuron populations is marked with *green arrows*. (C–E) Recording (visualized by Fluo-4 fluorescence as exemplary shown in (C)) and tracing (examples shown in (D)) of calcium spiking activity demonstrated correlated neuron activity inside the neuron population (upper graph in (E), measured for the cells marked with *blue circles* in (C)) confined in a dot after stimulation of one cell (marked with a *big gray/blue arrow* symbolizing the electrode pipet in (C)). Neuron populations connected by a neurite/axon bridge (see green bracket in (C)) showed also a higher level correlation (middle graph in (E), measured for the cells marked with *red circles* in (C)), compared to unconnected neuron populations (lower graph in (E), measured for cells marked with *yellow circles* in (C)). (F) Furthermore, temporally correlated Fluo-4 signals between neurons on the different dots were observed. Scale bar = 100 μm. *The figure was reproduced with permission from C. Schulte, J. Lamanna, A.S. Moro, C. Piazzoni, F. Borghi, M. Chighizola, et al., Neuronal cells confinement by micropatterned cluster-assembled dots with mechanotransductive nanotopography, ACS Biomater. Sci. Eng. (2018). https://doi.org/10.1021/acsbiomaterials.8b00916. Copyright 2018, American Chemical Society.*

promising potential as platforms to study neuronal communication and mechanotransductive influences on neuronal functioning. In addition, it is known that spatial confinement affects cytoskeletal tension, mechanotransductive responses, and cellular functioning in many cell types [57–59]. Therefore, also outside the mentioned neuroscience application, micropatterned substrates with nanotopographical surfaces fabricated by masked SCBD could be interesting as versatile tools for mechanobiological studies.

References

[1] A.S. Zeiger, B. Hinton, K.J. Van Vliet, Why the dish makes a difference: quantitative comparison of polystyrene culture surfaces, Acta Biomater. 9 (7) (2013) 7354–7361.

[2] M.J. Lerman, J. Lembong, S. Muramoto, G. Gillen, J.P. Fisher, The evolution of polystyrene as a cell culture material, Tissue Eng. Part B Rev. 24 (5) (2018) 359–372, https://doi.org/10.1089/ten.TEB.2018.0056.

[3] R.O. Hynes, The extracellular matrix: not just pretty fibrils, Science 326 (5957) (2009) 1216–1219.

[4] C. Frantz, K.M. Stewart, V.M. Weaver, The extracellular matrix at a glance, J. Cell Sci. 123 (Pt 24) (2010) 4195–4200.

[5] J.Z. Gasiorowski, C.J. Murphy, P.F. Nealey, Biophysical cues and cell behavior: the big impact of little things, Annu. Rev. Biomed. Eng. 15 (2013) 155–176.

[6] J.L. Young, A.W. Holle, J.P. Spatz, Nanoscale and mechanical properties of the physiological cell–ECM microenvironment, Exp. Cell Res. 343 (1) (2016) 3–6.

[7] P. Kollmannsberger, M.C. Bidan, C. Dunlop JW, P. Fratzl, The physics of tissue patterning and extracellular matrix organisation: how cells join forces, Soft Matter 7 (20) (2011) 9549–9560.

[8] P.M. Mendes, Cellular nanotechnology: making biological interfaces smarter, Chem. Soc. Rev. 42 (24) (2013) 9207–9218.

[9] M.J. Dalby, N. Gadegaard, R.O.C. Oreffo, Harnessing nanotopography and integrin-matrix interactions to influence stem cell fate, Nat. Mater. 13 (6) (2014) 558–569.

[10] W. Chen, Y. Shao, X. Li, G. Zhao, J. Fu, Nanotopographical surfaces for stem cell fate control: engineering mechanobiology from the bottom, Nano Today 9 (6) (2014) 759–784.

[11] S.W. Crowder, V. Leonardo, T. Whittaker, P. Papathanasiou, M.M. Stevens, Material cues as potent regulators of epigenetics and stem cell function, Cell Stem Cell 18 (1) (2016) 39–52.

[12] Y. Jin, J.S. Lee, J. Kim, S. Min, S. Wi, J.H. Yu, et al., Three-dimensional brain-like microenvironments facilitate the direct reprogramming of fibroblasts into therapeutic neurons, Nat. Biomed. Eng. 2 (7) (2018) 522–539.

[13] G.A. Abrams, S.L. Goodman, P.F. Nealey, M. Franco, C.J. Murphy, Nanoscale topography of the basement membrane underlying the corneal epithelium of the rhesus macaque, Cell Tissue Res. 299 (1) (2000) 39–46.

[14] A. Podestà, F. Borghi, M. Indrieri, S. Bovio, C. Piazzoni, P. Milani, Nanomanufacturing of titania interfaces with controlled structural and functional properties by supersonic cluster beam deposition, J. Appl. Phys. 118 (23) (2015) 234309.

[15] C. Schulte, S. Rodighiero, M.A. Cappelluti, L. Puricelli, E. Maffioli, F. Borghi, et al., Conversion of nanoscale topographical information of cluster-assembled zirconia surfaces into mechanotransductive events promotes neuronal differentiation, J. Nanobiotechnol. 14 (2016) 18.

[16] C. Schulte, A. Podestà, C. Lenardi, G. Tedeschi, P. Milani, Quantitative control of protein and cell interaction with nanostructured surfaces by cluster assembling, Acc. Chem. Res. 50 (2) (2017) 231–239.

[17] P. Piseri, A. Podestà, E. Barborini, P. Milani, Production and characterization of highly intense and collimated cluster beams by inertial focusing in supersonic expansions, Rev. Sci. Instrum. 72 (5) (2001) 2261–2267.

[18] K. Wegner, P. Piseri, H.V. Tafreshi, P. Milani, Cluster beam deposition: a tool for nanoscale science and technology, J. Phys. D: Appl. Phys. 39 (22) (2006) R439.

[19] C. Schulte, M. Ripamonti, E. Maffioli, M.A. Cappelluti, S. Nonnis, L. Puricelli, et al., Scale invariant disordered nanotopography promotes hippocampal neuron development and maturation with involvement of mechanotransductive pathways, Front. Cell. Neurosci. 10 (2016) 267.

[20] A. Galli, E. Maffioli, E. Sogne, S. Moretti, E.S. Di Cairano, A. Negri, et al., Cluster-assembled zirconia substrates promote long-term differentiation and functioning of human islets of Langerhans, Sci. Rep. 8 (1) (2018) 9979.

[21] N. Wang, J.D. Tytell, D.E. Ingber, Mechanotransduction at a distance: mechanically coupling the extracellular matrix with the nucleus, Nat. Rev. Mol. Cell Biol. 10 (1) (2009) 75–82.

[22] S.E. Winograd-Katz, R. Fässler, B. Geiger, K.R. Legate, The integrin adhesome: from genes and proteins to human disease, Nat. Rev. Mol. Cell Biol. 15 (4) (2014) 273–288.

[23] J.D. Humphries, N.R. Paul, M.J. Humphries, M.R. Morgan, Emerging properties of adhesion complexes: what are they and what do they do? Trends Cell Biol. 25 (7) (2015) 388–397.

[24] E.R. Horton, J.D. Humphries, J. James, M.C. Jones, J.A. Askari, M.J. Humphries, The integrin adhesome network at a glance, J. Cell Sci. 129 (22) (2016) 4159–4163.

[25] N.C. Gauthier, P. Roca-Cusachs, Mechanosensing at integrin-mediated cell–matrix adhesions: from molecular to integrated mechanisms, Curr. Opin. Cell Biol. 50 (2018) 20–26.

[26] J.Z. Kechagia, J. Ivaska, P. Roca-Cusachs, Integrins as biomechanical sensors of the microenvironment, Nat. Rev. Mol. Cell Biol. 20 (8) (2019) 457–473. https://doi.org/10.1038/s41580-019-0134-2.

[27] L.B. Case, M.A. Baird, G. Shtengel, S.L. Campbell, H.F. Hess, M.W. Davidson, et al., Molecular mechanism of vinculin activation and nanoscale spatial organization in focal adhesions, Nat. Cell Biol. 17 (7) (2015) 880–892.

[28] R. Changede, M. Sheetz, Integrin and cadherin clusters: a robust way to organize adhesions for cell mechanics, BioEssays. 39 (1) (2017) 1–12.

[29] R. Oria, T. Wiegand, J. Escribano, A. Elosegui-Artola, J.J. Uriarte, C. Moreno-Pulido, et al., Force loading explains spatial sensing of ligands by cells, Nature 552 (7684) (2017) 219.

[30] L.B. Case, C.M. Waterman, Integration of actin dynamics and cell adhesion by a three-dimensional, mechanosensitive molecular clutch, Nat. Cell Biol. 17 (8) (2015) 955–963.

[31] J. Liu, Y. Wang, W.I. Goh, H. Goh, M.A. Baird, S. Ruehland, et al., Talin determines the nanoscale architecture of focal adhesions, Proc. Natl. Acad. Sci. USA 112 (35) (2015) E4864–E4873.

[32] H.B. Schiller, C.C. Friedel, C. Boulegue, R. Fässler, Quantitative proteomics of the integrin adhesome show a myosin II-dependent recruitment of LIM domain proteins, EMBO Rep. 12 (3) (2011) 259–266.

[33] E.R. Horton, A. Byron, J.A. Askari, D.H.J. Ng, A. Millon-Frémillon, J. Robertson, et al., Definition of a consensus integrin adhesome and its dynamics during adhesion complex assembly and disassembly, Nat. Cell Biol. 17 (12) (2015) 1577–1587.

[34] P. Kanchanawong, G. Shtengel, A.M. Pasapera, E.B. Ramko, M.W. Davidson, H.F. Hess, et al., Nanoscale architecture of integrin-based cell adhesions, Nature 468 (7323) (2010) 580–584.

[35] T. Panciera, L. Azzolin, M. Cordenonsi, S. Piccolo, Mechanobiology of YAP and TAZ in physiology and disease, Nat. Rev. Mol. Cell Biol. 18 (12) (2017) 758–770.

[36] C. Uhler, G.V. Shivashankar, Chromosome intermingling: mechanical hotspots for genome regulation, Trends Cell Biol. 27 (11) (2017) 810–819.

[37] R. Carbone, I. Marangi, A. Zanardi, L. Giorgetti, E. Chierici, G. Berlanda, et al., Biocompatibility of cluster-assembled nanostructured TiO_2 with primary and cancer cells, Biomaterials 27 (17) (2006) 3221–3229.

[38] P.F. Manicone, P. Rossi Iommetti, L. Raffaelli, An overview of zirconia ceramics: basic properties and clinical applications, J. Dent. 35 (11) (2007) 819–826.

[39] Robert B. Heimann, Bioceramic Coatings for Medical Implants. https://www.bookdepository.com/Bioceramic-Coatings-for-Medical-Implants-Robert-B-Heimann.

[40] M. Belicchi, S. Erratico, P. Razini, M. Meregalli, A. Cattaneo, E. Jacchetti, et al., Ex vivo expansion of human circulating myogenic progenitors on cluster-assembled nanostructured TiO_2, Biomaterials 31 (20) (2010) 5385–5396.

[41] A.V. Singh, M. Ferri, M. Tamplenizza, F. Borghi, G. Divitini, C. Ducati, et al., Bottom-up engineering of the surface roughness of nanostructured cubic zirconia to control cell adhesion, Nanotechnology 23 (47) (2012) 475101.

[42] S. De Astis, I. Corradini, R. Morini, S. Rodighiero, R. Tomasoni, C. Lenardi, et al., Nanostructured TiO_2 surfaces promote polarized activation of microglia, but not astrocytes, toward a proinflammatory profile, Nanoscale 5 (22) (2013) 10963–10974.

[43] E.A. Cavalcanti-Adam, T. Volberg, A. Micoulet, H. Kessler, B. Geiger, J.P. Spatz, Cell spreading and focal adhesion dynamics are regulated by spacing of integrin ligands, Biophys. J. 92 (8) (2007) 2964–2974.

[44] E. Maffioli, C. Schulte, S. Nonnis, F. Grassi Scalvini, C. Piazzoni, C. Lenardi, et al., Proteomic dissection of nanotopography-sensitive mechanotransductive signalling hubs that foster neuronal differentiation in PC12 cells, Front. Cell. Neurosci. 11 (2017). https://doi.org/10.3389/fncel.2017.00417.

[45] N. Elad, T. Volberg, I. Patla, V. Hirschfeld-Warneken, C. Grashoff, J.P. Spatz, et al., The role of integrin-linked kinase in the molecular architecture of focal adhesions, J. Cell Sci. 126 (Pt 18) (2013) 4099–4107.

[46] C.I. Bargmann, E. Marder, From the connectome to brain function, Nat. Methods 10 (6) (2013) 483–490.

[47] T.N. Lerner, L. Ye, K. Deisseroth, Communication in neural circuits: tools, opportunities, and challenges, Cell 164 (6) (2016) 1136–1150.

[48] B.C. Wheeler, G.J. Brewer, Designing neural networks in culture, Proc. IEEE Inst. Electr. Electron Eng. 98 (3) (2010) 398–406.

[49] M. Shein-Idelson, E. Ben-Jacob, Y. Hanein, Engineered neuronal circuits: a new platform for studying the role of modular topology, Front. Neuroeng. 4 (2011).

[50] Y. Yi, J. Park, J. Lim, C.J. Lee, S.-H. Lee, Central nervous system and its disease models on a chip, Trends Biotechnol. 33 (12) (2015) 762–776.

[51] C. Tomba, C. Villard, Brain cells and neuronal networks: encounters with controlled microenvironments, Microelectron. Eng. 132 (Suppl. C) (2015) 176–191.

[52] M.J. Aebersold, H. Dermutz, C. Forró, S. Weydert, G. Thompson-Steckel, J. Vörös, et al., "Brains on a chip": towards engineered neural networks, TrAC Trends Anal. Chem. 78 (Suppl. C) (2016) 60–69.

[53] J.M. Stukel, R.K. Willits, Mechanotransduction of neural cells through cell–substrate interactions, Tissue Eng. B Rev. 22 (3) (2015) 173–182.

[54] Y. Kim, S.M. Meade, K. Chen, H. Feng, J. Rayyan, A. Hess-Dunning, et al., Nano-architectural approaches for improved intracortical interface technologies, Front. Neurosci. 12 (2018). https://doi.org/10.3389/fnins.2018.00456.

[55] C. Schulte, J. Lamanna, A.S. Moro, C. Piazzoni, F. Borghi, M. Chighizola, et al., Neuronal cells confinement by micropatterned cluster-assembled dots with mechanotransductive nanotopography, ACS Biomater. Sci. Eng. (2018). https://doi.org/10.1021/acsbiomaterials.8b00916.

[56] S. Weydert, S. Zürcher, S. Tanner, N. Zhang, R. Ritter, T. Peter, et al., Easy to apply polyoxazoline-based coating for precise and long-term control of neural patterns, Langmuir. (2017). https://doi.org/10.1021/acs.langmuir.7b01437.

[57] K.A. Kilian, B. Bugarija, B.T. Lahn, M. Mrksich, Geometric cues for directing the differentiation of mesenchymal stem cells, Proc. Natl. Acad. Sci. USA 107 (11) (2010) 4872–4877.

[58] M. Théry, Micropatterning as a tool to decipher cell morphogenesis and functions, J. Cell Sci. 123 (24) (2010) 4201–4213.

[59] E. D'Arcangelo, A.P. McGuigan, Micropatterning strategies to engineer controlled cell and tissue architecture in vitro, Biotechniques 58 (1) (2015) 13–23.

11

Antimicrobial nanostructured coating

Emanuele Cavaliere[1], Giulio Benetti[1,2], Francesco Banfi[3], Luca Gavioli[1,*]

[1]INTERDISCIPLINARY LABORATORIES FOR ADVANCED MATERIALS PHYSICS (I-LAMP) AND DIPARTIMENTO DI MATEMATICA E FISICA, UNIVERSITÀ CATTOLICA DEL SACRO CUORE, BRESCIA, ITALY; [2]DEPARTMENT OF PATHOLOGY AND DIAGNOSTICS—MEDICAL PHYSICS UNIT, UNIVERSITY HOSPITAL OF VERONA, P.LE STEFANI 1, VERONA, ITALY; [3]UNIVERSITÉ DE LYON, CNRS, UNIVERSITÉ CLAUDE BERNARD LYON 1, INSTITUT LUMIÉRE MATIÉRE, VILLEURBANNE, FRANCE

[*]CORRESPONDING AUTHOR

Chapter outline

11.1 Introduction

The application of nanostructured materials (NMs) [1–5] to different technological domains is favored by the increasing ability to tune the material physicochemical properties, such as electrical and thermal conductivity, chemical state, optical response and bactericidal functionalities [6–8]. The latter stems from the many different types of interaction of nanoparticles (NPs) with living organisms that have been mainly investigated for cells and bacteria in a solution containing the NPs [7–12].

The scientific interest in such research area has been further stimulated by the emergence of multidrug resistant pathogens, in particular in the hospital settings. These antibiotic-resistant strains are already causing a large socioeconomic impact, due to the

Cluster Beam Deposition of Functional Nanomaterials and Devices. https://doi.org/10.1016/B978-0-08-102515-4.00011-8

increase of infectious diseases directly generating from a) cross-contamination of nosocomial patients through infected **surfaces** [13,14], b) development of biofilm on the surfaces of medical implants such as catheters, dental or bone repair devices [15–18]. The transmission and proliferation of fungi, viruses, and bacteria, particularly by multidrug resistant bacteria, are causing infectious diseases for an estimated annual cost (direct and indirect) ranging from 6 to 60 billion US$ [19] in the United States only. Furthermore, very few new antibacterial compounds with good pharmacological profiles and low toxicity have been developed in the past decades [20] since, from an economic point of view, the pharmaceutical companies are more inclined in developing drugs for chronic conditions than for short-term treatments [21].

Taking into account the different mechanism of bactericidal action of NPs, the NMs would be less prone to promoting resistance in bacteria than antibiotics and could, in principle, provide a way to counteract the spreading of antibiotic-resistant strains [22]. Therefore a major research goal is finding appropriate materials to kill bacteria such as metal-based NPs [22–25], but is also to obtain functional nanostructured **surfaces** and/ or **thin films** [26,27] devised to limit the spread of bacteria through surfaces. The nanomaterial is required to have manyfold properites, such as (1) microbicidal activity against a wide number of multidrug resistant Gram negative (G−) and Gram positive (G+) pathogens; (2) tunable mechanical properties and adhesion, to tailor the NPs release and the film durability in different conditions; (3) cost-effective, environmental friendly production with high throughput. A major challenge recently brought up to attention is the accumulation of dead bacteria and organic debris that antibacterial surfaces might suffer [8]. The material accumulation on the surface might lead to screening of the active killing element (in particular for chemically functionalized surfaces), thereby reducing the bactericidal effect. Some works have hence been very recently devoted to investigate how to combine passive-defense antibacterial surfaces, which reduces the rate of bacteria adhesion, with bactericidal functionalities [28,29].

As it has been pointed out in a recent review [125], such attempts rely on the ability of combining materials at the nanoscale to tailor the properties of the coating and to devise the desired functionality. The latter, however, might be very different depending on the target of the application, i.e., on the type of surface where the antimicrobial agent has to be located, since morphological and chemical characteristics of a surface are determining the rate of microorganism adhesion [6,8,28,30]. In this framework, the knowledge of the major processes leading to the antibacterial effects of NPs becomes a fundamental asset. Although there are important points still to be clarified, a recent work [22] reviews in detail the NPs/ bacteria interaction mechanisms. These are NPs penetration of the bacterial cell membrane, induction of intracellular antibacterial effects (e.g., interactions with DNA and proteins), as well as disruption of the bacterial cell membrane or generation of reactive oxygen species (ROS) [22]. The killing mechanisms provided by the NPs depend on the size, composition, shape, and chemical state of the nanoobject, but also on the local morphology and availability at the interaction interface between the coating and the bacteria.

In this framework, many different strategies are explored, from deposition of polymeric films [6,8,27,31–33] with incorporated bactericidal agents [8,34–37], or metallic or oxide films [2,8,38,39], to metal decoration of nanostructures [40–42]. The strategies are based on wet synthesis (spin-coating [43], sol-gel [39,43,44]), photochemical [34], biological [45–47], biotechnological [11], and physical methods (laser ablation [48], magnetron sputtering [2,26,49–51], gas phase beams [52,53]). As a matter of fact, the literature available on nanostructured coatings obtained by physical methods is scarce.

This chapter is hence devoted to discuss some aspects of antibacterial coatings obtained with physical deposition techniques, shortly describing the kind of materials that can produce antibacterial effects, and then the issues related to the composition, mechanical properties, and antimicrobial properties of coatings, focusing on some recent works.

11.2 Role of material in the interaction with bacteria

The size, shape, and chemical composition of the NPs are fundamental in determining the pathway and effectiveness of bactericidal action. Hence, to devise and tailor antibacterial nanostructured coatings, it is important to shortly review some of the possible nanostructured coating interaction mechanism with bacteria that are summarized in Fig. 11.1. The scheme is listing the major interaction mechanism between NPs and the cell membrane and between NPs and the cytoplasmic components.

11.2.1 NPs interaction with cell barrier

NPs can release metal ions into the media surrounding the bacteria and bind with the negatively charged carboxyl and phosphate groups of the bacterial cell membrane [12,22]. The different components of the cell barrier of G+ and G− bacteria result in different adsorption pathways for NPs and therefore the antimicrobial action will depend on the specific bacterial cell composition. The lipoproteins and phospholipids composing the outer membrane of G- bacteria form a penetration barrier that allows the entrance of only macromolecules, resulting in a greater NP activity against G+ bacteria than against G- bacteria [12,52,54].

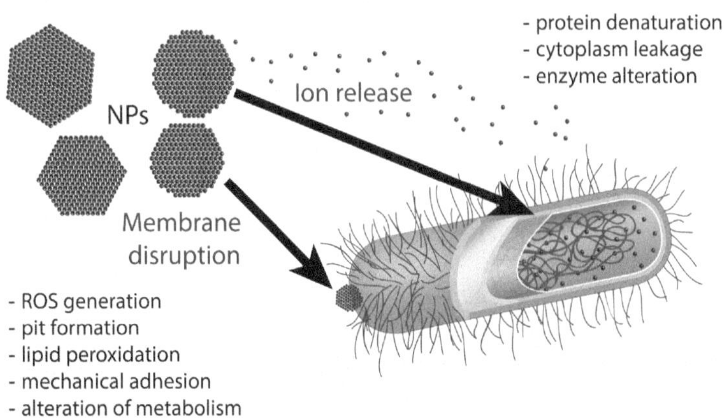

FIGURE 11.1 Schematic representation of a bacterium and of some NPs, with a list of the main NPs/bacterium interaction mechanisms. The NPs are not to scale with the bacterium size.

The lipopolysaccharide present in the G− wall can prevent the adhesion of NPs to the bacterial cell barrier and regulate the flow of ions in and out of the bacterial cell membrane [22,55], but ZnO NPs interact with the bacterial cell membrane generating strong lipid oxidative decomposition through the yield of H_2O_2, causing the leakage of intracellular content and bacterial death [55].

Nanodiamonds with different oxygen-containing surface groups are found to form covalent bonds with adjacent proteins and molecules on cell walls causing a disorder of the bacterial metabolism and, finally, cell death [56].

The photocatalytic action of TiO_2 NPs adhered to the surface of bacterial cells produces reactive ion species that damage the composition and structure of the cell membrane, causing leakage of cellular content and subsequent bacterial death [57]. Cytoplasmic leakage also results from the bacterial cell volume increase and oxidative decomposition of the bacterial membrane, induced by the photocatalytic activation of TiO_2 NP adhered to cell membrane [58].

Al_2O_3 NPs stimulate the formation of irregularly shaped pits and perforations in the cell membrane, which then interact with biological molecules in the cell. Attenuated total reflection analysis revealed that aluminum oxide NPs can interact with lipopolysaccharide through hydrogen binding and ligand exchange. Structural changes in the phospholipid lead to the loss of parent molecules, destruction of the cell membrane, and cytoplasmic leakage [59].

Ag NPs, through release of ions, have been shown to inactivate the respiratory activity of the bacterial cell membrane causing pit formation [60], while TiO_2 NPs were found to alter the potential of the cell membrane allowing diffusion of hydroxyl radicals into the bacteria [61]. CuO NPs are instead acting on the nitrogen metabolism and electron transfer on bacterial denitrification by causing a significant alteration of the expression of key proteins [62].

Since the outer membrane of the bacteria is charged, the interaction with oxidizing species has been recognized as an important antibacterial mechanism of NPs. The interaction of atoms or molecules with a strong positive redox potential with the bacterial wall is inducing an oxidative stress, leading to bacterial death [55,63–65].

Depending on the material, shape, and size of the NPS, different reactive ion species (ROS) can be produced by reducing oxygen molecules. CaO and MgO can produce superoxide radicals (O_2^-), ZnO might generate hydrogen peroxide (H_2O_2) and OH, while CUO NPs can also give rise to oxygen [22]. The generation of ROS can be obtained by photon absorption, employing the availability of surface states and defects to uptake the photogenerated electron hole pair to the surface of the NPs [66,67]. The absorbed photon generates an electron hole pair that, once it has reached the surface of the NP, can interact with molecules adsorbed on the NP surface [68]. For instance, in ZnO the holes on the surface can produce hydroxyl radical OH by interaction with OH^-, and the superoxide radical (O_2^-) by employing the photogenerated electron interaction with O_2.

Many experiments have shown that TiO_2-based NPs are more efficient in generating ROS species than ZnO-based materials [76]. However, the large band gap (3.2 eV), which does not allow the use of visible light to produce electron hole pairs [69–74] and the high recombination rate of the e-h pairs [65,72,75], resulted in a basically unexplored investigation of TiO_2 as an antibacterial coating.

11.2.2 Antibiofilm activity

Concerning the biofilm formation, the NP action has been investigated for Au [77,78], Ag [79], Mg [80], ZnO [24,78,81], CuO [78,82], CeO [78], Fe_3O_4 [83], YF [84], and TiO_2 [85], indicating that smaller NPs with high surface-to-volume ratio have a remarkable effect on biofilm destruction [86]. In human implantable devices (heart valves or dental implants) treated with antimicrobial coatings, it has been shown that a microporous TiO_2 coating loaded with NP of Ag, Ca, and Si is favoring osseointegration due to the roughness of about 300 nm [17] but also to prevent thrombosis and the occurrence of inflammation through the reduction of biofilm formation [16,17]. In dental implants, and particularly in root canal therapy, embedding nanodiamonds functionalized with amoxicillin, a broad-spectrum antibiotic led to improved mechanical properties of the filler and functionality of the drug [88]. In the case of partially implantable devices, such as catheters, the reduction of infection risk has been investigated by coating the catheter surface with Zn-doped CuO NPs to retard the growth of biofilms [18] and reduce the risk of bacterial infection and complications [87]. The catheters, coated on the outer and inner wall by sonochemical treatment, present a promising antibiofilm activity, biocompatibility, and absence of detectable irritation, although the antibacterial effect terminated after 7 days of use [18].

Atomic layer deposition was employed to coat maxillofacial prostheses with nano-TiO_2 thin films. The treated substrate presented a reduction of the incidence of tissue inflammation surrounding the prostheses and a rather good inhibition of a G+ and G− bacterial strains [85].

These results stimulated the use of various synthesis methods to provide antibacterial functionalities to objects employed in the biomedical field, such as wound dressings, and textiles [5]. The use of linens impregnated with CuO NPs was effective in reducing the occurrence of hospital-acquired microbial infections in healthcare facilities, hence bed sheets containing CuO NPs are now considered an interesting innovation in medical care [89].

The manifold interaction mechanisms between a coating and bacteria suggest that there is still wide potential for tailoring the coating's properties by modifying not only the individual NPs properties but also the composition, surface morphology, structure, and crystallinity of the coating as a whole. In the next section we will focus on recent developments of nanostructured coatings obtained by physical methods, highlighting the potential of such techniques to combine different elements with the aim of widening the bactericidal spectrum. Considering the potential applications described above, we will also consider some aspects of the mechanical properties of a coating to provide durability against use.

11.3 Nanostructured coatings

To limit the incidence of bacterial cross-contamination the type of surface where the antimicrobial agent has to be located should dictate the attack strategy, as the rate of microorganism adhesion is related to topological and chemical characteristics of a surface [6,8,28,30]. The most explored routes for the synthesis of nanostructured coatings with antimicrobial properties range from modification of surfaces through deposition of polymeric films [6,8,27,31–33] with incorporated bactericidal agents [8,34–37], or deposition of metallic or oxide films [2,8,38]. It has been found that Ag is improving the bactericidal activity of a TiO_2-based photocatalyst and may exploit the photooxidative mechanism of the oxide NPs to act against silver-resistant microorganisms [39]. The results led to an increased number of studies devoted to the Ag decoration of TiO_2 nanostructures [40–42], combining different synthesis routes even though physical methods to obtain nanostructured antibacterial coatings have not been in the major streamline of research.

Recently magnetron sputtering has been used to deposit coatings on textiles and catheters [90], implants [91], and food packaging [2,92]. The coatings may reach thicknesses up to few microns, and can be grown by combining different materials to obtain the target functionality [26,90,93]. However, magnetron sputtering presents some limitations to deposit NP with the same material flexibility to increase the surface-to-volume ratio and for ultrathin films, even more when core shell or Janus-like NPs might be needed to expose in a controlled way the active material. In particular, substrate temperature during deposition and resulting morphology of the film are two critical issues impacting on the applicability of the technique to a wide range of substrates.

Goderecci et al. [49] investigated the physical properties of 150 nm thick AgO/Ag_2O films on sapphire, measuring the Ag ion dissolution/elution rate by inductively coupled plasma mass spectroscopy (ICP-MS), and the coating bactericidal efficacy. The SEM and XRD data reported in Fig. 11.2 (A) and (B) indicate that the deposited films are polycrystalline with average grain size around 100 nm, and can be grown with a single AgO phase or mixed AgO/Ag_2O phases.

Since the main elution rates are measured in water while in-vitro experiments are executed in different media, ICP-MS was employed to test the environment effect on the Ag release rates. The results reported in Fig. 11.2 (C) show that Ag^+ ions elute at a faster rate in water as compared to the media used for bacterial growth experiments (LB) and for mammalian cell growth (DMDM). Interestingly, the overall trend of release rates is $H_2O > PBS > LB \approx DMEM$, indicating the need to measure elution rates in the media used for application of the coating. The results suggested the Ag ion release as the cause of the good bactericidal activity obtained by the AgO films. The continuous release was also inducing an increasing of the Ag concentration into the solution and hence toxicity effects against mammalian cells [49].

This work reveals the critical problem of dosing the bactericidal element to balance the antimicrobial action with the toxicity side effects, hence impacting on the coating applicability. In a closed system accumulation of Ag ions could lead to toxicity, a fact that

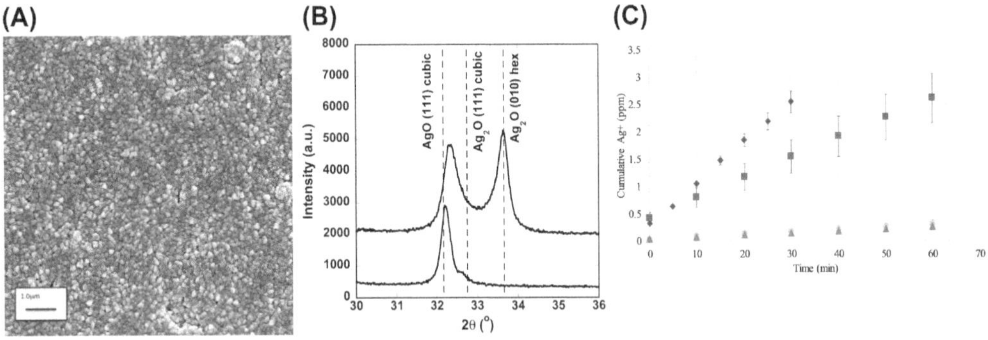

FIGURE 11.2 (A) SEM showing the typical surface microstructure of the silver oxide deposited at room temperature. The microstructure can be impacted by deposition pressure, deposition power, oxygen partial pressure, and coating thickness. (B) Plot of the X-ray diffraction intensity vs. 2θ showing single phase cubic AgO and mixed phase AgO and Ag_2O deposited at lower oxygen partial pressure. (C) Elution of Ag ions from AgxO coatings in water (diamonds), phosphate buffered saline (PBS) (squares), LB growth medium (circles) and DMEM growth medium (triangles) as measured by ICP-MS. *Adapted from S. Goderecci, E. Kaiser, M. Yanakas, Z. Norris, J. Scaturro, R. Oszust, C. Medina, F. Waechter, M. Heon, R. Krchnavek, L. Yu, S. Lofland, R. Demarest, G. Caputo, J. Hettinger, Silver oxide coatings with high silver-ion elution rates and characterization of bactericidal activity, Molecules 22 (9) (2017) 1487 under the Creative Commons Attribution License 4.0.*

might be avoided in an open system where released Ag ions can be transported away. Therefore, a permanent implant that remains surrounded by living cells might not be the target of such type of coating, while a wastewater purification system with a continuous liquid flow could be one.

The reported good adhesion of the Ag coating on hard and flexible substrates is introducing the issue of the mechanical response. The need to obtain killing activity against both G+ and G− strains has raised the issue of the coating mechanical stability related to the mixing into the coating of different active antimicrobial elements [26,38,94–96]. Musil et al. [26,96] suggested hundreds of nanometer thick films where the relative content of the most efficient antibacterial metals, Ag and Cu, needs to be in the range of 10%–30%. Such concentration resulted in a strong reduction of the film hardness and in poor mechanical stability, a major drawback for many practical applications requiring a long lifetime and a good resistance to wear [38,94,95]. Moreover, when such coatings are deposited on flexible substrates, they easily crack and/or delaminate due to the residual stress resulted from the growth mode.

The influence of Cu content on the mechanical and bactericidal properties of thin films prepared by reactive magnetron sputtering has been studied for Cr–Cu–O [26,96], Al–Cu–N [26,93], and Zr–Cu–N [97] compounds. The measured mechanical characteristics were the film hardness H, defined as its resistance to local plastic deformation [98], the elastic recovery We, defined as the fraction of a given deformation of a solid which behaves elastically [99], the Young's modulus E, and the effective Young's modulus $E^* = E(1-\nu^2)$, where ν = Poisson ratio obtained by mechanical indentation. The results obtained on the Zr–Cu–N coatings are summarized in Fig. 11.3A and B, reporting the mechanical properties as a function of Cu content in the Zr–Cu–N coating.

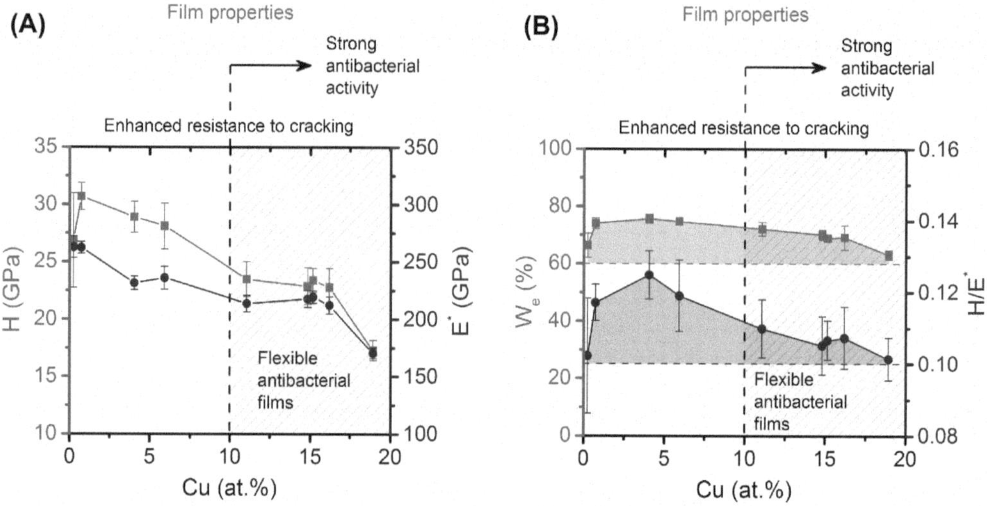

FIGURE 11.3 (A) Hardness H (*gray squares*), effective Young's modulus E* (*black circles*); and (B) elastic recovery We (*gray squares*) and H/E* ratio (*black circles*) of Zr–Cu–N coatings sputtered on Si (100) substrates as a function of Cu content. *Reprinted from J. Musil, M. Zítek, K. Fajfrlík, R. Čerstvý, Flexible antibacterial Zr–Cu–N thin films resistant to cracking, J. Vac. Sci. Technol. A, 34 (2) (2016) 021508, Copyright 2015, with permission from AIP Publishing LLC.*

The increasing Cu content determines the decreasing trend of H and E* of the film, while the elastic recovery We is ≥ 60%. The Zr–Cu–N coatings exhibit a ratio H/E* ≥ 0.1, a value indicating an enhanced resistance to cracking for all Cu content ranging from 0 to 19% Cu [97]. Moreover, the H values ranging from ~25 to ~17 GPa prevent the coating from being removed from the surface of a substrate by fretting (wear). The light gray shaded regions in the plots of Fig. 11.3 mark the Cu concentration for which the coatings exhibit a 100% killing efficacy of an *E. coli* bacteria, and set the lower limit for the Cu concentration. [97]. A major limitation of the reported results is the substrate deposition temperature of 450 °C, hampering the possibility to cover flexible substrates that would not withstand that temperature. Moreover, the use of precious metals like Ag to widen the bactericidal spectrum could result in an increase of the production costs, considering the total amount of the bactericidal metal needed to obtain a micron thick material over cm^2 areas.

Supersonic Cluster Beam Deposition (SCBD) provides a timely alternative to tackle the different problems discussed above [52,53,69,100–103]. The production of a pure, solvent-free supersonic beam of NPs is the peculiarity of this method [6,74,102,103]. The injection of an inert gas into the ablation chamber is synchronized to a high voltage discharge, thus triggering a plasma that ablates a target rod of the desired material. The condensed NPs are extracted through a nozzle thus forming a supersonic collimated beam directed on the substrate surface. The controlled pressure conditions of the beam synthesis and of the substrate allow the deposition of nanostructured film on virtually any substrate, either rigid or flexible [104], without the presence of solvents or other contaminants. The average kinetic energy of the NPs (about 0.2 eV/atom) [3,103] allows a soft landing of the NPs that maintains their original structure; hence, NPs are assembling on the substrate forming a film of desired thickness.

SCBD has been used to synthesize antibacterial coatings only in recent years [52,53]. The early work of Cavaliere et al. [52] described the physical and antibacterial properties of an Ag film deposited by SCBD on soda lime glass substrates at room temperature. The average NPs size was 7.8 ± 0.6 nm, and the correspondent film thickness was measured to be 8 nm, i.e., a single layer, with an NP density of $1.15 \pm 0.04 \times 10^{11}$ NPs/cm^2 [52]. The stability of the film with respect to time was recently reported by Benetti et al. [125], as described in detail in Fig. 11.4. The X-ray photoemission spectra of the O1s core level are plotted as a function of time after the film deposition (Fig. 11.4A, top panel). Very little changes are observed in the oxygen lineshape (Fig. 11.4A, bottom panel) suggesting that the film is stable against aging up to 15 days.

A least square fitting procedure decomposes the O1s peak in a main component at 531.6 eV binding energy, related to the Si-O bond in the substrate [105], and to the AgO feature at 530.2 binding energy [105,106] present in the film. The relative intensities of the two components as a function of time after deposition are reported in Fig. 11.4C, presenting a maximum change of 10% with respect to the initial value. This behavior is indicating major stability of the film against variation of the oxidation state.

Ag NP films were found to exert a broad-spectrum bactericidal activity, which was demonstrated both with reference strains and with a collection of clinical strains that exhibited extensively drugresistant phenotypes and/or belonged to high-risk hyperepidemic

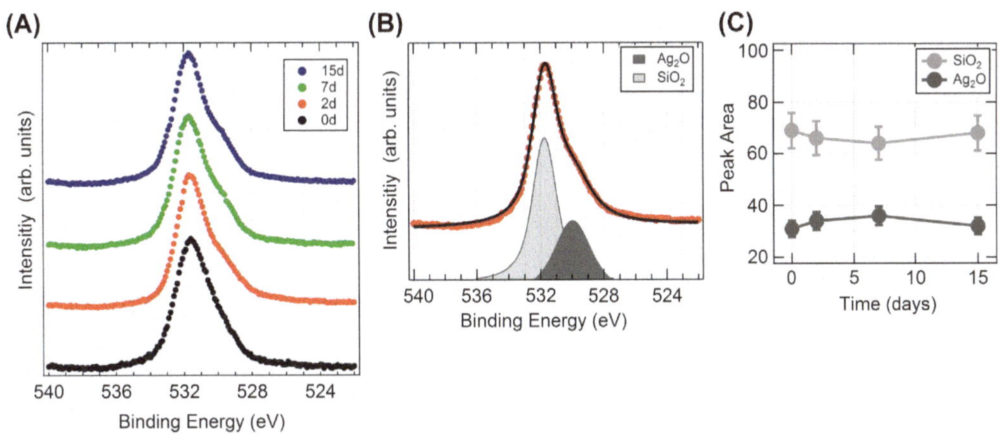

FIGURE 11.4 (A) Normalized O1s core level spectra obtained from the as-deposited Ag NPs film (curve 0d) and from the same film 2 days (curve 2d), 7 days (curve 7d), and 15 days (curve 15d) after the deposition. (B) O1s core level obtained from the Ag NPs film 2 days after deposition, with the peaks resulting from the least square fitting procedure. The AgO related peak (*dark gray*) is at 530.2 eV binding energy while the SiO_2 related peak (*light gray*) is at 531.6 eV binding energy. (C) intensity dependence of the relative area of the two peaks as a function of time. *Adapted from G. Benetti, E. Cavaliere, F. Banfi, L. Gavioli, Antimicrobial nanostructured coatings: a gas phase deposition and magnetron sputtering perspective, Materials 13 (2020) 784 under the Creative Commons Attribution License 4.0.*

clones, in particular clinical strains producing two of the most worrisome resistance mechanisms recently emerged in enterobacteria and capable of pandemic dissemination [107].

The interesting results obtained in the work of Cavaliere et al. point out that the film, being composed of Ag and AgO, has a limited efficacy on the film on some G+ bacteria. Moreover, the ultrathin coating morphology and thickness prevent the use of standard indentation techniques to measure the mechanical properties, even though a recent review reported on the attempts to unveil the hardness, elastic modulus, adhesion, and friction of single NPs [108]. Although the modeling of atomic force microscopy (AFM) interaction with a single nanoobject is well established, the same technique might not be applicable when the grown film is granular or porous. The NPs assembly into a coating might undergo plastic modifications under the load of the AFM tip, hampering a reliable extrapolation of the elastic costant from force distance curves.

A feasible solution to the problem has been recently proposed for a nanoporous granular system employing picosecond photoacoustic [109,110]. The technique exploits fs laser pulses to excite the film mechanical automodes, being sampled by detecting the variation in the reflectivity or transmissivity of a synchronized probe pulse. Considering an infinite two-dimensional film (the laser spot size is around 100 microns), the only automodes that are sampled are those perpendicular to the film surface. The data provides the period and the decay time and a simple model allows to determine the film elastic constant and the film adhesion to the substrate [109]. On pure Ag NPs films the density, longitudinal sound velocity, and elastic stiffness result to be 80% and the latter 50% of the respective values for bulk polycrystalline Ag, much lower than typical values obtained for coatings deposited by magnetron sputtering.

In recent years, the SCBD technique has been employed to grow NPs films of different materials (Ti [3,74], C [111], Pd [112]) and to dope TiO_2 NPs with Cr and N at high atomic concentrations [53,69,70,100,113]. The flexibility of the methods has been hence exploited to directly synthesize and deposit coatings with multielemental NPs. Recently Benetti et al. demonstrated the possibility to synthesize bi-element NPs of Ag and Ti with tunable elemental concentrations [53,101]. Two different rods has been used, the first (AgTi8020) with a nominal weight content of 80% Ag and 20% Ti, respectively, the second (AgTi5050) with a nominal weight content of 50% Ag and 50% Ti, respectively [53,101]. The structural characterization of the individual NPs obtained by high resolution scanning transmission electron microscope (HR-STEM) is reported in Fig. 11.5.

The observed lighter gray zone typically surrounding all the intense white spots of the image (Fig. 11.5A and D) is explained by the corresponding elemental maps (Fig. 11.5B and C): the contrast difference is because Ag (evidenced in red) and Ti (evidenced in green) are phase-separated into the NPs. The structural and chemical analysis of the NPs [53] indicates that such bi-element NPs are composed of Ag nanocrystals intermixed and embedded into amorphous TiO_2. This NPs structure is remarkably different from the Ag decoration of TiO_2 nanostructures [114, 115] obtained also by sol gel [39], solvothermal [40], hydrothermal [41], and hydrolyzed solution [116] methods. Moreover, the work of Benetti *et al.* clearly indicates a correspondence between the pristine rod and the

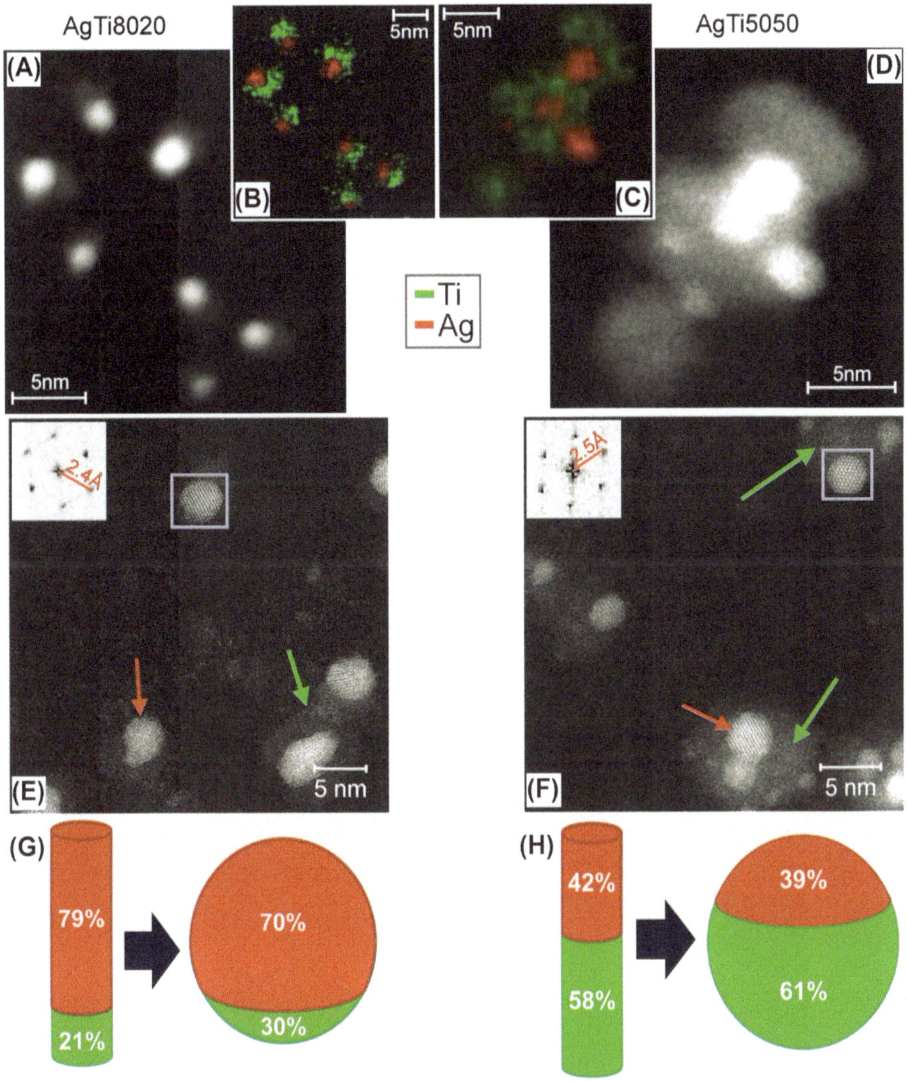

FIGURE 11.5 (A), (D) TEM images of AgTi8020 and AgTi5050 scattered NPs, respectively, with the relative elemental map plotted in panels (B) and (C), respectively. The data show that Ag and Ti are phase-separated into the NPs. (E), (F) HR-STEM images of the NPs, with the inset showing the FFT analysis of Ag crystalline structure of the zone in the *purple rectangle*. *Red arrows* are indicating small Ag NPs, and green arrows are pointing to the Ti part of the NPs. The data indicate that Ag is crystalline and Ti is amorphous. (G), (H) Schematic representation of the elemental weight in the initial rod and in the NPS, showing the good correspondence of the material concentration. *Adapted from G. Benetti, E. Cavaliere, A. Canteri, G. Landini, G.M. Rossolini, L. Pallecchi, M. Chiodi, M.J. Van Bael, N. Winckelmans, S. Bals, L. Gavioli, Direct synthesis of antimicrobial coatings based on tailored bi-elemental nanoparticles, Apl. Mater. 5 (3) (2017) 036105 under the Creative Commons Attribution License 4.0.*

synthesized bi-NPs stoichiometry [53,101], proving that SCBD is a powerful tool to tailor the NPs' chemical composition. Finally, the nanostructured Ag-Ti coating shows good adhesion to the substrate and exhibits an exceptional bactericidal effect against major Gram-negative nosocomial pathogens, remarkably similar to a pure Ag NPs coating [52], but with an 85% lower Ag mass content.

The remarkable activities against G- bacteria of Ag [52,117] and against G+ bacteria and yeasts of Cu suggest to combine these two elements to widen the microbicidal spectrum of a coating [118,119]. In order to keep the coating stability against wear, we opted to use as adhesion promoter the Mg, an element with good microbicidal activity [120] and an oxygen chemical affinity higher than that of Ti [121]. This choice has been also supported by the enhancement of the Mg microbicidal activity obtained with the presence of small quantities of Ag [122].

The rod employed in the experiment is composed of Mg, Ag, and Cu at nominal concentrations 20/50/30% in weight, respectively. The characterization of the multielement NPs synthesized by SCBD is presented in Fig. 11.6 [123]. In the TEM image of Fig. 11.6A, the heavier Ag and Cu appear as bright zones whereas Mg is visible as less bright areas of the nanocomposite system. The energy dispersive X-ray spectroscopy (EDX) map (Fig. 11.6B) shows that the NPs assume a cluster-in-cluster form, with a matrix of Mg (blue in Fig. 11.6B) partially embedding Ag (*green*) and Cu (*red*) clusters.

XPs data suggest that Ag and Cu are metallic, while Mg is oxidized to MgO. The relative weight content of the NPs as deduced from Rutherford backscattering spectrometry is Mg/Ag/Cu = 24/47/29. These values are in very good correspondence with the nominal weight concentration of the starting rod (20/50/30), indicating that the SCBD allows an almost one-to-one transfer of the elements to the nanostructured film, and hence the ability to finely tune the elemental composition of multielement nanoparticles. As for the Cu and Ag distribution in the brighter spots, one has to take into account the immiscible nature of Ag and Cu [124]; hence, we could expect the segregation of these two elements instead of an alloyed core.

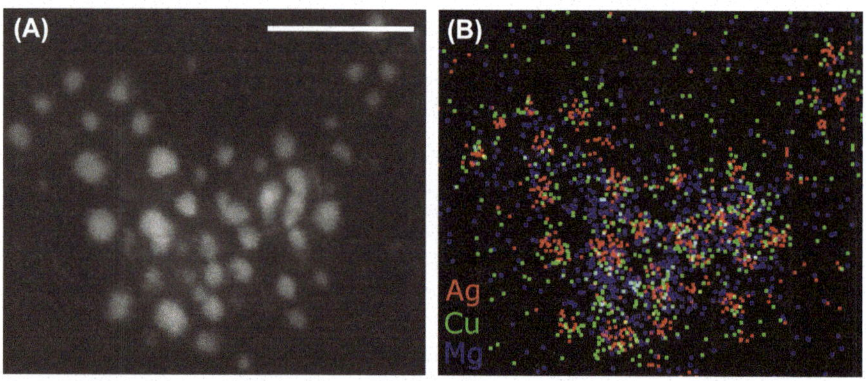

FIGURE 11.6 STEM and corresponding EDX elemental maps (Cu *red*; Ag *green*; Mg *blue*) for the Mg/Ag/Cu NPs. Scale bar 20 nm.

The antibacterial activity of the tri-element NPs has been tested by depositing a single layer (average thickness 6 nm) on soda-lime glasses. The activity has been tested against drug resistant G- and G+ reference strains (*Escherichia coli* ATCC 25,922 and *Staphylococcus aureus* ATCC 6538, respectively), using the procedure already employed for AgTi NPs [53]. When compared with the number of bacteria on bare substrates used as controls, the antimicrobial activity resulted higher than 99.999% in 3h for *E. coli* and higher than 99.99% in 24 h for *S. aureus.* In the case of *S. aureus*, the observed sterilization of the samples can be related with the presence of Cu and Ag, since it is was not observed in the case of pure Mg NPs.

It should be noted that for the coatings discussed in the present chapter there are no data on the mechanism leading to bacteria death, except for the work of Goderecci et al. [49], where the concentration of Ag ions released by the coating has been measured and suggested as the cause of the bacteria death. It will hence be desirable to understand how these innovative coatings are leading to the microbicidal effect. This knowledge, requiring an interdisciplinary effort, will allow better devising of the material component of the nanostructured coatings.

11.4 Summary and perspectives

Nanostructured coatings synthesized by physical methods have emerged as a potential way to limit the spread of bacteria in different environments, in particular, for surfaces related to medical devices. The solvent-free synthesis processes provide extreme control over the film purity, thickness, and relative element concentration, and hence the ability to tailor the coating functionalities. Moreover, magnetron sputtering has already reached the industrial level while SCBD is easily scalable. SCBD has the advantage of depositing already formed NPs in a soft landing regime, with an extreme flexibility in the choice of materials to tailor the coating for a specific application. The antimicrobial activity of the coatings is likely provided by the presence of materials such as Ag, Cu, and Ti. The coatings can in some cases be fabricated with good hardness and resistance to wear, and the adhesion to the substrate can be increased by using promoters such as TiO_2 or MgO.

Few considerations can be however made to address the future perspectives and challenges in the field. The first issue for antibacterial surfaces is the long-term stability and capability of a surface to maintain its property over a time scale of days, week, months, or even years. The time range of functioning of the antibacterial surface is of course strongly dependent on the application. On medical implants the antibacterial activity duration could be restricted to the period required by the body to accept the implant, while on medical devices used in surgery the antibacterial effect could be restricted to few hours. Hence, despite the successful bactericidal effect over short periods, the efficacy over time longer than few hours and after usage is still an open issue. Such open questions are also strictly related to the understanding of the main mechanism of action of the coating against bacteria, a task that is way far from being accomplished due to the almost total absence of studies addressing the bactericidal action of the coatings. A similar issue stems from the lack of knowledge of the coatings antibiofouling activity, since the wealth of research is performed mostly on paints. Both issues should be properly addressed in the future research to tailor the synthesis of nanostructured coatings.

The third issue is related to the use of standard strains to test the bactericidal activity as it is the case for most of the available literature on antibacterial coatings. This conduct is not taking into account the emergence of antibiotic resistant strains, against which the novel coatings should be tested.

As a final remark, we note that most of the work to date are proofs of concept of a coating antibacterial activity, tested on some model surfaces like silicon or glass. Since both magnetron sputtering and SCBD are already simple, cost-effective, environmentally friendly, and reproducible fabrication methods, it would be worth investigating the coating effect on surfaces of customer-oriented objects, for which the coating industrial scale-up should be envisaged. Tackling these important challenges will require collaborative efforts from researchers from the fields of surface chemistry, materials science, biomedical engineering, and biotechnology, and would also provide plenty of opportunities for innovation beyond antibacterial surfaces.

References

[1] J. Jeevanandam, A. Barhoum, Y.S. Chan, A. Dufresne, M.K. Danquah, Review on nanoparticles and nanostructured materials: history, sources, toxicity and regulations, Beilstein J. Nanotechnol. 9 (2018) 1050–1074.

[2] S. Calderon Velasco, A. Cavaleiro, S. Carvalho, Functional properties of ceramic-Ag nanocomposite coatings produced by magnetron sputtering, Prog. Mater. Sci. 84 (2016) 158–191.

[3] T. Mazza, E. Barborini, I.N. Kholmanov, P. Piseri, G. Bongiorno, S. Vinati, P. Milani, C. Ducati, D. Cattaneo, A. Li Bassi, C.E. Bottani, A.M. Taurino, P. Siciliano, Libraries of cluster-assembled titania films for chemical sensing, Appl. Phys. Lett. 87 (10) (2005) 103108.

[4] F. Baletto, R. Ferrando, Structural properties of nanoclusters: energetic, thermodynamic, and kinetic effects, Rev. Mod. Phys. 77 (1) (2005) 371–423.

[5] M. Grigore, E. Biscu, A. Holban, M. Gestal, A. Grumezescu, Methods of synthesis, properties and biomedical applications of CuO nanoparticles, Pharmaceuticals 9 (4) (2016) 75.

[6] A.V. Singh, V. Vyas, T.S. Salve, D. Cortelli, D. Dellasega, A. Podestà, P. Milani, W.N. Gade, Biofilm formation on nanostructured titanium oxide surfaces and a micro/nanofabrication-based preventive strategy using colloidal lithography, Biofabrication 4 (2) (2012) 025001.

[7] I. Banerjee, R.C. Pangule, R.S. Kane, Antifouling coatings: recent developments in the design of surfaces that prevent fouling by proteins, bacteria, and marine organisms, Adv. Mater. 23 (6) (2011) 690–718.

[8] T. Wei, Z. Tang, Q. Yu, H. Chen, Smart antibacterial surfaces with switchable bacteria-killing and bacteria-releasing capabilities, ACS Appl. Mater. Interfaces 9 (43) (2017) 37511–37523.

[9] B. Khodashenas, The influential factors on antibacterial behaviour of copper and silver nanoparticles, Indian Chem. Eng. 58 (3) (2016) 224–239.

[10] M. Ul-Islam, A. Shehzad, S. Khan, W.A. Khattak, M.W. Ullah, J.K. Park, Antimicrobial and biocompatible properties of nanomaterials, J. Nanosci. Nanotechnol. 14 (1) (2014) 780–791.

[11] K.S. Siddiqi, A. Husen, R.A.K. Rao, A review on biosynthesis of silver nanoparticles and their biocidal properties, J. Nanobiotechnol. 16 (1) (2018).

[12] Y.N. Slavin, J. Asnis, U.O. Häfeli, H. Bach, Metal nanoparticles: understanding the mechanisms behind antibacterial activity, J. Nanobiotechnol. 15 (1) (2017).

[13] E.I. Kritsotakis, F. Kontopidou, E. Astrinaki, M. Roumbelaki, E. Ioannidou, A. Gikas, Prevalence, incidence burden, and clinical impact of healthcare-associated infections and antimicrobial resistance: a national prevalent cohort study in acute care hospitals in Greece, Infect. Drug Resist. 10 (2017) 317–328.

[14] I. Wille, A. Mayr, P. Kreidl, C. Brühwasser, G. Hinterberger, A. Fritz, W. Posch, S. Fuchs, A. Obwegeser, D. Orth-Höller, C. Lass-Flörl, Cross-sectional point prevalence survey to study the environmental contamination of nosocomial pathogens in intensive care units under real-life conditions, J. Hosp. Infect. 98 (1) (2018) 90–95.

[15] K.A. Floyd, A.R. Eberly, M. Hadjifrangiskou, 3 - adhesion of bacteria to surfaces and biofilm formation on medical devices, in: Y. Deng, W. Lv (Eds.), Biofilms and Implantable Medical Devices, Woodhead Publishing, 2017, pp. 47–95.

[16] W. Xia, K. Grandfield, A. Hoess, A. Ballo, Y. Cai, H. Engqvist, Mesoporous titanium dioxide coating for metallic implants, J. Biomed. Mater. Res. B Appl. Biomater. 100B (1) (2012) 82–93.

[17] C. Della Valle, L. Visai, M. Santin, A. Cigada, G. Candiani, D. Pezzoli, C.R. Arciola, M. Imbriani, R. Chiesa, A novel antibacterial modification treatment of titanium capable to improve osseointegration, Int. J. Artif. Organs 35 (10) (2012) 864–875.

[18] Y. Shalom, I. Perelshtein, N. Perkas, A. Gedanken, E. Banin, Catheters coated with Zn-doped CuO nanoparticles delay the onset of catheter-associated urinary tract infections, Nano Res. 10 (2) (2017) 520–533.

[19] C.I. Michaelidis, M.J. Fine, C.J. Lin, J.A. Linder, M.P. Nowalk, R.K. Shields, R.K. Zimmerman, K.J. Smith, The hidden societal cost of antibiotic resistance per antibiotic prescribed in the United States: an exploratory analysis, BMC Infect. Dis. 16 (1) (2016) 655–662.

[20] M. Bassetti, E. Righi, Development of novel antibacterial drugs to combat multiple resistant organisms, Langenbeck's Arch. Surg. 400 (2) (2015) 153–165.

[21] J. Ziemska, A. Rajnisz, J. Solecka, New perspectives on antibacterial drug research, Open Life Sci. 8 (10) (2013).

[22] L. Wang, C. Hu, L. Shao, The antimicrobial activity of nanoparticles: present situation and prospects for the future, Int. J. Nanomed. 12 (2017) 1227–1249.

[23] A. Azam, A.S. Ahmed, M. Oves, M.S. Khan, S.S. Habib, A. Memic, Antimicrobial activity of metal oxide nanoparticles against Gram-positive and Gram-negative bacteria: a comparative study, Int. J. Nanomed. 7 (2012) 6003.

[24] M.J. Hajipour, K.M. Fromm, A. Akbar Ashkarran, D. Jimenez de Aberasturi, I.R. de Larramendi, T. Rojo, V. Serpooshan, W.J. Parak, M. Mahmoudi, Antibacterial properties of nanoparticles, Trends Biotechnol. 30 (10) (2012) 499–511.

[25] T.J. Webster, I. Seil, Antimicrobial applications of nanotechnology: methods and literature, Int. J. Nanomed. (2012) 2767.

[26] J. Musil, Flexible antibacterial coatings, Molecules 22 (5) (2017) 813.

[27] K. Yliniemi, M. Vahvaselkä, Y.V. Ingelgem, K. Baert, B.P. Wilson, H. Terryn, K. Kontturi, The formation and characterisation of ultra-thin films containing Ag nanoparticles, J. Mater. Chem. 18 (2) (2008) 199–206.

[28] Q. Gao, M. Yu, Y. Su, M. Xie, X. Zhao, P. Li, P.X. Ma, Rationally designed dual functional block copolymers for bottlebrush-like coatings: in vitro and in vivo antimicrobial, antibiofilm, and antifouling properties, Acta Biomater. 51 (2017) 112–124.

[29] G. Ye, J. Lee, F. Perreault, M. Elimelech, Controlled architecture of dual-functional block copolymer brushes on thin-film composite membranes for integrated "defending" and "attacking" strategies against biofouling, ACS Appl. Mater. Interfaces 7 (41) (2015) 23069–23079.

[30] M.L.W. Knetsch, L.H. Koole, New strategies in the development of antimicrobial coatings: the example of increasing usage of silver and silver nanoparticles, Polymers 3 (1) (2011) 340–366.

[31] A.V. Singh, Biotechnological applications of supersonic cluster beam-deposited nanostructured thin films: bottom-up engineering to optimize cell-protein-surface interactions, J. Biomed. Mater. Res. A 101 (10) (2013) 2994–3008.

[32] L. Duta, C. Ristoscu, G.E. Stan, M.A. Husanu, C. Besleaga, M.C. Chifiriuc, V. Lazar, C. Bleotu, F. Miculescu, N. Mihailescu, E. Axente, M. Badiceanu, D. Bociaga, I.N. Mihailescu, New bio-active, antimicrobial and adherent coatings of nanostructured carbon double-reinforced with silver and silicon by Matrix-Assisted Pulsed Laser Evaporation for medical applications, Appl. Surf. Sci. 441 (2018) 871–883.

[33] B.A. Camacho-Flores, O. Martínez-Álvarez, M.C. Arenas-Arrocena, R. Garcia-Contreras, L. Argueta-Figueroa, J. de la Fuente-Hernández, L.S. Acosta-Torres, Copper: synthesis techniques in nanoscale and powerful application as an antimicrobial agent, J. Nanomater. 2015 (2015) 1–10.

[34] F. Paladini, M. Pollini, A. Sannino, L. Ambrosio, Metal-based antibacterial substrates for biomedical applications, Biomacromolecules 16 (7) (2015) 1873–1885.

[35] M. Hoseinnejad, S.M. Jafari, I. Katouzian, Inorganic and metal nanoparticles and their antimicrobial activity in food packaging applications, Crit. Rev. Microbiol. 44 (2) (2018) 161–181.

[36] K. Vasilev, S.S. Griesser, H.J. Griesser, Antibacterial surfaces and coatings produced by plasma techniques, Plasma Process. Polym. 8 (11) (2011) 1010–1023.

[37] K. Vasilev, J. Cook, H.J. Griesser, Antibacterial surfaces for biomedical devices, Expert Rev. Med. Devices 6 (5) (2009) 553–567.

[38] L. Bai, R. Hang, A. Gao, X. Zhang, X. Huang, Y. Wang, B. Tang, L. Zhao, P.K. Chu, Nanostructured titanium–silver coatings with good antibacterial activity and cytocompatibility fabricated by one-step magnetron sputtering, Appl. Surf. Sci. 355 (2015) 32–44.

[39] E. Albert, P.A. Albouy, A. Ayral, P. Basa, G. Csík, N. Nagy, S. Roualdès, V. Rouessac, G. Sáfrán, á. Suhajda, Z. Zolnai, Z. Hórvölgyi, Antibacterial properties of Ag–TiO_2 composite sol–gel coatings, RSC Adv. 5 (73) (2015) 59070–59081.

[40] Y. Chen, Y. Deng, Y. Pu, B. Tang, Y. Su, J. Tang, One pot preparation of silver nanoparticles decorated TiO_2 mesoporous microspheres with enhanced antibacterial activity, Mater. Sci. Eng. C 65 (2016) 27–32.

[41] F. Liu, H. Liu, X. Li, H. Zhao, D. Zhu, Y. Zheng, C. Li, Nano-TiO_2@Ag/PVC film with enhanced antibacterial activities and photocatalytic properties, Appl. Surf. Sci. 258 (10) (2012) 4667–4671.

[42] H. Cheng, J. Ye, Y. Sun, W. Yuan, J. Tian, R.F. Bogale, P. Tian, G. Ning, Template-induced synthesis and superior antibacterial activity of hierarchical Ag/TiO_2 composites, RSC Adv. 5 (98) (2015) 80668–80676.

[43] S. Stankic, S. Suman, F. Haque, J. Vidic, Pure and multi metal oxide nanoparticles: synthesis, antibacterial and cytotoxic properties, J. Nanobiotechnol. 14 (1) (2016).

[44] D.P. Macwan, P.N. Dave, S. Chaturvedi, A review on nano-TiO_2 sol–gel type syntheses and its applications, J. Mater. Sci. 46 (11) (2011) 3669–3686.

[45] T.M. Abdelghany, A.M.H. Al-Rajhi, M.A. Al Abboud, M.M. Alawlaqi, A. Ganash Magdah, E.A.M. Helmy, A.S. Mabrouk, Recent advances in green synthesis of silver nanoparticles and their applications: about future directions. A review, BioNanoScience 8 (1) (2018) 5–16.

[46] P. Dheeban Shankar, S. Shobana, I. Karuppusamy, A. Pugazhendhi, V.S. Ramkumar, S. Arvindnarayan, G. Kumar, A review on the biosynthesis of metallic nanoparticles (gold and silver)

using bio-components of microalgae: formation mechanism and applications, Enzym. Microb. Technol. 95 (2016) 28–44.

[47] M. Manimaran, K. Kannabiran, Actinomycetes-mediated biogenic synthesis of metal and metal oxide nanoparticles: progress and challenges, Lett. Appl. Microbiol. 64 (6) (2017) 401–408.

[48] J.T.T. Angelina, S. Ganesan, T.M.R. Panicker, R. Narayani, M. Paul Korath, K. Jagadeesan, Pulsed laser deposition of silver nanoparticles on prosthetic heart valve material to prevent bacterial infection, Mater. Technol. 32 (3) (2017) 148–155.

[49] S. Goderecci, E. Kaiser, M. Yanakas, Z. Norris, J. Scaturro, R. Oszust, C. Medina, F. Waechter, M. Heon, R. Krchnavek, L. Yu, S. Lofland, R. Demarest, G. Caputo, J. Hettinger, Silver oxide coatings with high silver-ion elution rates and characterization of bactericidal activity, Molecules 22 (9) (2017) 1487.

[50] J. Musil, Flexible hard nanocomposite coatings, RSC Adv. 5 (74) (2015) 60482–60495.

[51] J. Wang, H. Zhou, G. Guo, T. Cheng, X. Peng, X. Mao, J. Li, X. Zhang, A functionalized surface modification with vanadium nanoparticles of various valences against implant-associated bloodstream infection, Int. J. Nanomed. 12 (2017) 3121–3136.

[52] E. Cavaliere, S. De Cesari, G. Landini, E. Riccobono, L. Pallecchi, G.M. Rossolini, L. Gavioli, Highly bactericidal Ag nanoparticle films obtained by cluster beam deposition, Nanomed. Nanotechnol. Biol. Med. 11 (6) (2015) 1417–1423.

[53] G. Benetti, E. Cavaliere, A. Canteri, G. Landini, G.M. Rossolini, L. Pallecchi, M. Chiodi, M.J. Van Bael, N. Winckelmans, S. Bals, L. Gavioli, Direct synthesis of antimicrobial coatings based on tailored bi-elemental nanoparticles, Apl. Mater. 5 (3) (2017) 036105.

[54] A. Sarwar, H. Katas, S.N. Samsudin, N.M. Zin, Regioselective sequential modification of chitosan via azide-alkyne click reaction: synthesis, characterization, and antimicrobial activity of chitosan derivatives and nanoparticles, PLoS One 10 (4) (2015) e0123084.

[55] J. Yu, W. Zhang, Y. Li, G. Wang, L. Yang, J. Jin, Q. Chen, M. Huang, Synthesis, characterization, antimicrobial activity and mechanism of a novel hydroxyapatite whisker/nano zinc oxide biomaterial, Biomed. Mater. 10 (1) (2014) 015001.

[56] J. Wehling, R. Dringen, R.N. Zare, M. Maas, K. Rezwan, Bactericidal activity of partially oxidized nanodiamonds, ACS Nano 8 (6) (2014) 6475–6483.

[57] H.A. Foster, I.B. Ditta, S. Varghese, A. Steele, Photocatalytic disinfection using titanium dioxide: spectrum and mechanism of antimicrobial activity, Appl. Microbiol. Biotechnol. 90 (6) (2011) 1847–1868.

[58] U. Joost, K. Juganson, M. Visnapuu, M. Mortimer, A. Kahru, E. Nõmmiste, U. Joost, V. Kisand, A. Ivask, Photocatalytic antibacterial activity of nano-TiO_2 (anatase)-based thin films: effects on *Escherichia coli* cells and fatty acids, J. Photochem. Photobiol. B Biol. 142 (2015) 178–185.

[59] M.A. Ansari, H.M. Khan, A.A. Khan, S.S. Cameotra, Q. Saquib, J. Musarrat, Interaction of Al_2O_3 nanoparticles with *Escherichia coli* and their cell envelope biomolecules, J. Appl. Microbiol. 116 (4) (2014) 772–783.

[60] I. Sondi, B. Salopek-Sondi, Silver nanoparticles as antimicrobial agent: a case study on *E. coli* as a model for Gram-negative bacteria, J. Colloid Interface Sci. 275 (1) (2004) 177–182.

[61] N. Nataraj, G.S. Anjusree, A.A. Madhavan, P. Priyanka, D. Sankar, N. Nisha, S.V. Lakshmi, R. Jayakumar, A. Balakrishnan, R. Biswas, Synthesis and anti-staphylococcal activity of TiO_2 nanoparticles and nanowires in ex vivo porcine skin model, J. Biomed. Nanotechnol. 10 (5) (2014) 864–870.

[62] Y. Su, X. Zheng, Y. Chen, M. Li, K. Liu, Alteration of intracellular protein expressions as a key mechanism of the deterioration of bacterial denitrification caused by copper oxide nanoparticles, Sci. Rep. 5 (1) (2015).

[63] Y. Li, W. Zhang, J. Niu, Y. Chen, Mechanism of photogenerated reactive oxygen species and correlation with the antibacterial properties of engineered metal-oxide nanoparticles, ACS Nano 6 (6) (2012) 5164–5173.

[64] S. Gurunathan, J. Woong Han, A. Abdal Daye, V. Eppakayala, J. Kim, Oxidative stress-mediated antibacterial activity of graphene oxide and reduced graphene oxide in *Pseudomonas aeruginosa*, Int. J. Nanomed. (2012) 5901.

[65] P.V. Laxma Reddy, B. Kavitha, P.A. Kumar Reddy, K.-H. Kim, TiO_2 - based photocatalytic disinfection of microbes in aqueous media: a review, Environ. Res. 154 (2017) 296–303.

[66] O. Carp, Photoinduced reactivity of titanium dioxide, Prog. Solid State Chem. 32 (1–2) (2004) 33–177.

[67] J.T. Yates, Photochemistry on TiO_2: mechanisms behind the surface chemistry, Surf. Sci. 603 (10–12) (2009) 1605–1612.

[68] S.G. Kumar, L.G. Devi, Review on modified TiO_2 photocatalysis under UV/visible light: selected results and related mechanisms on interfacial charge carrier transfer dynamics, J. Phys. Chem. A 115 (46) (2011) 13211–13241.

[69] M. Chiodi, C.P. Cheney, P. Vilmercati, E. Cavaliere, N. Mannella, H.H. Weitering, L. Gavioli, Enhanced dopant solubility and visible-light absorption in Cr–N codoped TiO_2 nanoclusters, J. Phys. Chem. C 116 (1) (2012) 311–318.

[70] C. Parks Cheney, P. Vilmercati, E.W. Martin, M. Chiodi, L. Gavioli, M. Regmi, G. Eres, T.A. Callcott, H.H. Weitering, N. Mannella, Origins of electronic band gap reduction in Cr/N codoped TiO_2, Phys. Rev. Lett. 112 (3) (2014).

[71] A.M. Márquez, J.J. Plata, Y. Ortega, J.F. Sanz, G. Colón, A. Kubacka, M. Fernández-García, Making photo-selective TiO_2 materials by cation–anion codoping: from structure and electronic properties to photoactivity, J. Phys. Chem. C 116 (35) (2012) 18759–18767.

[72] H. Pan, Y.-W. Zhang, V.B. Shenoy, H. Gao, Effects of H-, N-, and (H, N)-Doping on the photocatalytic activity of TiO_2, J. Phys. Chem. C 115 (24) (2011) 12224–12231.

[73] X. Ma, Y. Wu, Y. Lu, J. Xu, Y. Wang, Y. Zhu, Effect of compensated codoping on the photoelectrochemical properties of anatase TiO_2 photocatalyst, J. Phys. Chem. C 115 (34) (2011) 16963–16969.

[74] E. Barborini, A.M. Conti, I. Kholmanov, P. Piseri, A. Podestà, P. Milani, C. Cepek, O. Sakho, R. Macovez, M. Sancrotti, Nanostructured TiO_2 films with 2 eV optical gap, Adv. Mater. 17 (15) (2005) 1842–1846.

[75] M. D'Arienzo, N. Siedl, A. Sternig, R. Scotti, F. Morazzoni, J. Bernardi, O. Diwald, Solar light and dopant-induced recombination effects: photoactive nitrogen in TiO_2 as a case study, J. Phys. Chem. C 114 (42) (2010) 18067–18072.

[76] R.J. Barnes, R. Molina, J. Xu, P.J. Dobson, I.P. Thompson, Comparison of TiO_2 and ZnO nanoparticles for photocatalytic degradation of methylene blue and the correlated inactivation of gram-positive and gram-negative bacteria, J. Nanoparticle Res. 15 (2) (2013).

[77] Q. Yu, J. Li, Y. Zhang, Y. Wang, L. Liu, M. Li, Inhibition of gold nanoparticles (AuNPs) on pathogenic biofilm formation and invasion to host cells, Sci. Rep. 6 (1) (2016).

[78] J. Hou, Y. Yang, P. Wang, C. Wang, L. Miao, X. Wang, B. Lv, G. You, Z. Liu, Effects of CeO_2, CuO, and ZnO nanoparticles on physiological features of *Microcystis aeruginosa* and the production and composition of extracellular polymeric substances, Environ. Sci. Pollut. Control Ser. 24 (1) (2017) 226–235.

[79] A.R. Gillett, S.N. Baxter, S.D. Hodgson, G.C. Smith, P.J. Thomas, Using sub-micron silver-nanoparticle based films to counter biofilm formation by Gram-negative bacteria, Appl. Surf. Sci. 442 (2018) 288–297.

[80] J. Lellouche, A. Friedman, R. Lahmi, A. Gedanken, E. Banin, Antibiofilm surface functionalization of catheters by magnesium fluoride nanoparticles, Int. J. Nanomed. (2012) 1175.

[81] E. Hoseinzadeh, P. Makhdoumi, P. Taha, H. Hossini, J. Stelling, M. Amjad Kamal, G.M. Ashraf, A review on nano-antimicrobials: metal nanoparticles, methods and mechanisms, Curr. Drug Metabol. 18 (2) (2017) 120–128.

[82] L. Miao, C. Wang, J. Hou, P. Wang, Y. Ao, Y. Li, N. Geng, Y. Yao, B. Lv, Y. Yang, G. You, Y. Xu, Aggregation and removal of copper oxide (CuO) nanoparticles in wastewater environment and their effects on the microbial activities of wastewater biofilms, Bioresour. Technol. 216 (2016) 537–544.

[83] C. Chifiriuc, V. Grumezescu, A. Grumezescu, C. Saviuc, V. Lazăr, E. Andronescu, Hybrid magnetite nanoparticles/Rosmarinus officinalis essential oil nanobiosystem with antibiofilm activity, Nanoscale Res. Lett. 7 (1) (2012) 209.

[84] E. Banin, A. friedman, A. Gedanken, J. Lellouche, Antibacterial and antibiofilm properties of yttrium fluoride nanoparticles, Int. J. Nanomed. (2012) 5611.

[85] L. Liu, R. Bhatia, T. Webster, Atomic layer deposition of nano-TiO_2 thin films with enhanced biocompatibility and antimicrobial activity for orthopedic implants, Int. J. Nanomed. 12 (2017) 8711–8723.

[86] D.L. Slomberg, Y. Lu, A.D. Broadnax, R.A. Hunter, A.W. Carpenter, M.H. Schoenfisch, Role of size and shape on biofilm eradication for nitric oxide-releasing silica nanoparticles, ACS Appl. Mater. Interfaces 5 (19) (2013) 9322–9329.

[87] M.K.S. Ballo, S. Rtimi, C. Pulgarin, N. Hopf, A. Berthet, J. Kiwi, P. Moreillon, J.M. Entenza, A. Bizzini, In vitro and in vivo effectiveness of an innovative silver-copper nanoparticle coating of catheters to prevent methicillin-resistant *Staphylococcus aureus* infection, Antimicrob. Agents Chemother. 60 (9) (2016) 5349–5356.

[88] D.-K. Lee, S.V. Kim, A.N. Limansubroto, A. Yen, A. Soundia, C.-Y. Wang, W. Shi, C. Hong, S. Tetradis, Y. Kim, N.-H. Park, M.K. Kang, D. Ho, Nanodiamond–gutta percha composite biomaterials for root canal therapy, ACS Nano 9 (11) (2015) 11490–11501.

[89] A. Lazary, I. Weinberg, J.-J. Vatine, A. Jefidoff, R. Bardenstein, G. Borkow, N. Ohana, Reduction of healthcare-associated infections in a long-term care brain injury ward by replacing regular linens with biocidal copper oxide impregnated linens, Int. J. Infect. Dis. 24 (2014) 23–29.

[90] S. Rtimi, S. Giannakis, C. Pulgarin, Self-sterilizing sputtered films for applications in hospital facilities, Molecules 22 (7) (2017) 1074.

[91] M. Haenle, A. Fritsche, C. Zietz, R. Bader, F. Heidenau, W. Mittelmeier, H. Gollwitzer, An extended spectrum bactericidal titanium dioxide (TiO_2) coating for metallic implants: in vitro effectiveness against MRSA and mechanical properties, J. Mater. Sci. Mater. Med. 22 (2) (2011) 381–387.

[92] P. Carvalho, P. Sampaio, S. Azevedo, C. Vaz, J.P. Espinós, V. Teixeira, J.O. Carneiro, Influence of thickness and coatings morphology in the antimicrobial performance of zinc oxide coatings, Appl. Surf. Sci. 307 (2014) 548–557.

[93] J. Musil, J. Blažek, K. Fajfrlík, R. Čerstvý, Flexible antibacterial Al–Cu–N films, Surf. Coat. Technol. 264 (2015) 114–120.

[94] Y.-H. Chen, C.-C. Hsu, J.-L. He, Antibacterial silver coating on poly(ethylene terephthalate) fabric by using high power impulse magnetron sputtering, Surf. Coat. Technol. 232 (2013) 868–875.

[95] J.-H. Chu, J. Lee, C.-C. Chang, Y.-C. Chan, M.-L. Liou, J.-W. Lee, J.S.-C. Jang, J.-G. Duh, Antimicrobial characteristics in Cu-containing Zr-based thin film metallic glass, Surf. Coat. Technol. 259 (2014) 87–93.

[96] J. Musil, J. Blažek, K. Fajfrlík, R. Čerstvý, š. Prokšová, Antibacterial Cr–Cu–O films prepared by reactive magnetron sputtering, Appl. Surf. Sci. 276 (2013) 660–666.

[97] J. Musil, M. Zítek, K. Fajfrlík, R. Čerstvý, Flexible antibacterial Zr–Cu–N thin films resistant to cracking, J. Vac. Sci. Technol. A 34 (2) (2016) 021508.

[98] B. Jönsson, S. Hogmark, Hardness measurements of thin films, Thin Solid Films 114 (3) (1984) 257–269.

[99] B.R. Lawn, V.R. Howes, Elastic recovery at hardness indentations, J. Mater. Sci. 16 (10) (1981) 2745–2752.

[100] B.D. Fraters, E. Cavaliere, G. Mul, L. Gavioli, Synthesis of photocatalytic TiO_2 nano-coatings by supersonic cluster beam deposition, J. Alloy. Comp. 615 (2014) S467–S471.

[101] E. Cavaliere, G. Benetti, M. Van Bael, N. Winckelmans, S. Bals, L. Gavioli, Exploring the optical and morphological properties of Ag and Ag/TiO_2 nanocomposites grown by supersonic cluster beam deposition, Nanomaterials 7 (12) (2017) 442.

[102] P. Milani, P. Piseri, E. Barborini, A. Podesta, C. Lenardi, Cluster beam synthesis of nanostructured thin films, J. Vac. Sci. Technol. A 19 (4) (2001) 2025–2033.

[103] E. Barborini, I.N. Kholmanov, P. Piseri, C. Ducati, C.E. Bottani, P. Milani, Engineering the nano-crystalline structure of TiO_2 films by aerodynamically filtered cluster deposition, Appl. Phys. Lett. 81 (16) (2002) 3052–3054.

[104] C. Marega, J. Maculan, G. Andrea Rizzi, R. Saini, E. Cavaliere, L. Gavioli, M. Cattelan, G. Giallongo, A. Marigo, G. Granozzi, Polyvinyl alcohol electrospun nanofibers containing Ag nanoparticles used as sensors for the detection of biogenic amines, Nanotechnology 26 (7) (2015) 075501.

[105] J.F. Moulder, W.F. Stickle, P.E. Sobol, K.D. Bomben, J. Chastain, R.C. King Jr., Physical electronics, incorporation, in: Handbook of X-Ray Photoelectron Spectroscopy: A Reference Book of Standard Spectra for Identification and Interpretation of XPS Data, Physical Electronics, Eden Prairie, Minn., 1995.

[106] Y. Miura, H. Kusano, T. Nanba, S. Matsumoto, X-ray photoelectron spectroscopy of sodium borosilicate glasses, J. Non-Cryst. Solids 290 (1) (2001) 1–14.

[107] M. Michielsen, H.C. Comijs, E.J. Semeijn, A.T.F. Beekman, D.J.H. Deeg, J.J. Sandra Kooij, The comorbidity of anxiety and depressive symptoms in older adults with attention-deficit/hyperactivity disorder: a longitudinal study, J. Affect. Disord. 148 (2–3) (2013) 220–227.

[108] D. Guo, G. Xie, J. Luo, Mechanical properties of nanoparticles: basics and applications, J. Phys. D Appl. Phys. 47 (1) (2014) 013001.

[109] S. Peli, E. Cavaliere, G. Benetti, M. Gandolfi, M. Chiodi, C. Cancellieri, C. Giannetti, G. Ferrini, L. Gavioli, F. Banfi, Mechanical properties of Ag nanoparticle thin films synthesized by supersonic cluster beam deposition, J. Phys. Chem. C 120 (8) (2016) 4673–4681.

[110] G. Benetti, C. Caddeo, C. Melis, G. Ferrini, C. Giannetti, N. Winckelmans, S. Bals, M.J. Van Bael, E. Cavaliere, L. Gavioli, F. Banfi, Bottom-up mechanical nanometrology of granular Ag nanoparticles thin films, J. Phys. Chem. C 121 (40) (2017) 22434–22441.

[111] P. Milani, M. Ferretti, P. Piseri, C.E. Bottani, A. Ferrari, A. Li Bassi, G. Guizzetti, M. Patrini, Synthesis and characterization of cluster-assembled carbon thin films, J. Appl. Phys. 82 (11) (1997) 5793–5798.

[112] L. Ravagnan, G. Divitini, S. Rebasti, M. Marelli, P. Piseri, P. Milani, Poly(methyl methacrylate)–palladium clusters nanocomposite formation by supersonic cluster beam deposition: a method for microstructured metallization of polymer surfaces, J. Phys. D: Appl. Phys. 42 (8) (2009) 082002.

[113] M. Chiodi, E. Cavaliere, I. Kholmanov, M. de Simone, O. Sakho, C. Cepek, L. Gavioli, Nanostructured TiO_x film on Si substrate: room temperature formation of $TiSi_x$ nanoclusters, J. Nanoparticle Res. 12 (7) (2010) 2645–2653.

[114] D. Wang, Y. Li, Bimetallic nanocrystals: liquid-phase synthesis and catalytic applications, Adv. Mater. 23 (9) (2011) 1044–1060.

[115] N. Toshima, T. Yonezawa, Bimetallic nanoparticles—novel materials for chemical and physical applications, New J. Chem. 22 (11) (1998) 1179–1201.

[116] S.A.H. Jalali, A.R. Allafchian, S.S. Banifatemi, I. Ashrafi Tamai, The antibacterial properties of Ag/ TiO_2 nanoparticles embedded in silane sol–gel matrix, J. Taiwan Inst. Chem. Eng. 66 (2016) 357–362.

[117] X.-F. Zhang, Z.-G. Liu, W. Shen, S. Gurunathan, Silver nanoparticles: synthesis, characterization, properties, applications, and therapeutic approaches, Int. J. Mol. Sci. 17 (9) (2016) 1534.

[118] J.P. Ruparelia, A.K. Chatterjee, S.P. Duttagupta, S. Mukherji, Strain specificity in antimicrobial activity of silver and copper nanoparticles, Acta Biomater. 4 (3) (2008) 707–716.

[119] M. Valodkar, S. Modi, A. Pal, S. Thakore, Synthesis and anti-bacterial activity of Cu, Ag and Cu–Ag alloy nanoparticles: a green approach, Mater. Res. Bull. 46 (3) (2011) 384–389.

[120] T. Jin, Y. He, Antibacterial activities of magnesium oxide (MgO) nanoparticles against foodborne pathogens, J. Nanoparticle Res. 13 (12) (2011) 6877–6885.

[121] A.A. Marakushev, N.I. Bezmen, Chemical affinity of metals for oxygen and sulfur, Int. Geol. Rev. 13 (12) (1971) 1781–1794.

[122] Y. Cai, D. Wu, X. Zhu, W. Wang, F. Tan, J. Chen, X. Qiao, X. Qiu, Sol-gel preparation of Ag-doped MgO nanoparticles with high efficiency for bacterial inactivation, Ceram. Int. 43 (1) (2017) 1066–1072.

[123] G. Benetti, E. Cavaliere, M. van Bael, F. Banfi, R. Brescia, R. Ferrando, L. Pallecchi, L. Gavioli, Tailored multi-element nanoparticles for wide spectrum antibacterial coating, Nanoscale 11 (2019) 1626–1635.

[124] M.A. Jabbareh, F. Monji, Thermodynamic modeling of Ag – Cu nanoalloy phase diagram, Calphad 60 (2018) 208–213.

[125] G. Benetti, E. Cavaliere, F. Banfi, L. Gavioli, Antimicrobial nanostructured coatings: a gas phase deposition and magnetron sputtering perspective, Materials 13 (2020) 784.

12

Additive nano-manufacturing of 3D printed electronics using supersonic cluster beam deposition

Tommaso Santaniello, Paolo Milani

INTERDISCIPLINARY CENTRE FOR NANOSTRUCTURED MATERIALS AND INTERFACES (CIMAINA) AND DEPARTMENT OF PHYSICS, UNIVERSITÀ DEGLI STUDI DI MILANO, MILAN, ITALY

Chapter outline

12.1 Introduction

Additive Manufacturing (AM) refers to any technology relying on the fabrication of an object by joining materials, usually layer-upon-layer, directly from its three-dimensional virtual model [1]. AM-based processes are conceptually at the opposite pole of traditional top-down subtractive fabrication techniques, in which an object is realized by progressively removing parts off a starting material block. The possibility to promptly transform 3D data into physical components without using tools or molds is among the main advantages of AM. AM approaches also enable the fabrication of 3D freeform components with complex geometries; they can be applied to process a variety of materials (e.g., polymers, metals, ceramics, nanocomposites), and they increase design flexibility and product customization, reduce costs and raw material wastes [1–3].

Cluster Beam Deposition of Functional Nanomaterials and Devices. https://doi.org/10.1016/B978-0-08-102515-4.00012-X

3D printing (3DP) is nowadays the most widespread and extensively used subset of AM technologies for rapid prototyping, i.e., for the quick production of scale models of a part or finished product [4]. Recent advances in 3DP technologies in terms of hardware and software development suggest that they will be increasingly used in industry for small-scale production, and they are expected to reach a sufficient maturity for end products manufacturing and mass production by 2050 [5]. Despite the technical specifications and materials employed, a typical 3DP process can be summarized in four steps. First, the object is modeled using a computer-aided design (CAD) software and represented in a stereolithographic format as a solid. This is then "sliced" using a dedicated software that determines the cross-sections of the solid and enables the generation of the printer directive instructions. Finally, the object is additively fabricated according to its design and to the printing parameters set (e.g., layer height and printing speed) by processing the raw materials. Extrusion-based 3DP of polymers, such as fused filament fabrication (FFF) of thermoplastics, is the most employed and user-friendly AM technique for the fabrication of relatively small and complex plastic components and its use has now been democratized by the commercialization of low-cost consumer-grade machines [6].

The operational principle of FFF is based on the use of wires of thermoplastic polymers inserted into an extruder that heats up the material at a suitable temperature to be softened and then deposited by means of a nozzle onto a plate, where the extruded material cools and solidifies back, regaining its original physical properties [7]. The extruder is mounted on a motorized x/y/z stage and moves in the three axes during the printing process, following the directive instructions fed to the printer to reproduce the virtual model of the physical object. This takes place layer-by-layer, by depositing the melted filament in the form of strains, the characteristic diameter of which is essentially that of the nozzle [2]. Typical values of the deposited filament width ranges from 0.1 to 0.4 mm and determine the lateral resolution of the manufactured part. On the other hand, layer height can be typically controlled in a range going from 0.05 mm up to 0.5 mm [2,8]. A schematic of a standard FFF process is reported in Fig. 12.1.

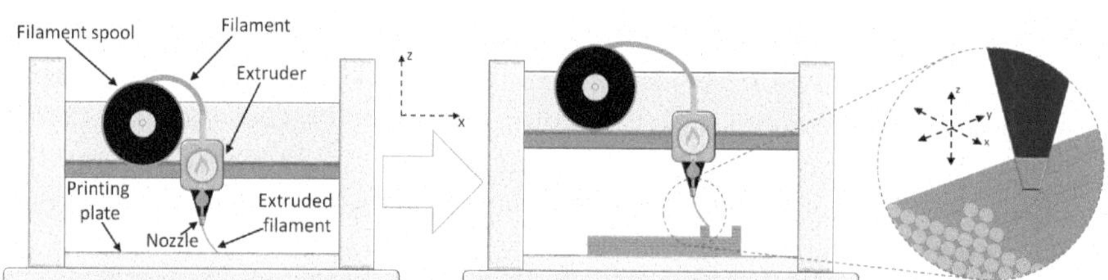

FIGURE 12.1 Schematic of a standard FFF process for the 3D printing of thermoplastic-based components. A thermoplastic filament is inserted into an extrusion head, mounted onto a motorized stage. The filament is heated up and extruded by the nozzle onto the printing plate and rapidly solidifies, while the extruder moves following the printer directive instructions. After the first layer is formed, the process iterates to additively generate the workpiece layer-upon-layer.

The surface quality of the printed piece, as well as the level of accuracy at which the manufactured object reproduces its virtual model, depends on a number of factors, including the type of material employed, the extrusion rate, printing speed and the specific printer hardware, and slicing software employed [9]. There is a broad variety of raw materials for FFF, among which the mostly used are polylactic acid (PLA), polycarbonate (PC), polystyrene (PS), and acrylonitrile butadiene styrene (ABS). These thermoplastic polymers are nontoxic and relatively inexpensive, and each of them presents different physical properties (e.g., glass transition temperature and melting point) that in turn affect the setting of the fabrication process parameters, such as extrusion and printing plate temperature, material flow rate into the nozzle and deposition rate that eventually determine the dimensional and structural integrity of the printed object. Beside rapid prototyping and tooling, FFF has been employed for the manufacturing of low volume consumer and industrial products in the automotive, aerospace, and biomedical medical area [2,5,6].

Recently, the development and fabrication of smart objects and functional devices with integrated sensing and actuation capabilities able to interact with the surrounding environment have been pushed by the rise of the Internet of Things (IoT) and by the paradigmatic model of Industry 4.0. This, in turn, led to a push of AM technologies toward multimaterial printing and printed electronics [10,11]. 3D printed electronics relies on the concept of integrating conductive structures, wires, passive electric and active electronic components into monolithic platforms, in order to confer the desired functional behavior to the object in a single fabrication process. This objective is far from accomplished, but state-of-the art machineries that implement the combined use of polymers and conductive materials in a single facility to embed conductive structures into freeform 3D substrates are emerging in the market [12]. This approach, which essentially combines FFF-based 3D printing of structural polymeric components with 2D printing of functional conductive lines in a multistep procedure, has several advantages, since it minimizes the assembly of the electronic object components, provides a direct packaging of the system, reduces the product size, and avoids the use of chemical treatments.

The 3DP of smart electronic objects predominantly exploits the use of nanocomposite functional inks for the production of the conductive parts. These are usually constituted by a liquid mixture containing high loadings of metal nanoparticles (usually silver or copper) dispersed in a suitable solvent and are managed by means of different printing strategies to directly write the conductive components onto the basis substrate through a nozzle, reproducing the 2D design and pattern of specific circuitries in an automated fashion. An effective route to process conductive inks for 3D printed electronics is the Aerosol Jet Printing (AJP) technique. AJP has been widely employed for the production of printed antennas, sensors, circuit boards, and passive electrical components onto freeform 3D plastic substrates and for the realization of embedded circuits in monolithic objects [13–15].

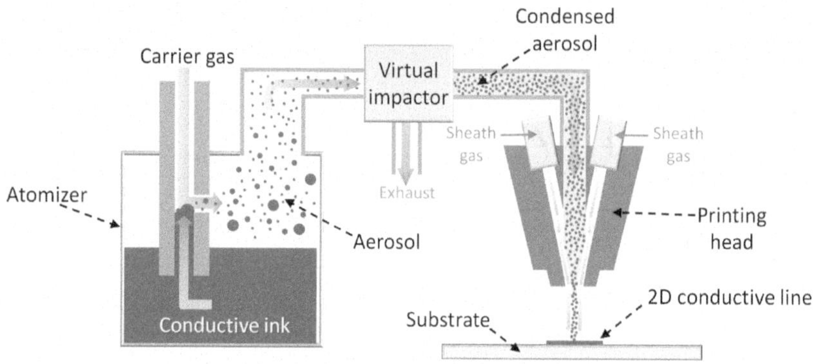

FIGURE 12.2 AJP working principle. Conductive lines can be printed onto flat or curved surfaces by exploiting a focused beam of a condensed aerosol carrying microdroplets of an electrically conductive ink.

AJP is a CAD controlled 2D printing process that is based on the spraying of micrometric droplets of a conductive ink at ambient pressure, focused onto the substrate using a carrier gas. The working principle of AJP is reported in Fig. 12.2, and can be summarized as follows.

The ink is kept into a pneumatic atomizer, constituted by a reservoir equipped with a fluidic inlet and outlet. An inert gas (usually nitrogen) is fluxed into the chamber and generates a mixture of gas and conductive ink droplets, with an average diameter of about few microns. While larger droplets (diameter $> 5\ \mu m$) are pushed downstream back to the ink bulk by gravity, smaller droplets are dragged by the carrier gas into the fluidic outlet, which directs the aerosol flow toward a virtual impactor. This device separates the aerosol stream into an exhaust gas line and a condensed aerosol, subsequently flowing to the printing head. Here, a second flux of nitrogen (namely, the sheath gas) compresses and focuses the aerosol, which is then ejected through a nozzle and directed onto the target substrate to print the desired pattern [13].

While pneumatic AJP is suitable to process high viscous inks, for low viscosity solutions (0.1–5 cP) an ultrasound atomizer is recommended: the latter approach also allows to obtain smaller droplet size, which can reach values down to fractions of microns. Droplet size, layer thickness (i.e., number of printed layers), ratio between the sheath gas and aerosol (focusing ratio), and printing speed are pivotal parameters to fabricate fine and highly resolved features ($<10\ \mu m$) using AJP. The surface wettability and roughness of the substrate with respect to the specific conductive ink formulation should also be carefully taken into account when dealing with these approaches, since they can also affect the printing resolution. The stand-off distance between the nozzle and the substrate is usually 1 –5 mm, and it also affects the printing lateral resolution to a great extent.

Considering that the aerosol beam can stay focused over a range of several millimeters (depending on process parameters, such as the sheath gas and aerosol ratio), AJP is suitable for conformal direct writing onto curved surfaces and is therefore promising for the deposition of conductive lines over 3D printed substrates, without altering the nozzle to substrate stand-off distance during printing. The flexibility and efficacy of AJP for 3D printed electronics were demonstrated in the framework of functional devices and sensors prototyping [16,17]. For example, finely resolved conductive tracks together with tens of micrometers-high pitches over 3D substrates were manufactured by means of AJP for the realization of interconnects and via holes in view of microelectronics packaging applications targeting at wire bonding replacement [14,18,19]. However, the interconnections between different layers of the fabricated devices are limited to vias with submillimeter dimensions and pitch heights around 60 μm. Furthermore, the feasibility of standard AJP to process 3D surfaces was assessed only by printing onto the so-called 2.5D surfaces, i.e., planar surfaces that are gently curved to a maximum angle of 45 degrees with respect to the aerosol beam direction. Different and more complex free-form substrates could only be processed by means of modified AJP facilities provided with 5-axes or 6-axes printing heads, enabling higher degrees of freedom to the aerosol beam motion during printing [20].

The main challenges regarding the use of traditional AJP and other direct writing techniques for 3D printed electronics are still related to the relatively low spatial resolution (minimum feature size achievable is around few micrometers) and to the poor control over the electrical resistance of the printed structures, which, given a specific ink, can be tailored only by setting the geometry of the lines. Other challenges are connected to the use of postprocess thermal treatments required to enhance the conductivity of the structures, which can limit the range of substrate materials, and to the high particles loading of the ink (>70% of the total weight), which can imply an increase in the overall process costs. Another major issue is that of adhesion between the conductive layer and the base polymer, a feature that is particularly critical for the fabrication of stretchable and wearable devices.

In the next section, the use of Supersonic Cluster Beam Deposition (SCBD) combined with FFF to print 3D monolithic smart objects with embedded electronics is presented and discussed. SCBD is an AM technology suitable for the fabrication of cluster-assembled conductive layers (few tens to few hundreds of nm thick) produced by directing supersonically accelerated beams of inert gas seeded with neutral metal nanoparticles onto a polymeric target [21,22]. Due to their low kinetic energies, clusters are able to penetrate into the polymer down to a characteristic depth and to progressively surface up the substrate to form a well-adhering conductive nanocomposite film, with electrical resistance depending only on the amount of nanoparticles deposited. SCBD is compatible with planar fabrication technologies and enables the patterning of highly resolved features using stencil masks, with lateral resolution down to one micron and no shadow effects, due to the low beam divergence. The combination of SCBD and FFF enables the fabrication of freeform functional devices with 3D polymer/metal nanocomposite lines, simplifying the topology of planar circuits by implementing bridge vias and out-of-plane electrical connections between different regions of the board [23]. Resistors with tailored electrical conductivity and sockets for standard electronic components fitting can be manufactured as well, minimizing the need to use bulky electrical components and soldering [23]. This integrated additive manufacturing approach was used to prototype an all-printed monolithic dark-activated light sensor.

The fabrication and characterization flexible and stretchable electronic devices, prototyped using top-down microfabrication techniques and SCBD printing, are also presented. First, SCBD printing of passive electrical components (resistors and planar capacitors) on plain paper is reported. Then, a stretchable keyboard composed of a molded elastomeric layer provided with SCBD electrodes and conductive paths is demonstrated and discussed. The soft nanocomposite keyboard is electrically connected to a touch sensor that can activate a matrix of LEDs printed on plain paper. The deformable device is also endowed with conductive vias, fabricated into the elastomer through SCBD, to electrically connect the top and bottom layers of the keyboard.

12.2 SCBD: operational principle

Supersonic Cluster Beam Deposition already proved to be an effective manufacturing tool for the fabrication of a number of functional polymer/metal nanocomposites for the production of stretchable and flexible electronic devices, sensors, and soft actuators [21,22,24–26]. A detailed description of the SCBD technology and of the deposition apparatus is reported in Ref. [27] and briefly discussed hereafter. A typical SCBD system is equipped with a Pulsed Microplasma Cluster Source (PMCS). Neutral metal clusters are produced by the PMCS and accelerated for deposition in a supersonic expansion. A PMCS schematically consists of a ceramic body with a cavity where a target metal rod, operating as a cathode, is sputtered by a localized electrical discharge ignited during the pulsed injection of Ar gas at high pressure (40 bar). The metal atoms sputtered from the target thermalize with the carrier gas and aggregate in the source cavity forming metal clusters, composed by several thousands of atomic units. The carrier gas-cluster mixture expands out of the PMCS through a nozzle into a low-pressure (10^{-3}–10^{-4} mbar) expansion chamber, thus producing a highly collimated supersonic beam with a divergence lower than 1 degree. The beam is focused by employing an aerodynamic focuser, constituted by a series of metal cylinders provided with aerodynamic lenses located at their terminal side [27,28]. These consist of metal discs having a central through-hole of 1 to 2 mm in diameter. The use of such lenses induces an abrupt change in direction of the flow lines along the beam passage through the focuser, by widening and narrowing the gas flow and selecting clusters with an average size ranging from 3 to 10 nm. The central part of the beam then enters a second vacuum chamber (deposition chamber, kept at a pressure of about 10^{-5} mbar) through a skimmer, and it impinges onto the target substrate [27,28]. This highly collimated supersonic beam carries neutral metallic clusters with a kinetic energy of about 0.5 $eV \cdot atom^{-1}$, avoiding fragmentation and significant deformation of the particles upon impact with the substrate (memory effect). Even if the kinetic energy is significantly lower than in ion implantation, neutral clusters are able to penetrate up to tens of nanometers into the polymeric target, progressively forming a conducting polymer/metal nanocomposite and avoiding electrical charging and carbonization of the polymeric base [22,29]. The deposition process takes place at room temperature and does not require the use of chemical treatments to enhance the adhesion between the polymeric and the metallic phase, since a significant fraction of the nanoparticles are physically embedded into the substrate, anchoring the conducting film to the polymer. By controlling the deposition time, it is possible to fabricate nanocomposite films emerging from the substrate surface with thickness ranging from tens to hundreds of nanometers [22].

Furthermore, since the beam is highly collimated, SCBD can be used to print microscale patterns with stencil masks, avoiding treating the polymer with chemical etching of the masks used for lift-off and ensuring minimization of shadows effects, guaranteeing high lateral resolution [30,31]. A schematic of the SCBD process and apparatus is reported in Fig. 12.3.

A motorized sample holder allows large area deposition moving the target substrate in the perpendicular plane with respect to the nanoparticles beam [22,31]. The deposition rate is typically tuned in a range from 0.1 nm/s to 0.7 nm/s. The amount of nanoparticles deposited onto the substrate is measured in terms of equivalent thickness, defined as the thickness of a film composed by an equivalent amount of nanoparticles deposited onto a rigid substrate [22], which usually consists of a partially masked Si wafer located nearby the substrate during the metallization process.

SCBD can be considered an AM technique, since it enables the printing of conductive structures onto polymers by additively generating cluster-assembled layers, joins nanoparticles together in a "pulse-by-pulse" fashion, and preserves their structure and functionality upon deposition. The physical properties of the clusters-seeded supersonic beam are that of a dry nanoparticle ink, focused onto the polymeric substrate due to its specific aerodynamic profile. As it will be explained in the next section, the electrical

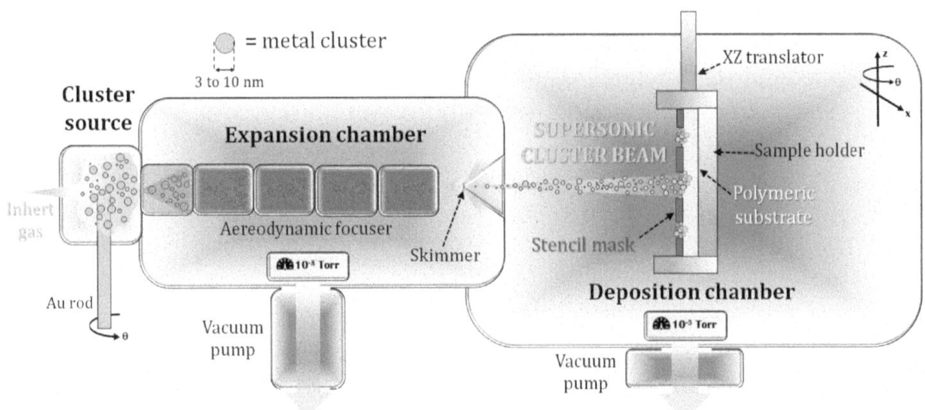

FIGURE 12.3 Schematic representation of a typical apparatus employed for SCBD printing.

resistance of the nanocomposite films produced by SCBD printing can be tailored by simply controlling the amount of clusters deposited (i.e., the deposition time), leading to the fabrication of layers with a conductivity behavior that is eventually independent from their geometry.

12.3 Controlling the electrical properties of SCBD printed polymer/metal nanocomposites

The incorporation of metal nanoparticles into polymers is a strategical solution to produce nanocomposite materials with controlled electrical properties. According to the filler volume fraction ϕ, i.e., the nanoparticles concentration in the host matrix, the transport properties of the composite can be varied from that of the pristine polymer to that of a conductive metallic film. The theory of percolation describes the evolution of the electrical features of nanocomposite materials by studying paths and connections formed by the nanoparticles and by the regions of the assembly. The main results of this theory are related to the characterization of the insulator-conductor transition in the immediate vicinity of the percolation threshold [32,33].

Three distinct conduction regimes can be identified in the nanocomposite formation, related to different values of φ and which are related to different electrical transport mechanisms in the material: (1) dielectric regime: at very low concentrations, the nanoparticles are isolated within the polymeric matrix and, in the case of insulating polymers, cannot therefore contribute to the conduction of the film. In this case, the clusters can significantly contribute to the variation of the dielectric constant of the polymer; (2) insulator-conductor transition regime: by implanting more metal particles, clusters begin to touch each other, leading to the formation of the first conductive paths. Once a critical volume fraction φ_C of the conductive phase is reached (namely at the percolation threshold), the amount of filler is enough to begin the formation of a continuous conductive network throughout the polymer matrix. The resistivity of the nanocomposite then undergoes a decrease of several orders of magnitude; iii) metallic regime: in this case ($\varphi \rightarrow 1$), the metal particles are well-connected with each other and form a metallic continuum with some dielectric inclusions. In this regime, the electrical transport is nearly similar to that of a thin metallic film, i.e., it is an ohmic behavior. Nevertheless, the electrical properties are rather modified due to strong electron scattering from the dielectric inclusions and the grain boundaries.

The electrical resistance of SCBD nanocomposites can be monitored online along the deposition process (Fig. 12.4A) and can be typically controlled in a range going from several GΩ (resistance of an insulating pristine polymer) to the ohmic regime ($\sim$10 Ω), solely depending on the amount of clusters implanted. The variation of resistivity of the nanocomposite materials, as the conductive filler content is increased, typically shows a classical curve as the one reported in Fig. 12.4B.

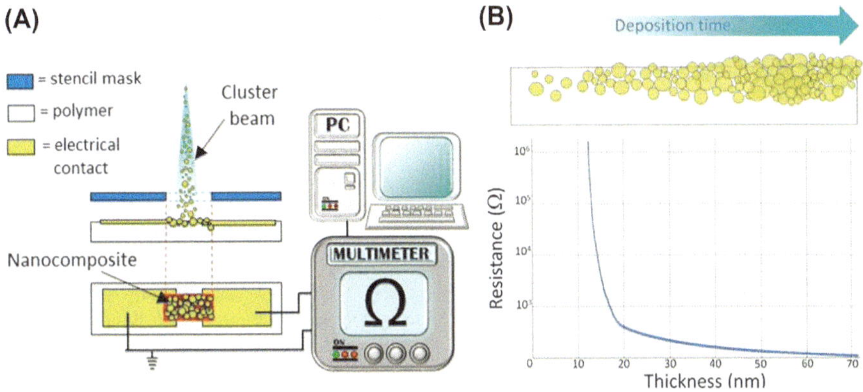

FIGURE 12.4 (A) Schematic of the experimental setup to monitor online the evolution of the electrical properties of an SCBD-printed nanocomposite. (B) Typical percolation curve.

12.4 Combining FFF and SCBD to manufacture 3D smart objects with embedded electronics

The combination of FFF and SCBD to produce all-printed monolithic functional devices with embedded electronics was recently demonstrated [23]. This integrated additive manufacturing approach consisted in a multistep procedure, alternating the two techniques to generate a 3D plastic object with integrated polymer/metal nanocomposite structures, such as 3D conductive lines, resistors with tailored electrical conductivity, 3D bridge vias, and sockets for standard electronic components fitting. The merging of FFF and SCBD offers new opportunities in printing 3D electronic circuits with simplified topology, enabling inter and intralayer connections between electrical components through the use of conducting vias and 3D bridging. Moreover, the presence of highly conductive socket slots avoid the use of soldering or conductive pastes to implement bulky electrical and electronic elements.

To demonstrate the feasibility of this fabrication route, a dark-activated light sensor was prototyped and validated. The sensor circuit layout is reported in Fig. 12.5. The operational principle of the device relies on the use of a photoresistor, the resistance (R_f) of which increases when the ambient light intensity drops below a certain threshold. The corresponding drop of the voltage imposed across the sensing element (V_{sens}) determines the switching on of one or two light emitting diodes (LEDs), whether V_{sens} falls below one or two fixed reference voltage signals (V_{ref1} and V_{ref2}), respectively. The system is therefore able to discriminate between two dark levels. The comparison between these analog voltages, generated using a 4.5 V power supply and two voltage dividers (R_1, R_2, R_3, and R_4), is done by means of two operational amplifiers (OP1 and OP2), connected to the LEDs. The circuit layout was translated into the printed smart object as schematized in Fig. 12.5. The device architecture comprehends three distinct levels, interconnected through SCBD printed oblique conductive paths: (1) a battery box for power supply, able

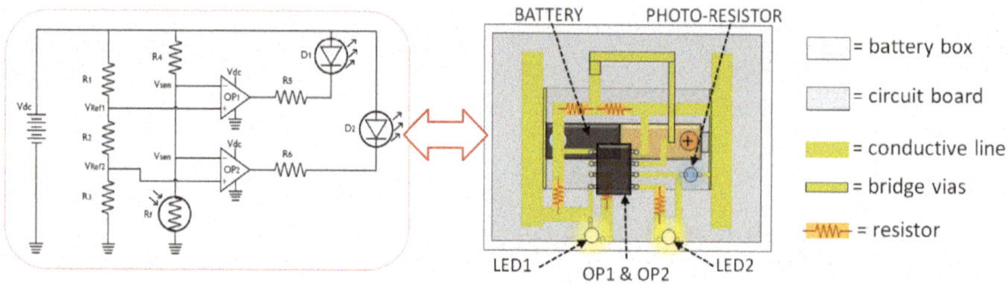

FIGURE 12.5 Circuit layout of the dark-activated switching device (*left*) and schematic representation of the 3D smart object produced by merging FFF and SCBD (*top view, right*). *Images readapted from A. Bellacicca, T. Santaniello, P. Milani, Embedding electronics in 3D printed structures by combining fused filament fabrication and supersonic cluster beam deposition, Addit. Manuf. 24 (1) (2018) 60–66.*

to host commercial batteries; (2) a circuit board provided with SCBD conductive lines and resistors, sockets, and active electronic components; (3) a 3D bridge vias for out-of-plane connections of the circuit board elements.

FFF was performed using acrylonitrile butadiene styrene (ABS, transparent, Orbitech, filament diameter 1.75 mm) as the basis material and employing a Delta Wasp 2040 3D printer. Simplify3D was used as the slicing software. The printing parameters (layer height from 0.05 to 0.1 mm, printing speed 10–15 mm/s, extrusion rate 110%) were empirically optimized to obtain an adequate surface smoothness and high dimensional accuracy. Moreover, an overprinting procedure was implemented to guarantee an optimal adhesion between the different levels of the object, avoiding thermal damages of the material when put in contact with the heated nozzle (ABS extrusion temperature was 230 degrees). During overprinting, dedicated frames, fabricated in ABS by mean of FFF, were glued to the printing plate and employed to firmly hold the workpiece. This optimized overprinting protocol also allowed bridging over empty areas, guaranteeing a proper flatness of the suspended printed structures.

SCBD was sequentially combined with FFF to print functional structures onto ABS. The metal employed for the deposition was gold. Prior device fabrication, the evolution of the electrical properties of the ABS/Au nanocomposites was monitored online during clusters deposition over flat ABS samples, in order to determine the percolation curve of the system. According to the amount of nanoparticles deposited, it was possible to print resistors with electrical resistance ranging from around 10 GΩ to 1 Ω.

A schematic of the step-by-step manufacturing process to print the light sensor is reported in Fig. 12.6. First, the battery box was printed using FFF (panel 1). Lateral oblique conductive lines linking the inner box walls to the top surface of the component were then printed using SCBD (panel 2), employing a stencil mask and by tilting the sample by an angle of 45 degrees with respect to the beam direction. These electrodes served to electrically connect the encased batteries to the upper circuit board. This level of the device, comprising the socket slots and interlayer vias, was subsequently overprinted on top of the battery box (panel 3). Conductive lines, resistors with tailored electrical conductivity, sockets, and the remaining 3D connections to the power supply

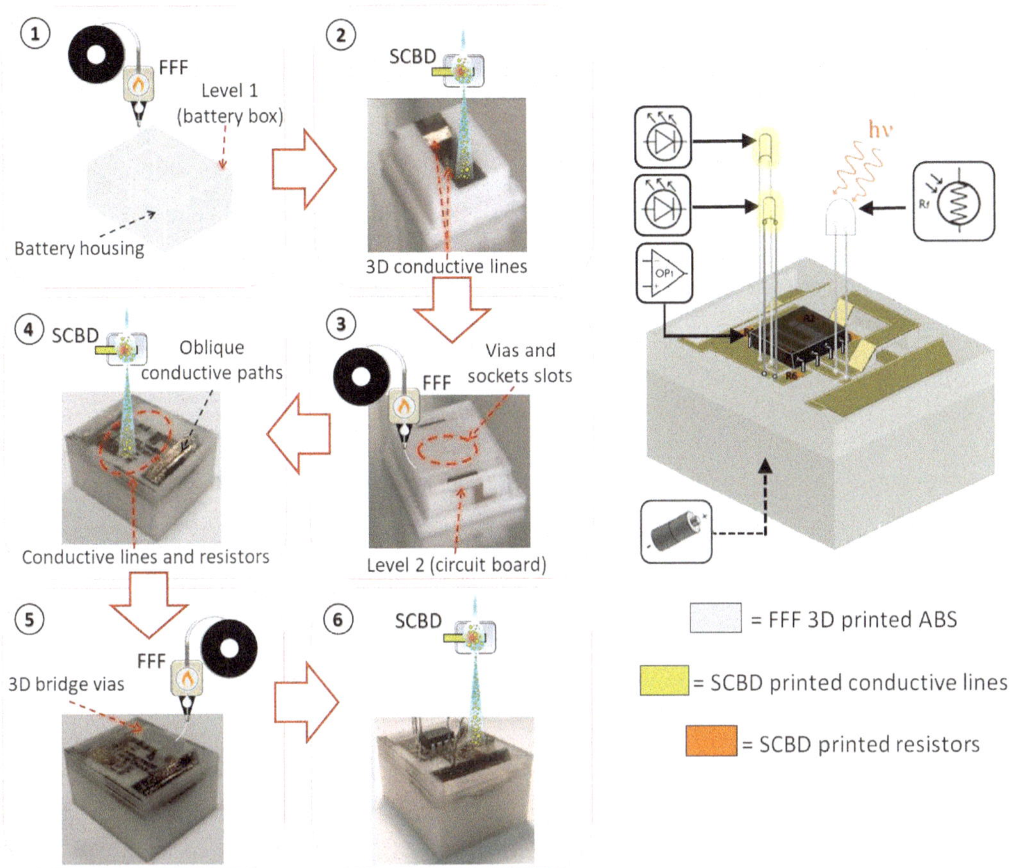

FIGURE 12.6 Schematic of the additive manufacturing approach combining FFF and SCBD for the realization of the monolithic light sensor. *Images readapted from A. Bellacicca, T. Santaniello, P. Milani, Embedding electronics in 3D printed structures by combining fused filament fabrication and supersonic cluster beam deposition, Addit. Manuf. 24 (1) (2018) 60–66.*

were then fabricated using SCBD (panel 4), using dedicated stencil masks. To uniformly print gold over the inner walls of the cylindrical socket slots, the sample was tilted at 30 degrees and 45 degrees with respect to the beam direction. At this stage, the overprinting of the bridge vias for the circuit out-of-plane connections was carried out (panel 5), followed by SCBD metallization of the structure (panel 6). To ensure low electrical resistance ($<10\ \Omega$), both 2D and 3D conductive lines and vias were fabricated with an equivalent thickness of 240 nm, while the four resistors were dimensioned in the range 1 kΩ to 6 kΩ by selecting the corresponding amount of nanoparticles. All of the masks employed for SCBD printing were also fabricated in ABS by means of FFF. In this case, the minimum feature size achievable was around 0.6 mm, but since the lateral resolution of the metallization process is determined only by the mask employed, the choice of

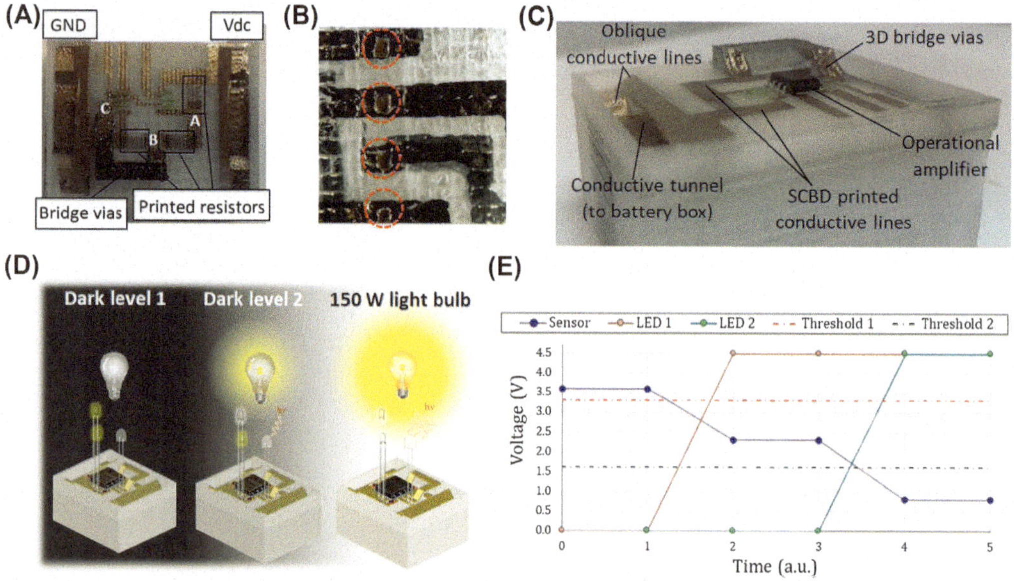

FIGURE 12.7 Photographs of the all-printed smart light sensor and device functionality assessment. (A) Top view of the system. (B) Closeup view of the conductive sockets for electronic components fitting. (C) Photograph of the 3D device. (D) Schematic of the sensor testing conditions to assess its functionality. (E) Voltage response of the system to different light intensities recorded over time. *Images readapted from A. Bellacicca, T. Santaniello, P. Milani, Embedding electronics in 3D printed structures by combining fused filament fabrication and supersonic cluster beam deposition, Addit. Manuf. 24 (1) (2018) 60–66.*

suitable stencils would lead to further miniaturization of the structures down to the microscale. The complete device (Fig. 12.6, right) was then provided with the operational amplifier, LEDs, photoresistors, and batteries.

Photographs of the 3D smart object are reported in Fig. 12.7. The device was designed with the 3D bridge vias (Fig. 12.7A and C) to transmit the threshold voltage V_{ref2} to the operational amplifier, bypassing the 2D conductive lines present in the circuit board, by connecting point B and C of Fig. 12.7A. This out-of-plane connection offered a substantial simplification of the circuit layout topology.

The device functionality was assessed by exposing the system to three different illumination conditions (dark, ambient light, and a 150 W light bulb) and by monitoring in real time the corresponding voltage response (V_{ref1}, V_{ref2}, V_{sens}, and the operational amplifier voltage outputs). The test conditions used and a graph showing the evolution over time of the voltage signals registered are reported in Fig. 12.7D and E, respectively. The device responded as expected, by switching the LEDs on according to the light intensities imposed, demonstrating that the electronic circuit layout of the dark-activated sensor could be successfully reproduced by the 3D object architecture [23].

The integration of FFF and SCBD into a single manufacturing process proved to be an effective solution for the additive fabrication of monolithic devices with embedded functional structures. Furthermore, passive electrical components and conductive lines could be directly printed from the virtual model of the object and circuit layout, and all of the fabrication steps exclusively involved the use of FFF and SCBD. This suggests that the two techniques have the potential to be merged in a single manufacturing facility.

12.5 SCBD printing of passive electrical components on paper

A polymeric substrate that is very promising for the fabrication of printed, flexible electronic devices and circuits is paper [34,35]: it is low cost and fully recyclable; it can be easily shaped and trimmed with scissors or cutters; and it can be used for complex self-standing 3D structures, fluidics, and electrochemical applications [36,37]. Although traditional papermaking still constitutes a robust industry for a number of end-products, 3D printing of plain paper using material extrusion-based approaches remains a technological challenge [38]. This is mainly due to the relatively poor thermal stability of cellulosic materials (e.g., lignocellulose, cotton pulp, and microfibrillated cellulose), since they tend to degrade before reaching the adequate rheology to be extruded [38,39]. Moreover, the rheological properties of cellulose-based inks at room temperature are also critical for syringe extrusion, since they depend on a number of interplaying factors, such as cellulose concentration, solvent, pH, the addition of thickening agents, ink preparation conditions [39,40]. Examples of 3D paper-based smart objects with embedded electronics and actuation capabilities were achieved using alternative additive manufacturing routes, relying on gluing trimmed paper sheets functionalized with conductive inks in a single printing facility [41].

Recently, an additive nanomanufacturing strategy enabling the fabrication of passive electrical components (resistors and capacitors) with controlled electrical properties on plain paper using SCBD coupled with shadow mask printing was reported [31]. Compared to standard deposition and AM technologies, SCBD allows for the rapid production of components with different shape and dimensions while controlling independently their electrical characteristics. This is achieved by exploiting two intrinsic features of this AM approach. On one hand, working in determined regions of the percolation curve, i.e., having control of the cluster-assembled film growth during printing, enables the onset of tailored conduction properties of the deposited metal layer, with fixed in-plane dimensions as determined by the stencil mask employed. On the other hand, the supersonic cluster beam profile typical of the SCBD process allows to print large batches of components with different electrical properties in a single fabrication step by a simple raster of the sample in one direction (along the z-axis in Fig. 12.3). In fact, as previously demonstrated [30], the distribution of the cluster density in the beam is that of a Gaussian function, which implies that the particle density decreases by moving from the central to the peripheral regions of the beam. As a result, a gradient in the dose of the deposited nanoparticles is naturally established in the direction perpendicular to that of the raster (y-axis in Fig. 3), while a homogeneous deposition is guaranteed along the raster direction, within a region of about 1 cm.

A batch of 27 resistors with electrical resistance ranging over three orders of magnitude was SCBD printed by depositing gold on plain paper over an area of 50 cm^2, with a lateral resolution below 1 μm, using a single fabrication step with a duration time of about 30 min. More specifically, a stencil mask consisting of three rows and seven to nine columns of rectangular-shaped through-holes was employed. The width and length of the through-holes were 30 × 5, 30 × 3, and 30 mm × 1 cm for each row, respectively. By displacing both the substrates and mask along the vertical direction with respect to the cluster beam, it was possible to print two sets of resistors having electrical resistance ranging from 100 Ω to 28 kΩ and from 270 Ω to 160 kΩ, respectively. Parallel plate capacitors having plain paper as dielectric were also additively manufactured using the same approach, by printing squared gold patterns on the opposite sides of the paper substrate. The capacitance values obtained were independent from the amount of clusters deposited (i.e., plate conductivity) and scaled linearly as a function of the plate area. By combining the printing of resistors and capacitors onto the same substrate, typical RC filters (R ~ 16.5 kΩ, C ~ 230 pF) with good stability over a wide range of frequencies were then produced and validated.

Moreover, coplanar microsupercapacitors (μ-SCaps) with nanostructured carbon electrodes and gold current collectors were printed by means of SCBD on plain paper substrates impregnated with an imidazolium-based ionic liquid electrolyte [42]. The as-prepared devices exhibited large electrochemical capacitance density (7 F/cm^3), operational voltage of 1.8 V, and good stability and repeatability upon charge/discharge cycling. The μ-SCaps (each electrode area was 0.24 cm^2) could also be encapsulated in PDMS using a simple overmolding procedure, without showing decrease in performance or variations in their operational features. A set of three devices connected in series was also printed on the same paper substrate. In this configuration, the operational voltage of three μ-SCaps connection increased to 5.4 V and the energy storage ability of the system was then validated by implementing the capacitors into a custom-designed device to drive a commercial temperature sensor [42].

The production of cluster-assembled passive electrical components and miniaturized energy storage devices suitable for low-power paper-based electronics renders SCBD printing a promising route for the additive manufacturing of flexible functional devices, where low cost, speed of fabrication, and the ability to integrate multiple, discrete components into the system are key parameters.

12.6 SCBD printing of a stretchable keyboard with multilayer electrical interconnects

The possibility to engineer by design both the mechanical and electrical properties of flexible and compliant conductive nanocomposite structures by controlling a single parameter (the equivalent thickness i.e., the deposition time) is particularly advantageous for the additive manufacturing of deformable electronic components, sensors, and actuators that can preserve the appropriate structural and functional properties when integrated into functional devices [43]. Although metal-elastomer nanocomposites are very convenient as substrates for stretchable electronics, their mechanical properties (Young's modulus) are critical for the interconnection with other devices, where the use of wires and rigid substrates is required.

Two major challenges are still associated with the fabrication of stretchable devices: (1) the limitation to integration density of the components when only a single layer of interconnects is used; (2) the connection of stretchable parts with different hard components or substrates: the matching between soft and rigid components in electrical or mechanical connections remains a major technical issue to be solved.

In this section, the fabrication and the characterization of a system consisting of a stretchable keyboard produced by means of SCBD printing is presented. The core of the device is a PDMS component having SCBD printed touchpads and conductive lines, connected with a touch sensor and an LED matrix printed on paper through paper-based interconnections. The stretchable keyboard is also provided with 3D cylindrical conductive vias to electrically connect the front and rear sides of the elastomer. Gold conductive films printed on paper are used to connect the keyboard with the electronic circuit that drives the LED matrix.

A schematic diagram of the system consisting of a stretchable keyboard, a touch sensor, an Arduino board, and an LED matrix is reported in Fig. 12.8. The stretchable

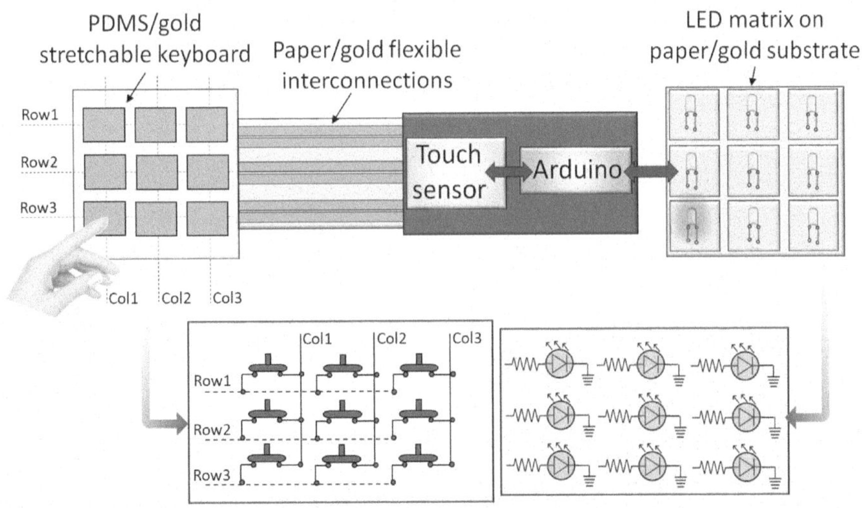

FIGURE 12.8 Schematic representation of a PDMS stretchable keyboard connected to a touch sensor with flexible conductive electrodes realized on a paper substrate. The touch sensor sends data to an Arduino controller that turns on or off the LEDs on a matrix printed on a paper substrate. The inset shows the electrical scheme of the keyboard: columns and rows are printed on the opposite sides of the PDMS slab and connected by vias.

keyboard is connected to a 12-key capacitive touch sensor MPR121 [44] through a flexible wiring made by electrically conductive gold lines printed on a paper substrate. An Arduino electronic board drives the capacitive sensors and runs the paper-based LED matrix according to the signals received from the touch sensor. The LED matrix is composed by nine LEDs powered by resistors printed on paper by SCBD, and it is connected to the Arduino board by standard LED-strip connectors.

The capacitance variation produced by touching a pad on the keyboard is transmitted to the rear side of the keyboard through conducting vias and then to the capacitive sensor connected by the metallized paper conductors. The sensor generates a set of univocal numbers to identify which pad has been touched, and it sends them to the Arduino board through an Inter-Integrated Circuit (I2C) communication channel. This triggers the Arduino board to turn on or off the corresponding LED fixed on the flexible paper substrate.

The stretchable keyboard is based on a matrix architecture printed on both sides of the PDMS substrate (Fig. 12.9). The bottom layer controls the columns of the matrix and it establishes the electrical connection between the keyboard and the paper connectors; the top layer integrates nine touchpads and electrical lines operating as bus, dedicated to control the rows of the matrix. The two layers of the keyboard are electrically connected by 12 conductive vias. Every touchpad is realized splitting the pad into two separate parts: one connected to the column-control bus and the other connected to the row-control bus. The keyboard and the paper interconnects are superimposed and bolted to a polylactic acid (PLA) plate to ensure electrical connection.

While the realization of conductive through-hole vias links is trivial for standard electronic circuits on hard substrates, the realization of conductive vias in elastomeric

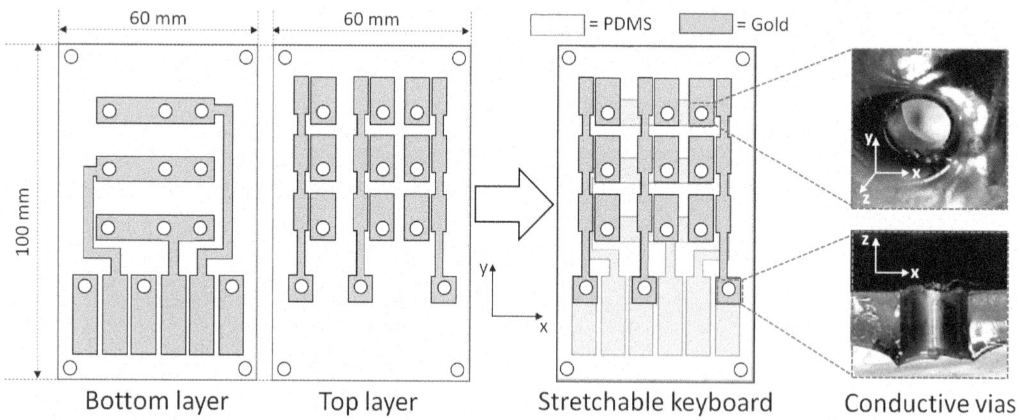

FIGURE 12.9 Keyboard design: 12 through-holes (3 mm diameter) were used as conductive vias to connect the electrodes on the top side of the keyboard with the electrodes on the bottom side. Bottom side with the six electrodes was used to send the capacitive signal to the external electronic circuit (*left*) and top side of the keyboard with the nine touchpads used as buttons (*right*). Insets show a side view and the cross-section of the PDMS vias metallized by SCBD.

substrates is still challenging when relying on metallization techniques [45,46]. By using SCBD-based additive manufacturing, it was possible to fabricate conductive vias in PDMS thick layers, since the direct and uniform metallization of the inner wall of the vias could be obtained by exploiting the high collimation of the beam typical of supersonic expansions in conjunction with an appropriate tilting of the target substrate [23].

The LED matrix was realized using SCBD coupled with shadow mask printing to deposit gold nanoparticles on a Xerox paper substrate (80 g/m^2) [31]. The electrical resistance of the conductive paths was tailored by measuring in situ the amount of nanoparticles deposited onto the paper. High power LEDs were glued on the paper and connected to the printed lines using a conductive silver paste to establish a good electrical contact (Fig. 12.10A and B). Exploiting the possibility to print resistors with well-defined electrical characteristic with SCBD, resistive conductive lines (RCL) were designed and additively fabricated to limit the amount of current that flows in the LEDs. The resistance of the gold elements printed of paper spanned from 100 to 250 Ω, obtained at an equivalent thickness of 180 nm (Fig. 12.11). These resistance values guarantee that the current sunk from the Arduino board does not exceed the maximum value, attested around 200 mA.

The connection between the electrodes of the bottom side of the PDMS keyboard and an external drive electronic circuit was fabricated by means of physical vapor deposition on a paper substrate, appropriately masked to define the electrical interconnects' shape (Fig. 12.10C).

The assembled device is shown in Fig. 12.11A. The touching of the touchpads causes the corresponding LED to turn on or off. This test was replicated while manually bending, twisting, and stretching the keyboard, with no observed delays or malfunctioning in the system response. The electrical connection between the keyboard and the electrodes printed on paper showed to be robust enough to pipe the capacitive signal from the keyboard to the touch sensor, withstanding the mechanical deformation applied to the keyboard. The electrical resistance of the vias was measured under

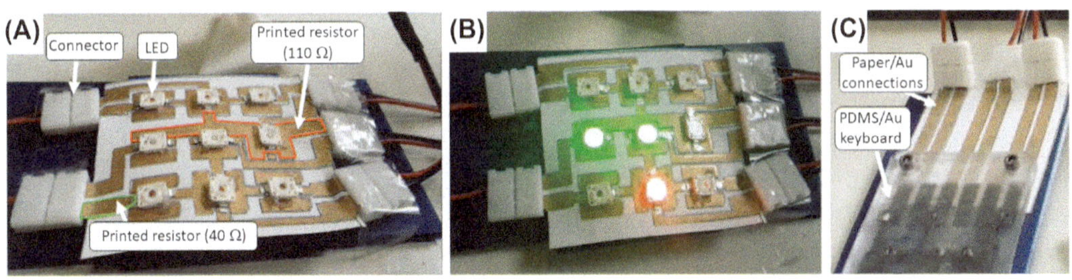

FIGURE 12.10 (A) Picture of the LED matrix showing the electrical circuit fabricated by SCBD printing of gold clusters on paper and the glued LEDS. (B) LEDs are turned on and off by touching the PDMS keyboard. (C) Conductive lines on paper connecting the PDMS keyboards with standard connectors and wires.

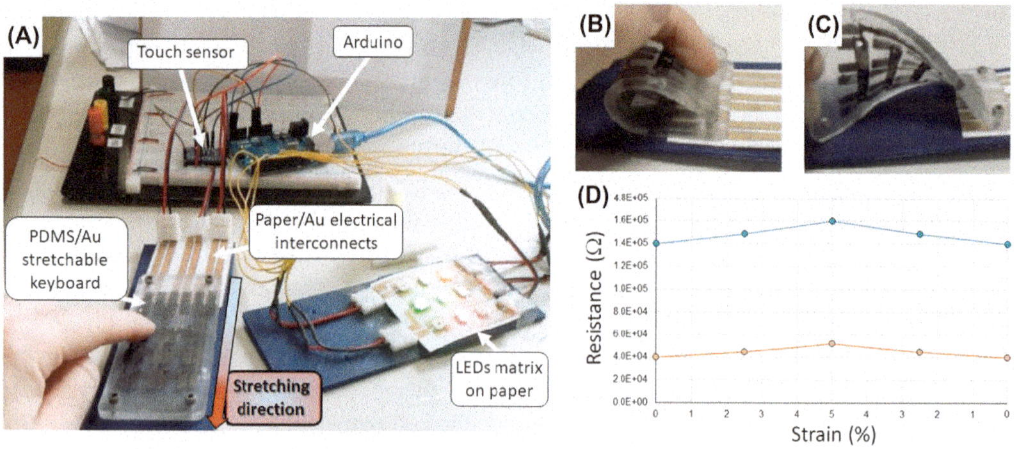

FIGURE 12.11 (A) Picture of the assembled device working under finger pressure. The electric signal is passed from the keyboard to the electronic circuit through gold/paper flexible electrodes. The Arduino board in response to the touch turns on or off an LED in the LED matrix made of flexible gold/paper circuit. The picture shows a test performed with the keyboard stretched by 5% along the direction indicated by the *arrow*. (B) and (C) stretching and bending of the PDMS keyboard. (D) Evolution of the resistance during a stretching cycle of two of the conductive lines present in the rear and front sides of the elastomer and connected by the vias. *Image readapted from A. Bellacicca, P. Milani, Supersonic Cluster Beam Fabrication of a Stretchable Keyboard with Multilayer Electrical Interconnects Integrated with Paper-based Flexible Devices, arXiv:1706.08586.*

stretching cycles (maximum number of cycles was 100) up to 5% of linear deformation. As showed in Fig. 12.11B and C the electrodes and vias were still conductive during the stretching phase, undergoing a small increase in the resistance values and restoring the initial value when the keyboard was reset to rest position (Fig. 12.11D).

12.7 Conclusions

SCBD is an additive fabrication technology enabling the printing of polymer/metal nanocomposites with controlled electrical conductivity and mechanical properties that can be tailored by regulating the amount of nanoparticles constituting the structures. This approach constitutes a promising solution for the manufacturing and rapid prototyping of 3D printed electronics and complex devices based on nonconventional substrates.

The merging of SCBD printing with standard FFF for the fabrication of freeform thermoplastic components enabled the production of smart monolithic objects with embedded electronics and sensing ability. This integrated manufacturing route is feasible for the realization of 3D conductive lines and vias for inter and intralayer electrical connections, simplifying the topology of planar circuits and minimizing the use of external bulky electronic components. Standard conductive socket slots were also realized. A dark-activated light sensor able to discriminate between two dark levels was prototyped and validated to demonstrate the suitability of this approach.

Moreover, the integration of nanoparticles and/or nanostructured layers in deformable and elastic polymers are the basis for the production of a novel class of systems that can be realized by combining top-down microfabrication strategies for polymer processing and the bottom-up additive assembling of nanoobjects that retain huge potential for soft robotics and deformable electronics applications.

References

[1] S.A.M. Tofail, E.P. Koumoulos, A. Bandyopadhyay, S. Bose, L. O'Donoghue, C. Charitidis, Additive manufacturing: scientific and technological challenges, market uptake and opportunities, Mater. Today 21 (1) (2018) 22–37.

[2] T.J. Horn, O.L.A. Harrysson, Overview of current additive manufacturing technologies and selected applications, Sci. Prog. 95 (3) (2012) 255–282.

[3] S. Singh, S. Ramakrishna, R. Singh, Material issues in additive manufacturing: a review, J. Manuf. Process. 25 (1) (2017) 185–200.

[4] R. Karania, D. Kazmer, Low volume plastics manufacturing strategies, J. Mech. Des. 129 (12) (2007) 1225–1233.

[5] M.J. Cotteleer, J. Joyce, 3D opportunity: additive manufacturing paths to performance, innovation, and growth. Deloitte review, in: SIMT Additive Manufacturing Symposium, 2014.

[6] K.V. Wong, A. Hernandez, A review of additive manufacturing, ISRN Mech. Eng. 2012 (1) (2012) 1–10.

[7] B.N. Turner, R. Strong, S.A. Gold, A review of melt extrusion additive manufacturing processes: I. Process design and modeling, Rapid Prototyp. J. 20 (3) (2014) 192–204.

[8] D.I. Wimpenny, P.M. Pandey, L. Jyothish Kumar (Eds.), Advances in 3D Printing and Additive Manufacturing Technologies, Springer, Singapore, 2016, https://doi.org/10.1007/978-981-10-0812-2.

[9] A. Boschetto, L. Bottini, Accuracy prediction in fused deposition modeling, Int. J. Adv. Manuf. Technol. 73 (5–8) (2014) 913–928.

[10] E. MacDonald, R. Salas, D. Espalin, M. Perez, E. Aguilera, D. Muse, et al., 3D printing for the rapid prototyping of structural electronics, IEEE Access 2 (2014) 234–242.

[11] D. Espalin, D.W. Muse, E. MacDonald, R.B. Wicker, 3D Printing multifunctionality: structures with electronics, Int. J. Adv. Manuf. Technol. 72 (5–8) (2014) 963–978.

[12] P.F. Flowers, C. Reyes, S. Ye, M.J. Kim, B.J. Wiley, 3D printing electronic components and circuits with conductive thermoplastic filament, Addit. Manuf. 18 (1) (2017) 156–163.

[13] A.A. Gupta, A. Bolduc, S.G. Cloutier, R. Izquierdo, Aerosol Jet Printing for printed electronics rapid prototyping, in: Proceedings - IEEE International Symposium on Circuits and Systems, 2016. https://doi.org/10.1109/ISCAS.2016.7527378.

[14] J.A. Paulsen, M. Renn, K. Christenson, R. Plourde, Printing conformal electronics on 3D structures with aerosol jet technology, in: FIIW 2012 - 2012 Future of Instrumentation International Workshop Proceedings, 2012. https://doi.org/10.1109/FIIW.2012.6378343.

[15] C. Goth, S. Putzo, J. Franke, Aerosol Jet printing on rapid prototyping materials for fine pitch electronic applications, in: Proceedings - Electronic Components and Technology Conference, 2011. https://doi.org/10.1109/ECTC.2011.5898664.

[16] B.-H. Lu, H.-B. Lan, H.-Z. Liu, Additive manufacturing frontier: 3D printing electronics, Opto-Electronic Adv. 1 (1) (2018) 17000401–17000410.

[17] J. Hoerber, J. Glasschroeder, M. Pfeffer, J. Schilp, M. Zaeh, J. Franke, Approaches for additive manufacturing of 3D electronic applications, in: Procedia CIRP, vol. 17 (1), 2014, pp. 806–811.

[18] M. Hedges, Aerosol Jetting of 3D interconnects & printed electronics, in: Fraunhofer EMFT Forum "Be-flexible" 2010, Flexible Electronic Systems on December, vol. 2, 9th International Workshop, 2010.

[19] M. Hedges, N. Services, M.T.P. Martin, 3D Printed Electronics via Aerosol Jet Printing, Neotech Services MTP, 2013.

[20] C. Kaestle, J. Hoerber, F. Oechsner, J. Franke, Prospects of wire bonding as an approach for contacting additive manufactured Aerosol Jet printed structures, in: European Microelectronics and Packaging Conference, 2015.

[21] M. Marelli, G. Divitini, C. Collini, L. Ravagnan, G. Corbelli, C. Ghisleri, et al., Flexible and biocompatible microelectrode arrays fabricated by supersonic cluster beam deposition on SU-8, J. Micromech. Microeng. 21 (4) (2011) 045013.

[22] G. Corbelli, C. Ghisleri, M. Marelli, P. Milani, L. Ravagnan, Highly deformable nanostructured elastomeric electrodes with improving conductivity upon cyclical stretching, Adv. Mater. 23 (39) (2011) 4504–4508.

[23] A. Bellacicca, T. Santaniello, P. Milani, Embedding electronics in 3D printed structures by combining fused filament fabrication and supersonic cluster beam deposition, Addit. Manuf. 24 (1) (2018) 60–66.

[24] T. Santaniello, L. Migliorini, F. Borghi, Y. Yan, S. Rondinini, C. Lenardi, et al., Spring-like electroactive actuators based on paper/ionogel/metal nanocomposites, Smart Mater. Struct. 27 (6) (2018), 065004-14.

[25] Y. Yan, T. Santaniello, L.G. Bettini, C. Minnai, A. Bellacicca, R. Porotti, et al., Electroactive ionic soft actuators with monolithically integrated gold nanocomposite electrodes, Adv. Mater. 29 (23) (2017), 1606109:1–9.

[26] L. Migliorini, T. Santaniello, S. Rondinini, P. Saettone, M. Comes Franchini, C. Lenardi, et al., Bioplastic electromechanical actuators based on biodegradable poly(3-hydroxybutyrate) and cluster-assembled gold electrodes, Sens. Actuators B Chem. 286 (1) (2019) 230–236.

[27] K. Wegner, P. Piseri, H.V. Tafreshi, P. Milani, Cluster beam deposition: a tool for nanoscale science and technology, J. Phys. D Appl. Phys. 39 (1) (2006) R439–R460.

[28] P. Piseri, H.V. Tafreshi, P. Milani, Manipulation of nanoparticles in supersonic beams for the production of nanostructured materials, Curr. Opin. Solid State Mater. Sci. 8 (1) (2004) 195–202.

[29] L. Ravagnan, G. Divitini, S. Rebasti, M. Marelli, P. Piseri, P. Milani, Poly(methyl methacrylate)-palladium clusters nanocomposite formation by supersonic cluster beam deposition: a method for microstructured metallization of polymer surfaces, J. Phys. D Appl. Phys. 42 (8) (2009), 082002-7.

[30] C. Ghisleri, F. Borghi, L. Ravagnan, A. Podestà, C. Melis, L. Colombo, et al., Patterning of gold-polydimethylsiloxane (Au-PDMS) nanocomposites by supersonic cluster beam implantation, J. Phys. D Appl. Phys. 47 (1) (2014) 015301–015311.

[31] F. Caruso, A. Bellacicca, P. Milani, F. Caruso, A. Bellacicca, P. Milani, High-throughput shadow mask printing of passive electrical components on paper by supersonic cluster beam deposition, Appl. Phys. Lett. 108 (1) (2016) 163501–163505.

[32] S. Kirkpatrick, Percolation and conduction, Rev. Mod. Phys. 45 (4) (1973) 574–588.

[33] F. Faupel, V. Zaporojtchenko, T. Strunskus, M. Elbahri, Metal-polymer nanocomposites for functional applications, Adv. Eng. Mater. 12 (12) (2010) 1177–1190.

[34] A.C. Siegel, S.T. Phillips, M.D. Dickey, N. Lu, Z. Suo, G.M. Whitesides, Foldable printed circuit boards on paper substrates, Adv. Funct. Mater. 20 (1) (2010) 28–35.

[35] R. Martins, I. Ferreira, E. Fortunato, Electronics with and on paper, Phys. Status Solidi Rapid Res. Lett. 5 (9) (2011) 332–335.

[36] J. Winfield, L.D. Chambers, J. Rossiter, J. Greenman, I. Ieropoulos, Urine-activated origami microbial fuel cells to signal proof of life, J. Mater. Chem. 3 (13) (2015) 7058–7065.

[37] M. Nogi, N. Komoda, K. Otsuka, K. Suganuma, Foldable nanopaper antennas for origami electronics, Nanoscale 5 (10) (2013) 4395–4399.

[38] L. Dai, T. Cheng, C. Duan, W. Zhao, W. Zhang, X. Zou, et al., 3D printing using plant-derived cellulose and its derivatives: a review, Carbohydr. Polym. 203 (1) (2019) 71–86.

[39] Q. Wang, J. Sun, Q. Yao, C. Ji, J. Liu, Q. Zhu, 3D printing with cellulose materials, Cellulose 25 (8) (2018) 4275–4301.

[40] C. Thibaut, A. Denneulin, S. Rolland du Roscoat, D. Beneventi, L. Orgéas, D. Chaussy, A fibrous cellulose paste formulation to manufacture structural parts using 3D printing by extrusion, Carbohydr. Polym. 212 (1) (2019) 119–128.

[41] H. Oh, T.D. Ta, R. Suzuki, M.D. Gross, Y. Kawahara, L. Yao, PEP (3D Printed Electronic papercrafts): an integrated approach for 3D sculpting paper-based electronic devices, in: Conference on Human Factors in Computing Systems - Proceedings, 2018. https://doi.org/10.1145/3173574.3174015.

[42] L.G. Bettini, A. Bellacicca, P. Piseri, P. Milani, Supersonic cluster beam printing of carbon microsupercapacitors on paper, Flex. Print. Electron. 2 (2) (2017) 025002.

[43] F. Borghi, C. Melis, C. Ghisleri, A. Podestà, L. Ravagnan, L. Colombo, et al., Stretchable nanocomposite electrodes with tunable mechanical properties by supersonic cluster beam implantation in elastomers, Appl. Phys. Lett. 106 (12) (2015) 121902–121907.

[44] S. Mirri, C. Prandi, P. Salomoni, L. Monti, Fitting like a GlovePi: a wearable device for deaf-blind people, in: 2017 14th IEEE Annual Consumer Communications and Networking Conference, CCNC 2017, 2017. https://doi.org/10.1109/CCNC.2017.7983285.

[45] L. Guo, S.P. DeWeerth, High-density stretchable electronics: toward an integrated multilayer composite, Adv. Mater. 22 (36) (2010) 4030–4033.

[46] A. Larmagnac, S. Eggenberger, H. Janossy, J. Vörös, Stretchable electronics based on Ag-PDMS composites, Sci. Rep. 4 (1) (2014) 1–7.

Index

'*Note:* Page numbers followed by "f" indicate figures and "t" indicates tables.'